VOLUME SEVENTY FOUR

CURRENT TOPICS IN MEMBRANES

Thermal Sensors

CURRENT TOPICS IN MEMBRANES, VOLUME 74

Series Editors

ROBERT BALABAN
National Heart, Lung and Blood Institute
National Institutes of Health
Bethesda, Maryland, USA

SIDNEY A. SIMON
Department of Neurobiology
Duke University Medical Centre
Durham, North Carolina, USA

VOLUME SEVENTY FOUR

CURRENT TOPICS IN MEMBRANES

Thermal Sensors

Edited by

LEÓN D. ISLAS

Department of Physiology, School of Medicine, National Autonomous University of Mexico (UNAM), Mexico City, Mexico

FENG QIN

Department of Physiology & Biophysics, State University of New York Buffalo, New York, USA

ELSEVIER

Amsterdam • Boston • Heidelberg • London
New York • Oxford • Paris • San Diego
San Francisco • Singapore • Sydney • Tokyo

Academic Press is an imprint of Elsevier

Academic Press is an imprint of Elsevier
225 Wyman Street, Waltham, MA 02451, USA
525 B Street, Suite 1800, San Diego, CA 92101-4495, USA
The Boulevard, Langford Lane, Kidlington, Oxford, OX5 1GB, UK
32 Jamestown Road, London NW1 7BY, UK
Radarweg 29, PO Box 211, 1000 AE Amsterdam, The Netherlands

First edition 2014

For information on all Academic Press publications visit our website at http://store.elsevier.com/

ISBN: 978-0-12-800181-3
ISSN: 1063-5823

CONTENTS

CONTRIBUTORS

Evan O. Anderson
Department of Cellular & Molecular Physiology, Yale School of Medicine, New Haven, CT, USA

David Baez
Centro Interdisciplinario de Neurociencia de Valparaíso, Facultad de Ciencias, Universidad de Valparaíso, Valparaíso, Chile

Sviatoslav N. Bagriantsev
Department of Cellular & Molecular Physiology, Yale University School of Medicine, New Haven, CT, USA

Matthew R. Cohen
Department of Pharmacology, Case Western Reserve University School of Medicine, Cleveland, OH, USA; Department of Physiology & Biophysics, Case Western Reserve University School of Medicine, Cleveland, OH, USA

Qin Feng
Department of Physiology & Biophysics, State University of New York, Buffalo, New York, USA

Gonzalo Ferreira
Laboratorio de Canales Iónicos, Departamento de Biofísica, Facultad de Medicina, Universidad de la República, Montevideo, Uruguay

Yuichiro Fujiwara
Laboratory of Integrative Physiology, Graduate School of Medicine, Osaka University, Suita, Japan

Carlos Gonzalez
Centro Interdisciplinario de Neurociencia de Valparaíso, Facultad de Ciencias, Universidad de Valparaíso, Valparaíso, Chile

Elena O. Gracheva
Department of Cellular and Molecular Physiology, Yale University School of Medicine, New Haven, CT, USA; Program in Cellular Neuroscience, Neurodegeneration and Repair, Yale University School of Medicine, New Haven, CT, USA

Enrique Hernández-García
Departamento de Neurodesarrollo y Fisiología, División de Neurociencias, Instituto de Fisiología Celular, Universidad Nacional Autónoma de México, Distrito Federal, México

Hongzhen Hu
The Center for the Study of Itch, Washington University Pain Center, The Department of Anesthesiology, Washington University School of Medicine in St. Louis, St. Louis, MO, USA

León D. Islas
Department of Physiology, School of Medicine, National Autonomous University of Mexico (UNAM), Mexico City, Mexico

Ramon Latorre
Centro Interdisciplinario de Neurociencia de Valparaíso, Facultad de Ciencias, Universidad de Valparaíso, Valparaíso, Chile

Willem J. Laursen
Department of Cellular and Molecular Physiology, Yale University School of Medicine, New Haven, CT, USA; Program in Cellular Neuroscience, Neurodegeneration and Repair, Yale University School of Medicine, New Haven, CT, USA

Jialie Luo
The Center for the Study of Itch, Washington University Pain Center, The Department of Anesthesiology, Washington University School of Medicine in St. Louis, St. Louis, MO, USA

Linlin Ma
Department of Physiology and Membrane Biology, University of California School of Medicine, Davis, CA, USA

Rodolfo Madrid
Departamento de Biología, Facultad de Química y Biología, Universidad de Santiago de Chile, Santiago, Chile

Vera Y. Moiseenkova-Bell
Department of Pharmacology, Case Western Reserve University School of Medicine, Cleveland, OH, USA; Department of Physiology & Biophysics, Case Western Reserve University School of Medicine, Cleveland, OH, USA

Yasushi Okamura
Laboratory of Integrative Physiology, Graduate School of Medicine, Osaka University, Suita, Japan

María Pertusa
Departamento de Biología, Facultad de Química y Biología, Universidad de Santiago de Chile, Santiago, Chile

Michael Pusch
Istituto di Biofisica, CNR, Genova, Italy

Natalia Raddatz
Centro Interdisciplinario de Neurociencia de Valparaíso, Facultad de Ciencias, Universidad de Valparaíso, Valparaíso, Chile; Centre for Plant Biotechnology and Genomics, Universidad Politécnica de Madrid, Pozuelo de Alarcón (Madrid), Spain

Tamara Rosenbaum
Departamento de Neurodesarrollo y Fisiología, División de Neurociencias, Instituto de Fisiología Celular, Universidad Nacional Autónoma de México, Distrito Federal, México

Eve R. Schneider
Department of Cellular & Molecular Physiology, Yale School of Medicine, New Haven, CT, USA

Jie Zheng
Department of Physiology and Membrane Biology, University of California School of Medicine, Davis, CA, USA

Giovanni Zifarelli
Istituto di Biofisica, CNR, Genova, Italy

PREFACE

Thermal biology of organisms is fundamentally determined by the activity of specialized receptors. Generally known as thermal receptors, these molecules are principally membrane-associated or integral membrane proteins. Among them, and as is found in many sensory systems, the main players are ion channels. Although the existence of thermally activated currents in sensory neurons was proposed in the decade of the 1980s, it was not until 1996 that it was demonstrated that sensory neurons poses heat activated currents, followed by the landmark discovery and cloning of the heat activated TRPV1 ion channel in 1997, which the field really took off.

Since then, researchers in the field have identified a large number of proteins that have the exotic property of being activated by temperature changes. These mainly belong to the relatively new family of polymodal TRP ion channels, and can be found in different flavors. Some are activated by cold, some by heat and there are variations in the ranges of temperature that they are responsive to. Some other ion channels, distinct from TRP channels, including proton channels, nonvoltage-dependent potassium channels and chloride channels have also been found to respond with exceptional sensitivity to changes in temperature and to contribute to thermal physiology.

Although very young, the field of thermal receptor biology and biophysics has attracted many talented investigators; thus, the number of insights into these proteins keeps increasing and the findings are ever more surprising.

This volume in the long-running series *Current Topics in Membranes* seeks to provide an up to date view of the basic concepts and main recent findings, from the pens of some of the leading workers in the field of thermal receptors.

The book is organized in such a way that it first presents the current knowledge and leading theories put forward to explain the mechanisms of thermal sensitivity at the molecular level. Later chapters explore insights obtained by modern structural experiments and explore the amazing diversity of functions and mechanisms displayed by these fascinating membrane proteins.

We would like to thank Sidney A. Simon for his generous invitation to participate in editing this volume and to all the authors of the chapters for their invaluable contributions and patience during this project.

Finally, our special recognition to Helene Kabes for her tolerance and constant support in compiling and editing the chapters.

León D. Islas
Mexico City

Feng Qin
Buffalo, NY

PREVIOUS VOLUMES IN SERIES

Current Topics in Membranes and Transport

Volume 23 Genes and Membranes: Transport Proteins and Receptors* (1985)
Edited by Edward A. Adelberg and Carolyn W. Slayman

Volume 24 Membrane Protein Biosynthesis and Turnover (1985)
Edited by Philip A. Knauf and John S. Cook

Volume 25 Regulation of Calcium Transport across Muscle Membranes (1985)
Edited by Adil E. Shamoo

Volume 26 Na^{+}–H^{+}Exchange, Intracellular pH, and Cell Function* (1986)
Edited by Peter S. Aronson and Walter F. Boron

Volume 27 The Role of Membranes in Cell Growth and Differentiation (1986)
Edited by Lazaro J. Mandel and Dale J. Benos

Volume 28 Potassium Transport: Physiology and Pathophysiology* (1987)
Edited by Gerhard Giebisch

Volume 29 Membrane Structure and Function (1987)
Edited by Richard D. Klausner, Christoph Kempf, and Josvan Renswoude

Volume 30 Cell **Volume** Control: Fundamental and Comparative Aspects in Animal Cells (1987)
Edited by R. Gilles, Arnost Kleinzeller, and L. Bolis

Volume 31 Molecular Neurobiology: Endocrine Approaches (1987)
Edited by Jerome F. Strauss, III, and Donald W. Pfaff

Volume 32 Membrane Fusion in Fertilization, Cellular Transport, and Viral Infection (1988)
Edited by Nejat Düzgünes and Felix Bronner

Volume 33 Molecular Biology of Ionic Channels* (1988)
Edited by William S. Agnew, Toni Claudio, and Frederick J. Sigworth

Volume 34 Cellular and Molecular Biology of Sodium Transport* (1989)
Edited by Stanley G. Schultz

Volume 35 Mechanisms of Leukocyte Activation (1990)
Edited by Sergio Grinstein and Ori D. Rotstein

*Part of the series from the Yale Department of Cellular and Molecular Physiology

Volume 36 Protein–Membrane Interactions* (1990)
Edited by Toni Claudio

Volume 37 Channels and Noise in Epithelial Tissues (1990)
Edited by Sandy I. Helman and Willy Van Driessche

Current Topics in Membranes

Volume 38 Ordering the Membrane Cytoskeleton Trilayer* (1991)
Edited by Mark S. Mooseker and Jon S. Morrow

Volume 39 Developmental Biology of Membrane Transport Systems (1991)
Edited by Dale J. Benos

Volume 40 Cell Lipids (1994)
Edited by Dick Hoekstra

Volume 41 Cell Biology and Membrane Transport Processes* (1994)
Edited by Michael Caplan

Volume 42 Chloride Channels (1994)
Edited by William B. Guggino

Volume 43 Membrane Protein–Cytoskeleton Interactions (1996)
Edited by W. James Nelson

Volume 44 Lipid Polymorphism and Membrane Properties (1997)
Edited by Richard Epand

Volume 45 The Eye's Aqueous Humor: From Secretion to Glaucoma (1998)
Edited by Mortimer M. Civan

Volume 46 Potassium Ion Channels: Molecular Structure, Function, and Diseases (1999)
Edited by Yoshihisa Kurachi, Lily Yeh Jan, and Michel Lazdunski

Volume 47 Amiloride-Sensitive Sodium Channels: Physiology and Functional Diversity (1999)
Edited by Dale J. Benos

Volume 48 Membrane Permeability: 100 Years since Ernest Overton (1999)
Edited by David W. Deamer, Arnost Kleinzeller, and Douglas M. Fambrough

Volume 49 Gap Junctions: Molecular Basis of Cell Communication in Health and Disease
Edited by Camillo Peracchia

Volume 50 Gastrointestinal Transport: Molecular Physiology
Edited by Kim E. Barrett and Mark Donowitz

Volume 51 Aquaporins
Edited by Stefan Hohmann, Søren Nielsen and Peter Agre

Volume 52 Peptide–Lipid Interactions
Edited by Sidney A. Simon and Thomas J. McIntosh

Volume 53 Calcium-Activated Chloride Channels
Edited by Catherine Mary Fuller

Volume 54 Extracellular Nucleotides and Nucleosides: Release, Receptors, and Physiological and Pathophysiological Effects
Edited by Erik M. Schwiebert

Volume 55 Chemokines, Chemokine Receptors, and Disease
Edited by Lisa M. Schwiebert

Volume 56 Basement Membranes: Cell and Molecular Biology
Edited by Nicholas A. Kefalides and Jacques P. Borel

Volume 57 The Nociceptive Membrane
Edited by Uhtaek Oh

Volume 58 Mechanosensitive Ion Channels, Part A
Edited by Owen P. Hamill

Volume 59 Mechanosensitive Ion Channels, Part B
Edited by Owen P. Hamill

Volume 60 Computational Modelling of Membrane Bilayers
Edited by Scott E. Feller

Volume 61 Free Radical Effects on Membranes
Edited by Sadis Matalon

Volume 62 The Eye's Aqueous Humor
Edited by Mortimer M. Civan

Volume 63 Membrane Protein Crystallization
Edited by Larry DeLucas

Volume 64 Leukocyte Adhesion
Edited by Klaus Ley

Volume 65 Claudins
Edited by Alan S. L. Yu

Volume 66 Structure and Function of Calcium Release Channels
Edited by Irina I. Serysheva

Volume 67 Advances in Adrenergic Receptor Biology
Edited by Qin Wang

Volume 68 Membrane Fusion
Edited by Leonid V. Chernomordik and Michael M. Kozlov

Volume 69 Metal Transporters
Edited by Svetlana Lutsenko and José M. Argüello

Volume 70 Co-Transport Systems
Edited by Mark O. Bevensee

Volume 71 Store-Operated Calcium Channels
Edited by Murali Prakriya

Volume 72 Functional Organization of Vertebrate Plasma Membrane
Edited by Vann Bennett

Volume 73 Exchangers
Edited by Mark O. Bevensee

CHAPTER ONE

Thermal Effects and Sensitivity of Biological Membranes

León D. Islas
Department of Physiology, School of Medicine, National Autonomous University of Mexico (UNAM), Mexico City, Mexico
E-mail: islas@liceaga.facmed.unam.mx

Contents

Abstract

Temperature is one of the key parameters that controlled the origin and evolution of life on earth and it continues to be a principal regulator of the functions of organisms. Some aspects of the response of simple and complex organisms to temperature variations are encoded in the physical properties of the cell components, with the all-important plasma membrane playing a principal role. Other responses to temperature are more specific and through evolution, specialized receptors with particular temperature sensitivities have appeared to mediate this signaling. While some of these receptors are ancient and can be found in very primitive organisms, it seems that the mechanisms used by prokaryotes and eukaryotes are very different, indicating that temperature sensitivity has evolved in more than one occasion during evolution.

1. INTRODUCTION

Ambient temperature is one of the most important parameters that make life on earth possible. At the same time, the existence of temperature

Current Topics in Membranes, Volume 74
ISSN: 1063-5823
http://dx.doi.org/10.1016/B978-0-12-800181-3.00001-4

gradients is also a determinant of an organisms' adaptive power and of biological diversity. At the scale of an individual, temperature changes are important clues for phenomena as varied as metabolic rate, migration, reproduction, harmful temperature avoidance, nutrient seeking behaviors, etc. (Angilletta, 2009).

In thermodynamic terms, temperature is a measure of the energy of a system. As such, it is a generalized property that by necessity affects all molecular processes (Feynman, Leighton, & Sands, 2013).

In the context of a cell, it is well known that temperature affects all metabolic reactions, through such mechanisms as changes in the diffusion constant of soluble components, control of enzymatic rates, changes in the resistivity of electrolytic solutions, the fluidity of the plasma membrane, just to name a few (Sharkey & Schrader, 2006; Willmer, Stone, & Johnston, 2009).

For these reasons it is of paramount importance that organisms, from single cells to the multicellular, be able to sense and respond to temperature gradients or steep changes, both external and internal. As a result, several mechanisms have evolved to function as thermal receptors, allowing organisms this fundamental interaction with their environment.

In this chapter, I will briefly review the response of model organisms to temperature changes and some of the general and fundamental effects of temperature in cellular parameters such as membrane composition and electrical properties and excitability of cells, and only mention when relevant, those processes involving specific thermal receptors, which are the main scope of this tome.

2. RESPONSE OF ORGANISMS TO CHANGES IN TEMPERATURE

2.1 Worms and flies

Most organisms, regardless of any ability to control their internal temperature, thrive in a preferred range of ambient temperatures. This is especially true in environments where this variable shows large fluctuations. For example, marine snails living in mud flats in tropical areas are subject to large daily variations in temperature between approximately 16 °C at night to 40 °C at midday. Accordingly, both in laboratory and natural conditions, these organisms exhibit temperature-stimulated motor behavior that drives them to seek a shelter region with their preferred temperature (Bertness & Schneider, 1976; Cameron, 1970; Cox & Rutherford, 2000). This temperature preference is generally related to other complex behaviors, such as feeding and egg laying and these are promoted when the organisms reach their preferred temperature ranges (McMahon & Russell-Hunter, 1977). In both vertebrates and complex invertebrates, the temperature seeking and avoidance behaviors are regulated by specialized sensory and motor neurons and their corresponding neuronal circuits (Wittenburg & Baumeister, 1999), and can be very complicated. For this reason, simpler organisms have been used as more tractable models in sensory physiology.

The nematode *Caenorhabditis elegans* has been extensively used to study these and other behaviors, because genetic mutants are readily available and neurons can be identified and selectively stimulated or disrupted (Li, Kang, Piggott, Feng, & Xu, 2011; Tobin & Bargmann, 2004; Wittenburg & Baumeister, 1999). It has been discovered that *C. elegans* responds quickly and efficiently to temperature changes even as small as a fraction of a degree centigrade. This organism has a preferred temperature of around 27 °C and displays a stereotyped heat avoidance behavior, orchestrated by several individual neurons acting as thermal receptors and which are located in the anterior, posterior, and mid-body regions of the worm. The use of mutant worms has identified some of the molecular regulators of this behavior mainly as thermally activated Transient Receptor Potential (thermoTRP) ion channels and glutamate receptors (Glauser, 2013; Glauser et al., 2011; Mohammadi, Byrne Rodgers, Kotera, & Ryu, 2013).

The fruit fly *Drosophila melanogaster* is another important model organism where similar insights have been garnered from physiological and genetic studies (Hoffmann, Sørensen, & Loeschcke, 2003; Patapoutian, Peier, Story, & Viswanath, 2003; Rosenzweig et al., 2005). *Drosophila* has a preferred temperature around 24 °C and shows remarkable heat avoidance responses that are initiated by specialized temperature sensitive organs located in the head antennae. Even in the larval stage this organism displays heat avoidance responses mediated by specialized thermosensitive neurons (Liu, Yermolaieva, Johnson, Abboud, & Welsh, 2003; Sayeed & Benzer, 1996), an indication of the importance of the role played by temperature in insect differentiation. Heat-sensitive channels also mediate these thermal responses in the fly. The molecules responsible for heat sensation are TRPA1 cation channels, which in the case of *D. melanogaster* are very diverse, especially through alternative splicing (Neely et al., 2011; Zhong et al., 2012).

Both *D. melanogaster* and *C. elegans* also respond to decreases in temperature. The response of organisms to low temperatures is less well understood, but at least in these model organisms it seems to also be mediated by specialized receptor ion channels. In the larval stage of the fly, the

TRP and TRPL channels have been implicated in the avoidance response to low temperatures (<18 °C) (Rosenzweig, Kang, & Garrity, 2008). Adult cold sensitivity seems to involve a completely different set of proteins, perhaps not even ion channels (Gallio, Ofstad, Macpherson, Wang, & Zuker, 2011).

2.2 Bacteria

The exquisite responses to temperature found in flies and worms are also present in much simpler organisms, such as bacteria. The Gram-negative *Escherichia coli* displays a response to temperature that is very similar to their chemotactic behavior. In general, *E. coli* displays two modes of motion, smooth, directed swimming and random tumbling. When the cells encounter a chemoattractant gradient, they bias their movement to favor swimming, while tumbling provides a better way to randomly sample as much space as possible when the chemoattractant is not present (Berg & Brown, 1972). The behavioral response to a temperature gradient is very similar and cells will swim towards and aggregate at a particular, preferred temperature in the gradient (Maeda, Imae, Shioi, & Oosawa, 1976). The change in the pattern of movement is mediated by a change in the rotation direction of the flagellar motor. Directed swimming occurs when the motor turns in a counter clockwise direction and the movement of the three or more flagella is coordinated, while tumbling occurs when the motor turns clockwise and the flagella become uncoordinated and propel the cell in random directions (Turner, Ryu, & Berg, 2000).

How *E. coli* modulates its motor behavior in relation to temperature has been explored both in populations of bacteria and in single, immobilized bacilli. Using this last approach, it was demonstrated that this bacteria respond to temperature changes by increasing the probability of directed swimming, by means of coordinating the movement of their multiple flagella, in a similar fashion to the chemotactic response. It was also observed that the response to a very brief pulse of temperature as well as chemoattractant concentration (the impulse response), has the same temporal course and is compatible with the bacteria performing a differential measurement of the stimulus, that is, the cell is actually sensing the rate of change of the stimulus, not its absolute value (Block, Segall, & Berg, 1982; Paster & Ryu, 2008; Segall, Block, & Berg, 1986).

In accordance with this shared behavioral response, a very interesting observation in *E. coli* and presumably several other prokaryotes is that the molecular substrate of temperature sensing and chemotaxis seems to be the same. That is, the same chemoattractant receptors function as thermoreceptors.

Two of the main amino acid receptors, Tar and Tsr (Falke & Hazelbauer, 2001; Sourjik & Wingreen, 2012), have been shown to also act as thermoreceptors and mediate a behavioral response that is the same for high temperature and high amino acid concentration. In these conditions cells will swim, while in the opposite condition cells will tumble (Maeda & Imae, 1979; Maeda et al., 1976). Other chemoreceptors such as Trg and Tap, responsible for sensing ribose and galactose, respectively, have also been shown to display thermal sensitivity, although their physiological role as thermoreceptors has not been demonstrated (Nara, Lee, & Imae, 1991). These membrane proteins have an extracellular (periplasmic) receptor domain, and transmembrane and cytoplasmic domains that are responsible for signaling. They generally function as arrays and their mechanism is probably highly cooperative (Hazelbauer, Falke, & Parkinson, 2008). It should be noted that the molecular mechanism responsible for the special thermal sensitivity of these receptor molecules is unknown, as is the mechanism of coupling to the flagellar motor.

3. GENERAL THERMAL DEPENDENCE OF MEMBRANE PROPERTIES

3.1 Role of membrane fluidity

As can be gleaned from the previous section, thermal responses in very different organisms, from bacteria to vertebrates, seem to be mainly mediated by the specific heat activation of specialized receptor proteins. However, temperature variations are felt not only by these receptors but also by the medium surrounding the cell, formed by both the plasma membrane and the cytoplasmic and extracellular electrolytes. So, an intriguing question is, what are the contributions of these to the thermoreceptive event?

Like all phospholipid bilayers, the plasma and intracellular membranes of cells undergo temperature-driven phase transitions between at least three states. At low temperatures, the bilayer can adopt a highly ordered structure known as the gel or L_o phase, which is characterized by reduced diffusion coefficients of lipids and by low fluidity and increased rigidity. At higher temperatures, the membrane is in a liquid, less ordered phase, L_L, which is the common state of biological membranes, since the L_L phase generally occurs at temperatures above 20 °C. At even higher temperatures >40 °C, the bilayer can exist in a nonlamellar hexagonal phase. The exact temperature (the transition temperature) at which each phase transition occurs is highly dependent on the lipid composition of the specific membrane. Since the survival and function of the cell is fundamentally dependent on the fluid state of the membrane, the lipid composition is strictly controlled to maintain the bilayer in the L_L state.

This phenomenon is readily observed in bacteria and plants. In *E. coli* and other bacteria of the *Bacillus* genus, it has been observed that a 10 °C reduction in the temperature from that at which cells are optimally grown reduces the fluidity of the membrane, which in these organisms is detected mainly by a membrane-associated histidine kinase denominated DesK (Albanesi et al., 2009; Cybulski et al., 2002; Mansilla, Albanesi, Cybulski, & de Mendoza, 2005; Mansilla, Cybulski, Albanesi, & de Mendoza, 2004). In plants, a similar mechanism for responses to lowered temperature has been proposed (Murata & Los, 1997). The activation of the histidine kinase results in the activation of enzymatic pathways that lead to unsaturation of the fatty acids in membrane phospholipids, which ultimately results in a decreased temperature for the L_o−L_L phase transition and in an increased fluidity of the membrane (Klein, Weber, & Marahiel, 1999).

The response of these organisms to an increased temperature is the exact opposite. At least in cyanobacteria, the degree of unsaturation of fatty acids of the membrane phospholipids is reduced at temperatures higher than the optimal growth temperature (Sato & Murata, 1980).

In poikilothermic eukaryotic organisms, similar biochemical responses have been observed. For example, in membranes from hepatocytes of the trout (*Oncorhynchus mykiss*) that has been cold acclimated to a temperature 15 °C below their normal growth temperature, the phase transitions to the L_o, ordered phase and to nonlamellar phases occur at lower temperatures than in fish grown at normal temperature, indicating that the composition of the plasma membrane has been enriched in fluidity-maintaining polyunsaturated phospholipids (Hazel, McKinley, & Gerrits, 1998; Hazel & Williams, 1990).

The cell membranes of animals have cholesterol, which in a high concentration, generally increases the order of the lipid bilayer and reduces its fluidity (Cooper, 1978; McIntosh, Magid, & Simon, 1989). The plasma membrane content of cholesterol has also been demonstrated to be regulated during prolonged temperature challenges and it has been shown to be reduced in cold acclimated organisms (Robertson & Hazel, 1995), with the consequence that membrane fluidity is maintained at lower temperatures and the functional characteristics of the membrane are preserved.

An immediate hypothesis derived from the previous description of the physical chemistry of the response of membranes to temperature, is the possibility that the fluid state of the membrane itself could act as a temperature sensor and transmit the information encoded in the fluid state of the bilayer to effector proteins. This mechanism has been documented in bacteria and plants, but not in animals. In *Bacillus subtilis,* the thermosensitive DesK histidine kinase senses changes in the thickness of the membrane, which are associated with the altered fluidity of the lipid bilayer and contributes to lipid content remodeling as an adaptation to cold temperatures (Cybulski, Martin, Mansilla, Fernandez, & de Mendoza, 2010; Mansilla et al., 2005). In rice, a calcium-dependent protein kinase related to DesK becomes autophosphorylated in response to cold and it is speculated that this activation occurs as a response to the altered fluidity of the membrane (Martin & Busconi, 2001).

In animals, especially endotherms, the physiological range of responses to temperature changes, coincides with the range of temperatures in which the membrane remains in its liquid state. This observation seems to indicate that phase transitions in the plasma membrane do not play a role in the perception of temperature changes at the cellular level. Even though drastic changes in the structure of the membrane may not play a role in temperature sensitivity, the fluid state of the membrane will change in the range of temperatures experienced by sensory cells. If and how this contributes to temperature perception remains unclear. Many temperature receptors are proteins modulated by direct interactions with lipids or by the local mechanical state of the membrane (Christensen & Corey, 2007; Hardie, 2007; Ramsey, Delling, & Clapham, 2006). The interplay between the physical state of the membrane, its lipid composition, and temperature promises to be a fertile area of research.

3.2 Membrane electrical properties and temperature

Ever since the first quantitative description of cellular excitability, researchers have been interested in the temperature dependence of electrical phenomena (Adrian, 1921; Lorente de No, 1947). Very early it was observed that there exists an optimal temperature for the firing of action potentials and that this temperature is highly correlated with the environment of the particular organism. In the case of the squid giant axon, the magnitude of the action potential increases at lower temperatures, being maximal at the preferred temperature of the squid. At higher temperatures, the magnitude of the action potential is reduced, perhaps because the extent and speed of sodium (Na^+) channel inactivation is increased with temperature (Hodgkin & Katz, 1949; Ishiko & Loewenstein, 1961).

The resting membrane electrical properties of excitable cells of exothermic or poikilothermic organisms show small but significant dependence on changes in temperature. Measurements in lizard muscle of membrane resistance, R_m, cell input resistance, R_{in}, cytoplasmic conductivity, R_a and membrane length constant, λ as well as membrane capacitance, C_m, have shown that as a general rule, resistive parameters tend to decrease with increasing temperature (Adams, 1987). Similar results have been obtained in muscle cells and synaptic terminal of a crustacean neuromuscular junction (Stephens & Atwood, 1982).

A practical way to express the temperature dependence of these quantities is the Q_{10}.

The fold-change in the value of any parameter P as a function of a 10 °C temperature change, can be expressed as the temperature coefficient or Q_{10} value, defined as:

$$Q_{10} = \left(\frac{P_2}{P_1}\right)^{10/(T_2-T_1)}$$

In the case of the membrane electrical parameters described above, all of them have measured Q_{10} values between 1 and 2 (Davenport, 1992; Schmidt-Nielsen, 1975). It should be noted that the Q_{10} of a diffusive process in water is approximately equal to 1 (Sidell & Hazel, 1987), which explains why quantities such as the membrane resistive parameters, which fundamentally depend on ionic diffusion in water, have similar Q_{10} values.

Contrary to resistances, the value of the membrane capacitance C_m increases with temperature in muscle (Adams, 1987) and in erythrocyte cell membranes (Bao, Davis, & Schmukler, 1992). Since the relationship between C_m and the membrane area A, thickness d, and dielectric constant ε is: $C_m = \varepsilon\varepsilon_o \frac{A}{d}$, the increased value of C_m at higher temperatures implies that heat reduces the thickness of the membrane or increases its dielectric constant, or both (but see below, Section 5).

In the case of mammals, the thermal sensitivity of passive electrical properties is more heterogeneous than in invertebrates and lower vertebrates. It has been shown that in some preparations, like rat diaphragm muscle, resting membrane resistance also decreases as a function of increased temperature, with a $Q_{10} \sim 1$ (Palade & Barchi, 1977), in a manner similar to reptilian muscle, while other muscles, like goat intercostal, show an opposite temperature dependence (Lipicky & Bryant, 1972). These differences might be due, at least in part, to different contributions from chloride (Cl^-) and leak potassium (K^+) channels to the resting conductance, because these channels have a large temperature dependence (see chapters by Pusch and Zifarelli and Qneibi et al. in this tome).

The described changes induced by high temperature in passive electrical properties have direct implications for the physiology of excitable cells; the reduced input and membrane resistances will tend to reduce the excitability by increasing the threshold for action potential firing and the magnitude of synaptic currents needed to excite the cell. Similarly, since the length constant, λ, is equal to:

$$\lambda = \sqrt{\frac{R_m}{R_a + R_o}}$$

where R_o is the extracellular bath resistance, the implication is that the length constant will be also smaller at higher temperature, contributing to the decrease in cell excitability.

The increased capacitance, on the other hand, will contribute to a reduction of the speed of propagation of an action potential, due to the increased voltage needed to charge the augmented capacitance and to slow down depolarization by increasing the time constant of the membrane (Adrian, 1975; Fernandez, Taylor, & Bezanilla, 1983).

It should be noted that the changes in electrical properties that we have reviewed in the previous section are observed for temperature variations that occur in a range in which the plasma membrane remains in its liquid phase, in such a way that its fluidity should not be a factor.

4. PYROELECTRICITY

Pyroelectricity is a characteristic of certain highly ordered polar crystals that have a polarization charge when whole crystals are dilated or contracted by heating or cooling, respectively. The result of this charge separation is the generation of a transient voltage (Whatmore, 1991). In pyroelectric crystals of minerals such as tourmaline, the voltages generated can be very large. Pyroelectric phenomena are generally observed in highly organized crystals with a permanent dipole moment, which also makes these crystals piezoelectric (Kepler, 1978).

Pyroelectricity has been demonstrated to be present in vertebrate bone and tendon and it is thought to arise as a consequence of the highly ordered mineral crystal arrays present in these tissues. Voltages generated by the pyroelectric effect in bone are very small, usually less than 1 mV (Lang, 1969; Williams & Breger, 1975). Pyroelectricity might play a role in the modulation of bone regeneration and growth by alternating magnetic fields and fixed electric fields (Bassett, 1987; Bassett, Pawluk, & Pilla, 1974).

Interestingly, the existence of pyroelectricity has been also demonstrated in the integument of insects. In the case of the cockroach, heating produces a negative voltage in the inner surface of the integument and the magnitude of the generated voltage transients is about −6 mV for a few degrees temperature change. These pyroelectric voltage transients are of potential interest for insect thermal receptors, since they can occur in the integument surrounding sensitive organs associated with "hairs" or vibrissae in the antennae (Athenstaedt & Claussen, 1981; French, 1985; Hoger & French, 1999), which have been shown to increase action potential firing frequency as a consequence of increased temperature. Temperature sensation in *Drosophila* and possibly other insects has been demonstrated to depend on thermally activated TRP channels (Hamada et al., 2008; Kang et al., 2012) and so it is unlikely that a pyroelectric effect is directly responsible for insect temperature sensitivity. Nonetheless, the pyroelectric effect could potentially make an important contribution to the overall response of thermal receptor neurons in these organisms. It should be noted that the potential contribution of pyrolectricity to thermal sensation in insects or other organisms is still in need of demonstration.

5. INFRA RED RADIATION AND CAPACITANCE

Heat can be delivered to matter in several ways. A very effective manner that does not require contact between a hot and a cold body is to provide energy in the form of radiation. A hot body will generally emit heat in the form of infrared radiation (IR), which is defined as the band in the electromagnetic spectrum between approximately 700 nm and 1 mm. Most molecules absorb infrared radiation through interaction of their dipole moment with the Poynting vector of the radiation and this leads to an increase of the rotational and vibrational moments of the molecule (Bohren & Huffman, 2008; Feynman et al., 2013). Heat absorption by water is important since it is a major constituent of biological systems. Water is mostly transparent to visible light but capable of absorbing red and IR radiation and responding by increasing its temperature (Falk & Ford, 1966).

Water IR absorption has been long used as a mechanism to produce local increases in temperature (Waynant, Ilev, & Gannot, 2001), but recently it has been shown to have important effects in membrane electrical properties by producing increases in the membrane capacitance, just as temperature was shown many years ago to increase the capacitance of nerve membranes (as discussed in section 3.2).

It is known that IR radiation is capable of producing excitatory responses in several tissues including hair cells, embryonic heart cells, auditory neurons, to name a few (Jenkins et al., 2010; Littlefield, Vujanovic, Mundi, Matic, & Richter, 2010; Rajguru et al., 2011). Several possibilities exist to explain the molecular origins of these responses and none are exclusive. Thermosensitive ion channels or transporters could be activated or deactivated by the increased temperature produced by the IR radiation, leading to ionic fluxes that can produce depolarization and action potential firing. This mechanism will be shown to be viable and will be briefly reviewed in the following section.

Another possibility is that IR could alter physical properties of the membrane. As it turns out is has been recently established that IR radiation increases the membrane capacitance, C_m of cells. Capacitive current in an idealized cell without ionic flux pathways is given by:

$$I_c = C_m(t)\frac{dV}{dt}$$

If the capacitance is a time-varying function and the membrane potential V is constant, we expect to see a time varying capacitive current. In the case of the cell types tested, the HEK 293 cell line and *Xenopus* frog oocytes, the reversal potential of this current is positive, leading to inward currents that produce membrane potential depolarizations (Shapiro, Homma, Villarreal, Richter, & Bezanilla, 2012). If we think of the plasma membrane as a classical capacitor, the charge in the charged "plates" is provided both by fixed charge present in the phospholipid head groups and other charge present in membrane proteins, and a diffuse double layer formed by the electrolyte solution bathing the membrane (McLaughlin, 1989). It has been suggested that the to 6–10% capacitance increase induced by IR heat changes, is due to perturbations of this double layer and not to changes in the dielectric constant or thickness of the lipid bilayer. The change in capacitance is physiologically relevant, since the capacitive current induced by the change in temperature is enough to produce a depolarization sufficiently large to elicit action potentials (Shapiro et al., 2012). A different group also working with HEK293 cells has reported the same changes in membrane capacitance elicited by IR radiation (Okunade & Santos-Sacchi, 2013).

More recently it was demonstrated that the capacitance change induced by IR heating can be seen in native *C. elegans* neurons and be very large, enough to produce sufficient depolarization to elicit action potential firing. It was also shown that the magnitude of the capacitive effect is proportional to the rate of change of temperature and not to the absolute temperature (Liu, Frerck, Holman, Jorgensen, & Rabbitt, 2014).

6. ACTIVATION OF SPECIFIC CHANNELS BY IR

It has been known for some time that electromagnetic radiation provided by laser emission is capable of exciting individual neurons in mollusks (Fork, 1971) and several isolated nerve preparations (Izzo et al., 2008). Apart from the direct effects of infrared radiation on the plasma membrane discussed above, which are ultimately a consequence of temperature changes, IR stimulation produces a global increase in temperature in the irradiated area. These changes in temperature can be very substantial, rising temperature beyond 60 °C (Liljemalm, Nyberg, & von Holst, 2013; Yao, Liu, & Qin, 2009). It is now very well established that increased temperatures activate several ion channels, especially some belonging to the thermoTRP variety. So it is not surprising that recent experimental work that has begun to unravel the mechanisms of IR-induced neuronal excitability, is showing that specific thermoTRPs are involved.

A recent study (Albert et al., 2012) demonstrated that the excitatory effects of IR radiation on sensory neurons is mediated by the ability to activate the thermoTRP ion channel TRPV4 by increasing the temperature around the cell. IR radiation-induced heating has been elegantly used as a tool to study thermoTRP channels. Yao et al. (Yao, Liu, & Qin, 2010; Yao et al., 2009) utilized an IR laser system to induce very rapid and constant temperature changes that are capable of activating the prototypical heat-activated TRPV1 channel, along with other heat-activated channels, like TRV3, TRPV2, and TRPV4 (Yao, Liu, & Qin, 2011).

Infrared-elicited temperature increases have a distinct physiological importance. It has been demonstrated that the special IR sensitivity of pit-bearing snakes is mediated by the activation of heat-activated TRPA1 channels; most likely due to an increased temperature caused by IR. In a similar fashion, heat-activated TRPV1 channels mediate the IR detection present in certain vampire bats (Gracheva et al., 2010, 2011).

It can be concluded from the last two sections that the effects of infrared radiation on excitable cells are all mediated by increases of temperature and are excitatory. On the one hand, increased temperature increases the capacitance, a fundamental property of membranes. Infrared produces substantial increases in temperature that result in specific activation of heat sensitive ion channels.

7. CONCLUSIONS

Thermal relationships of organisms with their environment are of fundamental importance. These interactions are mediated by a myriad of mechanisms, from simple temperature-dependence of physicochemical variables to the more sophisticated and specialized receptor/channel mediated signal transduction systems.

The research into the molecular mechanisms of temperature sensitivity has focused mainly on identifying molecules with especial temperature sensitivity. It is the aim of this chapter to provide a wider view of the mechanisms involved, with the hope of attaining a unified understanding of thermal sensation.

ACKNOWLEDGMENTS

I would like to thank the following agencies for financial support. CONACYT for grant No. 151297 and DGAPA-PAPIIT-UNAM for grant No. IN209209.

REFERENCES

Adams, B. A. (1987). Thermal dependence of passive electrical properties of lizard muscle fibres. *Journal of Experimental Biology, 133*, 169–182.

Adrian, E. D. (1921). The recovery process of excitable tissues: part II. *Journal of Physiology, 55*(3–4), 193–225.

Adrian, R. H. (1975). Conduction velocity and gating current in the squid giant axon. *Proceedings of the Royal Society of London, Series B: Biological Sciences, 189*(1094), 81–86.

Albanesi, D., Martin, M., Trajtenberg, F., Mansilla, M. C., Haouz, A., Alzari, P. M., et al. (2009). Structural plasticity and catalysis regulation of a thermosensor histidine kinase. *Proceedings of the National Academy of Sciences of the United States of America, 106*(38), 16185–16190.

Albert, E. S., Bec, J. M., Desmadryl, G., Chekroud, K., Travo, C., Gaboyard, S., et al. (2012). TRPV4 channels mediate the infrared laser-evoked response in sensory neurons. *Journal of Neurophysiology, 107*(12), 3227–3234.

Angilletta, M. J. (2009). *Thermal adaptation: A theoretical and empirical synthesis*. Oxford; New York: Oxford University Press.

Athenstaedt, H., & Claussen, H. (1981). Evidence for pyroelectric and piezoelectric sensory mechanisms in the insect integument. *Biophysical Journal, 35*(2), 365–374.

Bao, J. Z., Davis, C. C., & Schmukler, R. E. (1992). Frequency domain impedance measurements of erythrocytes. Constant phase angle impedance characteristics and a phase transition. *Biophysical Journal, 61*(5), 1427–1434.

Bassett, C. A. (1987). Electrical stimulation of repair of bone. *The Journal of Bone and Joint Surgery American Volume, 69*(7), 1110–1111.

Bassett, C. A., Pawluk, R. J., & Pilla, A. A. (1974). Augmentation of bone repair by inductively coupled electromagnetic fields. *Science, 184*(4136), 575–577.

Berg, H. C., & Brown, D. A. (1972). Chemotaxis in *Escherichia coli* analysed by three-dimensional tracking. *Nature, 239*(5374), 500–504.

Bertness, M. D., & Schneider, D. E. (1976). Temperature relations of puget sound thaids in reference to their intertidal distribution [Thais lamellosa; Thais emarginata]. *Veliger (United States), 19*(1).

Block, S. M., Segall, J. E., & Berg, H. C. (1982). Impulse responses in bacterial chemotaxis. *Cell, 31*(1), 215–226.

Bohren, C. F., & Huffman, D. R. (2008). *Absorption and scattering of light by small particles*. John Wiley & Sons.

Cameron, R. A. D. (1970). The effect of temperature on the activity of three species of helicid snail (Mollusca: Gastropoda). *Journal of Zoology, 162*(3), 303–315.

Christensen, A. P., & Corey, D. P. (2007). TRP channels in mechanosensation: direct or indirect activation? *Nature Reviews Neuroscience, 8*(7), 510–521.

Cooper, R. A. (1978). Influence of increased membrane cholesterol on membrane fluidity and cell function in human red blood cells. *Journal of Supramolecular Structure, 8*(4), 413–430.

Cox, T. J., & Rutherford, J. C. (2000). Thermal tolerances of two stream invertebrates exposed to diurnally varying temperature. *New Zealand Journal of Marine and Freshwater Research, 34*(2), 203–208.

Cybulski, L. E., Albanesi, D., Mansilla, M. C., Altabe, S., Aguilar, P. S., & de Mendoza, D. (2002). Mechanism of membrane fluidity optimization: isothermal control of the *Bacillus subtilis* acyl-lipid desaturase. *Molecular Microbiology, 45*(5), 1379–1388.

Cybulski, L. E., Martin, M., Mansilla, M. C., Fernandez, A., & de Mendoza, D. (2010). Membrane thickness cue for cold sensing in a bacterium. *Current Biology, 20*(17), 1539–1544.

Davenport, J. (1992). *Animal life at low temperature* (1st ed.). London; New York: Chapman & Hall.

Falke, J. J., & Hazelbauer, G. L. (2001). Transmembrane signaling in bacterial chemoreceptors. *Trends in Biochemical Sciences, 26*(4), 257–265.

Falk, M., & Ford, T. A. (1966). Infrared spectrum and structure of liquid water. *Canadian Journal of Chemistry, 44*(14), 1699–1707.

Fernandez, J. M., Taylor, R. E., & Bezanilla, F. (1983). Induced capacitance in the squid giant axon. Lipophilic ion displacement currents. *Journal of General Physiology, 82*(3), 331–346.

Feynman, R. P., Leighton, R. B., & Sands, M. (2013). Desktop Edition Volume I. *The Feynman lectures on physics* (Vol. 1). Basic Books.

Fork, R. L. (1971). Laser stimulation of nerve cells in aplysia. *Science, 171*(3974), 907–908.

French, A. S. (1985). The effects of temperature on action potential encoding in the cockroach tactile spine. *Journal of Comparative Physiology A, 156*(6), 817–821.

Gallio, M., Ofstad, T. A., Macpherson, L. J., Wang, J. W., & Zuker, C. S. (2011). The coding of temperature in the *Drosophila* Brain. *Cell, 144*(4), 614–624.

Glauser, D. A. (2013). How and why *Caenorhabditis elegans* uses distinct escape and avoidance regimes to minimize exposure to noxious heat. *Worm, 2*(4), e27285.

Glauser, D. A., Chen, W. C., Agin, R., Macinnis, B. L., Hellman, A. B., Garrity, P. A., et al. (2011). Heat avoidance is regulated by transient receptor potential (TRP) channels and a neuropeptide signaling pathway in *Caenorhabditis elegans*. *Genetics, 188*(1), 91–103.

Gracheva, E. O., Cordero-Morales, J. F., González-Carcacía, J. A., Ingolia, N. T., Manno, C., Aranguren, C. I., et al. (2011). Ganglion-specific splicing of TRPV1 underlies infrared sensation in vampire bats. *Nature, 476*(7358), 88–91.

Gracheva, E. O., Ingolia, N. T., Kelly, Y. M., Cordero-Morales, J. F., Hollopeter, G., Chesler, A. T., et al. (2010). Molecular basis of infrared detection by snakes. *Nature, 464*(7291), 1006–1011.

Hamada, F. N., Rosenzweig, M., Kang, K., Pulver, S. R., Ghezzi, A., Jegla, T. J., et al. (2008). An internal thermal sensor controlling temperature preference in *Drosophila*. *Nature, 454*(7201), 217–220.

Hardie, R. C. (2007). TRP channels and lipids: from *Drosophila* to mammalian physiology. *Journal of Physiology, 578*(1), 9–24.

Hazelbauer, G. L., Falke, J. J., & Parkinson, J. S. (2008). Bacterial chemoreceptors: high-performance signaling in networked arrays. *Trends in Biochemical Sciences, 33*(1), 9–19.

Hazel, J. R., McKinley, S. J., & Gerrits, M. F. (1998). Thermal acclimation of phase behavior in plasma membrane lipids of rainbow trout hepatocytes. *American Journal of Physiology, 275*(3 Pt 2), R861–R869.

Hazel, J. R., & Williams, E. E. (1990). The role of alterations in membrane lipid composition in enabling physiological adaptation of organisms to their physical environment. *Progress in Lipid Research, 29*(3), 167–227.

Hodgkin, A. L., & Katz, B. (1949). The effect of temperature on the electrical activity of the giant axon of the squid. *Journal of Physiology, 109*(1–2), 240–249.

Hoffmann, A. A., Sørensen, J. G., & Loeschcke, V. (2003). Adaptation of *Drosophila* to temperature extremes: bringing together quantitative and molecular approaches. *Journal of Thermal Biology, 28*(3), 175–216.

Hoger, U., & French, A. S. (1999). Temperature sensitivity of transduction and action potential conduction in a spider mechanoreceptor. *Pflugers Archiv, 438*(6), 837–842.

Ishiko, N., & Loewenstein, W. R. (1961). Effects of temperature on the generator and action potentials of a sense organ. *Journal of General Physiology, 45*, 105–124.

Izzo, A. D., Walsh, J. T., Jr., Jr., Ralph, H., Webb, J., Bendett, M., Wells, J., et al. (2008). Laser stimulation of auditory neurons: effect of shorter pulse duration and penetration depth. *Biophysical Journal, 94*(8), 3159–3166.

Jenkins, M. W., Duke, A. R., Gu, S., Doughman, Y., Chiel, H. J., Fujioka, H., et al. (2010). Optical pacing of the embryonic heart. *Nature Photonics, 4*(9), 623–626.

Kang, K., Panzano, V. C., Chang, E. C., Ni, L., Dainis, A. M., Jenkins, A. M., et al. (2012). Modulation of TRPA1 thermal sensitivity enables sensory discrimination in *Drosophila*. *Nature, 481*(7379), 76–80.

Kepler, R. G. (1978). Piezoelectricity, pyroelectricity, and ferroelectricity in organic materials. *Annual Review of Physical Chemistry, 29*(1), 497–518.

Klein, W., Weber, M. H., & Marahiel, M. A. (1999). Cold shock response of Bacillus subtilis: isoleucine-dependent switch in the fatty acid branching pattern for membrane adaptation to low temperatures. *Journal of Bacteriology, 181*(17), 5341–5349.

Lang, S. B. (1969). Thermal expansion coefficients and the primary and secondary pyroelectric coefficients of animal bone. *Nature, 224*(5221), 798–799.

Li, W., Kang, L., Piggott, B. J., Feng, Z., & Xu, X. Z. (2011). The neural circuits and sensory channels mediating harsh touch sensation in *Caenorhabditis elegans*. *Nature Communications, 2*, 315.

Liljemalm, R., Nyberg, T., & von Holst, H. (2013). Heating during infrared neural stimulation. *Lasers in Surgery and Medicine, 45*(7), 469–481.

Lipicky, R. J., & Bryant, S. H. (1972). Temperature effects on cable parameters and K efflux in normal and myotonic goat muscle. *American Journal of Physiology, 222*(1), 213–215.

Littlefield, P. D., Vujanovic, I., Mundi, J., Matic, A. I., & Richter, C. P. (2010). Laser stimulation of single auditory nerve fibers. *The Laryngoscope, 120*(10), 2071–2082.

Liu, Q., Frerck, M. J., Holman, H. A., Jorgensen, E. M., & Rabbitt, R. D. (2014). Exciting cell membranes with a blustering heat shock. *Biophysical Journal, 106*(8), 1570–1577.

Liu, L., Yermolaieva, O., Johnson, W. A., Abboud, F. M., & Welsh, M. J. (2003). Identification and function of thermosensory neurons in *Drosophila* larvae. *Nature Neuroscience, 6*(3), 267–273.

Lorente de No, R. (1947). A study of nerve physiology. *Studies from the Rockefeller Institute for Medical Research Reprints, 132*, 1–548.
Maeda, K., & Imae, Y. (1979). Thermosensory transduction in *Escherichia coli*: inhibition of the thermoresponse by L-serine. *Proceedings of the National Academy of Sciences of the United States of America, 76*(1), 91–95.
Maeda, K., Imae, Y., Shioi, J. I., & Oosawa, F. (1976). Effect of temperature on motility and chemotaxis of *Escherichia coli*. *Journal of Bacteriology, 127*(3), 1039–1046.
Mansilla, M. C., Albanesi, D., Cybulski, L. E., & de Mendoza, D. (2005). Molecular mechanisms of low temperature sensing bacteria. *Annals of Hepatology, 4*(3), 216–217.
Mansilla, M. C., Cybulski, L. E., Albanesi, D., & de Mendoza, D. (2004). Control of membrane lipid fluidity by molecular thermosensors. *Journal of Bacteriology, 186*(20), 6681–6688.
Martin, M. L., & Busconi, L. (2001). A rice membrane-bound calcium-dependent protein kinase is activated in response to low temperature. *Plant Physiology, 125*(3), 1442–1449.
McIntosh, T. J., Magid, A. D., & Simon, S. A. (1989). Cholesterol modifies the short-range repulsive interactions between phosphatidylcholine membranes. *Biochemistry, 28*(1), 17–25.
McLaughlin, S. (1989). The electrostatic properties of membranes. *Annual Review of Biophysics and Biophysical Chemistry, 18*(1), 113–136.
McMahon, R. F., & Russell-Hunter, W. D. (1977). Temperature relations of aerial and aquatic respiration in six littoral snails in respiration in six littoral snails in relation to their vertical zonation. *Biological Bulletin, 152*(2), 182–198.
Mohammadi, A., Byrne Rodgers, J., Kotera, I., & Ryu, W. S. (2013). Behavioral response of *Caenorhabditis elegans* to localized thermal stimuli. *BMC Neuroscience, 14*(1), 66.
Murata, N., & Los, D. A. (1997). Membrane fluidity and temperature perception. *Plant Physiology, 115*(3), 875–879.
Nara, T., Lee, L., & Imae, Y. (1991). Thermosensing ability of Trg and Tap chemoreceptors in *Escherichia coli*. *Journal of Bacteriology, 173*(3), 1120–1124.
Neely, G. G., Keene, A. C., Duchek, P., Chang, E. C., Wang, Q.-P., Aksoy, Y. A., et al. (2011). TrpA1 regulates thermal nociception in *Drosophila*. *PloS One, 6*(8), e24343.
Okunade, O., & Santos-Sacchi, J. (2013). IR laser-induced perturbations of the voltage-dependent solute carrier protein SLC26a5. *Biophysical Journal, 105*(8), 1822–1828.
Palade, P. T., & Barchi, R. L. (1977). Characteristics of the chloride conductance in muscle fibers of the rat diaphragm. *Journal of General Physiology, 69*(3), 325–342.
Paster, E., & Ryu, W. S. (2008). The thermal impulse response of *Escherichia coli*. *Proceedings of the National Academy of Sciences of the United States of America, 105*(14), 5373–5377.
Patapoutian, A., Peier, A. M., Story, G. M., & Viswanath, V. (2003). ThermoTRP channels and beyond: mechanisms of temperature sensation. *Nature Reviews Neuroscience, 4*(7), 529–539.
Rajguru, S. M., Richter, C. P., Matic, A. I., Holstein, G. R., Highstein, S. M., Dittami, G. M., et al. (2011). Infrared photostimulation of the crista ampullaris. *Journal of Physiology, 589*(6), 1283–1294.
Ramsey, I. S., Delling, M., & Clapham, D. E. (2006). An introduction to TRP channels. *Annual Review of Physiology, 68*, 619–647.
Robertson, J. C., & Hazel, J. R. (1995). Cholesterol content of trout plasma membranes varies with acclimation temperature. *American Journal of Physiology, 269*(5 Pt 2), R1113–R1119.
Rosenzweig, M., Brennan, K. M., Tayler, T. D., Phelps, P. O., Patapoutian, A., & Garrity, P. A. (2005). The *Drosophila* ortholog of vertebrate TRPA1 regulates thermotaxis. *Genes & Development, 19*(4), 419–424.
Rosenzweig, M., Kang, K., & Garrity, P. A. (2008). Distinct TRP channels are required for warm and cool avoidance in *Drosophila melanogaster*. *Proceedings of the National Academy of Sciences of the United States of America, 105*(38), 14668–14673.

Sato, N., & Murata, N. (1980). Temperature shift-induced responses in lipids in the blue-green alga, *Anabaena variabilis*: the central role of diacylmonogalactosylglycerol in thermo-adaptation. *Biochimica et Biophysica Acta, 619*(2), 353–366.

Sayeed, O., & Benzer, S. (1996). Behavioral genetics of thermosensation and hygrosensation in *Drosophila*. *Proceedings of the National Academy of Sciences of the United States of America, 93*(12), 6079–6084.

Schmidt-Nielsen, K. (1975). *Animal physiology: Adaptation and environment*. London; New York: Cambridge University Press.

Segall, J. E., Block, S. M., & Berg, H. C. (1986). Temporal comparisons in bacterial chemotaxis. *Proceedings of the National Academy of Sciences of the United States of America, 83*(23), 8987–8991.

Shapiro, M. G., Homma, K., Villarreal, S., Richter, C. P., & Bezanilla, F. (2012). Infrared light excites cells by changing their electrical capacitance. *Nature Communications, 3*, 736.

Sharkey, T. D., & Schrader, S. M. (2006). High temperature stress. In *Physiology and molecular biology of stress tolerance in plants* (pp. 101–129). Springer.

Sidell, B. D., & Hazel, J. R. (1987). Temperature affects the diffusion of small molecules through cytosol of fish muscle. *Journal of Experimental Biology, 129*, 191–203.

Sourjik, V., & Wingreen, N. S. (2012). Responding to chemical gradients: bacterial chemotaxis. *Current Opinion in Cell Biology, 24*(2), 262–268.

Stephens, P. J., & Atwood, H. L. (1982). Thermal acclimation in a crustacean neuromuscular system. *Journal of Experimental Biology, 98*, 39–47.

Tobin, D. M., & Bargmann, C. I. (2004). Invertebrate nociception: behaviors, neurons and molecules. *Journal of Neurobiology, 61*(1), 161–174.

Turner, L., Ryu, W. S., & Berg, H. C. (2000). Real-time imaging of fluorescent flagellar filaments. *Journal of Bacteriology, 182*(10), 2793–2801.

Waynant, R. W., Ilev, I. K., & Gannot, I. (2001). Mid–infrared laser applications in medicine and biology. *Philosophical Transactions of the Royal Society of London. Series A: Mathematical Physical and Engineering Sciences, 359*(1780), 635–644.

Whatmore, R. W. (1991). Piezoelectric and pyroelectric materials and their applications. In L. S. Miller, & J. B. Mullin (Eds.), *Electronic materials* (pp. 283–290). US: Springer.

Williams, W. S., & Breger, L. (1975). Piezoelectricity in tendon and bone. *Journal of Biomechanics, 8*(6), 407–413.

Willmer, P., Stone, G., & Johnston, I. (2009). *Environmental physiology of animals*. John Wiley & Sons.

Wittenburg, N., & Baumeister, R. (1999). Thermal avoidance in *Caenorhabditis elegans*: an approach to the study of nociception. *Proceedings of the National Academy of Sciences of the United States of America, 96*(18), 10477–10482.

Yao, J., Liu, B., & Qin, F. (2009). Rapid temperature jump by infrared diode laser irradiation for patch-clamp studies. *Biophysical Journal, 96*(9), 3611–3619.

Yao, J., Liu, B., & Qin, F. (2010). Kinetic and energetic analysis of thermally activated TRPV1 channels. *Biophysical Journal, 99*(6), 1743–1753.

Yao, J., Liu, B., & Qin, F. (2011). Modular thermal sensors in temperature-gated transient receptor potential (TRP) channels. *Proceedings of the National Academy of Sciences of the United States of America, 108*(27), 11109–11114.

Zhong, L., Bellemer, A., Yan, H., Honjo, K., Robertson, J., Hwang, R. Y., et al. (2012). Thermosensory and nonthermosensory isoforms of *Drosophila melanogaster* TRPA1 reveal heat-sensor domains of a thermoTRP channel. *Cell Reports, 1*(1), 43–55.

CHAPTER TWO

Temperature Sensing by Thermal TRP Channels: Thermodynamic Basis and Molecular Insights

Qin Feng
Department of Physiology and Biophysics, State University of New York, Buffalo, New York, USA

Contents

Abstract

All organisms need to sense temperature in order to survive and adapt. But how they detect and perceive temperature remains poorly understood. Recent discoveries of thermal Transient Receptor Potential (TRP) ion channels have shed light on the problem and unravel molecular entities for temperature detection and transduction in mammals. Thermal TRP channels belong to the large family of transient receptor potential channels. They are directly activated by heat or cold in physiologically relevant temperature ranges, and the activation is exquisitely sensitive to temperature changes. Thermodynamically, this strong temperature dependence of thermal channels occurs due to large enthalpy and entropy changes associated with channel opening. Thus understanding how the channel proteins obtain their exceptionally large energetics is central toward determining functional mechanisms of thermal TRP channels. The purpose of this chapter is to provide a comprehensive review on critical issues and challenges facing the problem, with emphases on underlying biophysical and molecular mechanisms.

Current Topics in Membranes, Volume 74
ISSN: 1063-5823
http://dx.doi.org/10.1016/B978-0-12-800181-3.00002-6

1. INTRODUCTION

The detection of ambient temperature is necessary for most organisms to seek preferred living temperatures and to avoid potentially damaging conditions. The detection of internal body temperature is also required for species capable of thermal regulation. In some species thermal sensation has even become the sixth sense. Snakes possess heat vision to detect a temperature difference between a moving prey and its surroundings on the scale of milliKelvins (Bulloc & Dieck, 1956). Fire-chasing beetles can sense infrared radiation produced by fires up to 130 km (Hart, 1998; Schmitz & Bleckmann, 1998). Despite such remarkable features, however, how temperature is detected, perceived, and regulated remains poorly understood in most organisms.

Thermal sensation in mammals involves peripheral sensory nerves innervating the skin and internal organs. The other end of the nerves enters the central nervous systems in the superficial dorsal horn of the spinal cord and end in the thalamus and somatosensory cortex where consciousness is made about what is happening on the surface—warm, cool, hot, or cold. As early as 1882, Blix discovered that a person's thermal sensations were associated with the stimulation of localized sensory spots on the skin (Blix, 1882). The modern pursuit of thermal sensation based on electrical recordings from skin-nerve preparations demonstrated unequivocally the existence of thermoreceptors (Hensel, 1974; Spray, 1986). On the basis of their conduction velocities, they are known to be small-diameter, slowly conducting unmyelinated C fibers and larger, more rapidly conducting, thinly myelinated Aδ fibers.

The sensations of temperature and pain are closely related. They both involve the C fibers, which are responsive to noxious thermal, mechanical, and chemical stimuli. The pain evoked by heat produces a sensation of burning (LaMotte & Campbell, 1978; Torebjork & Hallin, 1973), whereas the pain induced by cold can have various qualities including aching, burning, and pricking (Chery-Croze, 1983; Kreh et al., 1984; Lewis & Love, 1926; Rainville et al., 1992; Wahren, Torebjork, & Jorum, 1989; Wolf & Hardy, 1941; Yarnitsky & Ochoa, 1990). Temperatures that are normally innocuous can become noxious under pathological conditions (Julius & Basbaum, 2001; Levine et al., 1999; Sato et al., 2000; Takahashi, Sato, & Mizumura, 2003). Many forms of clinical pain are related to disorders of thermal sensation.

Although cutaneous thermal receptors had been implied decades ago, their molecular entities have only begun to emerge recently. Capsaicin-activated ion channels are well known for their roles in nociception and underlie the hallmark sensitivity of nociceptive neurons to chili peppers (Jancso, 1955; Jancso, Jancso-Gabor, & Szolcsanyi, 1967; Szolcsanyi & Jancso-Gabor, 1975, 1976; Wood, 1993). The seminal study of cloning and characterization of the channel leads to identification of the first molecular transducer in thermal sensation and nociception (Caterina et al., 1997). Subsequent searches for its homologs uncover a large number of related proteins in mammalian genomes. Collectively these channels fund the now rapidly growing transient receptor potential superfamily. Today, the TRP family contains 28 members and falls into seven main subfamilies (Clapham, 2003), TRPC (canonical), TRPV (vanilloid), TRPM (melastatin), TRPP (polycystin), TRPML (mucolipin), TRPA (ankyrin), and TRPN (NOMPC). Several members across different subfamilies have been found to be activated by temperature with distinct thermal properties. Their responsiveness ranges are correlated well with physiological temperatures causing the sensations of warm, cool, hot, and cold, thus supporting a general role for these thermal TRP channels in nociception and thermal sensation (Patapoutian et al., 2003). Plant-derived natural products that mimic temperature sensations also activate thermal TRP channels (Xu et al., 2006). The disruption of thermal TRP genes in mice results in deficient thermal sensitivity and reduced chemical and thermal hyperalgesia (Bautista et al., 2007; Caterina et al., 2000; Davis et al., 2000; Dhaka et al., 2007; Moqrich et al., 2005). Both pharmacological and behavioral characteristics support thermal TRP channels as key components on thermal and pain transduction pathways.

Thermal TRP channels are not only essential to acute nociception but are also substrates of chronic inflammatory mediators released in pathological pain states. It has been suggested that they contribute to such physiopathological conditions as inflammatory hyperalgesia, diabetic neuropathy, neuropathic pain associated with nerve lesion, etc. (Akbar et al., 2008; Cantero-Recasens et al., 2010; Engler et al., 2007; Meents, Neeb, & Reuter, 2010; Szallasi, 2002; Tsavaler et al., 2001; Valdes et al., 2011; Wondergem & Bartley, 2009; Wondergem et al., 2008; Yamamura et al., 2008). Owing to their significant roles in chronic pain, thermal TRP channels have become attractive targets for development of novel pain therapies that prevent generation and transduction of pain (Szallasi, Cruz, & Geppetti, 2006; TRPM, 2011).

Exquisite thermal sensitivity is a unique feature of thermal TRP channels and underlies their biological functions. All ion channels are sensitive to temperature, but few are directly activated by temperature and none has a sensitivity close to that of thermal TRP channels. Q_{10} is a common measurement of temperature dependence of a protein and describes the fold-change in response when temperature is increased by 10°. Whereas most ion channels have a Q_{10} in the range of 2–3 (DeCoursey & Cherny, 1998; Hille, 2001), thermal TRP channels reach a $Q_{10} > 100$ (Leffler et al., 2007; Yao, Liu, & Qin, 2010a, 2010b, 2011). This strong temperature dependence enables thermal TRP channels to discriminate small temperature gradients, but it also raises interesting questions on how these channels obtain the unusually strong thermal sensitivity. Presently, there is still a limited understanding of the issue owing to inherent complexity of the problem. Compared to voltage- or ligand-gated channels, the study of thermal TRP channels is still in its infancy. Below we will provide an overview on the status of the field, focusing on the biophysical and thermodynamic mechanisms and the molecular basis underlying the thermal sensitivity of the channels.

2. PRINCIPLES OF TEMPERATURE ACTIVATION

By Boltzmann equation the opening of an ion channel is determined by its free energy difference between the closed state and the open state (Hille, 2001):

$$P_o = \frac{1}{1 + e^{\frac{\Delta G}{RT}}}$$

where ΔG is the free energy change and R and T have their standard definitions. The free energy change is temperature dependent and can be represented explicitly in T by $\Delta G = \Delta H - T\,\Delta S$ where ΔH and ΔS are enthalpy change and entropy change, respectively. Thus the opening of a channel is related to temperature by

$$P_o = \frac{1}{1 + e^{\frac{\Delta H}{RT} - \frac{\Delta S}{R}}}.$$

This simple equation indicates that the thermal sensitivity of a channel lies in the enthalpy change during opening. The sign of the enthalpy change dictates the polarity of the thermal sensitivity: the channel is heat sensitive if

$\Delta H > 0$ and conversely cold sensitive. The opening becomes temperature independent if the net enthalpy change is vanished, which could occur if the opening and closing rates involve similar activation enthalpies. Both ΔH and ΔS could be also temperature dependent, for example, if the opening of the channel involves a heat capacity change. But over a narrow responsiveness range, ΔH and ΔS can be reasonably considered to be constant on a first order approximation.

Similar to the responsiveness curves of ligand- or voltage-gated channels, the temperature-evoked opening of thermal TRP channels also has a sigmoidal relationship with respect to the reciprocal of temperature. The two phenomenological parameters describing the curve are the midpoint and the slope sensitivity. The midpoint occurs at $\Delta G = 0$, i.e., when the entropic energy cancels the enthalpy, or $T_{1/2} = \Delta H/\Delta S$. The slope of the curve has the general relationship:

$$\frac{dP_o}{d\left(\frac{1}{T}\right)} = -P_o(1 - P_o)\frac{\Delta H}{R}$$

which becomes $\Delta H/4R$ at the midpoint. Thus the slope sensitivity at the midpoint of the responsiveness curve is proportional to the enthalpy change. The steepness of the curve, not the midpoint, measures the temperature dependence of thermal TRP channels.

Often the full-range temperature responses of thermal TRP channels cannot be resolved because the saturation temperature is either too high or too low beyond the experimentally accessible range. In such cases two other empirical parameters are useful quantifying the temperature dependence of the channels, including the activation threshold and the Q_{10} of the initial response. The activation threshold is an arbitrarily chosen value for a minimum activity that becomes visually resolvable above the background noise. Such a defined threshold has no simple mechanistic interpretation. Q_{10}, on the other hand, is usually defined for the rate of a process. When generalized to an equilibrium response, it may be applied to the initial activation where P_o is sufficiently small so that it can be approximated by:

$$P_o = e^{-\frac{\Delta H}{RT} + \frac{\Delta S}{R}}$$

Under this assumption, the Q_{10} of the response follows:

$$Q_{10} = e^{\frac{10\Delta H}{RT(T+10)}}$$

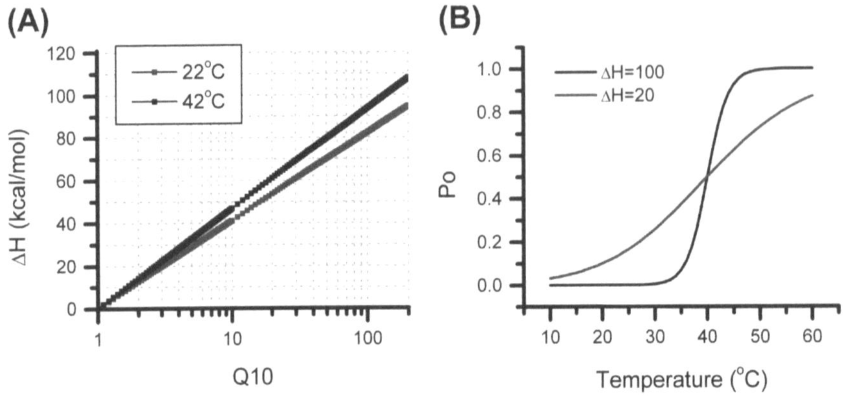

Figure 2.1 The temperature dependence of ion channel gating. (A) The correspondence of enthalpy change (ΔH) with Q_{10}. (B) Boltzmann relationship of P_o versus temperature for different values of enthalpy changes ($\Delta H = 20$ vs 100 kcal/mol).

and is explicitly related to the enthalpy change. This type of analysis of temperature dependence is analogous to the limiting slope analysis of voltage dependence for voltage-gated channels. Thus it suffers from the similar limitation that it is only applicable if no openings are present in the absence of a stimulus.

As a temperature transducer, thermal TRP channels need to have a responsiveness curve whose slope is as steep as possible. Below the activation threshold the channel is closed, but once temperature is above the threshold, the channel becomes maximally open. For most ion channels, the conformational change occurring during channel opening involves an enthalpy change in the range of 20–30 kcal/mol, corresponding to a Q_{10} value of 2–3 (Figure 2.1). As evident from Figure 2.1, the responsiveness of such a channel would span a wide temperature range (>50 °C), which is clearly inadequate as a thermal switch. Thus an intriguing question is what is the maximal thermal sensitivity a biological channel can have?

3. TEMPERATURE DEPENDENCE OF THERMAL TRP CHANNELS

The quantification of temperature dependence of thermal TRP channels is not as trivial as might be expected and depends on how it is measured among other experimental conditions. Typically, the temperature coefficient of the initial response on the rising phase of a temperature ramp is measured. The rate of heating or cooling is arbitrary and is usually set by the responsiveness time of the apparatus. The Q_{10} value is evaluated from the change of the current between two absolute temperatures, say T_1 and T_2, according to

$$Q_{10} = \left(\frac{I_2}{I_1}\right)^{\frac{10T_2}{(T_2-T_1)(T_1+10)}}$$

where I_1 and I_2 are absolute current values at T_1 and T_2, respectively. The enthalpy change results from the Q_{10} value according to

$$\Delta H = \frac{1}{10} RT_1(T_1 + 10)\ln Q_{10} \approx \frac{1}{10} RT_1^2 \ln Q_{10}.$$

The absolute current response can also be plotted on a log scale against the reciprocal of the absolute temperature. The enthalpy change is determined by the slope of the plot according to the Arrhenius equation $I = I_0 \exp(\Delta H/RT)$.

Table 2.1 summarizes temperature coefficients and enthalpy changes of thermal TRP channels (for comparison, a few other strongly temperature-dependent ion channels are also included at the end). Overall, thermal TRP channels have a temperature coefficient above 10, which is considered to be strongly temperature dependent. However, the measurements also show considerable variations. Even for the same channel, they may differ by many folds. Thus the estimates obtained from the initial temperature response need to be taken with a grain of salt. There are multiple reasons for the variations, such as the difficulty of experimental controls and the inherent gating complexity of thermal TRP channels.

Temperature control is always a major challenge in the study of thermal TRP channels. Conventional temperature controls are generally based on thermoelectric heating or cooling and have a slow time response, typically with a rate in the range of <1 °C/s (Welch, Simon, & Reinhart, 2000; Xu et al., 2002). The slow heating incurs technical challenges in experimentation, because it induces thermal stress on patch seals and compromises cell viability. The problem becomes further aggravated for channels with high activation thresholds. Heat-sensitive channels like TRPV1 and TRPV2 are activated by temperatures above 40–50 °C. Prolonged exposure to this high temperature can cause many problems including an increased leakage. The leak current, partly from the seal and partly from the cell membrane, is also temperature dependent and may obscure the currents of channels.

Table 2.1 Temperature dependence of thermal TRP channels

Channel	Q_{10}	References
TRPV1	21,26,40, >100	Liu et al. (2003), Vlachova, Simon, and Reinhart (2003), Welch et al. (2000), Yao et al. (2010a)
TRPV2	>100	Leffler et al. (2007), Yao et al. (2011)
TRPV3	6.6, 33	Peier et al. (2002), Xu et al. (2002)
TRPV4	10, 19	Guler et al. (2002), Watanabe et al. (2002)
TRPM8	24	Brauchi, Orio, & Latorre (2004)
TRPA1	8.8,13.7 (snake),9,116 (*Drosophila*)	Gracheva et al. (2010), Kang et al. (2012)
TRPC5	10	Zimmermann et al. (2011)
H^+	6–9	DeCoursey & Cherny (1998)
ClC-0 (*inactivation*)	40	Pusch, Ludewig, & Jentsch (1997)
Shaker K^+ (*inactivation*)	7.2	Nobile et al. (1997)

Thermal TRP channels are also polymodal receptors responsive to a variety of stimuli and cellular components. Prolonged thermal stress may alter cellular homeostasis which in turn alters the regulation of thermal TRP channels. The membrane potential is another complication factor. Thermal TRP channels are voltage-sensitive and can be activated by depolarization. The voltage-induced current may be low at room temperature but become significant when temperature becomes 20–40 °C higher. Furthermore, the relationship between temperature sensing and voltage sensing in thermal TRP channel is still not clear; if the two are coupled, the temperature dependence will be also voltage dependent.

Hysteresis of gating can also have an impact on the estimation. The temperature activation of thermal TRP channels appears to involve hysteresis, though the extent varies with different channels. For TRPV3, for example, repeated stimulation results in significant sensitization of the current. Both activation threshold and temperature dependence of these channels are dynamically dependent on the history of activation. The measurement of temperature coefficient for these channels will be use dependent. Consistent measurements will require precise temperature controls with defined stimulation protocols.

4. KINETICS AND ENERGETICS OF THERMAL CHANNELS

As to how the temperature coefficient (Q_{10}) of a process is defined, it would be more accurate to estimate the temperature dependence of thermal TRP channels based on activation rates. Unfortunately, the detection of the activation time course of thermal TRP channels has defied conventional temperature control methodologies. The measurement became possible only with the development of a fast temperature control based on laser irradiation using high power infrared laser diodes with appropriate wavelengths (Yao, Liu, & Qin, 2009). By coupling laser beams into an optic fiber and placing it near cells, the laser power is efficiently concentrated to a micro volume, so that the irradiation can induce an instant temperature jump. In practice, the heating can reach a rate of >100,000 °C/s, allowing for a sub-millisecond resolution for practically useful temperature jumps. Following temperature rise, the laser diode output is further modulated to clamp temperature at a constant level. Figure 2.2(A) illustrates electrical recordings of a family of temperature jumps produced by laser irradiation with a high power laser diode. Temperature was elevated from the room temperature (RT) to different target points within about 0.5 ms. Each temperature pulse lasts 100 ms, which considerably shortens the heating time over cells. As a result of reduced thermal stress, a much higher temperature range (>60 °C) becomes experimentally accessible than conventional temperature controls.

Presently TRPV1 is the only channel whose activation kinetics has been studied in relative detail (Yao et al., 2010a,b). Figure 2.2(B) illustrates responses of TRPV1 evoked by 100-ms-long temperature pulses at 0.5-ms resolution. The activation of the channel is surprisingly fast, with a time constant of a few milliseconds. TRPV1 is also activated by capsaicin and low pH. In contrast to heat activation, the activation by chemical agonists is much slower. Capsaicin activates the channel with a $t_{1/2}$ ~100 ms while $t_{1/2}$ for low pH is ~40 ms. Thus thermal TRP channels are truly good temperature detectors, not only having strong temperature dependence but also fast responsiveness.

The deactivation time course of TRPV1 has also been examined, but the measurement is technically more challenging and requires a combined use of

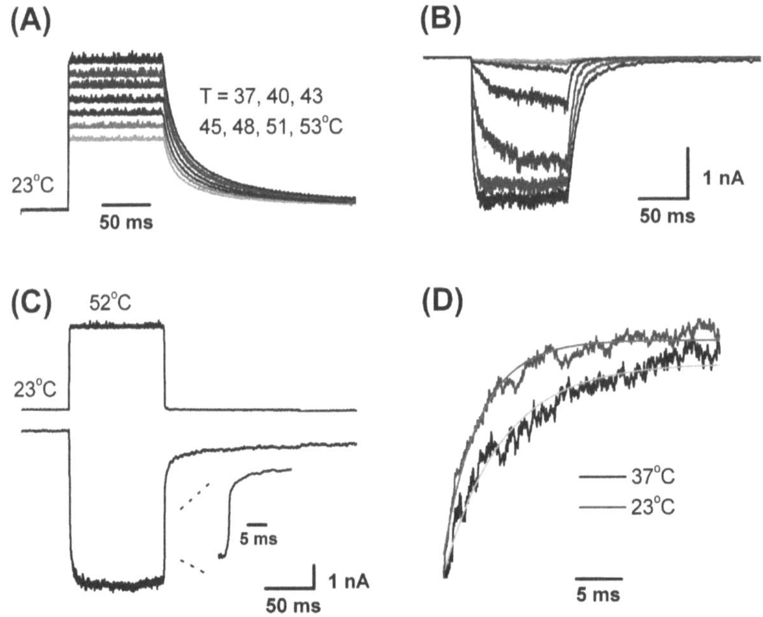

Figure 2.2 The time-resolved activation and deactivation of TRPV1. (A) A family of temperature jumps. (B) Current responses evoked by temperature jumps in (A). (C) Deactivation by rapid cooling. (D) Deactivation slowed by increasing temperature. (See the color plate.)

laser irradiation and rapid solution perfusion (Yao et al., 2010a). Immediately after laser heating is tuned off, the perfusion stream is rapidly moved towards the patch. This allows the deactivation of the channel at the room temperature to be resolved. For deactivation at higher temperatures, a second laser is used to raise the temperature of the perfusion stream. Figure 2.2(C) and (D) shows two deactivation time courses of TRPV1 at 23 °C and 37 °C, respectively. The comparison of the two time courses reveals some interesting features about the deactivation of thermal channels. First, the deactivation is also quite fast, on the order of milliseconds ($\tau = 4$ ms at 23 °C and <8 ms at 37 °C). Second, the deactivation at 37 °C is slower than the deactivation at 22 °C. This means that the rate of deactivation has negative temperature dependence; it is sensitive to cold instead of heat. Third, the change of the time constant between the two temperatures is moderate, less than two-fold over a 14 °C difference. Thus the deactivation of thermal channels has only nominal temperature dependence.

The nominal temperature dependence of deactivation implies that the strong temperature dependence of thermal TRP channels originates from

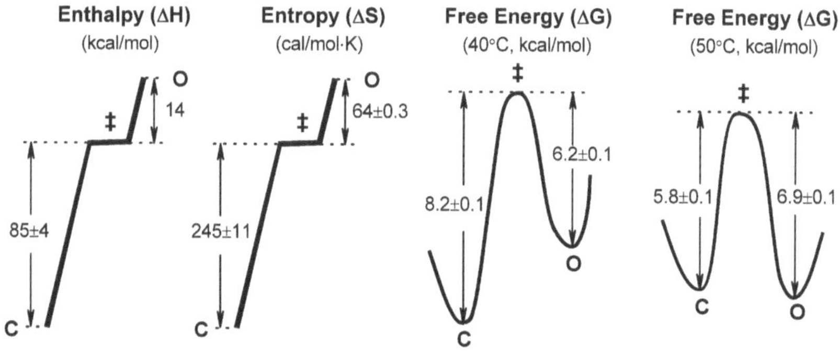

Figure 2.3 Energetic landscape of thermal activation of TRPV1. Large enthalpy and entropy changes are restricted to the activation step. Compared with enthalpy and entropy changes, free energy changes are small.

the activation rate. Indeed, the activation time course of TRPV1 is highly sensitive to temperature increases. Quantitatively, it can be fit by a simple two-state model in which temperature mainly drives the opening rate. The analysis results in an energy landscape for temperature activation of TRPV1 as shown in Figure 2.3. The opening of the channel involves a total enthalpy increase of about 100 kcal/mol. The activation enthalpy is about 78 kcal/mol, which accounts for most of the enthalpy change, while the deactivation involves only about 22 kcal/mol. This equates to a Q_{10} of ~50 at 40 °C for the activation rate. Despite the large enthalpy change, the free energy change is moderate, less than 10% of the enthalpy change.

An important feature of thermal TRP channels as implied by the energetics of TRPV1 is a tight compensation between enthalpy and entropy. This is manifested at least in two observations. One is the rapidness of activation and deactivation. The fast activation rate implies a small free energy change, which in turn requires the enthalpic and entropic energies to be well balanced. For other ion channels with a typical $\Delta H \sim 20$ kcal/mol, the enthalpy needs to be compensated by entropy within about 50% in order for activation to occur in a millisecond time scale. But thermal channels have $\Delta H \sim 100$ kcal/mol, so the compensation needs to be more precise to be within 10%. The other observation in support of enthalpy and entropy compensation is the robustness of temperature activation. With a large activation enthalpy, even a small fluctuation in enthalpy or entropy could result in a large free energy barrier, which would render the channel to be inactivatable by temperature. In contrast, the temperature activation of thermal TRP channels appears as robust as the activation by other stimuli; patches or cells showing agonist responses are always responsive to temperature and vice versa. Thus the large enthalpy and entropy changes involved in temperature activation in thermal TRP channels must be well correlated and balanced; either they are both very robust and resistant to fluctuations or the respective fluctuations can cancel each other.

5. HYSTERESIS OF TEMPERATURE-DEPENDENT GATING

The gating of ion channels is generally reversible. The activation of thermal TRP channels by temperature, however, appears to be different. Several thermal TRP channels have been reported to involve hysteresis though the extent varies. Hysteresis in temperature responses is often evident in experiments using conventional temperature controls for heating or cooling, which allows the current response of a channel to be continuously recorded during temperature rising and falling phases. When the absolute current is plotted against temperature, the activation pathway during heating is usually displaced from the deactivation pathway during cooling. Although there are multiple explanations for this displacement, one possibility is that the gating of the channel involves hysteresis.

The hysteresis of gating is probably most evident in the vanilloid receptor TRPV3. When first stimulated, either heat or a chemical agonist (e.g., 2-APB), TRPV3 produced only a small activity. But subsequent repeated stimulations cause its response to increase progressively (Figure 2.4). The final current can be many folds larger than the initial activity. This so-called sensitization was found to be sensitive to Ca^{2+} chelators; the sensitization of the current becomes accelerated when the fast chelator 1,2-Bis(2-aminophenoxy)ethane-N,N,N',N'-tetraacetic acid (BAPTA) was used in place of

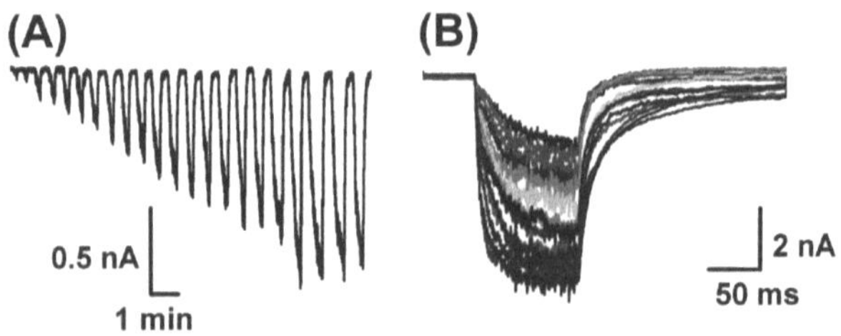

Figure 2.4 Hysteresis of TRPV3 responses evoked by repeated stimulation. (A) 2-APB responses. (B) Heat responses (53 °C).

Ethylene glycol tetraacetic acid (EGTA). This has led to the proposition of a regulatory mechanism dependent on intracellular Ca^{2+} release for the sensitization, such as Ca^{2+}-mediated CaM modulation of the channel. But latter findings show that the sensitization is an intrinsic property of the gating and occurs in a membrane-delimited manner (Liu, Yao, & Qin, 2011).

The hysteresis of thermal TRP channels usually results in a decrease in temperature dependence. For TRPV3, the change is dramatic. Upon initial stimulation, a high temperature >50 °C is required to elicit a detectable activity. Following sensitization, temperature around 30 °C suffices to activate the channel. The activation threshold drops by about 20 °C. The slope temperature dependence of the channel is also significantly affected. When activated by a family of increasing temperature jumps, the channel exhibits a much steeper sensitivity to temperature changes during the first sweep than during the subsequent ones. Interestingly, after hysteresis, TRPV3 becomes only nominally temperature-dependent (before sensitization: Q_{10} (50 °C) > 10; after sensitization: Q_{10} (30 °C) < 2).

The hysteresis of thermal TRP channels reveals that temperature may evoke irreversible structural changes. Presently, there is little known about the mechanism or the molecular basis underlying hysteresis of thermal TRP channels. In the case of TRPV3, the hysteresis may be conceived conceptually by a three state model: C_1- > C_2-O. The freshly synthesized channels in cells are folded into a structure corresponding to C_1. Stimulation of the channel drives it into an intermediate closed state C_2 before opening. The transition from C_1 to C_2 is practically irreversible, so that after opening, the channel closes to C_2 rather than C_1. The activation from C_1 to C_2 is slow compared to the opening and closing transitions between C_2 and O. Thus repeated stimulation progressively draws the channel from state C_1 into state C_2, and as the occupancy of C_2 is increased, the current to be activated becomes increasingly larger. The hysteresis occurs as a result of irreversible changes in the initial activation step (Figure 2.5).

The gating of thermal TRP channels is likely allosteric and involves multiple steps, so the hysteresis may take place in any step of the activation. But the profound change in the slope of temperature dependence of TRPV3 before and after sensitization suggests that the hysteresis is accompanied by large energetic changes. This seems to suggest that the structural change for hysteresis is correlated with the structure change for temperature sensing. Possibly, the activated temperature sensor of thermal TRP channels may become highly dynamic as implied by the large entropy increase during temperature activation. When temperature is decreased, the structure may

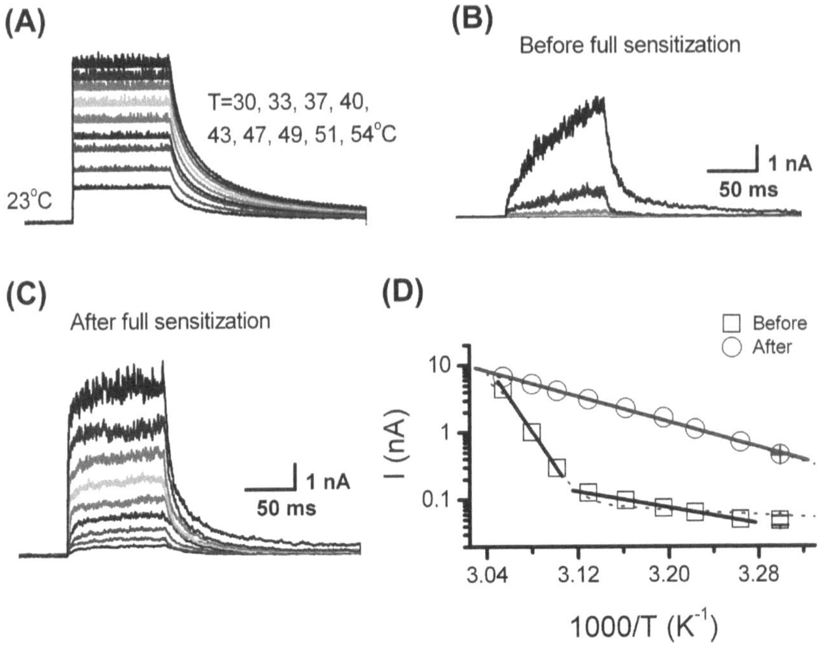

Figure 2.5 Temperature responses of TRPV3 evoked by a family of temperature jumps (A) before (B) and after sensitization (C). Slope sensitivity was decreased after sensitization (D). (See the color plate.)

relax to some intermediate fold different from the original native fold. In the future the elucidation of such structural changes will be important to understand the mechanism of the hysteresis, which in turn may provide insights on the mechanism of temperature sensing by thermal TRP channels.

6. HEAT CAPACITY THEORY

Although one may tend to associate a large energetic change with a large conformational change, the energetics itself provides no hint on the scale of conformational changes. Physically, an enthalpy change may result from two types of structural changes, those that involve a heat capacity change and those that do not. The energetics of thermal TRP channels as inferred from their temperature dependence is an apparent measurement and may arise from either change. A heat capacity enthalpy, $\Delta H = \Delta C_p(T - T_0)$, is increased proportionally with temperature changes. So in theory a large enthalpy change could also result from a heat capacity change. This is the case even for a small heat capacity change as long as the observation temperature T is far off from T_0. Thus the implication is that the large enthalpy change of thermal TRP channels may not be correlated with large conformational changes; instead it may be attributed to a heat capacity change, as recently suggested by Clapham and Miller (2011).

In proteins, the heat capacity change appears to result mainly from hydration of hydrophobic residues. Thus the role of heat capacity changes in temperature activation of thermal TRP channels may be understood with the model of hydrocarbon transfer between hydrophobic and aqueous environments (Baldwin, 1986). The transfer of liquid hydrocarbon to water is associated with a heat capacity change, which gives rise to an enthalpy change $\Delta H = \Delta C_p(T - T_0)$ where T_0 is ~22 °C. The entropy change follows as: $\Delta S = \Delta S_0 + \Delta C_p \ln(T/T_0)$. The free energy change, which determines the solubility of the hydrocarbon, is given by: $\Delta G = \Delta H - T\Delta S$. The equilibrium constant of the solvation follows:

$$lnK = \frac{\Delta S_0}{R} - \frac{\Delta C_p}{R}\left(1 - \frac{T_0}{T} + ln\frac{T_0}{T}\right).$$

An interesting property of the above equation is that the equilibrium constant does not simply increase with temperature in a monotonic manner. Instead, it has a bell shape relationship with respect to temperature. In other words, the heat capacity change causes the solubility of hydrocarbons to increase at both cold and warm temperatures. The minimum solubility is located at T_0. So another interpretation of T_0 is that it is the temperature where the solubility is lowest (Figure 2.6).

The above theory for hydrocarbon solvation also holds for the gating of thermal TRP channels assuming that the opening of the channels is accompanied with a heat capacity change. Instead of solubility, the free energy change determines the equilibrium constant of channel opening. One difference is that, besides a heat capacity enthalpy for temperature sensing, the gating of ion channels involves other enthalpy changes due to additional gating steps such as pore opening and allosteric coupling between temperature sensing and pore opening. These additional enthalpy changes, say ΔH_0, are associated with conformational changes with typical temperature dependence as occurring in other ion channels. Thus, they may be assumed to be temperature independent. The total enthalpy change for temperature activation of thermal TRP channels will be: $\Delta H = \Delta H_0 + \Delta C_p(T - T_0)$. Mathematically, ΔH_0 has an effect of shifting T_0 so that ΔH can be still represented by: $\Delta H = \Delta C_p(T - T_0')$ where $T_0' = \Delta H_0/\Delta C_p$. Thus, the

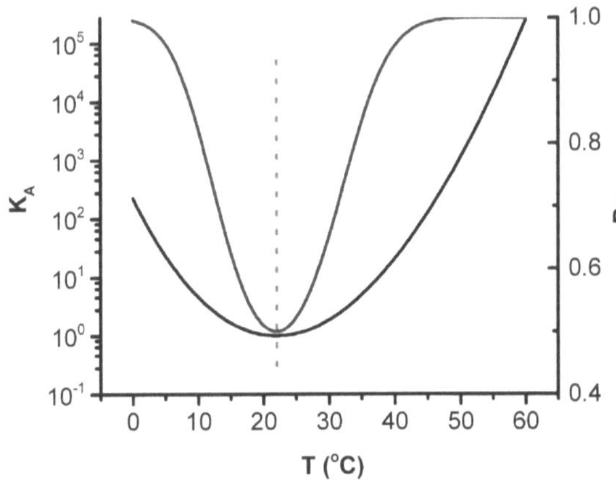

Figure 2.6 Bimodal temperature dependence of a heat capacity model (black for equilibrium constant and blue for solubility). (For interpretation of the references to color in this figure legend, the reader is referred to the online version of this book.)

opening of the channels would have similar bimodal temperature dependence as the solubility of liquid hydrocarbons.

The bimodal temperature dependence of the heat capacity model provides a potential explanation for both heat and cold sensitivity of thermal TRP channels. If T_0 is leftward-shifted to very cold temperatures, only the heat sensitive portion of the bimodal relationship will be observed within physiological temperatures. The channel behaves as a heat-sensitive receptor. On the other hand, if T_0 is rightward-shifted to very high temperatures, only the cold sensitive portion becomes accessible to physiological temperatures. The channel acts as a cold receptor. Thus it has been speculated that the molecular differences in thermal TRP channels have an effect on tuning T_0 to their respective temperature responsive ranges. The mechanism of temperature sensing is conserved among them.

The heat capacity model, while appealing in theory, still lacks experimental support. In fact the model appears to encounter some limitations when applied to thermal TRP channels. For example, according to the hydrocarbon transfer model, T_0 is around 22 °C. For TRPV1, the apparent enthalpy change is ~100 kcal/mol between 40 and 50 °C. Except the temperature sensing step, other gating steps of thermal TRP channels have nominal temperature dependence, indicating that the corresponding conformational changes are energetically similar to conformational changes occurring in other ion channels. This places ΔH_0 in the range of 20 kcal/mol. Thus the heat capacity change ΔC_p is estimated to be about 3.5 kcal/mol. This value of ΔC_p has two implications (Qin, 2013). One is a limited shift in T_0, which is less than 5 °C. As a result, TRPV1 would become cold sensitive below 17 °C. The other implication is that a large number of hydrophobic residues are needed to account for the heat capacity change. The heat capacity change is proportional to hydrophobic surface areas. On average, the heat capacity change per hydrophobic residue is about 30–40 cal/mol·K. Thus it would require a total number of 100 hydrophobic residues (or 25 residues per subunit) to undergo complete translocation between the hydrophobic core and the aqueous solution. This suggests that the heat capacity model would also necessitate a substantial conformational change.

Contrary to the model predictions, neither TRPV1 nor other thermal TRP channels are found to possess dual heat and cold sensitivity. Two orthologs channels reportedly have inverse temperature sensitivity. One is TRPA1, which was first shown to be activated by noxious cold in mammals. Later the channel in low vertebrates (flies and snakes) is found to be heat activated. However, the cold sensitivity of the mammalian TRPA1 is controversial. The other is TRPV3. The mammalian TRPV3 is activated by heat and is suggested to be a warm receptor (Patapoutian et al., 2003). The frog ortholog was recently reported to be activated by cold with a threshold at about 17 °C and was suggested to be a noxious cold receptor in frogs (Saito et al., 2011). This opposed thermal sensitivity of TRPV3 between mammals and frogs seems to fit the heat capacity model. However, our cloning and characterization of the frog TRPV3 show that the channel does not possesses the cold sensitivity as published. When heterogorously expressed in HEK cells, it is instead heat activated and have similar biophysical properties as its mammalian counterpart. Thus, it appears that there is presently no thermal TRP channel with the predicted dual heat and cold sensitivity of the heat capacity model. It remains an open question whether thermal TRP channels truly exploits heat capacity changes for temperature sensing or other possible mechanisms (Jara-Oseguera & Islas, 2013).

7. ORIGINS OF THERMAL SENSITIVITY

Since the finding of the first thermal TRP channel it has been a question whether the apparent temperature sensitivity of thermal TRP channels is intrinsic to channel proteins or their environments. Cell membranes are one major concern because lipid organizations are strongly temperature dependent. For example, the enthalpy change of TRPV1, which is about

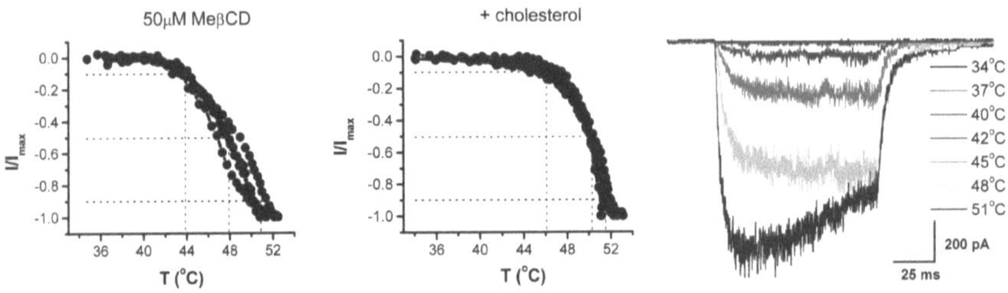

Figure 2.7 The heat sensitivity of TRPV1 is intrinsic. Either depletion or addition of cholesterol in cell membranes does not prevent heat activation of TRPV1 (A) and (B) (Liu et al., 2003). (C) Reconstituted TRPV1 are similarly activated by heat as in native membranes (Cao et al., 2013).

100 kcal/mol, is equivalent to melting of approximately 10 phospholipids (Heimburg, 2007). This is within the range of the number of annulus lipids around an ion channel protein. If these lipid molecules undergo a melting transition, they could result in an energy enough for activation of thermal TRP channels. The temperature where the phase transition of lipids occurs is also relevant and is close to activation thresholds of thermal TRP channels (e.g., TRPV1).

A first test for possible roles of lipids in thermal TRP channel functions is the study of cholesterol effects on heat activation of TRPV1 (Liu, Hui, & Qin, 2003). Cholesterol is known to modulate fluidity of lipid membranes; it can abolish temperature-dependent phase transitions in lipid bilayers. The level of cholesterol in cell membranes may be perturbed with cyclodextrin, a lipid molecule carrier that can shuffle cholesterol molecules out of membranes. Removal of cholesterol by cyclodextrin does not eliminate heat sensitivity of TRPV1. The channel remains activated by temperature with similar biophysical properties as in control (Figure 2.7). On the other hand, the introduction of additional cholesterol into cell membranes does not abrogate heat sensitivity either, though the activation threshold becomes elevated (Figure 2.7). Cholesterol also makes capsaicin more difficult to activate TRPV1 (Picazo-Juarez et al., 2011), suggesting that this effect of excess cholesterol is not specific to thermal sensitivity of the channel. Thus regardless of whether additional cholesterol is added or existing cholesterol molecules are depleted, TRPV1 remains sensitive to heat. The experiment provides the first clue that the membrane fluidity is not essential to the temperature sensitivity of thermal TRP channels.

Cellular components are another concern. Excised patch-clamp recording can provide some hint on whether the function of a channel depends on intracellular diffusive components. The heat sensitivity of thermal TRP channels is reproducible in either inside-out or outside-out patch configurations. The cold sensitivity of TRPM8 becomes lost after patch excision, but the problem is attributed to loss of membrane PIP2 and can be alleviated by supplying additional PIP2 or by inhibiting lipid phosphatases (Liu & Qin, 2005). Thus the excised patch clamp experiments suggest that the temperature sensitivity of thermal TRP channels is likely to be membrane delimited. More conclusive evidences on the issue came recently from the reconstitution of thermal TRP channels in artificial membranes. TRPM8 is among the first to be successfully reconstituted into chemically defined lipid bilayers (Zakharian, Cao, & Rohacs, 2010). The reconstituted TRPM8 is shown to be still activated by cold and the chemical agonist menthol. More recently, the heat sensitive channels such as TRPV1 have also been reconstituted in liposomes (Cao et al., 2013). In these experiments, the channels are purified from native membranes in multiple steps involving detergent solubilization, affinity chromatography and gel filtration. Thus the final proteins have a high homogeneity and purity, presumably devoid of regulatory components that would associate with the channels in cells. The reconstituted TRPV1 retains heat sensitivity in liposomes, and importantly, the activation exhibits nearly identical energetics as in native membranes. Together these studies provide strong support that the apparent heat or cold sensitivity of thermal TRP channels is an intrinsic property of the channels, independent of membrane lipids or intracellular regulators.

8. DISTRIBUTION OF THERMAL SENSITIVITY: GLOBAL OR LOCAL?

Knowing that the temperature sensitivity of thermal TRP channels is an intrinsic property, a next question is where it is located within the channels. There are two possible scenarios. Since temperature has global effects on protein dynamics, the thermal sensitivity may be integrated over the entire proteins. Alternatively it may be localized to some specific domain within the channels. The apparent large energetics of thermal TRP channels is obviously summed over the whole proteins. But this does not mean that all parts of proteins contribute equally. Some regions of the channels may make larger contributions than others or even dominate the overall energetics. In this case, it would suggest that thermal TRP channels involve a temperature sensor domain. Unfortunately, without fully understanding the gating at detailed structural levels, it is difficult to map out precise contributions of various molecular regions of the channels. However, some insights into the question may be drawn at the kinetic level.

Single-channel recording can unravel detailed gating events involved in the activation of an ion channel. If a stimulus has broad effects, one might expect that it will mediate multiple aspects of the gating process. Thus the analysis of single channel kinetics could provide insights on whether thermal sensitivity of TRP channels is localized to particular gating events. Single-channel analysis has been conducted for several thermal TRP channels (Fernandez et al., 2011; Liu et al., 2003; Zakharian et al., 2010). Temperature mainly influences the kinetics of gating. The unitary current has a temperature coefficient similar to that of other ion channels (Liu et al., 2003). The gating kinetics of thermal TRP channels is quite complex. For example, the opening of TRPV1 tends to occur in bursts (Figure 2.8). At low temperatures the opening bursts are sporadic, limiting the open probability. As temperature is increased, the bursts appear more frequently, resulting in an increase in P_O. In addition, the burst durations appear to also increase with temperature. Thus temperature has effects on both burst durations and durations of long closures (i.e., gaps) separating bursts, whose temperature coefficients are respectively about 32 and 7 at 40 °C. Both events are highly temperature dependent.

The opening bursts of TRPV1 are comprised of multiple openings and closures of relatively short durations. These events have fast kinetics,

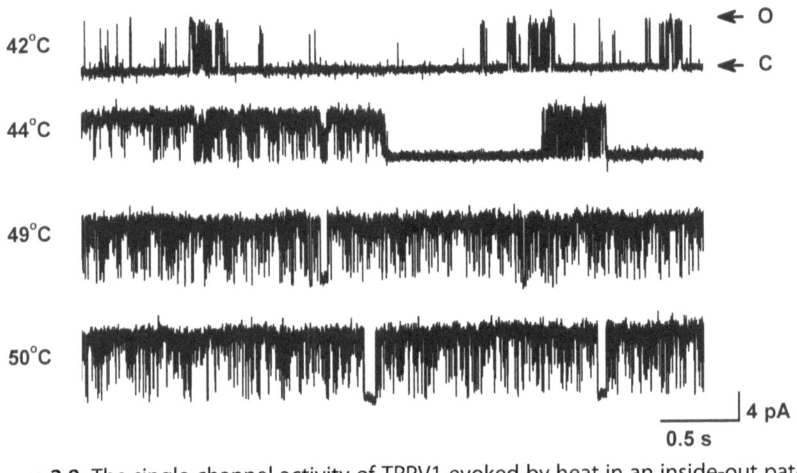

Figure 2.8 The single channel activity of TRPV1 evoked by heat in an inside-out patch (Liu et al., 2003).

occurring on a time scale of milliseconds. Analysis of the duration histograms suggests that there are at least three types of openings and three types of closures. They appear at both low and high temperatures. The Arrehenius plots of these events at different temperatures suggest that they all have a nominal $Q_{10} \sim 1$–2, some of which have negative temperature sensitivity. Thus the intraburst gating events are relatively independent of temperature as compared to the long closures between bursts.

Although the burst duration of TRPV1 is temperature dependent, an opening burst is not an elementary gating event; instead it is a consequence of asymmetry in gating kinetics. The formation of a burst requires a gating scheme of at least three states, e.g., $C_1 \underset{k_{21}}{\overset{k_{12}}{\rightleftharpoons}} C_2 \underset{k_{32}}{\overset{k_{23}}{\rightleftharpoons}} O_3$, where the final opening and closing steps occur on a fast time scale relative to transitions between C_1 and C_2. In such a case, the duration of a burst is determined by $t = k_{21}^{-1}(1 + k_{23}/k_{32})$. Since the intraburst events are temperature independent, the temperature dependence of the burst duration would result from the closing step from C_2 to C_1. However, from the macroscopic measurements we know that temperature mainly drives the opening, not the closing of TRPV1 (Yao et al., 2010a,b). This suggests that the temperature sensitivity of burst durations in single channel recording arise from alternative mechanisms. A possible explanation is that there are multiple types of bursts with distinct durations. The long bursts are favored by high temperatures while the short bursts are evoked by low temperatures. The apparent temperature dependence of the burst durations results from population changes of distinct bursts. In support of multiple types of bursts, the intraburst activity of TRPV1 indeed changes with temperature, mainly as a result of changes in relative occupancy of different openings and closures within bursts. Thus, it is hypothesized that the gating of thermal TRP channels involves multiple steps of sensor activation each of which leads to a burst of openings and closures. Presumably the multiple activation steps occur because the channel has multiple subunits and each subunit activation can lead to channel opening. As more subunits are activated, the opening bursts become more stabilized.

In summary, at the single channel level temperature has localized effects on gating kinetics of TRPV1. Although the gating of TRPV1 involves as many as seven components, temperature mainly influences the durations of long closures and bursts. The change of burst durations may reflect different levels of subunit activation, while the shortening of long closures by temperatures is likely associated with temperature sensor activation.

The intraburst openings and closures, which are relatively insensitive to stimulus, occur probably after sensor activation. These differential effects of temperature on different gating events are in agreement with the asymmetric effects of temperature on activation and deactivation of TRPV1 at the macroscopic level. Consequently, both single-channel and macroscopic kinetics of thermal TRP channels favor a model in which temperature sensitivity is localized rather than distributed across whole channels.

9. IDENTIFICATION OF MOLECULAR BASIS OF THERMAL SENSITIVITY

TRP channels have a membrane topology similar to that of voltage-gated channels, comprising six transmembrane domains with a reentrant pore loop between the last two transmembrane segments S5 and S6 (Figure 2.9). The N terminus of TRP channels is relatively large, accounting for approximately half of the total size. The sequence of the N terminus also diverges across TRP subfamilies. At the molecular level, several regions of thermal TRP channels have been implicated in temperature activation. For example, by random mutagenesis screening, residues at the external S6 linker were found to be crucial for heat activation of TRPV3 (Grandl et al., 2008). Mutations of these residues preferentially abrogated heat responses of the channel. Swapping of the C terminus between TRPV1 and TRPM8 was reportedly to reverse their heat and cold sensitivity, causing the TRPV1 chimera to be responsive to cold while the TRPM8 chimera sensitive to heat (Brauchi et al., 2006). The pore turret is another region found to be involved in heat activation of TRPV1 (Yang et al, 2010) but see Yao et al. (2010b). Finally, exchanges of N terminal domains can successfully transfer heat sensitivity to temperature-insensitive homologs in both mammalian vanilloid receptors (Yao et al., 2011) and nonmammalian TRPA1 (Cordero-Morales, Gracheva, & Julius, 2011; Kang et al., 2012; Zhong et al., 2012). Thus the N terminus, C terminus, and pore domain all appear to play a role in temperature regulation of thermal TRP channels.

At first glance, the disparate locations of these identified regions would suggest that the temperature sensitivity of thermal TRP channels is broadly distributed within channels. However, it should be realized that the gating of an ion channel is a multistep process, involving stimulus sensor activation, pore opening and allosteric coupling between them. A change in any activation step may alter the apparent response of the channel. This is also likely

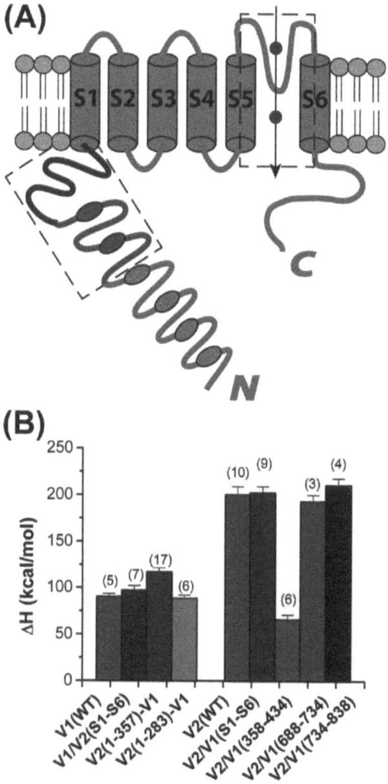

Figure 2.9 Proximal N terminus mediates heat sensitivity of vanilloid receptors. (A) Membrane topology of vanilloid receptors. The proximal N terminus meading heat sensitivity is shown in red. (See the color plate.) (B) Enthalpy changes upon exchanges of various regions in TRPV1 and TRPV2. Only exchanges of the proximal N terminus had significantly impacts on the energetics.

to be the case for temperature activation of thermal TRP channels (Jara-Oseguera & Islas, 2013; Liu et al., 2003; Yao et al., 2010a). Thus a region that is able to affect temperature activation of thermal TRP channels is not necessarily responsible for temperature sensing, but instead it may mediate gating events occurring after temperature sensor activation. Distinguishing the specific effect of a perturbation on the gating process requires thorough understanding of the underlying gating mechanism. Unfortunately, this is rarely the case for most thermal TRP channels. Most studies measure the responses of thermal TRP channels in semiequilibrium at moderate temperatures, and such measurements generally cannot unmask gating details. As a consequence, how a molecular region specifically mediates temperature activation of thermal TRP channels is usually not determined.

Thus, the finding of multiple regions involved in the gating by temperature is not necessarily contradictory with a localized temperature sensing mechanism for thermal TRP channels.

Below we take TRPV channels as an example to discuss possible roles of N terminal regions in thermal sensitivity of TRP channels (Yao et al., 2011). The TRPV family contains several members with different temperature phenotypes. In particular, the fast temperature jump experiments show that rat TRPV1 and TRPV2 have distinct activation kinetics and energetics. The activation of TRPV1 can be approximated by a two state model and involves an enthalpy change of ~100 kcal/mol. The time course of TRPV2 requires a three state model, and more importantly, its activation enthalpy is nearly twice that of TRPV1. This enthalpy difference between TRPV1 and TRPV2 is considerably larger than the enthalpy change associated with gating events with nominal temperature dependence, suggesting that it is pertinent to temperature sensing. Thus, the characterization of the energetic difference and identification of the underlying molecular domains would provide a way to unravel the molecular determinants of temperature sensitivity of thermal TRP channels.

Protein sequences of these channels provide no direct hint on possible candidate regions. The differences in residues scatter throughout entire sequences, thus the whole proteins would need to be scrutinized in order to locate the molecular basis of the energetics. By extensive mutagenesis it was found that the whole channels may be divided up at strategic positions so that individual fragments can be exchanged between TRPV1 and TRPV2. These fragments include the transmembrane segments, the ankyrin repeats, the membrane proximal linker region between ankyrin repeats and the first transmembrane segment, and the proximal and distal C termini. Exchanges of most of these regions have effects on temperature activation properties, consistent with the notion that multiple domains are involved in the activation of thermal TRP channels. But what is interesting is that most of them impact the temperature threshold and the time course kinetics of activation. Only the N terminal domains can significantly perturb the enthalpy of the channels. These include the last two ankyrin repeats and the linker region connecting the ankyrin repeats to the first transmembrane segment. Between the two regions, the proximal linker domain appears to have a dominant role; its exchange alone could account for the energetic difference between TRPV1 and TRPV2. While the last two ankyrin repeats are less effective, they dictate the multistate kinetics of TRPV2 activation.

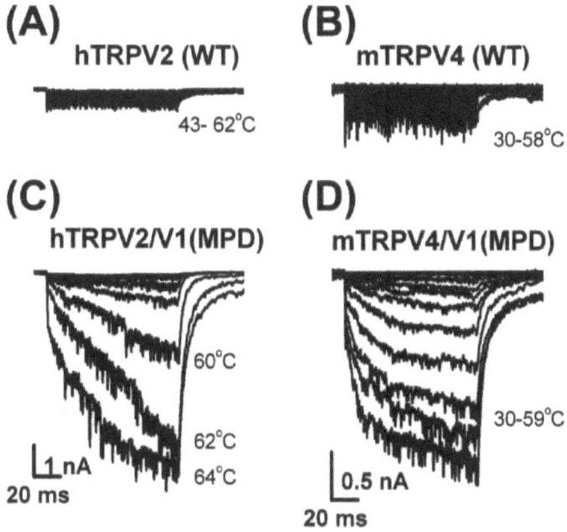

Figure 2.10 The replacement of the proximal N terminus suffices to restore the heat sensitivity of temperature insensitive TRPV homologs.

Exchange of both regions could transform the activation of TRPV2 to resemble TRPV1.

The proximal N terminal regions identified from TRPV1 and TRPV2 are found to be also crucial for other TRPV members. Some TRPVs are insensitive to temperature (Figure 2.10). For example, human TRPV2 produces no heat-evoked activity for temperature up to 60 °C. The mouse TRPV4 also shows only a small current at a temperature as high as 58 °C. When the proximal N-terminal linker region of these channels is replaced by the counter part of a functional channel (e.g., TRPV1), the resultant chimeras become vigorously responsive to heat. The wild-type TRPV3, albeit responsive to heat, has a high activation threshold (>50 °C) and undergoes strong hysteresis. After replacement of its N-terminal linker domain, the chimeric TRPV3 becomes activated by warm temperature with a threshold around 30 °C. The hysteresis of activation also becomes diminished in the chimeric channel. These experiments indicate that the proximal N terminus mediates heat sensitivity of all TRPVs.

The N terminus is also implicated in thermal sensitivity of nonmammalian TRPA1 channels. The heat sensitive TRPA1 from snake and fly has a long N terminus containing nearly twice as many ankyrin repeats as mammalian TRPVs. The overall homology of these channels is quite diverse. But chimeric exchanges of the N terminus could also transfer heat sensitivity to heat-insensitive TRPA1 homologs (Cordero-Morales et al., 2011; Kang et al., 2012; Zhong et al., 2012). Within the N terminus, the regions that can impact temperature dependence of heat responses are also located towards the C-terminal end of the N terminus, including the proximal ankyrin repeats and the linker domain. Consequently, although TRPV1 and TRPA1 differ considerably in their N terminus, the overall distribution of the heat sensitivity appears to be related, suggesting that the molecular mechanisms of temperature sensing by these channels may have been conserved between mammalian and nonmammalian orthologs.

In general, the gain-of-function of a chimeric channel does not speak to the particular gating function of the molecular region being exchanged. However, the profound change in the energetics as observed in TRPV channels suggest that the changes are intimate to the temperature sensing process. The alteration of other gating steps could perturb the energetics, but their contributions would be rather limited as compared to the observed changes in TRPV1 and TRPV2. Furthermore, the energetic analysis of TRPV chimeras is based on fitting activation time courses directly with state-dependent models. While the analysis takes account of multiple gating steps, changes in activation energetics are specifically required in order to fit the change in activation kinetics. Thus, the energetic differences of these channels most likely lie in differences in temperature sensing elements, and consequently, the identified N terminal regions that can account for the energetic differences stand as strong candidates for temperature sensing in thermal TRP channels (Figure 2.11).

It is still unknown how the proximal N terminus confers the large energetics of thermal TRP channels. The new high-resolution structure of TRPV1 reveals that the linker region connecting the ankyrin repeats to the transmembrane domains has a delicate, complex fold. On one end, it forms a loop interacting with a hydrophobic pocket on the ankyrin repeats of the next subunit. On the other end, it folds into three short helices forming a grove to support the TRP box helix coming out of S6. The loop region between the first two helices appears to be also in close proximity to the unresolved S2–S3 loop, which is known to be functionally important (Numazaki et al., 2002). Thus, the proximal linker domain along with the adjacent ankyrin repeats fold into a pivotal structural assembly hosting complex interactions in TRPV channels. This arrangement is not anticipated from the tetramerization domain of potassium channels at similar positions, but seems to be consistent with the crucial functions of the region as found in TRPVs and homologs. It is likely that some interactions within the assembly are highly endothermic thereby providing the energetics required for temperature activation of thermal TRP channels.

(A)

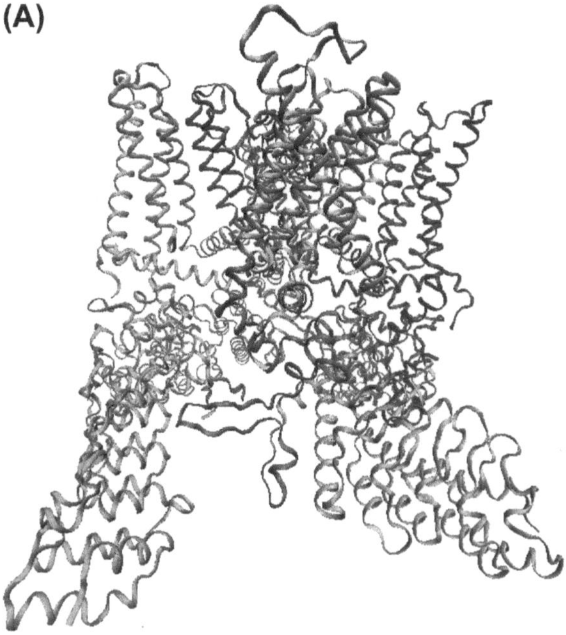

(B)

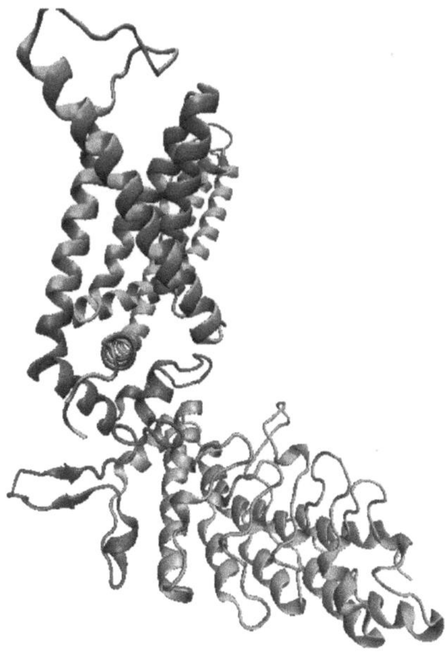

Figure 2.11 Cyo EM structure of TRPV1. (A) Tetramer. (B) Monomer. Regions colored in red corresponds to the linker domain connecting ankyrin repeats to S1. Regions in yellow are the last two ankyrin repeats (Liao et al., 2013). (See the color plate.)

10. SUMMARY

In essence, thermo TRP channels are molecular machines built perhaps with the largest activation energetics. Although it has been more than 15 years since the discovery of the first thermal TRP channel, it remains unclear how these proteins attain their unusual thermodynamic properties. But progress has been made and has been accelerating recently. Thermodynamic basis of channel functions is being understood. The time-resolved detection of temperature activation is becoming possible. Molecular sites mediating temperature activation are being identified. High-resolution structures of the channels are emerging. The present findings support that the channel proteins, rather than their environments, are intrinsically sensitive to temperature and that the thermal sensitivity possibly involves specialized molecular structures, though exactly what region in the channels is responsible has not yet reached a consensus. The problem underscores the need of a thorough understanding of biophysical mechanisms of temperature gating and capability to properly determine the energetic contribution of molecular regions. The large energetics of thermal TRP channels implies that temperature sensing may involve large structure changes. But temperature evoked conformational changes have remained largely elusive. Recent advances on high-resolution structures and functional reconstitution in artificial lipids open new possibility for detecting temperature-dependent conformational changes. Molecular dynamic simulation could potentially provide atomic insights on structural basis of temperature gating at both high temporal and spatial resolutions. Solutions to these problems may be still a long way off, but the current progress has set the stage for future investigations at detailed molecular and structural levels, which will eventually unravel the physiochemical mechanisms conferring the large energetics of thermal TRP channels.

REFERENCES

Akbar, A., Yiangou, Y., Facer, P., Walters, J. R., Anand, P., Ghosh, S., et al. (2008). Increased capsaicin receptor TRPV1-expressing sensory fibres in irritable bowel syndrome and their correlation with abdominal pain. *Gut, 57*(7), 923–929.

Baldwin, R. L. (1986). Temperature dependence of the hydrophobic interaction in protein folding. *Proceedings of the National Academy of Sciences of the United States of America, 83*(21), 8069–8072.

Bautista, D. M., Siemens, J., Glazer, J. M., Tsuruda, P. R., Basbaum, A. I., Stucky, C. L., et al. (2007). The menthol receptor TRPM8 is the principal detector of environmental cold. *Nature, 448*(7150), 204–208.

Blix, M. (1882). Experimentela bidrag till losning av fragan om hudnervernas specifika energi. *Uppsala LakFor Forh, 18*, 87–102.

Brauchi, S., Orta, G., Salazar, M., Rosenmann, E., & Latorre, R. (2006). A hot-sensing cold receptor: C-terminal domain determines thermosensation in transient receptor potential channels. *Journal of Neuroscience, 26*(18), 4835–4840.

Brauchi, S., Orio, P., & Latorre, R. (2004). Clues to understanding cold sensation: thermodynamics and electrophysiological analysis of the cold receptor TRPM8. *Proceedings of the National Academy of Sciences of the United States of America, 101*(43), 15494–15499.

Bulloc, T. H., & Dieck, F. P. (1956). Properties of an infra-red receptor. *Journal of Physiology, 134*(1), 47–87.

Cantero-Recasens, G., Gonzalez, J. R., Fandos, C., Duran-Tauleria, E., Smit, L. A., Kauffmann, F., et al. (2010). Loss of function of transient receptor potential vanilloid 1 (TRPV1) genetic variant is associated with lower risk of active childhood asthma. *The Journal of Biological Chemistry, 285*(36), 27532–27535.

Cao, E., Cordero-Morales, J. F., Liu, B., Qin, F., & Julius, D. (2013). TRPV1 channels are intrinsically heat sensitive and negatively regulated by Phosphoinositide lipids. *Neuron, 77*(4), 667–679.

Caterina, M. J., Schumacher, M. A., Tominaga, M., Rosen, T. A., Levine, J. D., & Julius, D. (1997). The capsaicin receptor: a heat-activated ion channel in the pain pathway. *Nature, 389*(6653), 816–824.

Caterina, M. J., Leffler, A., Malmberg, A. B., Martin, W. J., Trafton, J., Petersen-Zeitz, K. R., et al. (2000). Impaired nociception and pain sensation in mice lacking the capsaicin receptor. *Science, 288*(5464), 306–313.

Chery-Croze, S. (1983). Relationship between noxious cold stimuli and the magnitude of pain sensation in man. *Pain, 15*(3), 265–269.

Clapham, D. E. (2003). TRP channels as cellular sensors. *Nature, 426*(6966), 517–524.

Clapham, D. E., & Miller, C. (2011). A thermodynamic framework for understanding temperature sensing by transient receptor potential (TRP) channels. *Proceedings of the National Academy of Sciences of the United States of America, 108*(49), 19492–19497.

Cordero-Morales, J. F., Gracheva, E. O., & Julius, D. (2011). Cytoplasmic ankyrin repeats of transient receptor potential A1 (TRPA1) dictate sensitivity to thermal and chemical stimuli. *Proceedings of the National Academy of Sciences of the United States of America, 108*(46), E1184–E1191.

Davis, J. B., Gray, J., Gunthorpe, M. J., Hatcher, J. P., Davey, P. T., Overend, P., et al. (2000). Vanilloid receptor-1 is essential for inflammatory thermal hyperalgesia. *Nature, 405*(6783), 183–187.

DeCoursey, T. E., & Cherny, V. V. (1998). Temperature dependence of voltage-gated H^+ currents in human neutrophils, rat alveolar epithelial cells, and mammalian phagocytes. *The Journal of General Physiology, 112*(4), 503–522.

Dhaka, A., Murray, A. N., Mathur, J., Earley, T. J., Petrus, M. J., & Patapoutian, A. (2007). TRPM8 is required for cold sensation in mice. *Neuron, 54*(3), 371–378.

Engler, A., Aeschlimann, A., Simmen, B. R., Michel, B. A., Gay, R. E., Gay, S., et al. (2007). Expression of transient receptor potential vanilloid 1 (TRPV1) in synovial fibroblasts from patients with osteoarthritis and rheumatoid arthritis. *Biochemical and Biophysical Research Communications, 359*(4), 884–888.

Fernandez, J. A., Skryma, R., Bidaux, G., Magleby, K. L., Scholfield, C. N., McGeown, J. G., et al. (2011). Voltage- and cold-dependent gating of single TRPM8 ion channels. *The Journal of General Physiology, 137*(2), 173–195.

Gracheva, E. O., Ingolia, N. T., Kelly, Y. M., Cordero-Morales, J. F., Hollopeter, G., Chesler, A. T., et al. (2010). Molecular basis of infrared detection by snakes. *Nature, 464*(7291), 1006–1011.

Grandl, J., Hu, H., Bandell, M., Bursulaya, B., Schmidt, M., Petrus, M., et al. (2008). Pore region of TRPV3 ion channel is specifically required for heat activation. *Nature Neuroscience.*

Guler, A. D., Lee, H., Iida, T., Shimizu, I., Tominaga, M., & Caterina, M. (2002). Heat-evoked activation of the ion channel, TRPV4. *Journal of Neurophysiology, 22*(15), 6408–6414.

Hart, S. (1998). Beetle mania: an attraction to fire. *Bioscience, 48*(1), 3–5.

Heimburg, T. (2007). *Thermal biophysics of membranes.* Weinheim: Wiley-VCH, Verlag GmbH & Co. KGaA.

Hensel, H. (1974). Thermoreceptors. *Annual Review of Physiology, 36*, 233–249.

Hille, B. (2001). *Ionic channels of excitable membranes* (3rd ed.). Sunderland, MA: Sinauer Associates.

Jancso, N. (1955). *Speicherung. Stoffanreicherung im Retikuloendothel und in der Niere* (Budapest).

Jancso, N., Jancso-Gabor, A., & Szolcsanyi, J. (1967). Direct evidence for neurogenic inflammation and its prevention by denervation and by pretreatment with capsaicin. *British Journal of Pharmacology, 31*(1), 138–151.

Jara-Oseguera, A., & Islas, L. D. (2013). The role of allosteric coupling on thermal activation of thermo-TRP channels. *Biophysical Journal, 104*(10), 2160–2169.

Julius, D., & Basbaum, A. I. (2001). Molecular mechanisms of nociception. *Nature, 413*(6852), 203–210.

Kang, K., Panzano, V. C., Chang, E. C., Ni, L., Dainis, A. M., Jenkins, A. M., et al. (2012). Modulation of TRPA1 thermal sensitivity enables sensory discrimination in *Drosophila. Nature, 481*(7379), 76–80.

Kreh, A., Anton, F., Gilly, H., & Handwerker, H. O. (1984). Vascular reactions correlated with pain due to cold. *Experimental Neurology, 85*(3), 533–546.

LaMotte, R. H., & Campbell, J. N. (1978). Comparison of responses of warm and nociceptive C-fiber afferents in monkey with human judgments of thermal pain. *Journal of Neurophysiology, 41*(2), 509–528.

Leffler, A., Linte, R. M., Nau, C., Reeh, P., & Babes, A. (2007). A high-threshold heat-activated channel in cultured rat dorsal root ganglion neurons resembles TRPV2 and is blocked by gadolinium. *European Journal of Neuroscience, 26*(1), 12–22.

Levine, J. D., & Reichling, D. B. (1999). Peripheral mechanisms of inflammatory pain. In P. D. Wall, & R. Melzack (Eds.), *Textbook of pain* (pp. 59–73). New York: Churchill-Livingstone.

Lewis, T., & Love, W. S. (1926). Vascular reactions of the skin to injury. Part III. Some effects of freezing, of cooling and of warming. *Heart, 13*, 27–60.

Liao, M., Cao, E., Julius, D., & Cheng, Y. (2013). Structure of the TRPV1 ion channel determined by electron cryo-microscopy. *Nature, 504*(7478), 107–112.

Liu, B., Hui, K., & Qin, F. (2003). Thermodynamics of heat activation of single capsaicin ion channels VR1. *Biophysical Journal, 85*(5), 2988–3006.

Liu, B., & Qin, F. (2005). Functional control of cold- and menthol-sensitive TRPM8 ion channels by phosphatidylinositol 4,5-bisphosphate. *Journal of Neurophysiology, 25*(7), 1674–1681.

Liu, B. L., Yao, J., & Qin, F. (2011). Hysteresis of gating underlines sensitization of TRPV3 channels. *The Journal of General Physiology, 138*(5), 509–520.

Meents, J. E., Neeb, L., & Reuter, U. (2010). TRPV1 in migraine pathophysiology. *Trends in Molecular Medicine, 16*(4), 153–159.

Moqrich, A., Hwang, S. W., Earley, T. J., Petrus, M. J., Murray, A. N., Spencer, K. S., et al. (2005). Impaired thermosensation in mice lacking TRPV3, a heat and camphor sensor in the skin. *Science, 307*(5714), 1468–1472.

Nobile, M., Olcese, R., Toro, L., & Stefani, E. (1997). Fast inactivation of Shaker K^+ channels is highly temperature dependent. *Experimental Brain Research, 114*(1), 138–142.

Numazaki, M., Tominaga, T., Toyooka, H., & Tominaga, M. (2002). Direct phosphorylation of capsaicin receptor VR1 by PKCe and identification of two target serine residues. *The Journal of Biological Chemistry, 277*, 13375–13378.

Patapoutian, A., Peier, A. M., Story, G. M., & Viswanath, V. (2003). ThermoTRP channels and beyond: mechanisms of temperature sensation. *Nature Reviews Neuroscience, 4*(7), 529–539.

Peier, A. M., Reeve, A. J., Andersson, D. A., Moqrich, A., Earley, T. J., Hergarden, A. C., et al. (2002). A heat-sensitive TRP channel expressed in keratinocytes. *Science, 296*(5575), 2046–2049.

Picazo-Juarez, G., Romero-Suarez, S., Nieto-Posadas, A., Llorente, I., Jara-Oseguera, A., Briggs, M., et al. (2011). Identification of a binding motif in the S5 helix that confers cholesterol sensitivity to the TRPV1 ion channel. *The Journal of Biological Chemistry, 286*(28), 24966–24976.

Pusch, M., Ludewig, U., & Jentsch, T. J. (1997). Temperature dependence of fast and slow gating relaxations of ClC-0 chloride channels. *The Journal of General Physiology, 109*(1), 105–116.

Qin, F. (2013). Demystifying thermal channels: driving a channel both forwards and backwards with a single gear? *Biophysical Journal, 104*(10), 2118–2120.

Rainville, P., Feine, J. S., Bushnell, M. C., & Duncan, G. H. (1992). A psychophysical comparison of sensory and affective responses to four modalities of experimental pain. *Somatosensory and Motor Research, 9*(4), 265–277.

Saito, S., Fukuta, N., Shingai, R., & Tominaga, M. (2011). Evolution of vertebrate transient receptor potential vanilloid 3 channels: opposite temperature sensitivity between mammals and western clawed frogs. *PLoS Genetics, 7*(4), e1002041.

Sato, J., Morimae, H., Takanari, K., Seino, Y., Okada, T., Suzuki, M., et al. (2000). Effects of lowering ambient temperature on pain-related behaviors in a rat model of neuropathic pain. *Experimental Brain Research, 133*(4), 442–449.

Schmitz, H., & Bleckmann, H. (1998). The photomechanic infrared receptor for the detection of forest fires in the beetle Melanophila acuminata (Coleoptera : Buprestidae). *Journal of Comparative Physiology A – Sensory, Neural, and Behavioral Physiology, 182*(5), 647–657.

Spray, D. C. (1986). Cutaneous temperature receptors. *Annual Review of Physiology, 48*, 625–638.

Szallasi, A. (2002). Vanilloid (capsaicin) receptors in health and disease. *American Journal of Clinical Pathology, 118*(1), 110–121.

Szallasi, A., Cruz, F., & Geppetti, P. (2006). TRPV1: a therapeutic target for novel analgesic drugs? *Trends in Molecular Medicine, 12*(11), 545–554.

Szolcsanyi, J., & Jancso-Gabor, A. (1975). Sensory effects of capsaicin congeners I. Relationship between chemical structure and pain-producing potency of pungent agents. *Arzneimittelforschung., 25*(12), 1877–1881.

Szolcsanyi, J., & Jancso-Gabor, A. (1976). Sensory effects of capsaicin congeners. Part II: Importance of chemical structure and pungency in desensitizing activity of capsaicin-type compounds. *Arzneimittelforschung., 26*(1), 33–37.

Takahashi, K., Sato, J., & Mizumura, K. (2003). Responses of C-fiber low threshold mechanoreceptors and nociceptors to cold were facilitated in rats persistently inflamed and hypersensitive to cold. *Neuroscience Research, 47*(4), 409–419.

Torebjork, H. E., & Hallin, R. G. (1973). Perceptual changes accompanying controlled preferential blocking of A and C fibre responses in intact human skin nerves. *Experimental Brain Research, 16*(3), 321–332.

TRPM8 channel modulators. (2011). *Nature Review Drug Discovery, 10*, 569.

Tsavaler, L., Shapero, M. H., Morkowski, S., & Laus, R. (2001). Trp-p8, a novel prostate-specific gene, is up-regulated in prostate cancer and other malignancies and shares high homology with transient receptor potential calcium channel proteins. *Cancer Research, 61*(9), 3760–3769.

Valdes, A. M., De, W. G., Doherty, S. A., Lories, R. J., Vaughn, F. L., Laslett, L. L., et al. (2011). The Ile585Val TRPV1 variant is involved in risk of painful knee osteoarthritis. *Annals of the Rheumatic Diseases, 70*(9), 1556–1561.

Vlachova, V., Teisinger, J., Suankova, K., Lyfenko, A., Ettrich, R., & Vyklicky, L. (2003). Functional role of C-terminal cytoplasmic tail of rat vanilloid receptor 1. *Journal of Neuroscience, 23*(4), 1340–1350.

Wahren, L. K., Torebjork, E., & Jorum, E. (1989). Central suppression of cold-induced C fibre pain by myelinated fibre input. *Pain, 38*(3), 313–319.

Watanabe, H., Vriens, J., Suh, S. H., Benham, C. D., Droogmans, G., & Nilius, B. (2002). Heat-evoked activation of TRPV4 channels in a HEK293 cell expression system and in native mouse aorta endothelial cells. *The Journal of Biological Chemistry, 277*(49), 47044–47051.

Welch, J. M., Simon, S. A., & Reinhart, P. H. (2000). The activation mechanism of rat vanilloid receptor 1 by capsaicin involves the pore domain and differs from the activation by either acid or heat. *Proceedings of the National Academy of Sciences of the United States of America, 97*(25), 13889–13894.

Wolf, S., & Hardy, J. D. (1941). Studies on pain. Observations on pain due to local cooling and on factors involved in the 'cold pressor' effect. *Journal of Clinical Investigation, 20*, 521–533.

Wondergem, R., Ecay, T. W., Mahieu, F., Owsianik, G., & Nilius, B. (2008). HGF/SF and menthol increase human glioblastoma cell calcium and migration. *Biochemical and Biophysical Research Communications, 372*(1), 210–215.

Wondergem, R., & Bartley, J. W. (2009). Menthol increases human glioblastoma intracellular Ca2+, BK channel activity and cell migration. *Journal of Biomedical Science, 16*, 90.

Wood, J. N. (Ed.). (1993). *Capsaicin in the study of pain*. San Diego: Academic Press.

Xu, H., Delling, M., Jun, J. C., & Clapham, D. E. (2006). Oregano, thyme and clove-derived flavors and skin sensitizers activate specific TRP channels. *Nature Neuroscience, 9*(5), 628–635.

Xu, H. X., Ramsey, I. S., Kotecha, S. A., Moran, M. M., Chong, J. H. A., Lawson, D., et al. (2002). TRPV3 is a calcium-permeable temperature-sensitive cation channel. *Nature, 418*(6894), 181–186.

Yamamura, H., Ugawa, S., Ueda, T., Morita, A., & Shimada, S. (2008). TRPM8 activation suppresses cellular viability in human melanoma. *American Journal of Physiology Cell Physiology, 295*(2), C296–C301.

Yang, F., Cui, Y., Wang, K., & Zheng, J. (2010). Thermosensitive TRP channel pore turret is part of the temperature activation pathway. *Proceedings of the National Academy of Sciences of the United States of America, 107*(15), 7083–7088.

Yao, J., Liu, B., & Qin, F. (2009). Rapid temperature jump by infrared diode laser irradiation for patch-clamp studies. *Biophysical Journal, 96*(9), 3611–3619.

Yao, J., Liu, B. L., & Qin, F. (2010). Kinetic and energetic analysis of thermally activated TRPV1 channels. *Biophysical Journal, 99*(6), 1743–1753.

Yao, J., Liu, B. L., & Qin, F. (2010). Pore turret of thermal TRP channels is not essential for temperature sensing. *PNAS, 107*(32), E125.

Yao, J., Liu, B., & Qin, F. (2011). Modular thermal sensors in temperature-gated transient receptor potential (TRP) channels. *Proceedings of the National Academy of Sciences of the United States of America, 108*, 11109–11114.

Yarnitsky, D., & Ochoa, J. L. (1990). Release of cold-induced burning pain by block of cold-specific afferent input. *Brain, 113*(Pt 4), 893–902.

Zakharian, E., Cao, C., & Rohacs, T. (2010). Gating of transient receptor potential melastatin 8 (TRPM8) channels activated by cold and chemical agonists in planar lipid bilayers. *Journal of Neuroscience, 30*(37), 12526–12534.

Zhong, L., Bellemer, A., Yan, H., Honjo, K., Robertson, J., Hwang, R., et al. (2012). Thermosensory and nonthermosensory isoforms of *Drosophila melanogaster* TRPA1 reveal heat-sensor domains of a thermoTRP Channel. *Cell Reports, 1*, 1–13.

Zimmermann, K., Lennerz, J. K., Hein, A., Link, A. S., Kaczmarek, J. S., Delling, M., et al. (2011). Transient receptor potential cation channel, subfamily C, member 5 (TRPC5) is a cold-transducer in the peripheral nervous system. *Proceedings of the National Academy of Sciences of the United States of America, 108*(44), 18114–18119.

CHAPTER THREE

Gating of Thermally Activated Channels

David Baez[1], Natalia Raddatz[1,2], Gonzalo Ferreira[3], Carlos Gonzalez[1] and Ramon Latorre[1,*]
[1]Centro Interdisciplinario de Neurociencia de Valparaíso, Facultad de Ciencias, Universidad de Valparaíso, Valparaíso, Chile
[2]Centre for Plant Biotechnology and Genomics, Universidad Politécnica de Madrid, Pozuelo de Alarcón (Madrid), Spain
[3]Laboratorio de Canales Iónicos, Departamento de Biofísica, Facultad de Medicina, Universidad de la República, Montevideo, Uruguay
*Corresponding author: E-mail: ramon.latorre@uv.cl

Contents

Abstract

A class of ion channels that belongs to the transient receptor potential (TRP) superfamily and is present in specialized neurons that project to the skin has evolved as temperature detectors. These channels are classified into subfamilies, namely canonical (TRPC), melastatin (TRPM), ankyrin (TRPA), and vanilloid (TRPV). Some of these channels are activated by heat (TRPM2/4/5, TRPV1-4), while others by cold (TRPA1, TRPC5, and TRPM8). The general structure of these channels is closely related to that of the voltage-dependent K^+ channels, with their subunits containing six transmembrane segments that form tetramers. Thermal TRP channels are polymodal receptors. That is, they can be activated by temperature, voltage, pH, lipids, and agonists. The high temperature sensitivity in these thermal TRP channels is due to a large enthalpy change (~100 kcal/mol), which is about five times the enthalpy change in voltage-dependent gating. The characterization of the macroscopic currents and

Current Topics in Membranes, Volume 74
ISSN: 1063-5823
http://dx.doi.org/10.1016/B978-0-12-800181-3.00003-8

single-channel analysis demonstrated that gating by temperature is complex and best described by branched or allosteric models containing several closed and open states. The identification of molecular determinants of temperature sensitivity in TRPV1, TRPA1, and TRPV3 strongly suggest that thermal sensitivity arises from a specific protein domain.

1. INTRODUCTION

Considering that temperature-sensitive ion channels come in all kind of different flavors and that temperature is an intensive thermodynamic parameter, to write a chapter on their gating is like driving into a slippery road with brakes problems. Thus, since temperature is an intensive measure of the average speed of atoms and molecules, everything is subject to thermal fluctuations and this makes the search of the molecular structures in charge of sensing temperature (i.e., the "temperature sensor") in this type of channels something similar to be looking for the Holy Grail. In general, the temperature-sensitive ion channels behave as polymodal receptors (e.g., Latorre, Brauchi, Orta, Zaelzer, & Vargas, 2007) meaning that they not only respond to changes in temperature but also to a multiplicity of other stimuli (voltage, agonists, lipids, ions, pressure). It is now not difficult to envision how the electrical energy contained in the electric field, or how the chemical energy produced by the binding of an agonist is transformed into mechanical energy (the pore opening). It is still a mystery, however, how the thermal energy leads to channel opening with such exquisite temperature sensitivity as in the case of some TRP channels. Yao et al. (Yao, Liu, & Qin, 2010a) pointed out that the enthalpy change between the open and closed conformations in TRPV1 amounts to about 100 kcal/mol, which is equivalent to an electrical energy moving 71 unit charges across 60 mV! It is most probable that unlike the stereotyped nature of the voltage sensor in voltage-dependent ion channels, the structures in charge of sensing temperature in temperature-sensitive ion channels do not follow a single pattern. It is likely that these structures will be proved to be as diverse as the number of different temperature-sensitive channels we know at present.

Early studies on thermal sensation showed that thermoreceptors are located in the cutaneous nerve endings and, although more sparsely, in the mouth and nasal cavity. The same studies classified these receptors into two groups of cold and warm receptors (Darian-Smith et al., 1979; Hensel, Andres, & von During, 1974). The molecular nature of these receptors remained a mystery until a Pandora box was opened with the cloning of the transient receptor potential vanilloid1, TRPV1 channel (Caterina et al.,

1997). Seven families comprise the TRP channel superfamily (Figure 3.1): classical or canonical subfamily (TRPC) with seven members; the vanilloid (TRPV) family having six members; melastatin-related (TRPM) subfamily containing eight members; TRPA (ankyrin) with only one member, TRPA1. TRPN (NO-mechano potential); TRPP (polycystin) subfamily formed by the polycystic kidney disease protein or polycystins and containing three members; and lastly TRPML (mucolipin) subfamily consisting of channels that appear to be only present in intracellular compartments (for reviews on the TRP superfamily of ion channels the reader should consult Latorre, Zaelzer, and Brauchi (2009); Montell (2005); Ramsey, Delling, and Clapham (2006)). Temperature-sensitive TRP channels (usually dubbed thermoTRP channels) are spread in the TRPV (TRPV1-4), TRPM (TRPM2-5 and TRPM8), TRPA1, and in the TRPC subfamily (TRPC5) (Figure 3.1).

TRP channels are nonselective cationic channels that show, in several cases, a high Ca^{2+} permeability relative to monovalent cations (Owsianik, Talavera, Voets, & Nilius, 2006). Structurally, TRP channels are tetramers formed by proteins containing, as in voltage-dependent K^+ (Kv) channels,

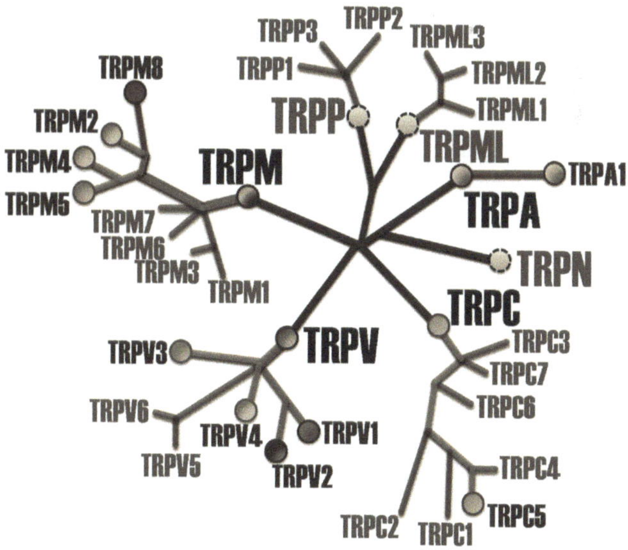

Figure 3.1 TRP superfamily dendogram. Seven subfamilies has been identified and the identity between them is low (~20%). Canonical or classic (TRPC), vanilloid (TRPV), ankyrin (TRPA), nonmechanoreceptor potential (TRPN), mucolipins (TRPML), polycystins (TRPP), and melastatin (TRPM). ThermoTRPs in the TRPM, TRPV, TRPA, and TRPC subfamilies are identified in colors. (See the color plate.)

six (S1–S6) transmembrane segments. Until very recently we only had low-resolution structures at 19, 35, and 16 Å for TRPV1, TRPV4, and TRPA1, respectively, obtained using electron microscopy (Cvetkov, Huynh, Cohen, & Moiseenkova-Bell, 2011; Moiseenkova-Bell, Stanciu, Serysheva, Tobe, & Wensel, 2008; Shigematsu, Sokabe, Danev, Tominaga, & Nagayama, 2010). However, this situation has been radically changed with the recent publication of two reports by Julius and colleagues (Cao, Liao, Cheng, & Julius, 2013; Liao, Cao, Julius, & Cheng, 2013). In the first article, they were able to solve the structure of TRPV1 at 3.4 Å using cryoelectron microscopy (Liao et al., 2013) and in the second paper the TRPV1 structure was revealed in the presence of toxins and agonists (Cao et al., 2013). Interestingly, all these structures conserve some of the general aspects of Kv channel structure. They all contain a tetrameric assembly of subunits residing in the lipid bilayer and a cytoplasmic "hanging gondola" (Kobertz, Williams, & Miller, 2000) formed by the N and C termini containing four large windows. These findings show that, as Kv channels, TRP channels are modular proteins from which we can distinguish an N-terminal, S1–S4, a pore (S5-P domain-S6), and a C-terminal module. The structures reported by Julius et al. show the presence of a cytoplasmic gate where the narrowest part is determined by a hydrophobic plug conformed by four isoleucines at position 679. The second gate is located at the selectivity filter being flanked by glycine 643 and methionine 644. The simultaneous binding of the spider toxin double-knot toxin (DkTx) and resiniferatoxin activates both gates, thereby greatly increasing the pore diameter, but the binding of capsaicin alone opens the pore at the level of the cytoplasmic gate only.

The different modules in thermoTRP channels serve as binding sites (sensors) for different agonists. Thus, capsaicin in TRPV1 and menthol in TRPM8 bind to the S1–S4 module (Bandell et al., 2006; Cao et al., 2013; Gavva et al., 2004; Jordt & Julius, 2002; Voets, Owsianik, Janssens, Talavera, & Nilius, 2007). Electrophilic compounds in TRPA1 and TRPV1 bind to the N terminal module (Hinman, Chuang, Bautista, & Julius, 2006; Macpherson et al., 2007; Salazar et al., 2008). PIP_2 in TRPM8 and TRPV1 binds to the C-terminal module (Rohacs, Lopes, Michailidis, & Logothetis, 2005; Ufret-Vincenty, Klein, Hua, Angueyra, & Gordon, 2011). The S4, on the other hand, appears to play a role in voltage sensing in TRPM8 channels (Voets et al., 2007). However, the molecular determinants of temperature sensing have been found to be, depending on the thermoTRP channel under study, spread in different modules (see below).

This chapter provides the reader with a critical appraisal of the information we possess at present regarding activation by temperature in thermoTRP channels. We first give an overview of the diversity of this type of channels. Because the TRP channel-forming proteins are essentially thermodynamic machines, we discuss next what we know about the thermodynamic parameters that modulate thermoTRP channel gating and discuss the most appropriate kinetic models that can account for their polymodal receptor behavior. The last section of this chapter is dedicated to a detailed description of the molecular determinants of temperature sensitivity that have been identified so far.

2. TEMPERATURE-SENSITIVE CHANNEL DIVERSITY

TRPV1 was found to be activated by depolarizing voltages, capsaicin (the active compound in chili peppers), resiniferatoxin (RTX) an extremely potent toxin obtained from the *Euphorbia* plant, fatty acids, anandamide, heat, external cations (Na^+, Mg^{2+}, and Ca^{2+}), and protons and to be inhibited by capzasepine (Ahern, 2003; Ahern, Brooks, Miyares, & Wang, 2005; Caterina et al., 1997; Chavez, Chiu, & Castillo, 2010; Jordt, Tominaga, & Julius, 2000; Tominaga et al., 1998). TRPV1 is a cation-selective channel with a high Ca^{2+} permeability ($P_{Ca}/P_K \sim 10$) and is activated by temperature above 40 °C at negative voltages. See Figure 3.2 for a schematic representation of the temperature range in which the different thermoTRPs are activated. TRPV1 is expressed in small to medium size primary sensory neurons in dorsal root and trigeminal ganglion, the "pain" pathway (Caterina et al., 1997; Helliwell et al., 1998; Tominaga et al., 1998). The TRPV1 knockout mouse showed that this channel was involved not only in painful thermal sensation but also in thermal hyperalgesia (Caterina et al., 2000). Some compounds thought to be specific for the noxious cold receptor TRPA1 (see below) proved to activate TRPV1. This is the case of the pungent compound derived from onions, allicin (Macpherson et al., 2005; Salazar et al., 2008). The activation of TRPV1 is also produced by a fraction of the venom of the tarantula, *Psalmopoeus cambridgei*, that contains three cysteine knot (Kremeyer et al., 2010) peptides, dubbed vanillotoxins (Siemens et al., 2006). In particular, the vanillotoxin called DkTx, which consists of two independent toxin domains connected by a short linker, traps TRPV1 in its open state almost irreversibly (Bohlen et al., 2010). Whether PIP_2 activates TRPV1 is still controversial (cf. Cao, Cordero-Morales, Liu, Qin, & Julius, 2013; Lukacs, Rives, Sun, Zakharian, & Rohacs, 2013; Ufret-Vincenty et al., 2011).

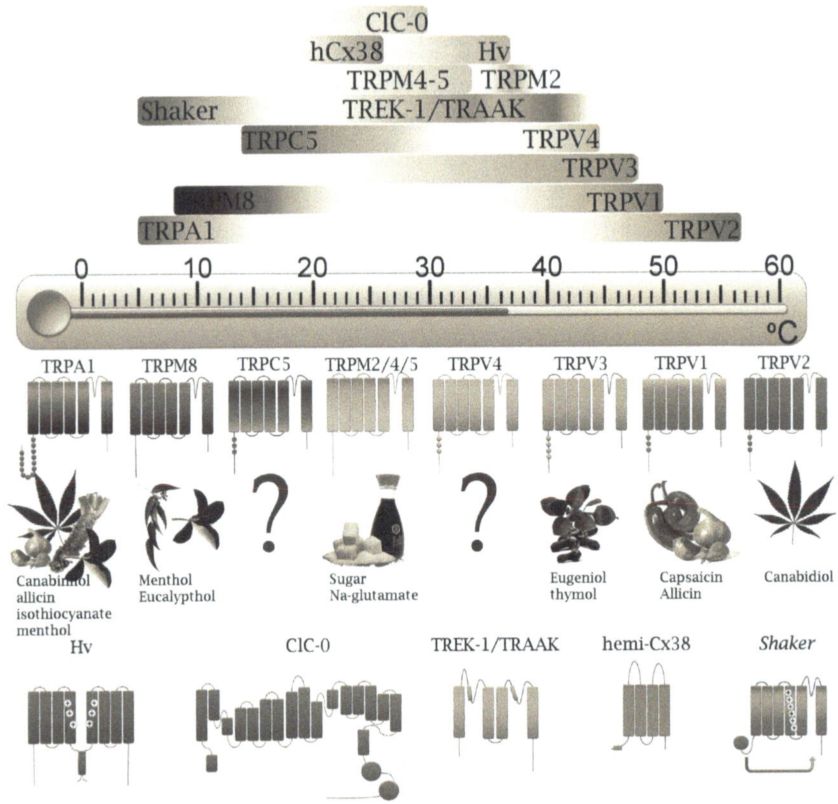

Figure 3.2 The temperature sensitivity of thermoTRPs and other temperature-dependent ion channels. The open probability (P_o) of the different temperature-activated channels increases in different temperature ranges. In this figure the higher the color intensity the larger the P_o. ThermoTRPs share a common organization. They are tetramers where the permeation pathway is located in the center of symmetry of the channel. ThermoTRP are activated by different agonist obtained from different plants. Other highly sensitive ion channels that do not share a common structure are shown in the bottom of the figure. Proton channels (H_v), chloride channel (ClC-0), and connexin 38 (hemi-Cx38). The green ball in the case of the Shaker channel is the inactivation gate that once the channel opens binds to the channel internal vestibule. (See the color plate.)

As discussed in the Introduction, the structure of TRPV1 revealed two gates that expand upon binding of RTX and DkTx allowing the passage of cations (Cao, et al., 2013). Two gates were also detected using cysteine scanning mutagenesis and accessibility to thiol-modifying agents and the Ag^+ ion (Salazar et al., 2009). Leucine 681 located towards the distal part of S6 was found to hinder the flow of large cations (5.8 Å). In the closed state tyrosine 671 impedes the access of the small Ag^+ ion. The position of these gates differs from that revealed in the TRPV1 structure (Liao et al., 2013). It remains to be explained why Ag^+ in the experiments of Salazar et al. (2009) is able to reach deep into the pore overcoming the energy barrier imposed by the cytoplasmic gate located at isoleucine 679 when the channel is closed.

In medium to large diameter human sensory neurons Caterina et al. (Caterina, Rosen, Tominaga, Brake, & Julius, 1999) identified a Ca^{2+} permeable ($P_{Ca}/P_{Na} \sim 2.9$) cationic channel activated by high temperatures ($\sim$52 °C) now known as TRPV2. TRPV2 was found to be insensitive to capsaicin and pH. It should be noted here that a mouse TRPV2 ortholog was cloned at the same time (Kanzaki et al., 1999) and was found to be localized mainly in intracellular pools and translocated to the plasma membrane upon the stimulation of cells with insulin-like growth factor. This channel proved to be rather insensitive to temperature, pH, and capsaicin but once in the plasma membrane it was able to increase cell Ca^{2+} entry. The vanilloid family of TRP channels has two more temperature-sensitive channels TRPV3 and TRPV4, which are thermosensors for nonnoxious heat (>33 °C) (Guler et al., 2002; Peier, et al., 2002; Smith et al., 2002; Xu et al., 2002). Mainly expressed in keratinocytes, TRPV3 is activated by camphor, and other components of plants such as carvacrol, eugenol, and thymol (Xu, Delling, Jun, & Clapham, 2006). Paradoxically, menthol, a well-known activator of the cold receptor TRPM8 (see below), also activates TRPV3 (Macpherson et al., 2006). TRPV3 channels are discussed extensively in the chapter by Hu in this volume. TRPV4, was initially described as hypo-osmolarity-activated (Liedtke et al., 2000; Strotmann, Harteneck, Nunnenmacher, Schultz, & Plant, 2000; Wissenbach, Bodding, Freichel, & Flockerzi, 2000) and its heat-evoked responses are greater in hypo-osmotic solutions (Guler et al., 2002). TRPV4 is expressed in a variety of different tissues including epithelial cells of the renal distal convoluted tubule, trachea, and in autonomic nerve fibers (Delany et al., 2001). Because TRPV4 temperature-mediated activation has been observed only in intact cells e.g., (Watanabe et al., 2002), it has been hypothesized that this channel lacks an intrinsic temperature sensor. However, in a recent report, Garcia-Elias et al. (Garcia-Elias et al., 2013) demonstrated that TRPV4 is a bona fide temperature-sensitive channel but the channel activation by temperature requires the presence of PIP_2. TRPV4 is activated also by the synthetic phorbol ester 4α-phorbol 12,13-didecanoate (Watanabe et al., 2002), by the active compound of an extract of the plant *Andrographis paniculata*, bisandrographolide A (Smith, Maloney, Pothen, Clardy, & Clapham, 2006) and by compounds derived from the arachidonic acid (Watanabe et al., 2003).

Potential temperature-sensitive channels are also present in the melastatin subfamily (TRPM2-5 are heat receptors and TRPM8 is a cold receptor) and in the ankyrin subfamily, TRPA1 a cold receptor. TRPM2, a Ca^{2+} permeable cation channel first designated as NUDT9 because it contains a Nudix hydrolase (NudT9-H), was found to be gated by free ADP-ribose (Perraud et al., 2001). TRPM2 is activated by warm temperatures (>35 °C) and regulates Ca^{2+} entry into pancreatic β-cells (Togashi et al., 2006). However, this channel is not present in primary sensory neurons and a role in thermosensation has not been reported (reviewed in Jiang, Gamper, and Beech (2011)). As TRPM2, TRPM3 also shows a high Ca^{2+} permeability (P_{Ca}/P_{Na} ~ 1.6) and hypotonic external solutions increase Ca^{2+} entry in cells expressing this channel (Grimm, Kraft, Sauerbruch, Schultz, & Harteneck, 2003; Lee et al., 2003). Five different splice variants of TRPM3 have been identified, one of these splice sites is located in a putative pore region and their biophysical characterization showed that the two well-studied variants differ mainly in their divalent cation selectivity (Oberwinkler, Lis, Giehl, Flockerzi, & Philipp, 2005). TRPM3 is expressed in a subset of sensory neurons and is activated by the neurosteroid pregnenolone sulfate (Rosenzweig et al., 2005) and heat (>35 °C) (Wagner et al., 2008). TRPM3-deficient mice exhibit deficits in their avoidance response to noxious heat (Vriens et al., 2011). Interestingly, the largest population of heat positive neurons responds to PS and capsaicin, suggesting the coexpression of TRPM3 and TRPV1, although there are two other populations of neurons, one responding only to PS and the other only to capsaicin. TRPM4 is a Ca^{2+}-activated nonselective cation channel, which is also activated by depolarizing voltages (Launay et al., 2002; Nilius et al., 2003) and modulated by PIP_2 (Nilius et al., 2006). PIP_2 produces a leftward shift of the channel conductance–voltage curve and increases the apparent Ca^{2+} sensitivity. TRPM4 behaves as a heat receptor in the temperature range between 15 °C and 35 °C (Talavera et al., 2005). Similar to TRPM4, TRPM5 is activated by depolarizing voltages and is heat-activated (>15 °C) playing an important role in vertebrate taste transduction (Perez et al., 2002; Talavera et al., 2005; Zhang et al., 2003). Heat potentiates the gustatory nerve responses to sweet compounds in wild type animals but not in TRPM5 knockout mice, a finding that shows this channel contributes to the modulation by temperature of different taste modalities. Similar to TRPM4, PIP_2 reverses desensitization, partially restoring channel activity (Liu & Liman, 2003). Unlike the other temperature-sensitive channels in the TRPM subfamily that are heat receptors, TRPM8 is a cold receptor activated by temperatures <25 °C. This channel, identified first as an upregulated protein in prostate cancer (Tsavaler, Shapero, Morkowski, & Laus, 2001), was cloned and characterized simultaneously by two groups (McKemy, Neuhausser, & Julius, 2002; Peier et al., 2002). Cooling compounds such as menthol and the super cooling agent icilin, PIP_2, and depolarizing voltages activated TRPM8 channels (Liu & Qin, 2005; McKemy et al., 2002; Peier et al., 2002; Rohacs et al., 2005). However, we should mention here that camphor, an agonist of TRPV3 channels, activates and potentiates TRPM8 responses evoked by cooling or icilin but inhibits the TRPM8 channel activation by menthol (Selescu, Ciobanu, Dobre, Reid, & Babes, 2013).

From the moment TRPA1 was cloned from vertebrates, the temperature sensitivity of this channel has been the subject of heated debate. The human TRPA1 (hTRPA1) was found to be temperature insensitive (Jordt et al., 2004) whereas in the case of mouse TRPA1 noxious cold (<17 °C) elicited Ca^{2+} influx and cationic currents (Story et al., 2003). From then on, the TRPA1 world was divided in two since several studies confirmed that TRPA1 was a receptor of noxious cold (Bandell et al., 2004; Karashima et al., 2009; Kremeyer et al., 2010; Sawada, Hosokawa, Hori, Matsumura, & Kobayashi, 2007) while others reported that cold did not activate TRPA1 (Bautista et al., 2006; Cordero-Morales, Gracheva, & Julius, 2011; Jordt et al., 2004; Zurborg, Yurgionas, Jira, Caspani, & Heppenstall, 2007). Some of this confusion may have risen from the fact that TRPA1 is also activated by internal Ca^{2+} (Zurborg et al., 2007) because in some cases Ca^{2+} was used in the external solutions (Bandell et al., 2004; Kremeyer et al., 2010). Apparently, this controversy has been recently partially settled in a detailed study by Chen et al. (2013) who found that there is an evolutionary divergence in the thermal sensitivity of TRPA1 (see also Saito et al., 2012). hTRPA1, as well as the rhesus monkey TRPA1, is insensitive to temperature. Mosquito, fruit fly, and rattle snake TRPA1, like *Drosophila* TRPA1, is a heat receptor but in mouse and rat it is activated by noxious cold. However, evolutionary divergence is probably part but not the whole story since some *Drosophila* TRPA1 isoforms are insensitive to temperature (Zhong et al., 2012; see below).

TRPA1 functions as a receptor of a plethora of noxious and electrophilic compounds such as the pungent principle of wasabi, ally isothiocyanate, thiosulfinate from onions, and α,β-unsaturated aldehyde compounds from cinnamon and other aldehydes (Hinman et al., 2006; Jordt et al., 2004; Macpherson et al., 2007). Menthol, a compound that potently activates

TRPM8, shows a bimodal action on TRPA1. At low concentrations (<10 μM) menthol activates TRPA1 but at higher concentrations it behaves as a TRPA1 blocker. A TRPA1 gain-of-function point mutation (N855S) by causing a fivefold increase in current at the resting potential is linked to an autosomal-dominant familial episodic pain syndrome (Kremeyer et al., 2010).

Last but not the least, TRPC5 present in neurons of the human and rat dorsal root ganglia behaves as a cold receptor of innocuous temperatures (Zimmermann et al., 2011). TRPC5 is activated in the temperature range 37–25 °C, the cold-induced activation is potentiated by phospholipase C-coupled receptor activation.

More diversity of thermoreceptors can be obtained through the formation of heteromeric channels as shown for the case of TRPV1-4 channels (Cheng, Yang, Takanishi, & Zheng, 2007) and in vivo experiments that provide evidence that TRPV1 and TRPV2 interact in rat cerebral cortex cells (Hellwig, Albrecht, Harteneck, Schultz, & Schaefer, 2005). Heteromeric TRPV1/TRPV3 channels with altered heat sensitivity have been recently reported, thus showing that heteromeric channels have a reduced heat sensitivity ($Q_{10} = 4$) compared to TRPV1 ($Q_{10} = 20$) (Cheng et al., 2012).

From the above discussions of the temperature-sensitive channels in the TRP channel superfamily, it is clear that the gating of these channels is extremely promiscuous. They are gated by temperature, voltage, agonists, lipids, and ions. This level of diversity of gating makes it more complex and difficult to explore, and thus forcing investigators to focus on simplified situations in which a few variables are at play (e.g., temperature and voltage or temperature and agonists, etc.), when trying to construct reasonable gating kinetic schemes. A word of caution: we also showed above that compounds that initially appeared to be specific, turned out to be effective on more than one temperature-sensitive TRP channel (e.g., menthol; Figure 3.2).

For the sake of completeness, we mention here other ion channels highly temperature dependent. These include voltage-dependent proton channels (DeCoursey & Cherny, 1998), chloride channels (Pusch, Ludewig, & Jentsch, 1997), the two pore channels TREK1 and TRAAK (Noel et al., 2009), the hemichannel formed by connexin38 (Ebihara, 1996), and the fast inactivation process in voltage-dependent Shaker channels (Nobile, Olcese, Toro, & Stefani, 1997) (Figure 3.2). The good is that most of these channels will be discussed in other sections of this volume. However, some of the properties of their temperature-dependent gating are useful to illustrate various aspects of the mechanisms governing the transformation of thermal energy into mechanical energy (the pore opening) and they will be discussed accordingly.

3. ENERGETICS OF TEMPERATURE-SENSITIVE CHANNELS

Temperature-sensitive TRP channels are polymodal receptors but detailed biophysical studies of their gating are scarce. In this regard, voltage- and temperature-dependent gating of TRPM8 and TRPV1 (they are outward rectifiers) are among the best characterized, and we will take advantage of what is known in this regard as the basis of our general discussion on the gating of temperature-sensitive channels (Brauchi, Orio, & Latorre, 2004; Fernandez et al., 2011; Latorre et al., 2007; Liu, Hui, & Qin, 2003; Matta & Ahern, 2007; Voets et al., 2004).

Activation in temperature-sensitive TRP channels is characterized by very large enthalpy changes that are compensated by similarly large entropy changes (Table 3.1). For example, enthalpy changes of about 100 kcal/mol, 200 kcal/mol, and −60 kcal/mol are involved in the opening of TRPV1, TRPV2, and TRPM8, respectively. Because of its simplicity, these parameters have been calculated assuming a two-state model, closed (C)–open (O) for channel gating (Brauchi et al., 2004; Karashima et al., 2009; Liu et al., 2003).

Table 3.1 Thermodynamic parameters of different thermally-activated ion channels

Ion channel	ΔS (cal/mol K)	ΔH (kcal/mol)	Q_{10}	References
TRPV1	260	90	27	Yao et al. (2011)
TRPV2	586	200	>100	Yao et al. (2011)
TRPV3	–	54	17	Xu et al. (2002)
TRPV4	–	57	19	Guler et al. (2002)
TRPC5	–	−40	10	Zimmermann et al. (2011)
TRPM8	−384	−112	24	Brauchi et al. (2004)
TRPM2	–	75	44	Togashi et al. (2006)
TRPA1	−140	−40	10	Karashima et al. (2009)
ClC-0	–	62	40	Pusch et al. (1997)
Hv1	–	40	9	DeCoursey and Cherny (1998)
Shaker	–	−16	7	Nobile et al. (1997)

$$C \overset{\alpha(T,V)}{\underset{\beta(T,V)}{\leftrightarrow}} O$$

The equilibrium constant is in this case $K_{eq} = \alpha(T, V)/\beta(T, V) = K(0)$ $exp(zFV/RT)$ where $K(0)$ is the equilibrium constant in the absence of applied potential, V; z is the apparent number of gating charges involved in channel opening, and F, R, and T have their usual meanings. In this case the free energy difference (ΔG) between the two conformations is:

$$\Delta G = \Delta H - T\Delta S + zFV = -RT \ln K_{eq} \tag{1}$$

where ΔH and ΔS are the enthalpy and entropy changes, respectively and T is the absolute temperature. ΔG is related to the channel open probability, P_o by

$$P_o = \frac{1}{1 + K_{eq}} = \frac{1}{1 + K(0)\exp(zF(V - V_{1/2})/RT)} \tag{2}$$

where $V_{1/2}$ is the voltage at which P_o = ½ and is given by the expression:

$$V_{1/2} = \frac{\Delta G}{zF} = \frac{\Delta H - T\Delta S}{zF} \tag{3}$$

In the case $V_{1/2}$ is measured at two different temperatures, T_1 and T_2, from Eqn (3, #18) we have that:

$$V_{1/2}(T_2) - V_{1/2}(T_1) = -\frac{\Delta S(T_2 - T_1)}{zF} \tag{4}$$

The P_o versus V data are obtained from single channel or macroscopic current data and are fitted using the Boltzmann equation (Eqn (2)). It has been argued that the weak voltage dependency of thermoTRP channels is an important factor in determining the large leftward shifts in the P_o–V curves by different stimuli (Nilius et al., 2005). It is clear from Eqn (4) that the smaller the value of z, the larger the difference between the $V_{1/2}$ measured at two different temperatures. However, it is important to mention here that $V_{1/2}$ is related to the free energy that defines $K(0)$ and that the voltage dependence of the channel relates to the steepness of the P_o–V curve. Thus, the shifts in the P_o–V curve promoted by temperature represent only modifications of the energy that separates closed from open states at zero voltage. From Eqn (4) it is also clear that if the P_o–V curve is displaced to the left for a temperature decrease, it implies a negative entropic change during channel opening (Correa, Bezanilla, & Latorre, 1992). This is the case for the cold receptors TRPM8 and TRPA1. The two-state model demands that $V_{1/2}$ is related linearly with temperature (Eqn (3)). In thermoTRP channels $z \sim 1$ therefore $zFV \sim 2$ kcal/mol at 100 mV, which is much smaller than that of ΔH (Table 3.1; see above). The temperature dependence of $K(0)$ can be obtained by measuring P_o at different temperatures determined either from single channel data (Liu et al., 2003) or from macroscopic currents combined with noise analysis (Brauchi et al., 2004). With the provision that the contribution of zFV to ΔG is small, Eqn (1) leads to the van't Hoff relation:

$$\ln K_{eq} = -\frac{\Delta H}{RT} + \frac{\Delta S}{R} \tag{5}$$

$$\frac{d \ln K_{eq}}{d(1/T)} = \frac{\Delta H}{R} \tag{6}$$

Therefore, if ΔH is independent of temperature a plot of $\ln K_{eq}$ versus $1/T$ will be linear with slope yielding ΔH.

The changes in ΔH and ΔS are negative in the case of cold receptors and positive for heat receptors and they compensate each other bringing as a consequence a small ΔG, which allows the reversibility of the closed–open reaction. The sign of ΔH implies that ΔG increases in the case of cold receptors and decreases for heat receptors (Figure 3.3).

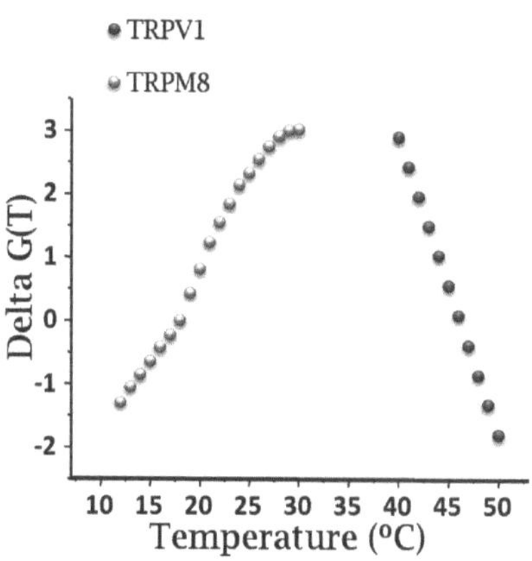

Figure 3.3 The free energy associated to temperature activation in thermoTRP. The free energy (DeltaG(T)) decreases as temperature is reduced in the case of TRPM8, a cold receptor (green solid circles). (DeltaG(T)) in the case of TRPV1, a heat receptor, decreases as temperature is augmented (red solid circles). (For interpretation of the references to color in this figure legend, the reader is referred to the online version of this book.)

Rate constants α and β are given by the following equations:

$$\alpha = Ae^{-\frac{\Delta G_{\alpha}^{\mp}}{RT}} = Ae^{-\frac{\left[\Delta H_{\alpha}^{\mp} + T\Delta S_{\alpha}^{\mp} + z_{\alpha}FV\right]}{RT}} \quad (7)$$

$$\beta = Ae^{-\frac{\Delta G_{\beta}^{\mp}}{RT}} = Ae^{-\frac{\left[\Delta H_{\beta}^{\mp} + T\Delta S_{\beta}^{\mp} + z_{\beta}FV\right]}{RT}} \quad (8)$$

where A is the frequency factor that for protein folding is on the order of $10^6\ s^{-1}$ (Chakrapani & Auerbach, 2005; Yang & Gruebele, 2003), $\Delta G_{\alpha}^{\mp}$, $\Delta H_{\alpha}^{\mp}$, and $\Delta S_{\alpha}^{\mp}$ are the activation free energy, activation enthalpy, and activation entropy, respectively, and z_{α} is the apparent gating charge associated with the forward reaction. In Eqn (7) the same parameters are associated with the deactivation process. The rate constants, α and β, can be obtained from macroscopic current measurements by determining $P_o = \alpha/(\alpha + \beta)$ and the time constant $\tau = 1/(\alpha + \beta)$. In thermoTRPs, it appears as a rule that in heat receptors (e.g., TRPV1) only the channel activation shows high activation enthalpies ($\Delta H_{\alpha}^{\mp}$), whereas in cold receptors the temperature-sensitive parameter is the deactivation time constant associated to $\Delta H_{\beta}^{\mp}$ (Brauchi et al., 2004; Voets et al., 2004) (see Table 3.1). In general, $\Delta H^{\mp} < \Delta H$ meaning that the channel temperature dependence is not completely contained in the activation or deactivation rates a fact difficult to explain based only on the two-state model since in this case $P_o = \alpha/(\alpha + \beta)$ and P_o should have the same T dependence of either α or β. For TRPM8, the deactivation process has two exponential components but the fastest process was claimed to be temperature and voltage insensitive (Brauchi et al., 2004).

Yao et al. (Yao et al., 2010a) have directly measured the activation and deactivation kinetics of TRPV1 using optically generated submilliseconds temperature steps in the 22 °C to >60 °C range. The time course of the currents elicited by the different temperature jumps follows an exponential course. The plot of the activation time constants ($\tau_a = 1/\alpha$) versus $1/T$ (Arrhenius plot) yields a $\Delta H_{\alpha}^{\mp} = 85$ kcal/mol and a $\Delta S_{\alpha}^{\mp} = 245$ cal/mol K. On the other hand, the steady-state current (I) was fit to a Boltzmann equation:

$$I = I_L e^{-\frac{\Delta H_L}{RT}} + \frac{I_{\max} e^{-\frac{\Delta H_i}{RT}}}{1 + e^{-\frac{\Delta G}{RT}}} \quad (9)$$

where ΔH_L is the enthalpy change of the leak current and the second term accounts for the TRPV1 channel activity, ΔH_i is the enthalpy change involved in the closed-open reaction and $\Delta G = \Delta H_i - T\Delta S$. Using a -60 mV holding potential, a fit to the current-$1/T$ data using Eqn (9) made it possible to estimate the enthalpy and entropy changes between the TRPV1 closed and open configurations. The resulting values were $\Delta H = 101$ kcal/mol and $\Delta S = 315$ cal/mol K. The large entropy value indicates that the channel transits between a highly ordered closed state and a strongly disordered open state. A highly ordered closed state is also predicted from the leftward shift of the $P_o(V)$ as temperature is increased (Eqn (4)). These values are in reasonable agreement with those obtained previously by measuring the effect of temperature on the single channel kinetics (Liu et al., 2003) and the value of $\Delta H_{\alpha}^{\ddagger}$ is close to the equilibrium enthalpy change (ΔH) between open and closed states. It was found, however, that the TRPV1 energy landscape changes with the applied voltage because as voltage is made more depolarizing ΔH and ΔS decrease in parallel. For example at 60 mV $\Delta H = 65$ kcal/mol and $\Delta H = 204$ kcal/mol K. A voltage dependence of ΔH reveals a certain degree of coupling between the temperature and voltage sensors but the fact that temperature is more effective in opening TRPV1 channels at hyperpolarizing voltages also suggest that temperature alone can open the channel. These temperature jump experiments, as it was also demonstrated in voltage jump experiments (Voets et al., 2004), also showed that temperature mainly drives the channel opening (α in the two-state model) and that the closing rate, β, has a very low temperature dependence. Intriguingly, the TRPV1 closing rate has a negative temperature dependence (β increase as temperature is decreased), a result at odds with that from modeling of thermal sensitivity of voltage responses (Yao et al., 2010a).

On the basis of these and other results (Brauchi et al., 2004; Karashima et al., 2009; Voets et al., 2004) it is possible to construct the enthalpy, entropy, and free energy landscapes of temperature gating for TRPV1, TRPM8, and TRPA1 noticing that for heat receptors there are large and positive ΔH and ΔS changes during channel activation, whereas for cold receptors the reverse situation occurs (Figure 3.4). Thus gating evolves from a low enthalpy ordered closed states to high enthalpy disordered open state in heat receptors and from a high enthalpy disordered closed state to a low enthalpy ordered open state in the case of cold receptors.

Of course all the previous analysis is adequate only if ΔH is a temperature-independent parameter, an assumption that has been challenged by recalling that given the large and positive value in molar heat capacity (ΔC_p) observed for the overall unfolding of a polypeptide chain (Dill, 1990; Privalov & Gill, 1988), ΔH and ΔS must both vary with T (Clapham & Miller, 2011). Accordingly, the thermal stability of proteins is dominated by concave energy dependence with temperature. In this case, a plot of K_{eq} versus T

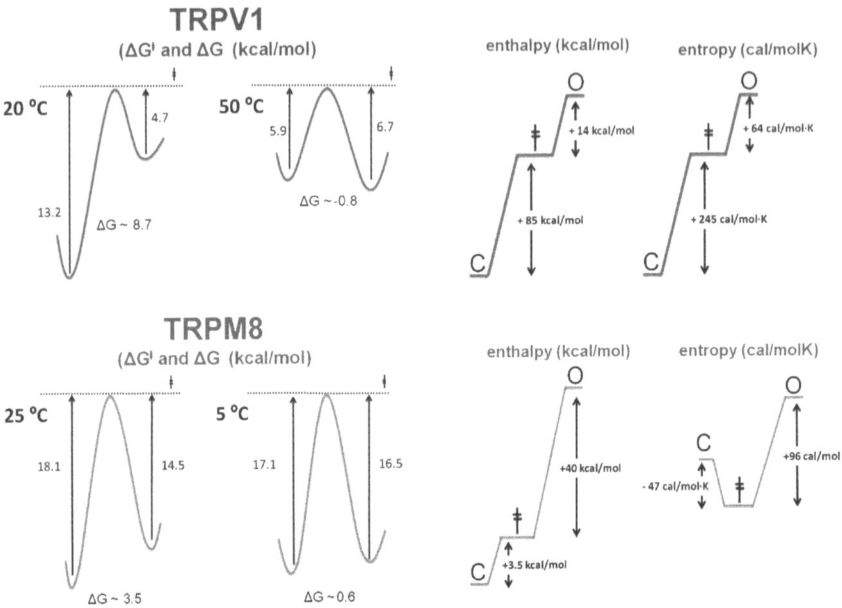

Figure 3.4 Energy landscape for TRPV1 and TRPM8. Increasing the temperature from 20 °C to 50 °C in TRPV1, changes ΔG from 8.7 kcal/mol to −0.8 kcal/mol. Thus, the channel opening reaction at 50 °C becomes spontaneous. This can be explained as large entropy and enthalpy changes associated to the activation rate constant (assuming a two-state model). In the case of TRPM8, a cold receptor, cooling from 25 °C to 5 °C, changes ΔG from 3.5 kcal/mol to 0.6 kcal/mol. In this case large enthalpy and entropy changes are associated to the deactivation rate constant.

should yield a "U"-shaped curve where the steepness of its arms is determined by the value ΔC_p. In other words, the Gibbs–Helmholtz free energy follows Eqn (10) for a two-state system (Privalov & Gill, 1988).

$$\Delta G = \Delta H(T_0)\left(1 - \frac{T}{T_0}\right) + \Delta C_p\left(T - T_0 - T\ln\frac{T}{T_0}\right) \tag{10}$$

where $\Delta H(T_0)$ is the enthalpy of the reaction at the temperature T_0 at which $\Delta G = 0$. The Eqn (10) predicts that all temperature-sensitive channels with a large ΔC_p can, in principle, behave as both cool and heat receptors (see also Privalov & Gill, 1988). We will return to this point at the end of this chapter when discussing thermal sensing in ion channels.

3.1 Voltage sensor

Although temperature sensing is the main topic of the present chapter and since we have discussed that thermoTRPs are also voltage dependent, we thought convenient to say a few words regarding the present knowledge we have about thermoTRPs' voltage sensor. The weak voltage dependence of these channels (see Nilius et al., 2005) could occur if many charges are moving very little across the electric field, that the number of gating charges is small, or that the z obtained from the fit of the $P_o(V)$ data is underestimated (Bezanilla & Stefani, 1994). Given that these channels are tetramers, a $z \sim 1$, means about 0.25 e_0 charges moving across the electric field in each channel subunit. The charge associated to channel opening in the case that there is direct coupling between channel voltage sensor activation and channel opening, can be obtained using the limiting slope procedure by measuring the slope of the $P_o(V)$ curve at voltages where $\ln(P_o)$ reaches a constant value (Almers, 1978; Sigg & Bezanilla, 1997). In this case:

$$\lim_{V \to -\infty} RT \frac{d \ln P_o}{dV} = Q_t \quad (11)$$

where Q_t is the total charge associated to the channel opening. In the case of TRPM8, $Q_t = 0.89\ e_0$ and decreases to about 0.6 e_0 when either R842 in S4 or K856 in the linker S4–S5 were neutralized (Voets et al., 2007). Thus, although these positively charged amino acids appear to be sensing the membrane voltage, there is still about 50% of the total gating charge unaccounted for. The most plausible explanation for the weak voltage sensitivity in TRPM8 channels is that the number of gating charges is small and/or they move only a small fraction of the total electric field. There are, however, some caveats to this method since it is not valid for models with charge movement between open states and in some types of branched models.

4. GATING KINETICS IN THERMOTRP CHANNELS

Voets (Voets, 2012) has discussed in detail the different gating kinetic models proposed to explain thermoTRP channel activation by temperature and voltage, and the reader should consult this comprehensive review if interested in the detailed predictions of these gating schemes. Below we discuss two types of models based on an analysis of the single-channel kinetics and from macroscopic currents, branched, and allosteric models.

A detailed analysis of the TRPV1 single channel gating kinetics showed the existence of two to six closed states and two to five open states depending on temperature (Liu et al., 2003). These single channel current records show well-defined bursts of activity consisting of short openings separated by long closures (gaps). Increasing temperature has the effect of shortening the gaps ($Q_{10} = 7$, $\Delta H^{\mp} = 40$ kcal/mol) and prolonging the bursts ($Q_{10} = 32$, $\Delta H^{\mp} = -70$ kcal/mol) with the individual kinetic components having only a nominal temperature dependence ($Q_{10} < 3$). Therefore, the TRPV1 temperature dependence resides mainly in the burst duration, a result that can be interpreted assuming that the closing rate is highly temperature dependent. However, if this is the case, it conflicts with the macroscopic current results that show that temperature mainly affects the activation rate (see above). A closer look at the single-channel activity showed the existence of short and long bursts. Increasing temperature has the effect of shifting the equilibrium form short to long ones, thus leading to an apparent increase in burst duration (Liu et al., 2003). We should mention here that a TRPV1 agonist such as capsaicin, at constant temperature, also decreases gap duration and greatly elongates burst duration (Hui, Liu, & Qin, 2003). No channel activity was found in the absence of capsaicin and the compound addition reveals at least five closed states and three open states showing strong correlation between short closing intervals and long openings and between long closings and short openings. No detailed gating kinetic model was proposed to account for the temperature dependence of channel gating but again as in the case of channel activation mediated by capsaicin, short closings were coupled to long openings (Liu et al., 2003). The fact that closing intervals are correlated with open components makes linear kinetic schemes untenable.

Several closed (5) and open states (2) are also needed to account for the TRPM8 single channel kinetics, so a seven-state kinetic model was identified (Fernandez et al., 2011; Figure 3.5). In this model there is also a correlation between closed and open states in that the brief openings are connected to C_4 whereas long openings emerge from C_5. In this model, transitions between closed states (C–C transitions) and between closed and open states (C–O transitions) are voltage dependent and depolarizing voltages increase the forward rates with no effect on the backwards rates. The same model was used to explain the increase in channel activity when temperature is decreased from 30 °C to 20 °C. Cooling accelerates forward transitions and decreases backward rates between closed and open sates. The thermodynamic details of TRPM8 temperature dependence were not given in Fernández et al. (Fernandez et al., 2011) study, but that study did state that temperature accelerates forward rates among closed states, which is at odds with the results that show that only

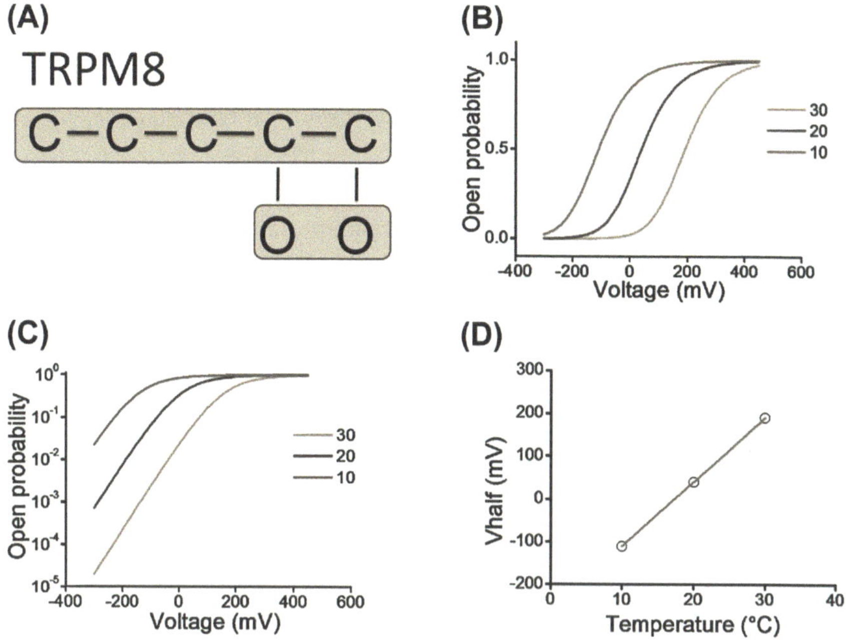

Figure 3.5 Kinetic model proposed for TRPM8 on the basis of single-channel analysis. (A) The kinetic model proposed by Fernandez et al. (2011) corresponds to five sequential closed states with two open states connected to the last two closed states. (B) Model prediction for the channel open probability as a function of temperature; note that the V_{half} is shifted to the left when the system is cooled down. The model predicts a maximum $P_o = 1$ at all temperatures. (C) The limiting slope method gave at the different temperatures gives, unambiguously, the number of gating charges per channel. (D) The V_{half}-temperature data is well described by a linear relationship.

deactivation (backwards) rates have high activation enthalpy in TRPM8 channels. This model predicts that the P_o–V curve should reach a constant and well-defined slope at extreme hyperpolarization (Figure 3.5(C)) and that there is a linear relationship between $V_{1/2}$ and temperature (Figure 3.5(D)).

Given that thermoTRPs can be viewed as modular proteins and polymodal receptors, we would like to discuss the main experimental evidence for models that assume that voltage and temperature sensors in different structures and that they communicate between each other allosterically (Brauchi et al., 2004; Latorre et al., 2007; Matta & Ahern, 2007).

The most robust observations that indicate that different structures are defined for the different sensors contained in thermoTRP channels are those obtained using mutagenesis (see below). For instance, high-throughput mutagenesis of TRPV3 and TRPV1 showed that channel temperature sensitivity can be greatly modified without appreciable changes in voltage dependence (z) (Grandl et al., 2008, 2010). Chimeric TRPV1/TRPV2 channels (Yao, Liu, & Qin, 2011) as well as some TRPV1 deletions (Yang, Cui, Wang, & Zheng, 2010) confirmed that temperature dependence can be either switched by exchanging certain channel regions or obliterated, respectively, without perturbing channel voltage dependence. It is difficult to conciliate these results with gating schemes that consider that voltage and temperature sensors are strictly coupled as is the case of linear models or branched models that assume common activation pathways for channel activation by temperature and voltage (e.g., Fernandez et al., 2011). On the other hand, Latorre et al. (2007) have pointed out that in the case of a model with n identical sensors that are highly temperature and voltage dependent, there will be n identical equilibrium constants with the form of Eqn (1). In this case, as in the two-state model, the $V_{1/2}$ should be a linear function in the whole range of temperature (see Eqn (3)). However, this is not the case of TRPM8 where $V_{1/2}$ deviates from a linear function at low and high temperatures, a behavior explained by a simple allosteric model assuming structural independence of voltage and temperature sensors (Figure 3.6(B)) (Fernandez et al., 2011; Latorre et al., 2007). Unlike linear (strictly coupled) models, allosteric models can also account for the observations that agonists of TRPV1 and TRPM8 (capsaicin, resineferatoxin and menthol) at high concentrations promote an essentially voltage-independent gating (Matta & Ahern, 2007). The fact that voltage is a partial activator of TRPV1 and TRPM8 can be explained on the basis of gating kinetic models by making the closed to open transition voltage independent (Figure 3.6(B)). In that case, the maximum open probability will be given by the equilibrium constant that define the open-closed transition (Figure 3.6(A)). We note here that the allosteric model depicted in Figure 3.6(A) predicts that at very negative voltages and extreme high temperatures (TRPM8) or extreme low temperatures (TRPV1) temperature and voltage sensors are at rest and the scheme simplifies to:

$$C \overset{L}{\leftrightarrow} O$$

In this case the limiting slope method does not give the total number of gating charges associated to channel opening but rather the voltage dependence (or lack of it) of the equilibrium constant L. This is a strict

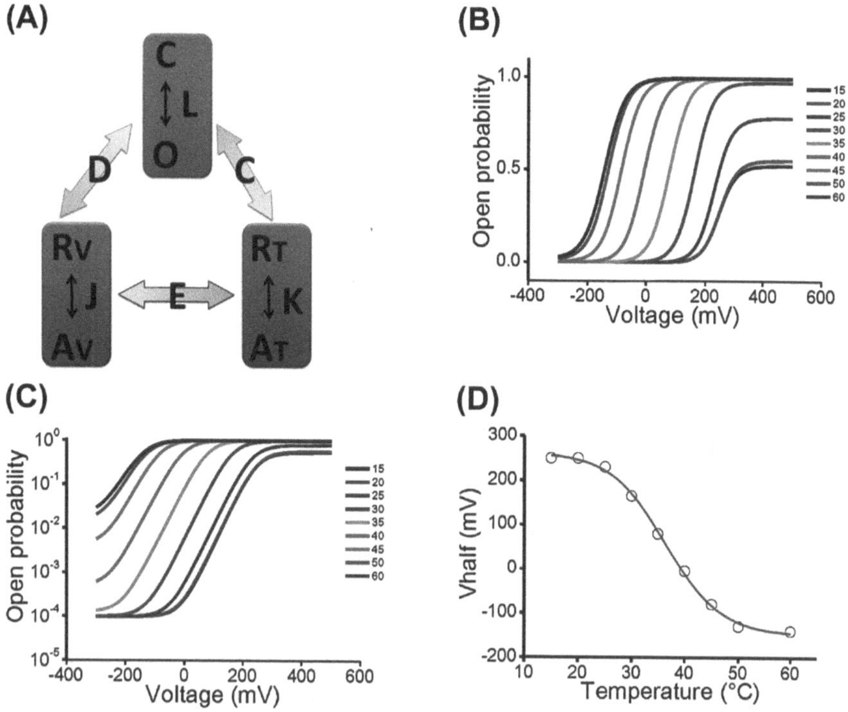

Figure 3.6 Allosteric kinetic model proposed for heat receptor. (A) Schematic representation for the model where the equilibrium constant *L* governs the open-closed transitions, J the voltage sensor activation and K the temperature sensor activation. Pore opening, voltage and temperature sensor activation are coupled between them through the allosteric constants D, C, and E. D and C connect the pore module with the voltage and temperature sensor, respectively and E represents the strength of the coupling between both sensors. When both sensors are in the model is reduced to the C–O transition only, that is, the channel can open in the absence of activated sensors. When all sensors are activated the equilibrium constant that defines the C–O transition is LDC, where LDC >> *L*. (B) The allosteric model predicts that an increase in temperature shifts the P_o–V curves to the left along the voltage axis and if the C–O transition is voltage-independent, voltage becomes a partial channel activator (P_o does not reach one). (C) The model predicts that the limiting slope gives the number of apparent gating charges of the C–O reaction, which in this case has been made voltage-independent. Notice that the model also predictas that temperature alone can increase P_o by more than two orders of magnitude. (D) V_{half} saturates at low and high temperatures. (See the color plate.)

prediction of this type of models and will also serve to test whether or not the channel can open when all temperature and voltage sensors are at rest and if thermoTRP channels can be activated by temperature alone (Figure 3.6(C)).

5. MOLECULAR DETERMINANTS OF TEMPERATURE SENSING IN TRP CHANNELS

5.1 TRPV1 and TRPV2

The molecular determinants of temperature sensing in TRPV1 have been located in the N terminal domain (Yao et al., 2011) in the pore turret (Yang et al., 2010), in the outer pore region (Grandl et al., 2010), and in the C terminal (Brauchi, Orta, Salazar, Rosenmann, & Latorre, 2006). TRPV1 and TRPV2, described above, are heat receptors with different temperature thresholds and share an amino acid identity of about 70% (Caterina et al., 1999) with activation enthalpies (ΔH) of ~90 and 200 kcal/mol, respectively (Yao et al., 2010a, 2011) showing that TRPV2 is more sensitive to temperature than TRPV1 (Table 3.1). Using a systematic chimeric analysis, Yao et al. (Yao et al., 2011) found that the N-terminal region that connects ARs to S1 (amino acid residues 358–434 defined as the membrane proximal domain, (MPD); Figure 3.7(A) and (B)) induced in TRPV2 a TRPV1 ΔH phenotype (ΔH ~ 66 kcal/mol). The MPD also produced TRPV1-like phenotypes in TRPV3 and TRPV4, making it tempting to suggest that it may correspond to a modular thermal sensor. However, although insensitive to capsaicin and protons, TRPV1b, a splice variant of TRPV1 missing 60 amino acid residues in the N-terminal domain (350–409; equivalent to about 70% of the MPD), was reported to be activated by temperature (>47 °C) (Lu, Henderson, Liu, Reinhart, & Simon, 2005). This finding was not confirmed in a subsequent study that convincingly demonstrated that TRPV1b is resistant to thermal stimuli up to 50 °C (Vos et al., 2006).

In the KcsA channel, Doyle et al. (Doyle et al., 1998) defined the linker between S5 and the pore helix as the pore turret. When TRPV1 amino acid residues from the putative pore turret are replaced with an artificial sequence the temperature response is eliminated (Yang et al., 2010; Figure 3.7(C)). This replacement, however, did not modify the channel activation by capsaicin or depolarizing voltages. On the other hand, fluorescent resonance energy transfer experiments showed that temperature induces a rearrangement of the TRPV1 pore turret. Increasing temperature promoted a decrease in the distances that separate the turrets contained in the different subunits. The involvement of the pore turret as a molecular determinant of channel temperature sensitivity has been questioned by experiments in which TRPV1 channels containing deletion of a 15-residue pore turret fragment remained activated by temperature with

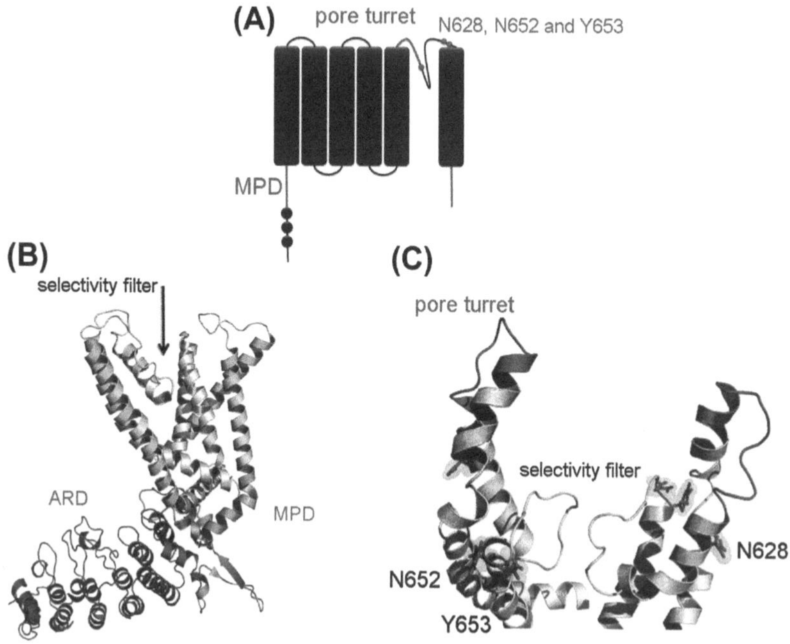

Figure 3.7 TRPV1 molecular determinants associated to temperature gating. (A) Schematic representation of TRPV1 monomer, highlighted in red are the different structures which has an impact on temperature gating. (B) Image of the recently TRPV1 structure determined by crioEM technique (PDB:3J5P) showing the molecular determinants located in the N-terminus claimed to be the temperature sensor structure. MPD is the connection between the ARD and the S1 in TRPV1. This portion of the protein is able to transfer the temperature sensitivity from TRPV1 to TRPV2 (see text). (C) Homology model (see Brauchi et al., 2007) showing the pore turret and the amino acid residues that when mutated reduce TRPV1 temperature sensitivity. (See the color plate.)

the same temperature sensitivity as the wild type channel (Yao, Liu, & Qin, 2010b, 2011). A more detailed study in which deletions of the pore turret were performed showed that the same deletion mutant used by Qin et al. (Yao et al., 2010b, 2011) was able to respond to temperature albeit with a much lower temperature sensitivity and that larger than 15-residue deletions or deletion of the seven distal amino acid residues of the pore turret completely disrupted channel function (Cui et al., 2012). This controversy is far from being settled since Julius et al. (Liao et al., 2013) reported that a TRPV1 deletion mutant lacking the whole turret region (residues 604–626) responded to heat.

Using random and targeted mutagenesis screens in the outer pore region of the TRPV1 channels, it was possible to identify point mutations that impaired temperature activation (Grandl et al., 2010). Point mutations N628K, N652T, and Y653T located at the distal part of the pore turret and in the linker between the filter and S6 (Figure 3.7(C)), respectively, severely decrease the TRPV1 temperature sensitivity. The triple mutant N628K/N652T/Y653T is essentially temperature independent and the leftward shift of the conductance–voltage curve promoted by an increase in temperature from 20 °C to 40 °C is merely −7 mV compared to the −100 to −180 mV reported for the wild-type TRPV1 (Grandl et al., 2010; Voets et al., 2004).

The C-terminus of TRPV1 has also been implicated in temperature sensing because the exchange of this domain with that of TRPM8 reverses the hot to cold sensitivity in this channel (Brauchi et al., 2006). Additionally, a region located distal to the TRP domain and comprising the TRPV1 amino acids Q727 and W752 is able to turn TRPM8 into a heat receptor (Brauchi et al., 2007). These results are at odds with those of Yao et al. (Yao et al., 2011) who observed that the exchange of the C-terminal between TRPV1 and TRPV2 did not produce a significant change in temperature dependence of the channels. As pointed out by Yao et al. (Yao et al., 2011), it is possible that the C-terminal interacts with the MPD of the TRPV1 N-terminal to mediate the transduction of thermal energy sensed by the N-terminus.

In summary, on the basis of the present information, it is difficult to decide whether the temperature-sensing structures have been identified in TRPV1. The molecular determinants of temperature sensing in TRPV1 described above can be or can be part of the coupling machinery that connects allosterically the actual temperature sensor with pore opening. From the thermodynamic point of view the more convincing results are those that show that only the exchange of specific regions of the N-terminal between TRPV1 and TRPV2 is able to alter the enthalpy of gating especially in view that the channel temperature sensitivity is determined by this parameter (Latorre et al., 2007; Liu et al., 2003).

5.2 TRPV3

In the case of TRPV3, from a mutant library of 14,000 clones obtained using high-throughput mutagenesis, three mutants (I644S, N647Y, and Y661C) with greatly decreased heat sensitivity were discovered (Grandl et al., 2008). All these mutants are located in S6 and the adjacent extracellular

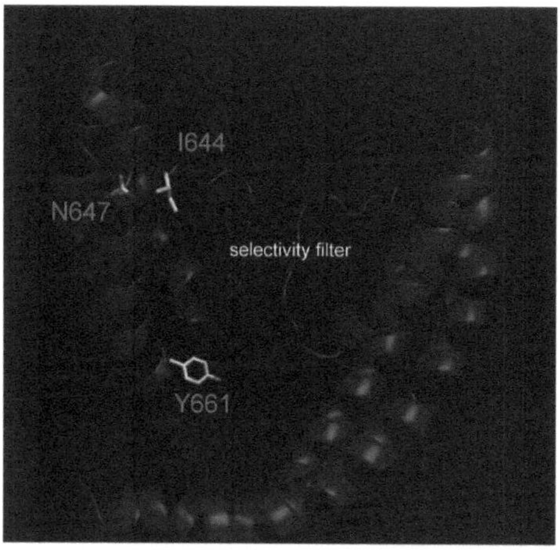

Figure 3.8 TRPV3 amino acid residues involved in temperature detection. The residues I644, N647, and Y661 located in the neighborhood of the pore domain have been identified as the molecular determinants involved in temperature detection with a high throughput mutagenesis technique. The triple mutant in TRPV3 channel abolishes the temperature sensitivity of this channel. (See the color plate.)

loop (Figure 3.8). Noticeable, the three mutants have a normal response to voltage, to the agonist 2-aminoethoxydiphenyl borate (2-APB), and unaltered ion selectivity. Since these clone libraries are not saturating, 45 more mutants in the region between S5 and S6 were made, finding two other clones in S6 with decreased temperature sensitivity (F645S and L647E). Molecular modeling suggests that F654S, L657E, and Y661C (Figure 3.8) are located in a periodic pattern probably aligned on the lipid-facing side of the α-helix (Grandl et al., 2008).

5.3 TRPA1

Using well-controlled experimental conditions, Chen et al. (2013) have confirmed previous results indicating that cold activates rat and mouse TRPA1 but not human TRPA1, and they further showed that the TRPA1 channel from rhesus monkey, behaves as hTRPA1. Taking advantage of the differences in temperature sensitivity between hTRPA1 and rTRPA1, Chen et al. (2013) also demonstrated that introducing the S5 and S5–S6 linker from hTRPA1 into rTRPA1 abolished cold activation. Surprisingly, changing the glycine878 in rTRPA1 to valine875 residing in an equivalent position in hTRPA1, obliterates cold activation (Figure 3.9(A)). Apparently, this glycine in rTRPV1 is not the only

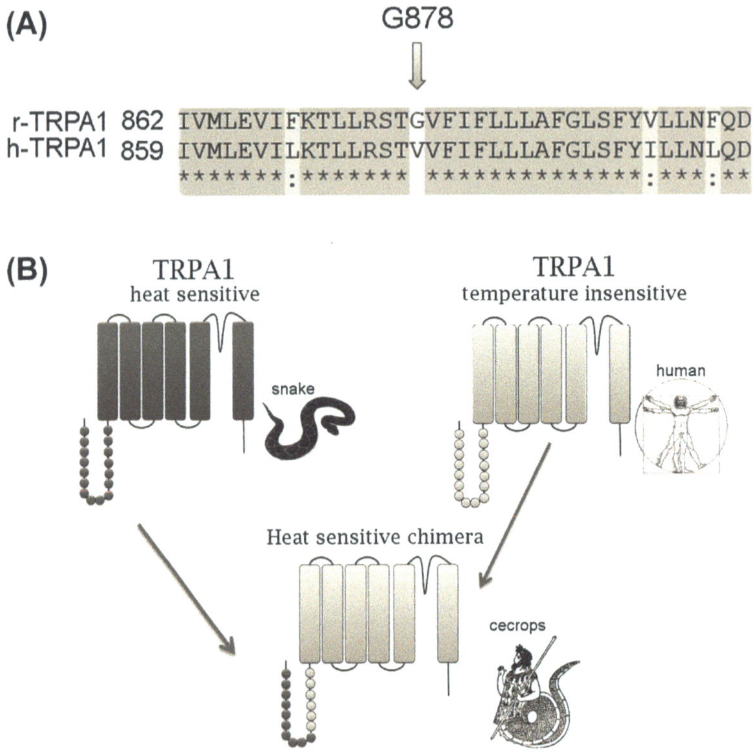

Figure 3.9 The molecular nature of temperature dependence of TRPA1. (A) Alignment between human and rat TRPA1, where hTRPA1 is temperature insensitive and rTRPA1 is cold activated, the residue G878 plays an important role in stimuli transduction mutation in this residue disrupts temperature sensitivity in rTRPA1. (B) In rattlesnakes TRPA1 is a heat-activated channel. Replacing the first 10 of the 15 ARDs from rsTRPA1 into hTRPA1 conveys heat activation to the chimeric ion channel, with a $Q_{10} \sim 14$.

requirement for cold activation since the reverse mutation V875G in hTRPA1 could not confer cold activation to this channel.

We recall here that *Drosophila* and snake TRPA1 orthologs are heat activated. In some species of snakes (e.g., rattle snakes) the TRPA1 channel is highly enriched in the pit organ and is used to detect infrared radiation (prey detection, predator avoidance; Gracheva et al., 2010). In insects the thermal sensitivity of the channel contributes to thermal nociception and to thermal avoidance behaviors (Hamada et al., 2008; Zhong et al., 2012). Swapping the first 10 ankyrin domains (AD) of the N-terminal between snake and human TRPA1 gave rise to an hTRPA1 activated by heat and allyl isothiocyanate (AITC) (Cordero-Morales et al., 2011) (Figure 3.9(B)). Moreover, elimination of the electrophile modification sites (3 cysteines and 1 lysine) contained in the N-terminal linker region renders this chimeric channel insensitive to AITC but not to temperature. This is one more example indicating that the structures involved in agonist and temperature channel activation are different and that channel gating is better explained using allosteric kinetic models. The N-terminal of TRPA1 contains 16–17 ankyrin repeats (AR) and introducing ARs 3 to 8 from the N-terminal of rattle snake into hTRA1 produces channels that activate with a $Q_{10} \sim 5$ but hTRPA1 containing ARs 10 to 15 from rsTRPA1 shows a $Q_{10} \sim 11$, a result that shows that the N-terminal of rattle snake contains two transferable regions able to induce heat-sensitivity in hTRPA1. However, transfer of ARs 10 to 15 into hTRPA1 results in temperature sensitivity more similar to rsTRPA1 ($Q_{10} \sim 14$).

From the work done using TRPA1 human/rat chimeras and TRPA1 snake/human chimeras it is tempting to suggest that the molecular determinants for heat sensing in rsTRPA1 are placed in the N-terminal and those for cold sensing in rTRPA1 in the neighborhood of the conduction pore. Four *Drosophila* TRPA1 isoforms have been identified, named dTRPA1-A, -B, -C, and -D. From these channels only TRPA1-A and -B are heat sensors. A detailed study using the four dTRPA1 isoforms showed that the 37 amino acids located between the last ankyrin repeat and S1 play a critical role in temperature sensing (Zhong et al., 2012). These results tend to support the hypothesis that the heat responsive element in TRPA1 is located in the N-terminal. However, a recent report using TRPA1 *Drosophila*/human chimeras has established that dTRPA1 heat sensitivity is reduced by introducing either the S6 or the pore helix from hTRPA1 into dTRPA1 (Wang, Schupp, Zurborg, & Heppenstall, 2013).

5.4 Lipids are not the answer

Lipids can undergo phase transitions and there is a possibility the temperature sensitivity of thermoTRPs is a consequence of changes in physical properties of the bulk lipids or the so-called boundary lipids with temperature. The fact that TRPM8 and TRPV1 are fully functional in reconstituted systems strongly speaks against lipids as the energy source for temperature gating. TRPM8 was reconstituted in planar lipid bilayers and in this alien lipid milieu can be activated by PIP_2, menthol, icilin, and temperature (Zakharian, Cao, & Rohacs, 2010). As it was found in cells, the reconstituted TRPM8 is highly temperature dependent ($Q_{10} \sim 40$). Purified TRPV1 reconstituted into artificial liposomes can be gated by capsaicin, protons, and temperature (Q_{10} ~ 20) (Cao, Cordero-Morales, Liu, Qin, & Julius, 2013). These experiments are consistent with the idea that thermoTRPs are intrinsically thermosensitive.

6. CODA

Despite having three-dimensional reconstructions of TRPV1 to resolutions of less than 4Å (Cao et al., 2013; Liao et al., 2013) it is still difficult to decide which gating mechanisms are more appropriate to describe how these channels gate when confronted with changes in voltage and temperature. The problem is that although they are built as tetramers implying the existence of four voltage sensors as in Kv channels, we not only do not know how many temperature sensors there are but we also lack a clear definition of the temperature-sensing structures. The TRPV1 channel is the perfect example of how confusing the situation is given the fact that four different channel regions, the N-terminus, the C-terminus, the turret and some pore residues have been proposed as potential temperature sensors. Most probably, some of these structures will prove to be part of the coupling system in charge of communicating the sensor with the pore gate. The best we can do at present is to picture the temperature sensor as a protein structure contained in the channel that undergoes a process equivalent to denaturation since, as in thermoTRP channels, large enthalpy and large entropy changes are involved in protein denaturation. In particular, cold denaturation of proteins is a consequence of the large ΔC_p value originated from the exposure of hydrophobic residues to the solvent upon protein unfolding (Dill, 1990; Privalov, 1990). As discussed above the Gibbs–Helmholtz equation for a two-state system predicts two thermal unfolding transitions for a protein—cold unfolding and heat unfolding (Privalov, 1990). However, cold denaturation appears to involve a different mechanism than the one mediating protein unfolding by heat (e.g., Richardson, Lemaire, Jacquot, & Makhatadze, 2000) and it is possible that in temperature sensitive channels only one type of process prevail (cold or heat denaturation). Large intervals of temperature have been tested for TRPM8 and TRPV1 and in no case the concave free-energy dependence with temperature has been observed. Moreover, the large and positive entropy value involved in TRPV1 gating indicates that the channel transits between a highly ordered closed state and a strongly disordered open state and that the reverse situation is true in the case of TRPM8. This implies a process of protein "folding" in TRPM8 and "unfolding" in TRPV1 when channels transit from the closed to the open configuration. Without knowing the molecular characteristics of the temperature sensor there is at present no experimental strategy that enables us to determine its ΔC_p. Therefore, the determination of the thermodynamic parameters using the van't Hoff relation is still the best way to get a glimpse at the nature of the thermoTRP channels temperature dependence. However, we should mention here that by considering thermoTRP channels as allosteric proteins, it is also possible that there is not a well-defined temperature sensor and that temperature sensitivity is given by the sum of small temperature sensitivities of the allosteric coupling processes (Jara-Oseguera & Islas, 2013).

As discussed above, the fast inactivation process in Shaker K^+ channels is highly temperature dependent (Nobile et al., 1997) ($Q_{10} = 7$), while the recovery from inactivation has only a $Q_{10} = 1.6$. This process represents another example of thermal gating worthwhile discussing since it can give us some hints about possible molecular mechanisms involved in this kind of temperature-dependent processes. The group of Aldrich (Hoshi, Zagotta, & Aldrich, 1990; Zagotta, Hoshi, & Aldrich, 1990) showed that the first 20 amino acids of the Shaker K^+ channel interact with the open channel to cause inactivation and that a synthetic peptide of the N-terminal inactivation domain was able to restore inactivation in a Shaker noninactivating mutant. The effect of the inactivating peptide can be interpreted in terms of an open channel block and, in agreement with Nobile et al. (Nobile et al., 1997) results, it was found that temperature highly affected the blockade association rate constant ($Q_{10} = 5$) with a dissociation rate constant rather temperature insensitive ($Q_{10} = 1.1$; Murrell-Lagnado & Aldrich, 1993a,b). Here we have a case in which the "temperature sensor," the inactivation peptide, is structurally well defined (Schott, Antz, Frank, Ruppersberg, & Kalbitzer, 1998; Wissmann et al., 1999). In aqueous solution, the structure of the inactivating peptide approached that of a random coil but existed in fast equilibrium of multiple weakly structured states. The most parsimonious way to explain the temperature-dependent characteristics of the peptide-blocking reaction is to assume that increasing temperature increases the fraction of time that the peptide spends in the correct conformation for binding. This mechanism is an alternative to the two-state denaturation process in that it explains the high enthalpy change processes since in this case from several preexistent structures temperature only increases the probability of finding the sensor in the right conformation.

In general, given the large enthalpy and entropy changes involved in the closed-opening reaction, even larger than those involved in the denaturation of whole proteins (e.g., Privalov, 1979), we are of the opinion that activation by temperature of thermoTRP channels involves large conformational changes of specialized protein domains.

ACKNOWLEDGMENTS

This work was supported by FONDECYT grants 1110430 (RL), 1120802 (Carlos Gonzalez); ANILLO grant ACT1104 (Carlos Gonzalez). CSIC p944, the CSIC International Cooperation Program, UdelaR-CINV-UV (GF and CG). The *Centro Interdisciplinario de Neurociencia de Valparaíso* is a Millennium Institute supported by the Millennium Scientific Initiative of the Chilean Ministry of Economy, Public Works, and Tourism.

REFERENCES

Ahern, G. P. (2003). Activation of TRPV1 by the satiety factor oleoylethanolamide. *The Journal of Biological Chemistry, 278*, 30429–30434.

Ahern, G. P., Brooks, I. M., Miyares, R. L., & Wang, X. B. (2005). Extracellular cations sensitize and gate capsaicin receptor TRPV1 modulating pain signaling. *Journal of Neuroscience, 25*, 5109–5116.

Almers, W. (1978). Gating currents and charge movements in excitable membranes. *Reviews of Physiological, Biochemistry & Pharmacology, 82*, 96–190.

Bandell, M., Dubin, A. E., Petrus, M. J., Orth, A., Mathur, J., Hwang, S. W., et al. (2006). High-throughput random mutagenesis screen reveals TRPM8 residues specifically required for activation by menthol. *Nature Neuroscience, 9*, 493–500.

Bandell, M., Story, G. M., Hwang, S. W., Viswanath, V., Eid, S. R., Petrus, M. J., et al. (2004). Noxious cold ion channel TRPA1 is activated by pungent compounds and bradykinin. *Neuron, 41*, 849–857.

Bautista, D. M., Jordt, S. E., Nikai, T., Tsuruda, P. R., Read, A. J., Poblete, J., et al. (2006). TRPA1 mediates the inflammatory actions of environmental irritants and proalgesic agents. *Cell, 124*, 1269–1282.

Bezanilla, F., & Stefani, E. (1994). Voltage-dependent gating of ionic channels. *Annual Review of Biophysics and Biomolecular Structure, 23*, 819–846.

Bohlen, C. J., Priel, A., Zhou, S., King, D., Siemens, J., & Julius, D. (2010). A bivalent tarantula toxin activates the capsaicin receptor, TRPV1, by targeting the outer pore domain. *Cell, 141*, 834–845.

Brauchi, S., Orio, P., & Latorre, R. (2004). Clues to understanding cold sensation: thermodynamics and electrophysiological analysis of the cold receptor TRPM8. *Proceedings of the National Academy of Sciences of the United States of America, 101*, 15494–15499.

Brauchi, S., Orta, G., Mascayano, C., Salazar, M., Raddatz, N., Urbina, H., et al. (2007). Dissection of the components for PIP_2 activation and thermosensation in TRP channels. *Proceedings of the National Academy of Sciences of the United States of America, 104*, 10246–10251.

Brauchi, S., Orta, G., Salazar, M., Rosenmann, E., & Latorre, R. (2006). A hot-sensing cold receptor: C-terminal domain determines thermosensation in transient receptor potential channels. *Journal of Neuroscience, 26*, 4835–4840.

Cao, E., Cordero-Morales, J. F., Liu, B., Qin, F., & Julius, D. (2013). TRPV1 channels are intrinsically heat sensitive and negatively regulated by phosphoinositide lipids. *Neuron, 77*, 667–679.

Cao, E., Liao, M., Cheng, Y., & Julius, D. (2013). TRPV1 structures in distinct conformations reveal activation mechanisms. *Nature, 504*, 113–118.

Caterina, M. J., Leffler, A., Malmberg, A. B., Martin, W. J., Trafton, J., Petersen-Zeitz, K. R., et al. (2000). Impaired nociception and pain sensation in mice lacking the capsaicin receptor. *Science, 288*, 306–313.

Caterina, M. J., Rosen, T. A., Tominaga, M., Brake, A. J., & Julius, D. (1999). A capsaicin-receptor homologue with a high threshold for noxious heat. *Nature, 398*, 436–441.

Caterina, M. J., Schumacher, M. A., Tominaga, M., Rosen, T. A., Levine, J. D., & Julius, D. (1997). The capsaicin receptor: a heat-activated ion channel in the pain pathway. *Nature, 389*, 816–824.

Chakrapani, S., & Auerbach, A. (2005). A speed limit for conformational change of an allosteric membrane protein. *Proceedings of the National Academy of Sciences of the United States of America, 102*, 87–92.

Chavez, A. E., Chiu, C. Q., & Castillo, P. E. (2010). TRPV1 activation by endogenous anandamide triggers postsynaptic long-term depression in dentate gyrus. *Nature Neuroscience, 13*, 1511–1518.

Chen, J., Kang, D., Xu, J., Lake, M., Hogan, J. O., Sun, C., et al. (2013). Species differences and molecular determinant of TRPA1 cold sensitivity. *Nature Communications, 4*, 2501.

Cheng, W., Yang, F., Liu, S., Colton, C. K., Wang, C., Cui, Y., et al. (2012). Heteromeric heat-sensitive transient receptor potential channels exhibit distinct temperature and chemical response. *Journal of Biological Chemistry, 287*, 7279–7288.

Cheng, W., Yang, F., Takanishi, C. L., & Zheng, J. (2007). Thermosensitive TRPV channel subunits coassemble into heteromeric channels with intermediate conductance and gating properties. *Journal of General Physiology, 129*, 191–207.

Clapham, D. E., & Miller, C. (2011). A thermodynamic framework for understanding temperature sensing by transient receptor potential (TRP) channels. *Proceedings of the National Academy of Sciences of United States of America, 108*, 19492–19497.

Cordero-Morales, J. F., Gracheva, E. O., & Julius, D. (2011). Cytoplasmic ankyrin repeats of transient receptor potential A1 (TRPA1) dictate sensitivity to thermal and chemical stimuli. *Proceedings of the National Academy of Sciences of United States of America, 108*, E1184–E1191.

Correa, A. M., Bezanilla, F., & Latorre, R. (1992). Gating kinetics of batrachotoxin-modified Na^+ channels in the squid giant axon. Voltage and temperature effects. *Biophysical Journal, 61*, 1332–1352.

Cui, Y., Yang, F., Cao, X., Yarov-Yarovoy, V., Wang, K., & Zheng, J. (2012). Selective disruption of high sensitivity heat activation but not capsaicin activation of TRPV1 channels by pore turret mutations. *Journal of General Physiology, 139*, 273–283.

Cvetkov, T. L., Huynh, K. W., Cohen, M. R., & Moiseenkova-Bell, V. Y. (2011). Molecular architecture and subunit organization of TRPA1 ion channel revealed by electron microscopy. *Journal of Biological Chemistry, 286*, 38168–38176.

Darian-Smith, I., Johnson, K. O., LaMotte, C., Shigenaga, Y., Kenins, P., & Champness, P. (1979). Warm fibers innervating palmar and digital skin of the monkey: responses to thermal stimuli. *Journal of Neurophysiology, 42*, 1297–1315.

DeCoursey, T. E., & Cherny, V. V. (1998). Temperature dependence of voltage-gated H^+ currents in human neutrophils, rat alveolar epithelial cells, and mammalian phagocytes. *Journal of General Physiology, 112*, 503–522.

Delany, N. S., Hurle, M., Facer, P., Alnadaf, T., Plumpton, C., Kinghorn, I., et al. (2001). Identification and characterization of a novel human vanilloid receptor-like protein, VRL-2. *Physiological Genomics, 4*, 165–174.

Dill, K. A. (1990). Dominant forces in protein folding. *Biochemistry, 29*, 7133–7155.

Doyle, D. A., Morais Cabral, J., Pfuetzner, R. A., Kuo, A., Gulbis, J. M., Cohen, S. L., et al. (1998). The structure of the potassium channel: molecular basis of K^+ conduction and selectivity. *Science, 280*, 69–77.

Ebihara, L. (1996). Xenopus connexin38 forms hemi-gap-junctional channels in the nonjunctional plasma membrane of Xenopus oocytes. *Biophysical Journal, 71*, 742–748.

Fernandez, J. A., Skryma, R., Bidaux, G., Magleby, K. L., Scholfield, C. N., McGeown, J. G., et al. (2011). Voltage- and cold-dependent gating of single TRPM8 ion channels. *Journal of General Physiology, 137*, 173–195.

Garcia-Elias, A., Mrkonjic, S., Pardo-Pastor, C., Inada, H., Hellmich, U. A., Rubio-Moscardo, F., et al. (2013). Phosphatidylinositol-4,5-biphosphate-dependent rearrangement of TRPV4 cytosolic tails enables channel activation by physiological stimuli. *Proceedings of the National Academy of Sciences of United States of America, 110*, 9553–9558.

Gavva, N. R., Klionsky, L., Qu, Y., Shi, L., Tamir, R., Edenson, S., et al. (2004). Molecular determinants of vanilloid sensitivity in TRPV1. *Journal of Biological Chemistry, 279*, 20283–20295.

Gracheva, E. O., Ingolia, N. T., Kelly, Y. M., Cordero-Morales, J. F., Hollopeter, G., Chesler, A. T., et al. (2010). Molecular basis of infrared detection by snakes. *Nature, 464*, 1006–1011.

Grandl, J., Hu, H., Bandell, M., Bursulaya, B., Schmidt, M., Petrus, M., et al. (2008). Pore region of TRPV3 ion channel is specifically required for heat activation. *Nature Neuroscience, 11*, 1007–1013.

Grandl, J., Kim, S. E., Uzzell, V., Bursulaya, B., Petrus, M., Bandell, M., et al. (2010). Temperature-induced opening of TRPV1 ion channel is stabilized by the pore domain. *Nature Neuroscience, 13*, 708–714.

Grimm, C., Kraft, R., Sauerbruch, S., Schultz, G., & Harteneck, C. (2003). Molecular and functional characterization of the melastatin-related cation channel TRPM3. *Journal of Biological Chemistry, 278*, 21493–21501.

Guler, A. D., Lee, H., Iida, T., Shimizu, I., Tominaga, M., & Caterina, M. (2002). Heat-evoked activation of the ion channel, TRPV4. *Journal of Neuroscience, 22*, 6408–6414.

Hamada, F. N., Rosenzweig, M., Kang, K., Pulver, S. R., Ghezzi, A., Jegla, T. J., et al. (2008). An internal thermal sensor controlling temperature preference in *Drosophila*. *Nature, 454*, 217–220.

Helliwell, R. J., McLatchie, L. M., Clarke, M., Winter, J., Bevan, S., & McIntyre, P. (1998). Capsaicin sensitivity is associated with the expression of the vanilloid (capsaicin) receptor (VR1) mRNA in adult rat sensory ganglia. *Neuroscience Letters, 250*, 177–180.

Hellwig, N., Albrecht, N., Harteneck, C., Schultz, G., & Schaefer, M. (2005). Homo- and heteromeric assembly of TRPV channel subunits. *Journal of Cell Science, 118*, 917–928.

Hensel, H., Andres, K. H., & von During, M. (1974). Structure and function of cold receptors. *Pflugers Arch, 352*, 1–10.

Hinman, A., Chuang, H. H., Bautista, D. M., & Julius, D. (2006). TRP channel activation by reversible covalent modification. *Proceedings of the National Academy of Sciences of United States of America, 103*, 19564–19568.

Hoshi, T., Zagotta, W. N., & Aldrich, R. W. (1990). Biophysical and molecular mechanisms of Shaker potassium channel inactivation. *Science, 250*, 533–538.

Hui, K., Liu, B., & Qin, F. (2003). Capsaicin activation of the pain receptor, VR1: multiple open states from both partial and full binding. *Biophysical Journal, 84*, 2957–2968.

Jara-Oseguera, A., & Islas, L. D. (2013). The role of allosteric coupling on thermal activation of thermo-TRP channels. *Biophysical Journal, 104*, 2160–2169.

Jiang, L. H., Gamper, N., & Beech, D. J. (2011). Properties and therapeutic potential of transient receptor potential channels with putative roles in adversity: focus on TRPC5, TRPM2 and TRPA1. *Current Drug Targets, 12*, 724–736.

Jordt, S. E., Bautista, D. M., Chuang, H. H., McKemy, D. D., Zygmunt, P. M., Hogestatt, E. D., et al. (2004). Mustard oils and cannabinoids excite sensory nerve fibres through the TRP channel ANKTM1. *Nature, 427*, 260–265.

Jordt, S. E., & Julius, D. (2002). Molecular basis for species-specific sensitivity to "hot" chili peppers. *Cell, 108*, 421–430.

Jordt, S. E., Tominaga, M., & Julius, D. (2000). Acid potentiation of the capsaicin receptor determined by a key extracellular site. *Proceedings of the National Academy of Sciences of the United States of America, 97*, 8134–8139.

Kanzaki, M., Zhang, Y. Q., Mashima, H., Li, L., Shibata, H., & Kojima, I. (1999). Translocation of a calcium-permeable cation channel induced by insulin-like growth factor-I. *Nature Cell Biology, 1*, 165–170.

Karashima, Y., Talavera, K., Everaerts, W., Janssens, A., Kwan, K. Y., Vennekens, R., et al. (2009). TRPA1 acts as a cold sensor in vitro and in vivo. *Proceedings of the National Academy of Sciences of the United States of America, 106*, 1273–1278.

Kobertz, W. R., Williams, C., & Miller, C. (2000). Hanging gondola structure of the T1 domain in a voltage-gated K(+) channel. *Biochemistry, 39*, 10347–10352.

Kremeyer, B., Lopera, F., Cox, J. J., Momin, A., Rugiero, F., Marsh, S., et al. (2010). A gain-of-function mutation in TRPA1 causes familial episodic pain syndrome. *Neuron, 66*, 671–680.

Latorre, R., Brauchi, S., Orta, G., Zaelzer, C., & Vargas, G. (2007). ThermoTRP channels as modular proteins with allosteric gating. *Cell Calcium, 42*, 427–438.

Latorre, R., Zaelzer, C., & Brauchi, S. (2009). Structure-functional intimacies of transient receptor potential channels. *Quarterly Reviews of Biophysics, 42*, 201–246.

Launay, P., Fleig, A., Perraud, A. L., Scharenberg, A. M., Penner, R., & Kinet, J. P. (2002). TRPM4 is a Ca^{2+}-activated nonselective cation channel mediating cell membrane depolarization. *Cell, 109*, 397–407.

Lee, N., Chen, J., Sun, L., Wu, S., Gray, K. R., Rich, A., et al. (2003). Expression and characterization of human transient receptor potential melastatin 3 (hTRPM3). *Journal of Biological Chemistry, 278*, 20890–20897.

Liao, M., Cao, E., Julius, D., & Cheng, Y. (2013). Structure of the TRPV1 ion channel determined by electron cryo-microscopy. *Nature, 504*, 107–112.

Liedtke, W., Choe, Y., Marti-Renom, M. A., Bell, A. M., Denis, C. S., Sali, A., et al. (2000). Vanilloid receptor-related osmotically activated channel (VR-OAC), a candidate vertebrate osmoreceptor. *Cell, 103*, 525–535.

Liu, B., Hui, K., & Qin, F. (2003). Thermodynamics of heat activation of single capsaicin ion channels VR1. *Biophysical Journal, 85*, 2988–3006.

Liu, D., & Liman, E. R. (2003). Intracellular Ca^{2+} and the phospholipid PIP_2 regulate the taste transduction ion channel TRPM5. *Proceedings of National Academy of Sciences of the United States of America, 100*, 15160–15165.

Liu, B., & Qin, F. (2005). Functional control of cold- and menthol-sensitive TRPM8 ion channels by phosphatidylinositol 4,5-bisphosphate. *The Journal of Neuroscience: The Official Journal of the Society for Neuroscience, 25*, 1674–1681.

Lu, G., Henderson, D., Liu, L., Reinhart, P. H., & Simon, S. A. (2005). TRPV1b, a functional human vanilloid receptor splice variant. *Molecular Pharmacology, 67*, 1119–1127.

Lukacs, V., Rives, J. M., Sun, X., Zakharian, E., & Rohacs, T. (2013). Promiscuous activation of transient receptor potential vanilloid 1 (TRPV1) channels by negatively charged intracellular lipids: the key role of endogenous phosphoinositides in maintaining channel activity. *Journal of Biological Chemistry, 288*, 35003–35013.

Macpherson, L. J., Dubin, A. E., Evans, M. J., Marr, F., Schultz, P. G., Cravatt, B. F., et al. (2007). Noxious compounds activate TRPA1 ion channels through covalent modification of cysteines. *Nature, 445*, 541–545.

Macpherson, L. J., Geierstanger, B. H., Viswanath, V., Bandell, M., Eid, S. R., Hwang, S., et al. (2005). The pungency of garlic: activation of TRPA1 and TRPV1 in response to allicin. *Current Biology, 15*, 929–934.

Macpherson, L. J., Hwang, S. W., Miyamoto, T., Dubin, A. E., Patapoutian, A., & Story, G. M. (2006). More than cool: promiscuous relationships of menthol and other sensory compounds. *Molecular Cellular Neuroscience, 32*, 335–343.

Matta, J. A., & Ahern, G. P. (2007). Voltage is a partial activator of rat thermosensitive TRP channels. *Journal of Physiology, 585*, 469–482.

McKemy, D. D., Neuhausser, W. M., & Julius, D. (2002). Identification of a cold receptor reveals a general role for TRP channels in thermosensation. *Nature, 416*, 52–58.

Moiseenkova-Bell, V. Y., Stanciu, L. A., Serysheva, I. I., Tobe, B. J., & Wensel, T. G. (2008). Structure of TRPV1 channel revealed by electron cryomicroscopy. *Proceedings of National Academy of Sciences of the United States of America, 105*, 7451–7455.

Montell, C. (2005). The TRP superfamily of cation channels. *Science's STKE, 2005*, re3.

Murrell-Lagnado, R. D., & Aldrich, R. W. (1993). Energetics of Shaker K channels block by inactivation peptides. *Journal of General Physiology, 102*, 977–1003.

Murrell-Lagnado, R. D., & Aldrich, R. W. (1993). Interactions of amino terminal domains of Shaker K channels with a pore blocking site studied with synthetic peptides. *Journal of General Physiology, 102*, 949–975.

Nilius, B., Mahieu, F., Prenen, J., Janssens, A., Owsianik, G., Vennekens, R., et al. (2006). The Ca^{2+}-activated cation channel TRPM4 is regulated by phosphatidylinositol 4,5-biphosphate. *EMBO Journal, 25*, 467–478.

Nilius, B., Prenen, J., Droogmans, G., Voets, T., Vennekens, R., Freichel, M., et al. (2003). Voltage dependence of the Ca^{2+}-activated cation channel TRPM4. *Journal of Biological Chemistry, 278*, 30813–30820.

Nilius, B., Talavera, K., Owsianik, G., Prenen, J., Droogmans, G., & Voets, T. (2005). Gating of TRP channels: a voltage connection? *Journal of Physiology, 567*, 35–44.

Nobile, M., Olcese, R., Toro, L., & Stefani, E. (1997). Fast inactivation of Shaker K^+ channels is highly temperature dependent. *Experimental Brain Research, 114*, 138–142.

Noel, J., Zimmermann, K., Busserolles, J., Deval, E., Alloui, A., Diochot, S., et al. (2009). The mechano-activated K^+ channels TRAAK and TREK-1 control both warm and cold perception. *EMBO Journal, 28*, 1308–1318.

Oberwinkler, J., Lis, A., Giehl, K. M., Flockerzi, V., & Philipp, S. E. (2005). Alternative splicing switches the divalent cation selectivity of TRPM3 channels. *Journal of Biological Chemistry, 280*, 22540–22548.

Owsianik, G., Talavera, K., Voets, T., & Nilius, B. (2006). Permeation and selectivity of TRP channels. *Annual Reviews of Physiology, 68*, 685–717.

Peier, A. M., Moqrich, A., Hergarden, A. C., Reeve, A. J., Andersson, D. A., Story, G. M., et al. (2002). A TRP channel that senses cold stimuli and menthol. *Cell, 108*, 705–715.

Peier, A. M., Reeve, A. J., Andersson, D. A., Moqrich, A., Earley, T. J., Hergarden, A. C., et al. (2002). A heat-sensitive TRP channel expressed in keratinocytes. *Science, 296*, 2046–2049.

Perez, C. A., Huang, L., Rong, M., Kozak, J. A., Preuss, A. K., Zhang, H., et al. (2002). A transient receptor potential channel expressed in taste receptor cells. *Nature Neuroscience, 5*, 1169–1176.

Perraud, A. L., Fleig, A., Dunn, C. A., Bagley, L. A., Launay, P., Schmitz, C., et al. (2001). ADP-ribose gating of the calcium-permeable LTRPC2 channel revealed by Nudix motif homology. *Nature, 411*, 595–599.

Privalov, P. L. (1979). Stability of proteins: small globular proteins. *Advances in Protein Chemistry, 33*, 167–241.

Privalov, P. L. (1990). Cold denaturation of proteins. *Critical Review in Biochemistry and Molecular Biology, 25*, 281–305.

Privalov, P. L., & Gill, S. J. (1988). Stability of protein structure and hydrophobic interaction. *Advances in Protein Chemistry, 39*, 191–234.

Pusch, M., Ludewig, U., & Jentsch, T. J. (1997). Temperature dependence of fast and slow gating relaxations of ClC-0 chloride channels. *Journal of General Physiology, 109*, 105–116.
Ramsey, I. S., Delling, M., & Clapham, D. E. (2006). An introduction to TRP channels. *Annual Review of Physiology, 68*, 619–647.
Richardson, J. M., 3rd, Lemaire, S. D., Jacquot, J. P., & Makhatadze, G. I. (2000). Difference in the mechanisms of the cold and heat induced unfolding of thioredoxin h from *Chlamydomonas reinhardtii*: spectroscopic and calorimetric studies. *Biochemistry, 39*, 11154–11162.
Rohacs, T., Lopes, C. M., Michailidis, I., & Logothetis, D. E. (2005). PI(4,5)P_2 regulates the activation and desensitization of TRPM8 channels through the TRP domain. *Nature Neuroscience, 8*, 626–634.
Rosenzweig, M., Brennan, K. M., Tayler, T. D., Phelps, P. O., Patapoutian, A., & Garrity, P. A. (2005). The *Drosophila* ortholog of vertebrate TRPA1 regulates thermotaxis. *Genes & Development, 19*, 419–424.
Saito, S., Nakatsuka, K., Takahashi, K., Fukuta, N., Imagawa, T., Ohta, T., et al. (2012). Analysis of transient receptor potential ankyrin 1 (TRPA1) in frogs and lizards illuminates both nociceptive heat and chemical sensitivities and coexpression with TRP vanilloid 1 (TRPV1) in ancestral vertebrates. *Journal of Biological Chemistry, 287*, 30743–30754.
Salazar, H., Jara-Oseguera, A., Hernandez-Garcia, E., Llorente, I., Arias, O., II, Soriano-Garcia, M., et al. (2009). Structural determinants of gating in the TRPV1 channel. *Nature Structural & Molecular Biology, 16*, 704–710.
Salazar, H., Llorente, I., Jara-Oseguera, A., Garcia-Villegas, R., Munari, M., Gordon, S. E., et al. (2008). A single N-terminal cysteine in TRPV1 determines activation by pungent compounds from onion and garlic. *Nature Neuroscience, 11*, 255–261.
Sawada, Y., Hosokawa, H., Hori, A., Matsumura, K., & Kobayashi, S. (2007). Cold sensitivity of recombinant TRPA1 channels. *Brain Research, 1160*, 39–46.
Schott, M. K., Antz, C., Frank, R., Ruppersberg, J. P., & Kalbitzer, H. R. (1998). Structure of the inactivating gate from the Shaker voltage gated K^+ channel analyzed by NMR spectroscopy. *European Biophysics Journal, 27*, 99–104.
Selescu, T., Ciobanu, A. C., Dobre, C., Reid, G., & Babes, A. (2013). Camphor activates and sensitizes transient receptor potential melastatin 8 (TRPM8) to cooling and icilin. *Chemical Senses, 38*, 563–575.
Shigematsu, H., Sokabe, T., Danev, R., Tominaga, M., & Nagayama, K. (2010). A 3.5-nm structure of rat TRPV4 cation channel revealed by Zernike phase-contrast cryoelectron microscopy. *Journal of Biological Chemistry, 285*, 11210–11218.
Siemens, J., Zhou, S., Piskorowski, R., Nikai, T., Lumpkin, E. A., Basbaum, A. I., et al. (2006). Spider toxins activate the capsaicin receptor to produce inflammatory pain. *Nature, 444*, 208–212.
Sigg, D., & Bezanilla, F. (1997). Total charge movement per channel. The relation between gating charge displacement and the voltage sensitivity of activation. *Journal of General Physiology, 109*, 27–39.
Smith, G. D., Gunthorpe, M. J., Kelsell, R. E., Hayes, P. D., Reilly, P., Facer, P., et al. (2002). TRPV3 is a temperature-sensitive vanilloid receptor-like protein. *Nature, 418*, 186–190.
Smith, P. L., Maloney, K. N., Pothen, R. G., Clardy, J., & Clapham, D. E. (2006). Bisandrographolide from *Andrographis paniculata* activates TRPV4 channels. *Journal of Biological Chemistry, 281*, 29897–29904.
Story, G. M., Peier, A. M., Reeve, A. J., Eid, S. R., Mosbacher, J., Hricik, T. R., et al. (2003). ANKTM1, a TRP-like channel expressed in nociceptive neurons, is activated by cold temperatures. *Cell, 112*, 819–829.

Strotmann, R., Harteneck, C., Nunnenmacher, K., Schultz, G., & Plant, T. D. (2000). OTRPC4, a nonselective cation channel that confers sensitivity to extracellular osmolarity. *Nature Cell Biology, 2*, 695–702.

Talavera, K., Yasumatsu, K., Voets, T., Droogmans, G., Shigemura, N., Ninomiya, Y., et al. (2005). Heat activation of TRPM5 underlies thermal sensitivity of sweet taste. *Nature, 438*, 1022–1025.

Togashi, K., Hara, Y., Tominaga, T., Higashi, T., Konishi, Y., Mori, Y., et al. (2006). TRPM2 activation by cyclic ADP-ribose at body temperature is involved in insulin secretion. *EMBO J, 25*, 1804–1815.

Tominaga, M., Caterina, M. J., Malmberg, A. B., Rosen, T. A., Gilbert, H., Skinner, K., et al. (1998). The cloned capsaicin receptor integrates multiple pain-producing stimuli. *Neuron, 21*, 531–543.

Tsavaler, L., Shapero, M. H., Morkowski, S., & Laus, R. (2001). Trp-p8, a novel prostate-specific gene, is up-regulated in prostate cancer and other malignancies and shares high homology with transient receptor potential calcium channel proteins. *Cancer Research, 61*, 3760–3769.

Ufret-Vincenty, C. A., Klein, R. M., Hua, L., Angueyra, J., & Gordon, S. E. (2011). Localization of the PIP_2 sensor of TRPV1 ion channels. *Journal of Biological Chemistry, 286*, 9688–9698.

Voets, T. (2012). Quantifying and modeling the temperature-dependent gating of TRP channels. *Reviews of Physiology, Biochemistry and Pharmacology, 162*, 91–119.

Voets, T., Droogmans, G., Wissenbach, U., Janssens, A., Flockerzi, V., & Nilius, B. (2004). The principle of temperature-dependent gating in cold- and heat-sensitive TRP channels. *Nature, 430*, 748–754.

Voets, T., Owsianik, G., Janssens, A., Talavera, K., & Nilius, B. (2007). TRPM8 voltage sensor mutants reveal a mechanism for integrating thermal and chemical stimuli. *Nature Chemical Biology, 3*, 174–182.

Vos, M. H., Neelands, T. R., McDonald, H. A., Choi, W., Kroeger, P. E., Puttfarcken, P. S., et al. (2006). TRPV1b overexpression negatively regulates TRPV1 responsiveness to capsaicin, heat and low pH in HEK293 cells. *Journal of Neurochemistry, 99*, 1088–1102.

Vriens, J., Owsianik, G., Hofmann, T., Philipp, S. E., Stab, J., Chen, X., et al. (2011). TRPM3 is a nociceptor channel involved in the detection of noxious heat. *Neuron, 70*, 482–494.

Wagner, T. F., Loch, S., Lambert, S., Straub, I., Mannebach, S., Mathar, I., et al. (2008). Transient receptor potential M3 channels are ionotropic steroid receptors in pancreatic beta cells. *Nature Cell Biology, 10*, 1421–1430.

Wang, H., Schupp, M., Zurborg, S., & Heppenstall, P. A. (2013). Residues in the pore region of *Drosophila* transient receptor potential A1 dictate sensitivity to thermal stimuli. *Journal of Physiology, 591*, 185–201.

Watanabe, H., Davis, J. B., Smart, D., Jerman, J. C., Smith, G. D., Hayes, P., et al. (2002). Activation of TRPV4 channels (hVRL-2/mTRP12) by phorbol derivatives. *Journal of Biological Chemistry, 277*, 13569–13577.

Watanabe, H., Vriens, J., Prenen, J., Droogmans, G., Voets, T., & Nilius, B. (2003). Anandamide and arachidonic acid use epoxyeicosatrienoic acids to activate TRPV4 channels. *Nature, 424*, 434–438.

Watanabe, H., Vriens, J., Suh, S. H., Benham, C. D., Droogmans, G., & Nilius, B. (2002). Heat-evoked activation of TRPV4 channels in a HEK293 cell expression system and in native mouse aorta endothelial cells. *Journal of Biological Chemistry, 277*, 47044–47051.

Wissenbach, U., Bodding, M., Freichel, M., & Flockerzi, V. (2000). Trp12, a novel Trp related protein from kidney. *FEBS Letters, 485*, 127–134.

Wissmann, R., Baukrowitz, T., Kalbacher, H., Kalbitzer, H. R., Ruppersberg, J. P., Pongs, O., et al. (1999). NMR structure and functional characteristics of the hydrophilic N terminus of the potassium channel beta-subunit Kvbeta1.1. *Journal of Biological Chemistry, 274*, 35521–35525.

Xu, H., Delling, M., Jun, J. C., & Clapham, D. E. (2006). Oregano, thyme and clove-derived flavors and skin sensitizers activate specific TRP channels. *Nature Neuroscience, 9*, 628–635.

Xu, H., Ramsey, I. S., Kotecha, S. A., Moran, M. M., Chong, J. A., Lawson, D., et al. (2002). TRPV3 is a calcium-permeable temperature-sensitive cation channel. *Nature, 418*, 181–186.

Yang, F., Cui, Y., Wang, K., & Zheng, J. (2010). Thermosensitive TRP channel pore turret is part of the temperature activation pathway. *Proceedings of National Academy of Sciences of the United States of America, 107*, 7083–7088.

Yang, W. Y., & Gruebele, M. (2003). Folding at the speed limit. *Nature, 423*, 193–197.

Yao, J., Liu, B., & Qin, F. (2010). Kinetic and energetic analysis of thermally activated TRPV1 channels. *Biophysics Journal, 99*, 1743–1753.

Yao, J., Liu, B., & Qin, F. (2010). Pore turret of thermal TRP channels is not essential for temperature sensing. *Proceedings of National Academy of Sciences of the United States of America, 107*, E125. author reply E126–127.

Yao, J., Liu, B., & Qin, F. (2011). Modular thermal sensors in temperature-gated transient receptor potential (TRP) channels. *Proceedings of National Academy of Sciences of the United States of America, 108*, 11109–11114.

Zagotta, W. N., Hoshi, T., & Aldrich, R. W. (1990). Restoration of inactivation in mutants of Shaker potassium channels by a peptide derived from ShB. *Science, 250*, 568–571.

Zakharian, E., Cao, C., & Rohacs, T. (2010). Gating of transient receptor potential melastatin 8 (TRPM8) channels activated by cold and chemical agonists in planar lipid bilayers. *Journal of Neuroscience, 30*, 12526–12534.

Zhang, Y., Hoon, M. A., Chandrashekar, J., Mueller, K. L., Cook, B., Wu, D., et al. (2003). Coding of sweet, bitter, and umami tastes: different receptor cells sharing similar signaling pathways. *Cell, 112*, 293–301.

Zhong, L., Bellemer, A., Yan, H., Ken, H., Jessica, R., Hwang, R. Y., et al. (2012). Thermosensory and nonthermosensory isoforms of *Drosophila melanogaster* TRPA1 reveal heat-sensor domains of a thermoTRP Channel. *Cell Reports, 1*, 43–55.

Zimmermann, K., Lennerz, J. K., Hein, A., Link, A. S., Kaczmarek, J. S., Delling, M., et al. (2011). Transient receptor potential cation channel, subfamily C, member 5 (TRPC5) is a cold-transducer in the peripheral nervous system. *Proceedings of National Academy of Sciences of the United States of America, 108*, 18114–18119.

Zurborg, S., Yurgionas, B., Jira, J. A., Caspani, O., & Heppenstall, P. A. (2007). Direct activation of the ion channel TRPA1 by Ca^{2+}. *Nature Neuroscience, 10*, 277–279.

CHAPTER FOUR

TRPA1 Channels: Chemical and Temperature Sensitivity

Willem J. Laursen[1,2], Sviatoslav N. Bagriantsev[1,*] and Elena O. Gracheva[1,2,*]
[1]Department of Cellular and Molecular Physiology, Yale University School of Medicine, New Haven, CT, USA
[2]Program in Cellular Neuroscience, Neurodegeneration and Repair, Yale University School of Medicine, New Haven, CT, USA
*Corresponding author: E-mail: elena.gracheva@yale.edu, sviatoslav.bagriantsev@yale.edu

Contents

Abstract

Transient receptor potential ankyrin 1 (TRPA1) is a polymodal excitatory ion channel found in sensory neurons of different organisms, ranging from worms to humans. Since its discovery as an uncharacterized transmembrane protein in human fibroblasts, TRPA1 has become one of the most intensively studied ion channels. Its function has been linked to regulation of heat and cold perception, mechanosensitivity, hearing, inflammation, pain, circadian rhythms, chemoreception, and other processes. Some of these proposed functions remain controversial, while others have gathered considerable experimental support. A truly polymodal ion channel, TRPA1 is activated by various stimuli, including electrophilic chemicals, oxygen, temperature, and mechanical force, yet the molecular mechanism of TRPA1 gating remains obscure. In this review, we discuss recent advances in the understanding of TRPA1 physiology, pharmacology, and molecular function.

Current Topics in Membranes, Volume 74
ISSN: 1063-5823
http://dx.doi.org/10.1016/B978-0-12-800181-3.00004-X

1. INTRODUCTION

TRPA1 (transient receptor potential ankyrin 1, also known as ANKTM1) is a polymodal ion channel that belongs to a functionally and structurally diverse group of ion channels known as transient receptor potential (TRP) channels. It was originally discovered as an obscure molecule in cultured human fibroblasts (Jaquemar, Schenker, & Trueb, 1999), and later received considerable attention as a putative cold sensor in rodents (Story et al., 2003). TRPA1 is expressed in a specific subset of somatosensory neurons of trigeminal, dorsal root, and nodose ganglia, and exhibits pronounced promiscuity in terms of activation: the channel can interact with an increasingly long list of identified ligands, including the electrophilic compounds in pungent spices, industrial pollutants, and various external irritants (Julius, 2013; Nilius, Appendino, & Owsianik, 2012). TRPA1 is also activated by endogenous inflammatory mediators and plays a significant role in transducing nociceptive signals associated with tissue damage and inflammation (for reviews see Bautista, Pellegrino, & Tsunozaki (2013), Julius (2013)). Indeed, TRPA1 is expressed specifically in peptidergic C-fibers, which release an endogenous neurogenic inflammatory cocktail, containing CGRP (Calcitonin gene-related peptide), substance P, and neurokinins, resulting in the initiation of cellular responses and tissue inflammation.

A growing number of studies implicate TRPA1 in the development of migraine headache (Benemei, Fusi, Trevisan, & Geppetti, 2013). This notion is supported by the fact that umbellulone (the active compound of the so-called "headache tree") activates TRPA1 channels within the trigeminal system (Zhong et al., 2011), thereby initiating CGRP release, leading to significantly increased meningeal blood flow (Kunkler, Ballard, Oxford, & Hurley, 2011). Recently, additional support was given to the idea of TRPA1 as a mediator of pain in humans when a gain-of-function mutation was identified and found to be associated with a familial episodic pain syndrome (Kremeyer et al., 2010).

Itch (pruritus) is a very important protective physiological response to different irritants, including insect bites, poisonous plants, and pollutant allergens. Not long ago, TRPA1was shown to be involved in the itch pathway and proposed to act as an integrator of multiple histamine-independent signaling cascades (Bautista, Wilson, & Hoon, 2014; Oh et al., 2013; Wilson et al., 2011).

In some invertebrate and vertebrate species, including fruit flies, *Caenorhabditis elegans*, pit-bearing snakes, frogs, and lizards, the TRPA1 channel evolved as a temperature sensor (Chatzigeorgiou et al., 2010; Gracheva et al., 2010; Kang et al., 2012; Saito et al., 2012; Viswanath et al., 2003). It is interesting to note that, in these animals, sensitivity of TRPA1 to temperature often appears to come at the expense of chemical activation, with thermo-activated orthologues displaying decreased chemosensitivity (Cordero-Morales, Gracheva, & Julius, 2011; Gracheva et al., 2010; Saito et al., 2012) (see Table 4.1).

Nonetheless, its role as a polymodal sensor of such a variety of stimuli including temperature and noxious irritants—combined with its described role in episodic pain syndrome, inflammation and migraine—makes TRPA1 an attractive target for analgesic intervention. A thorough understanding of channel activation mechanisms will therefore be crucial for the development of effective and specific therapeutics.

2. ACTIVATION AND REGULATION OF TRPA1 BY CHEMICAL COMPOUNDS

TRPA1 is a nonselective cation channel that contains six transmembrane domains, a huge cytoplasmic N terminus (~720 amino acids), and a cytoplasmic C terminus. It is distinguishable from other members of the TRP channel family by the presence of numerous regulatory ankyrin repeats within the N terminus (Jaquemar et al., 1999; Story et al., 2003) (Figure 4.1). Ankyrin repeats are 33-amino-acid-long structural motifs involved in protein–protein interactions. This region of the TRPA1 channel is thought to play a key role in the regulation of gating and integration of multiple stimuli, including temperature, multiple classes of chemicals, and cytoplasmic calcium (Cordero-Morales et al., 2011; Nilius, Prenen, & Owsianik, 2011; Zurborg, Yurgionas, Jira, Caspani, & Heppenstall, 2007).

2.1 Chemical activation of TRPA1 by covalent modification

The most well-characterized mechanism of TRPA1 activation occurs through covalent modification of the thiol groups of conserved cysteine residues that are located in the "linker" region of the channel that connects the ankyrin repeats with the transmembrane domain (Hinman, Chuang, Bautista, & Julius, 2006; Macpherson, Dubin, et al., 2007) (Figure 4.1). There are several pungent culinary compounds that activate TRPA1 via electrophilic attack on these cysteines, including allyl isothiocyanate

Table 4.1 Temperature and chemical sensitivity data for TRPA1 orthologues from different animal species

Temperature and chemical sensitivity of TRPA1 orthologues					
Species	**Cold/heat**	**Threshold (°C)**	**Q_{10}**	**AITC (EC_{50})**	**References**
Human (*Homo sapiens*)	Cold?	≤17	N.D.	Sensitive (62 μM)	Story et al. (2003), Cordero-Morales et al. (2011)
Rhesus macaque (*Macaca mulatta*)	Insensitive	–	–	Sensitive (75.5 μM)	Chen et al. (2013)
Rat (*Rattus norvegicus*)	Cold?	≤15	N.D.	Sensitive (11 μM)	Story et al. (2003), Jordt et al. (2004)
Mouse (*Mus musculus*)	Cold?	≤18	N.D.	Sensitive (0.39 μM)	Story et al. (2003), Sawada et al. (2007), Zhou et al. (2013)
Chicken (*Gallus gallus domesticus*)	Heat	39.4 ± 1.1	N.D.	Sensitive (N.D.)	Saito et al. (2014)
Rattlesnake (*Crotalus atrox*)	Heat	27.6 ± 0.9	13.7	Sensitive (>2000 μM)	Gracheva et al. (2010)
Rat Snake (*Elaphe obsoleta lindheimeri*)	Heat	37.2 ± 0.7	8.8	Sensitive (>500 μM)	Gracheva et al. (2010)
Python (*Python regius*)	Heat	32.7 ± 1.3	N.D.	Sensitive (>500 μM)	Gracheva et al. (2010)

Boa (*Corallus hortulanus*)	Heat	29.6 ± 0.7	N.D.	Sensitive (>500 μM)	Gracheva et al. (2010)
Anole (*Anolis carolinensis*)	Heat	33.9 ± 0.8	45.71	Sensitive (N.D.)	Saito et al. (2012)
Frog (*Xenopus tropicalis*)	Heat	39.7 ± 0.7	59.24	Sensitive (N.D.)	Saito et al. (2012)
Zebrafish (*Danio rerio*)	Insensitive	–	–	Sensitive (N.D.)	Prober et al. (2008)
Fruit fly (*Drosophila melanogaster*)				Sensitive (277.6 μM)	Cordero-Morales et al. (2011), Kang et al. (2010), Kang et al. (2012), Zhong et al. (2012)
TRPA1(A)	Heat	29.7 ± 0.3	9		
TRPA1(B)	Heat	27.8 ± 0.4	116		
TRPA1(C)	Insensitive	–	–		
TRPA1(D)	Heat	≥34	N.D.		
Mosquito (*Anopheles*	Heat	34.2 ± 1.8	4	Sensitive (N.D.)	Wang et al. (2011), Kang et al. (2012)
gambiae)		25.2 ± 0.9	200		
TRPA1(A)					
TRPA1(B)					
Nematode (*Caenorhabditis elegans*)	Cold?	≤18	N.D.	Insensitive	Chatzigeorgiou et al. (2010), Kindt et al. (2007)

N.D., Not Defined.

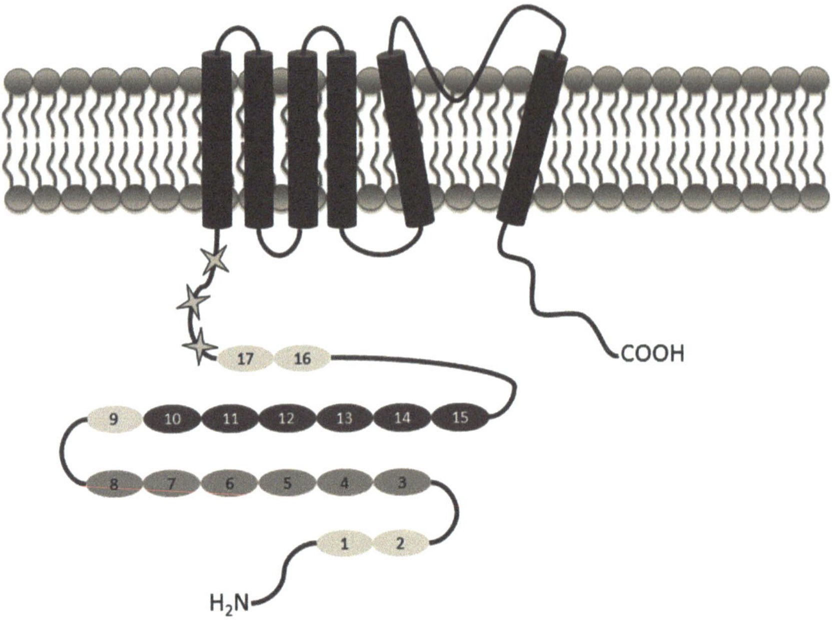

Figure 4.1 Topological organization of a snake TRPA1 channel. Ankyrin repeats are shown as numbered ovals. The primary and enhancer modules important for determining heat sensitivity correspond to repeats number 10–15 and 3–8, respectively. Stars in the linker region between the transmembrane domains and ankyrin repeat domain depict conserved cysteine residues involved in channel activation by electrophiles. (See the color plate.)

(AITC; found in horseradish, wasabi, and mustard) (Bandell et al., 2004; Jordt et al., 2004), allicin and diallyl disulfide (from raw garlic) (Bautista et al., 2005; Macpherson et al., 2005), and cinnamaldehyde (cinnamon) (Bandell et al., 2004), among others (Figure 4.2). These molecules are widely used in the laboratory to explore physiological roles of TRPA1 and to determine structure–function relationships.

Electrophilic compounds are capable of forming reversible adducts with thiol groups (Hinman et al., 2006; Macpherson, Dubin, et al., 2007). Structure–function analyses identified several key conserved cysteine residues (Cys619, Cys639, and Cys663 for human TRPA1) in the linker region between the cytoplasmic N terminus and the transmembrane domains of the channel that are necessary prerequisites for proper channel activation by electrophiles (Hinman et al., 2006) (Figure 4.1). However, different TRPA1 channel orthologues, such as those from human and rattlesnake, exhibit significantly different sensitivities to AITC, despite the presence of all three conserved cysteine residues (Gracheva et al., 2010). This

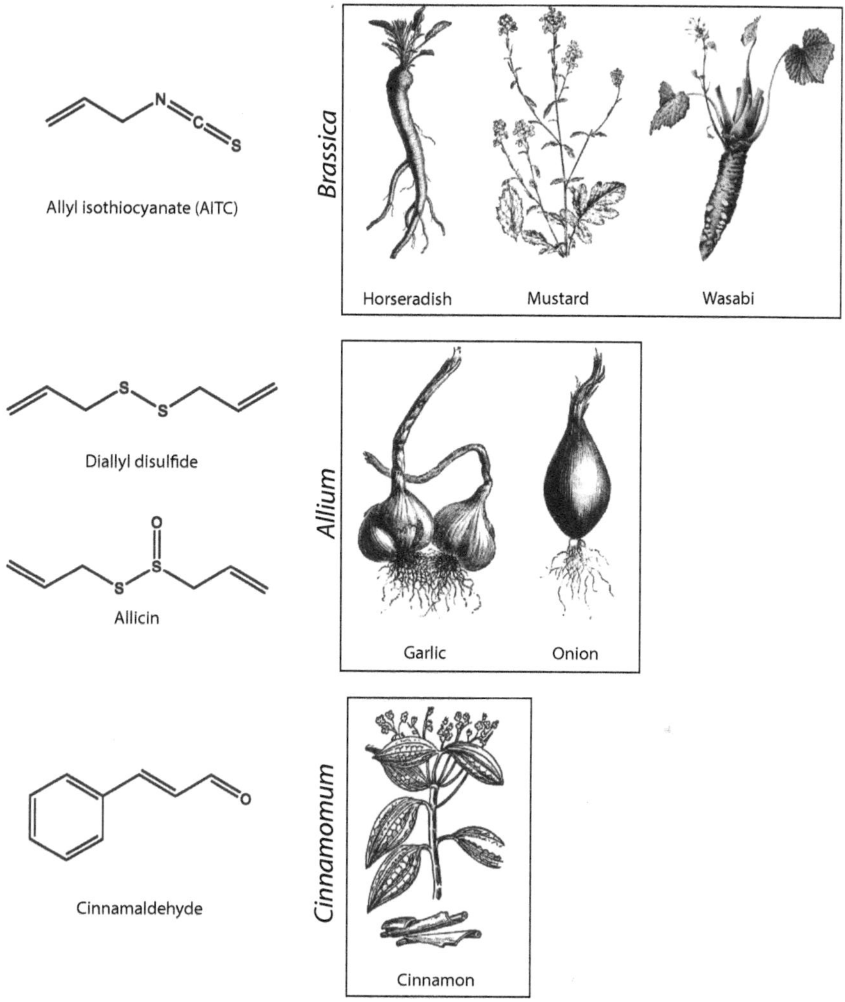

Figure 4.2 Examples of electrophilic activators of TRPA1 from three different plant genera.

observation suggests that other regulatory elements may be involved in determining sensitivity to electrophilic ligands.

In addition, a number of other plant-based compounds have been identified as TRPA1 agonists: gingerol (ginger), thymol (thyme), oleocanthal (olive oil), eugenol (cloves), methyl salicylate (wintergreen), Δ^9-tetrahydrocannabinol (psychoactive compound in marijuana), and carvacrol (oregano, marjoram)—just to name a few (Bandell et al., 2004; Jordt et al., 2004; Peyrot des Gachons et al., 2011; Xu, Delling, Jun, & Clapham, 2006).

Interestingly, some of these chemicals, like carvacrol and oleocanthal, do not act through the reactive cysteine residues, suggesting the presence of an alternative activation mechanism (Peyrot des Gachons et al., 2011). Nevertheless, all these compounds are associated with a spicy, chemesthetic sensation upon ingestion owing to activation of TRPA1. It is believed that the evolution of TRPA1 agonists in these plants may have arisen as self-defense strategy in the molecular arms race against attacks by pests and consumption by herbivores.

From a physiological point of view, TRPA1 plays an important protective role because it can detect the presence of less-than-appetizing (i.e., potentially harmful) environmental pollutants such as toluene diisocyanate, hypochlorite, and H_2O_2 (industrial pollutants); H_2S; as well as α, β-aldehydes from smoke (acrolein, crotonaldehyde, etc.) (Andersson, Gentry, Moss, & Bevan, 2008; Bautista et al., 2006; Bessac, Sivula, von Hehn, Escalera, & Cohnet, 2008; Taylor-Clark, Kiros, Carr, & McAlexander, 2009). Like AITC and other organosulfur activators, these electrophiles covalently interact with the conserved linker cysteines.

Moreover, TRPA1 appears to mediate the irritating effects of exposure to formalin, another widespread environmental pollutant (Macpherson, Xiao, et al., 2007; McNamara et al., 2007; Yonemitsu et al., 2013). TRPA1-knockout animals display significantly attenuated pain responses when injected with a dilute formalin solution into the hindpaw (Macpherson, Xiao, et al., 2007; McNamara et al., 2007). Inhalation of these environmental irritants activates TRPA1 and leads to depressed respiration rates, coughing, and bronchial contraction (Bessac & Jordt, 2010). Therefore, the ability to detect the presence of these hazardous chemicals, even at low concentrations, can be beneficial to limit injurious exposure.

The military has exploited this protective physiological function of TRPA1 as a noxious chemosensor for decades. 1-chloroacetophenone (Mace®), dibenz[b,f][1,4]-oxazepine, 2-chlorobenzylidene malononitrile, and chloropicrin are extremely effective tear gases and riot control agents. These electrophiles are also some of the most potent TRPA1 agonists identified (Bessac et al., 2009; Brone et al., 2008). With half-maximal activation (EC_{50}) values in the pico- to nanomolar range for heterologously expressed TRPA1, these alkylating agents lead to profuse lachrymation, pain, coughing, and respiratory distress. Genetic deletion of the channel, mutation of the reactive cysteine residues, or treatment with TRPA1 antagonists can significantly reduce pain behavior in vivo, suggesting a potential treatment strategy for exposed patients (Bessac et al., 2009; Brone et al., 2008).

In addition to external irritants, endogenous signals, such as mediators of inflammation, also make use of the reactive thiol moieties to activate TRPA1 nociceptors (Andersson et al., 2008; Bautista et al., 2013; Takahashi et al., 2008; Taylor-Clark, Ghatta, Bettner, & Undem, 2009; Taylor-Clark et al., 2008). Following injury, activation of the cyclooxygenase pathway leads to the production of inflammatory mediators known as prostaglandins. The prostaglandin 15-deoxy-$\Delta^{12,14}$-prostaglandin J_2 (15d-PGJ_2), one of the molecules whose production leads to hyperalgesia associated with the carrageenan-induced laboratory model of inflammation, was shown to activate a specific subset of dissociated dorsal root ganglion cells (Takahashi et al., 2008). It was subsequently demonstrated that this molecule could activate TRPA1 via the conserved cysteines in the N terminus. Specifically, the α,β-unsaturated moieties of 15d-PGJ_2 are believed to alkylate TRPA1 via a Michael addition reaction (Andersson et al., 2008). When assessed by calcium imaging, mutations in specific N-terminal cysteine residues of human TRPA1 are able to significantly reduce activation by 15d-PGJ_2 (Takahashi et al., 2008).

Other compounds associated with the inflammatory milieu include NO, H_2O_2, and 4-hydroxynonenal, a product of lipid peroxidation due to release of reactive oxygen species (Basbaum & Woolf, 1999; Schneider, Porter, & Brash, 2008). All these molecules have been found to activate TRPA1, albeit via a slightly different mechanism than 15d-PGJ_2 (Andersson et al., 2008; Macpherson, Xiao, et al., 2007; Takahashi et al., 2008; Trevisani et al., 2007). Unlike15d-PGJ_2, the interactions of these molecules with the N-terminal cysteines are not inhibited by the presence of the reducing agent dithiothreitol. This is consistent with nitrosylation or oxidation of the sulfhydryl group, as opposed to alkylation by 15d-PGJ_2 (Figure 4.3).

2.2 Noncovalent activation of TRPA1

As is the case for the monoterpenic phenol carvacrol and several of the agonists mentioned earlier in the text, TRPA1 can be activated by mechanisms independent of covalent modifications. For instance, a decrease in intracellular pH can activate the channel. Studies have shown that weak acids such as acetic acid, formic acid, and others can cause activation (Wang et al., 2011). This represents another mechanism for activation of TRPA1 at areas of inflammation due to local acidosis near the injury site. Sensitivity of TRPA1 to acidic pH has also been demonstrated to mediate the stinging, pungent effects of carbonated drinks. The carbon dioxide (CO_2) in these

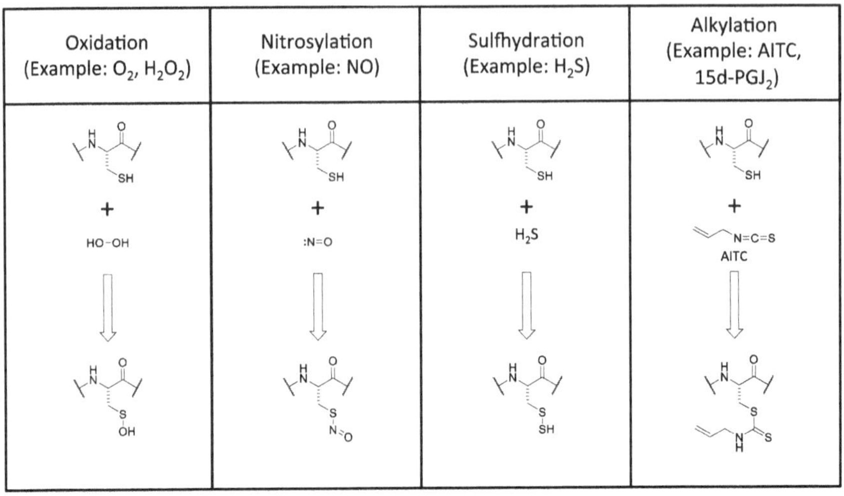

Figure 4.3 Mechanisms of cysteine modification by oxidation, nitrosylation, sulfhydration, and alkylation.

beverages is able to diffuse into the cell, thereby lowering intracellular pH and activating TRPA1 (Wang, Chang, & Liman, 2010).

Oxygen (O_2) is another gas for which TRPA1 is suggested to serve as a sensor (Takahashi et al., 2011). Because it is required for mitochondrial respiration, oxygen levels must be carefully monitored. Signaling by neurons in sensory and vagus nerves alerts the animal to hypoxic conditions. On the other hand, elevated levels of oxygen (hyperoxia) can have toxic effects and cells will respond to deviations from normoxia in either direction. Genetic mutation of TRPA1 in mice prevents signaling from sensory and vagal nerves in response to hypoxia or hyperoxia. During states of normoxia, TRPA1 activation is prevented by the actions of prolyl hydroxylases (PHDs). At near atmospheric oxygen levels, the action of these O_2-dependent enzymes is believed to result in the hydroxylation of a conserved proline residue within the N terminus (Pro394 of human TRPA1), thus inhibiting channel activation during normoxia. When O_2 levels drop, activity of PHDs decrease and the channel is freed from inhibition. In contrast, hyperoxia has been shown to override PHD-mediated inhibition to activate TRPA1 via direct oxidation of reactive cysteine residues by O_2 (Takahashi et al., 2011) (Figure 4.3).

Chronic metal toxicity has become a very hot topic in sensory physiology and pathology. Over the past decade, a number of studies suggested a role for TRPA1 in this condition, with results showing direct activation and modulation of channel function by divalent cations, including zinc, copper, cadmium, and calcium (Andersson, Gentry, Moss, & Bevan, 2009; Gu & Lin, 2010; Hu, Bandell, Petrus, Zhu, & Patapoutian, 2009; Miura et al., 2013; Zurborg et al., 2007). Animals lacking TRPA1 display attenuated irritation and nociception after zinc chloride injection (Gu & Lin, 2010). Single amino acid mutagenesis studies identified N-terminal Cys614, C-terminal Cys1021, and His983 as important prerequisites for activation by zinc. Although it is not clear how the last of these residues is involved, it is interesting to note that cysteine/histidine residues are commonly found in zinc interaction sites of other proteins (Hu et al., 2009).

2.3 Receptor-operated activation of TRPA1

Like many other TRP channels, TRPA1 activation can result as a downstream consequence of signaling cascades generated by other receptors. Coexpression of TRPA1 with phospholipase-C (PLC)-coupled receptors like the bradykinin receptor (B_2R) (Bandell et al., 2004; Wang, Dai, et al., 2008) and the M1 muscarinic acetylcholine receptor (Jordt et al., 2004) show typical receptor-operated channel behavior. In heterologous expression systems, activation of either of these proteins alone results in minimal noticeable currents on the plasma membrane. However, when these receptors are coexpressed with TRPA1, the magnitude of the signal increases dramatically, an effect that can be reversed by treatment with PLC inhibitors. However, the exact mechanism(s) by which PLC signaling leads to TRPA1 activation remains controversial and activation of this pathway results in the formation of a number of products that have been proposed to regulate TRPA1 function (Dai et al., 2007; Nilius, Owsianik, & Voets, 2008; Rohacs, Thyagarajan, & Lukacs, 2008; Wang, Chang, Waters, McKemy, & Liman, 2008; Zurborg et al., 2007).

Phosphatidylinositol-4,5-bisphosphate (PIP_2) is a membrane phospholipid that acts as a substrate for cleavage by PLC. PIP_2 can also function as an anchor for a variety of proteins. Of particular relevance to this review, PIP_2 has been suggested to elicit both stimulatory and inhibitory effects on members of the TRP channel family (Qin, 2007). Although not as well understood or as well studied as the interaction of PIP_2 with the Transient receptor potential, subfamily V, member 1 channel, there is still some evidence to suggest that PIP_2 may display dual regulation over TRPA1. One study found that desensitization in response to agonist stimulation was delayed by addition of exogenous PIP_2. Furthermore, reduction of PIP_2 levels by application of neomycin increased the rate of desensitization (Nilius et al., 2008). On the other hand, another group found that application of PIP_2 led to inhibition of TRPA1 potentiation by bradykinin (Dai et al., 2007). Similarly, a third group found that application of PIP_2 to inside-out patches of mouse TRPA1-expressing HeLa cells did not activate the channel and, in fact, when applied subsequently to AITC, actually inhibited TRPA1 currents. Removal of PIP_2 either through polylysine treatment or through the use of an antibody led to increased AITC-activated current in whole-cell configuration (Kim, Cavanaugh, & Simkin, 2008).

At the same time, it is well known that PLC signaling can lead to increases in cytosolic calcium. Ca^{2+} itself has been suggested to play major roles in both activation and desensitization of TRPA1 (Wang, Chang, Waters, McKemy, & Liman, 2008; Zuborg et al., 2007). Moreover, it has been proposed that cold-induced activation of TRPA1 reported for some heterologous systems may be indirect, resulting from Ca^{2+}-induced activation (Zuborg et al., 2007). TRPA1 contains a putative intracellular EF-hand motif suggested to mediate the effects of Ca^{2+}. This hypothesis has proven to be somewhat controversial, however, as different groups have reported conflicting results when this region is mutated (Wang, Chang, et al., 2008). Although the precise mechanisms remain to be elucidated, it does appear that Ca^{2+}-mediated potentiation/desensitization are mediated by cytosolic rather than extracellular Ca^{2+}, as mutations that limit TRPA1 Ca^{2+} permeability are shown to abrogate both these processes, while application of Ca^{2+} to the cytosolic face of excised patches was able to restore them (Wang, Chang, et al., 2008).

In the past few years, PLC-independent receptor-operated activation pathways for TRPA1 were also described. As mentioned earlier, TRPA1 has been linked to the pathophysiology of histamine-independent itch (Wilson et al., 2011). Several converging pathways are capable of activating TRPA1 to mediate this response. While some of these signals occur through PLC pathways, activation of Mas-related G protein-coupled receptor A3 (MrgprA3), the receptor for the infamously pruritogenic antimalarial drug chloroquine (CQ), leads to TRPA1 activation independent of PLC. Interestingly, signaling by the Gβγ subunit has been proposed to mediate the effects of CQ receptor activation and treatment with subunit chelators or small molecule inhibitors prevent TRPA1 activation.

By acting as a downstream target of different signaling cascades, TRPA1 is able to further diversify the list of ligands to which it effectively responds. This role also endows the channel with an important function as an amplifier and integrator of signals from a wide range of both internal and external sources.

To understand the role of TRPA1 in vivo, it will be of utmost importance to develop a thorough knowledge of activation mechanisms for the plethora of proposed ligands. This knowledge will be critical to control for potentially physiologically undesirable effects resulting from channel activation by unidentified cellular components in the nonuniform heterologous expression systems currently used to investigate TRPA1 function.

3. TEMPERATURE SENSITIVITY OF TRPA1

Temperature is a potent regulator of TRPA1, but unlike sensitivity to electrophilic compounds—conserved in all known TRPA1 orthologues except nematodes—temperature sensitivity of TRPA1 varies significantly between species.

3.1 TRPA1 in mammals

In the somatosensory system, TRPA1 is expressed in a subset of TRPV1-positive neurons (Kobayashi et al., 2005; Mishra & Hoon, 2010; Story et al., 2003) and initially the channel was suggested as a cold sensor in the sub-Transient receptor potential, subfamily M, member 8 range (Kwan et al., 2006; Story et al., 2003). Since then, this hypothesis has been refined (McKemy, 2013), leading to the prevalent notion that TRPA1 contributes to cold sensitivity in injury-evoked or pathological conditions (del Camino et al., 2010; Knowlton, Bifolck-Fisher, Bautista, & McKemy, 2010).

In an attempt to understand the potential contributions of TRPA1 as a cold sensor in vivo, knockout mice were independently developed by two groups (Bautista et al., 2006; Kwan et al., 2006). Despite the fact that the animals were created using similar genetic schemes, there is disagreement about the resulting phenotype. Notably, one group reported the absence—even in WT (wild type) mice—of menthol-insensitive, cold-activated trigeminal neurons that fit TRPA1 pharmacological profiles (Bautista et al., 2006). The same group also found no significant differences between TRPA1-knockout and WT animals in terms of either the latency to lift the paw in cold plate assays, flinching due to acetone-induced cooling of the paw (Bautista et al., 2006), or by temperature preference assay (Bautista et al., 2007). These results led them to conclude that TRPA1 does not behave as a cold sensor in vivo. In contrast, Kwan et al. (2006) reported dramatically decreased paw lifting in TRPA1-knockout mice exposed to a 0 °C cold plate as well as decreased behavioral responses to acetone application. A number of other conflicting reports have emerged either supporting or contradicting the notion of TRPA1 as a noxious cold sensor (del Camino et al., 2010; Fajardo, Meseguer, Belmonte, & Viana, 2008; Karashima et al., 2009; Knowlton et al., 2010; Kwan & Corey, 2009; Kwan, Glazer, Corey, Rice, & Stucky, 2009). Of note, a recent study showed that specific ablation of TRPV1-positive neurons in adult mice leads to disappearance of TRPA1 neurons, but does not abolish animal sensitivity to painful cold, further arguing against a role of TRPA1 in nonpathological detection of noxious temperatures (Pogorzala, Mishra, & Hoon, 2013).

Interestingly, while there remains considerable debate as to the role of TRPA1 as a noxious cold sensor, the channel does not appear to be involved in thermoregulation. This is demonstrated by the lack of change in core body temperature induced by genetic knockout (Bautista et al., 2007) or treatment with channel-specific antagonist (Chen et al., 2011). An effect on thermoregulation is also not observed in cold-exposed TRPA1-knockout animals or antagonist-treated rats when core body temperatures are decreased below the reported TRPA1 activation threshold of 17 °C (Oliveira et al., 2014). These findings are in stark contrast to other thermo-TRPs such as TRPV1, inhibition of which results in hyperthermia (Gawa et al., 2007).

Perhaps the most controversial data exist with regard to temperature activation of mammalian TRPA1 orthologues in heterologous systems by cold (Caspani & Heppenstall, 2009). A number of studies have reported potent activation of mammalian TRPA1 at ~17 °C (del Camino et al., 2010; Karashima et al., 2009; Kremeyer et al., 2010; Sawada et al., 2007; Story et al., 2003), while others claimed the absence of a response (Cordero-Morales et al., 2011; Jordt et al., 2004; Knowlton et al., 2010; Nagata, Duggan, Kumar, & Garcia-Anoveros, 2005; Zurborg et al., 2007). The reason for this discrepancy may reside in the differences in experimental conditions, such as expression system, expression level, ionic composition of recording solution, etc. In addition, much of the controversy may be due to species-specific properties of TRPA1 orthologues. In other temperature-sensitive TRP channels, species differences are not as pronounced. For example, all tested mammalian TRPV1 and TRPM8 orthologues are, respectively, heat and cold sensitive, and a similar behavior, perhaps erroneously, was expected from TRPA1.

Along these lines, a recent report documented a side-by-side comparison of mouse, rat, human, and rhesus monkey TRPA1 in identical experimental conditions (Chen et al., 2013). The study showed that while all four molecules are potentiated by AITC, cold activates only mouse and rat orthologues, while primate channels are insensitive to temperature changes in the 8–24 °C range.

3.2 TRPA1 in insects and worms

As a heat sensor, TRPA1 was originally identified in *Drosophila melanogaster* (Viswanath et al., 2003). It was demonstrated that insect TRPA1 is a polymodal ion channel that can be activated by temperature and electrophilic compounds (Kang et al., 2010). Later, the same group identified two isoforms of TRPA1 that play distinct physiological roles (Kang et al., 2012). The longer TRPA1(A) isoform is less temperature sensitive than TRPA1(B), with reported Q_{10} values of 9 and 116, respectively. The latter is localized to specific thermosensory neurons and is involved in temperature discrimination, whereas TRPA1(A) is expressed in chemosensory neurons and plays a key role in chemical sensitivity rather than heat perception. Mosquitoes have similar gene organization and use the same molecular strategies to produce two different isoforms of TRPA1 with distinct temperature activation profiles, suggesting a common evolutionary trend in insects for modulating TRPA1 function (Kang et al., 2012). Moreover, two additional isoforms (TRPA1(C) and TRPA1(D)) were cloned from *Drosophila* larvae (Zhong et al., 2012). These two channels are expressed in nociceptors and activated by AITC. However, TRPA1(C) is not temperature sensitive, whereas TRPA1(D) can be activated by heat at temperatures $\geq$34 °C.

In contrast, nematode TRPA1 is proposed to function as both a mechanosensor and cold sensor in a distinct population of sensory neurons (Chatzigeorgiou et al., 2010; Kindt et al., 2007). *Caenorhabditis elegans* TRPA1 is a polymodal ion channel, but compared to other orthologues, it is not activated by chemical compounds such as AITC (see Table 4.1). Moreover, recent studies revealed an additional role for TRPA1: worm TRPA1 is proposed to detect thermal changes in the environment and, as a result, promote longevity at cold temperatures (Xiao et al., 2013).

3.3 TRPA1 in fish, birds, reptiles, and amphibians

Zebrafish (*Danio rerio*) contain two genes that encode TRPA1a and TRPA1b. Both these molecules are expressed specifically in sensory neurons. Unlike the invertebrate channel, TRPA1a and TRPA1b are activated only by chemical compounds, not temperature (Prober et al., 2008). At the physiological level, TRPA1b (but not TRPA1a) is implicated in nocifensive responses to chemical irritants.

A recent report presented the first characterization of an avian TRPA1 channel (Saito et al., 2014). TRPA1 cloned from chickens (*Gallus gallus domesticus*) is activated by heat with an average threshold of ~39.4 °C. Furthermore, the channel is responsive to classic noxious electrophilic TRPA1 agonists as well as sensitive to the agricultural bird repellant methyl anthranilate. This finding is especially interesting because it represents the first characterization of a TRPA1 orthologue from a nonmammalian homeotherm. The fact that the channel responds selectively to heat as opposed to activation by cold, as is proposed for some mammalian TRPA1 channels, suggests that the directionality of thermal-sensing properties of the channel did not change coincident with the evolution of homeothermy. Along these lines, it should be noted that, based on amino acid sequence, chicken TRPA1 displays closer similarity to green anole lizard (*Anolis carolinensis*) TRPA1 (82%) than to human and mouse TRPA1 (64% and 65%, respectively).

In frogs, lizards, and snakes, TRPA1 serves as a detector of noxious heat and chemical irritants (Gracheva et al., 2010; Saito et al., 2012). In more specialized species such as pit-bearing snakes (boas, pythons, and rattlesnakes), TRPA1 plays a crucial role in the detection of infrared radiation emitted by warm-blooded animals (Gracheva et al., 2010). This enhanced thermal sensitivity comes at the cost of chemical sensitivity to electrophilic compounds with these heat-activated orthologues displaying significantly reduced sensitivity to electrophiles (Table 4.1). Taking into consideration that insect TRPA1s are also heat sensitive, this may suggest that heat sensitivity evolved and adjusted multiple times during evolution to support adaptability to environmental conditions as well as to unique feeding habits. The ability to separate modalities (temperature/chemical/mechanical activation) in order to prevent conflicting sensory input may be the key to the adaptive significance and functional diversification that TRPA1 has attained throughout the course of evolution.

3.4 TRPA1: Molecular mechanism of temperature sensitivity

Little is known about the molecular mechanism of temperature sensitivity in TRPA1. As discussed above, some TRPA1 orthologues are activated by heat, and others by cold, which at first sight may suggest profound differences in the temperature-sensing apparatus of these channels. However, as counterintuitive as it may seem, activation by heat and cold may represent similar conformational changes that come hand in hand when considered from the point of view of thermodynamics of heat- and cold-induced protein folding.

Basic thermodynamic considerations dictate that an ion channel with a nonzero change in heat capacity between open and closed conformations will be active at the extreme sides of the temperature spectrum. With regard to temperature-gated ion channels, it is expected that a cold-activated ion channel will also be activated by heat. However, practical limitations prevent us from assessing the full temperature range, so we often observe only one arm of the U-shaped activation profile, and, based on the results, call the channel either a cold or a heat sensor. While these considerations (for details, see (Clapham & Miller, 2011)) await experimental support (Chowdhury, Jarecki, & Chanda, 2014), we note here that this thermodynamic framework may provide an elegant explanation as to why different TRPA1 orthologues exhibit seemingly incompatible temperature activation properties. For example, rattlesnake TRPA1 is activated by heat (Gracheva et al., 2010), while the ~60% identical mouse orthologue is activated by cold (Chen et al., 2013; Story et al., 2003). It was suggested that the point of minimum in the U-shaped activation curve can be left or right shifted by slightly changing the amino acid composition without significantly affecting heat capacity of the channel's temperature-sensing element (Clapham & Miller, 2011). As a result, the practically testable temperature window will shift, leaving only the cold- or heat-activated "arm" of the U-curve accessible for experimental analysis. In practical terms, this means that alterations in the amino acid composition may lead to the apparent reversal of temperature properties, manifested as cold or heat activation in different TRPA1 orthologues.

Alternatively, temperature-dependent activation of TRPA1 may be explained using modular allosteric gating and coupling models (Jara-Oseguera & Islas, 2013; Qin, 2013; Salazar, Moldenhauer, & Baez-Nieto, 2011; Voets, 2012). Regulation of channel gating in these scenarios takes place via allosteric coupling to a heat sensor module. In this way, temperature-induced channel activation may occur without relying on changes in heat capacity. Interestingly, when temperature-dependent coupling parameters are included in their model, Oseguera and Islas (2013) point out that a channel exhibiting allosteric coupling may also display a U-shaped temperature activation profile, similar to predictions from the heat capacity model.

How temperature is translated into opening of the TRPA1 pore is a matter of intense research. It is of great interest to determine whether the temperature-sensing module of TRPA1 has defined structural boundaries, or whether different parts of the molecule collectively contribute to the formation of such a sensor. If the temperature-sensing module has defined boundaries, does it include gating elements? As of this writing, these questions have not been definitively answered.

The fact that both cold- and heat-activated thermo-TRP channels retain their temperature responses when reconstituted in artificial bilayers suggests that temperature responses and directionality are not determined by membrane lipid composition (Cao, Cordero-Morales, Liu, Qin, & Julius, 2013; Zakharian, Cao, & Rohacs, 2010). Chimeras and mutagenesis experiments in TRP channels, including TRPA1, provide further support for this idea. Transposition of ankyrin repeats from the N terminus of fly or rattlesnake TRPA1 confers heat sensitivity onto the human orthologue without changing AITC responses (Cordero-Morales et al., 2011). Additionally, a random mutagenesis screen in mouse TRPA1 identified three single-point mutations in ankyrin repeat 6 that are reported to confer warmth-induced activation on the channel while maintaining sensitivity to chemical responses (Jabba et al., 2014). If we assume that heat and AITC open the channel at the same molecular gate, then these findings suggest that the ankyrin repeats of the heat-sensitive TRPA1s contain functional prerequisites for heat activation. The existence of a heat "module" within the same ankyrin repeats indicate that such evolutionarily distant species as flies and snakes undertook similar molecular strategies for developing heat sensitivity in TRPA1 (see Figure 4.1). These observations suggest that the N terminus of the temperature-insensitive human TRPA1 does not contain the heat module. However, substitution of the entire N and C termini in fly TRPA1 with human domains fails to abolish heat sensitivity (Wang, Schupp, Zurborg, & Heppenstall, 2013), indicating that regions beyond the ankyrin repeats also contribute to temperature activation. Indeed, in many insects, such as flies, mosquitoes, and lice, alternative splicing of the extreme N-terminal region was shown to be critical for defining the steepness of temperature responses (Kang et al., 2012). Collectively, these data strongly support the existence of heat-sensing modules in the N terminus of TRPA1, separate from the regions sensing electrophilic compounds, such as AITC.

Single or multiple amino acid substitutions in the pore region of TRPA1 can obliterate temperature responses without affecting chemical activation (Chen et al., 2013; Wang et al., 2013). These data clearly establish a critical role of the pore region in temperature gating. It remains unclear, however, if the gate only responds to temperature-induced conformational changes in the N terminus, or if temperature causes structural rearrangements in various parts of the channel, including the cytosolic and transmembrane domains, which then leads to opening of the channel gate.

To complicate the matter, the location of a functional temperature-activated gate in TRPA1 has not been clearly defined. In the absence of structural data, it is reasonable to assume that, by analogy with voltage-gated potassium channels, the region between transmembrane helices 5 and 6 of TRPA1 will form the ion-conducting pore. In many potassium channels, the pore domain contains two gates: the extracellular selectivity filter-based gate and the intracellular gate formed by the inner helices, topologically analogous to transmembrane segment 6 in TRPs (Mathie, Al-Moubarak, & Veale, 2010). Whether or not any of these regions, or both, are opened by changes in temperature, is unclear. This aspect is critical for understanding the molecular mechanism of TRPA1 gating by temperature and chemical compounds.

A thorough understanding of the biophysics of temperature-induced TRPA1 activation mechanism(s) may be critical for understanding and manipulating channel function in vivo. It will be fascinating to see if the same (or similar) mechanisms apply to other temperature-activated ion channels. We look forward to progress on this topic in the near future.

ACKNOWLEDGMENTS

This work was supported by fellowships from the Beckman Foundation, Rita Allen Foundation, and Alfred P. Sloan Foundation to E.O.G.; by a grant from American Heart Association (14SDG17880015) to S.N.B.W.J.L. was supported by a grant from National Institutes of Health (T32 HG-3198-10). Correspondence should be addressed to E.O.G. (elena.gracheva@yale.edu) or S.N.B (sviatoslav.bagriantsev@yale.edu).

REFERENCES

Andersson, D. A., Gentry, C., Moss, S., & Bevan, S. (2008). Transient receptor potential A1 is a sensory receptor for multiple products of oxidative stress. *Journal of Neuroscience, 28*(10), 2485–2494.

Andersson, D. A., Gentry, C., Moss, S., & Bevan, S. (2009). Clioquinol and pyrithione activate TRPA1 by increasing intracellular Zn^{2+}. *Proceedings of the National Academy of Sciences of the United States of America, 106*(20), 8374–8379.

Bandell, M., Story, G. M., Hwang, S. W., Viswanath, V., Eid, S. R., Petrus, M. J., et al. (2004). Noxious cold ion channel TRPA1 is activated by pungent compounds and bradykinin. *Neuron, 41*(6), 849–857.

Basbaum, A. I., & Woolf, C. J. (1999). Pain. *Current Biology, 9*(12), R429–R431.

Bautista, D. M., Jordt, S. E., Nikai, T., Tsuruda, P. R., Read, A. J., Poblete, J., et al. (2006). TRPA1 mediates the inflammatory actions of environmental irritants and proalgesic agents. *Cell, 124*(6), 1269–1282.

Bautista, D. M., Movahed, P., Hinman, A., Axelsson, H. E., Sterner, O., Hogestatt, E. D., et al. (2005). Pungent products from garlic activate the sensory ion channel TRPA1.

Proceedings of the National Academy of Sciences of the United States of America, 102(34), 12248–12252.

Bautista, D. M., Pellegrino, M., & Tsunozaki, M. (2013). TRPA1: a gatekeeper for inflammation. *Annual Review of Physiology, 75*, 181–200.

Bautista, D. M., Siemens, J., Glazer, J. M., Tsuruda, P. R., Basbaum, A. I., Stucky, C. L., et al. (2007). The menthol receptor TRPM8 is the principle detector of environmental cold. *Nature, 448*, 204–208.

Bautista, D. M., Wilson, S. R., & Hoon, M. A. (2014). Why we scratch an itch: the molecules, cells and circuits of itch. *Nature Neuroscience, 17*(2), 175–182.

Benemei, S., Fusi, C., Trevisan, G., & Geppetti, P. (2013). The TRPA1 channel in migraine mechanism and treatment. *British Journal of Pharmacology, 171*(10), 2552–2567.

Bessac, B. F., & Jordt, S. E. (2010). Sensory detection and responses to toxic gases: mechanisms, health effects, and countermeasures. *Proceedings of the American Thoracic Society, 7*(4), 269–277.

Bessac, B. F., Sivula, M., von Hehn, C. A., Caceres, A. I., Escalera, J., Jordt, S. E., et al. (2009). Transient receptor potential ankyrin 1 antagonists block the noxious effects of toxic industrial isocyanates and tear gases. *FASEB Journal, 23*(4), 1102–1114.

Bessac, B. F., Sivula, M., von Hehn, C. A., Escalera, J., Cohn, L., Jordt, S. E., et al. (2008). TRPA1 is a major oxidant sensor in murine airway sensory neurons. *Journal of Clinical Investigation, 118*(5), 1899–1910.

Brone, B., Peeters, P. J., Marrannes, R., Mercken, M., Nuydens, R., Meert, T., et al. (2008). Tear gasses CN, CR, and CS are potent activators of the human TRPA1 receptor. *Toxicology and Applied Pharmacology, 231*(2), 150–156.

Cao, E., Cordero-Morales, J. F., Liu, B., Qin, F., & Julius, D. (2013). TRPV1 channels are intrinsically heat sensitive and regulated by phosphoinositide lipids. *Neuron, 77*(4), 667–679.

Caspani, O., & Heppenstall, P. A. (2009). TRPA1 and cold transduction: an unresolved issue? *Journal of General Physiology, 133*(3), 245–249.

Chatzigeorgiou, M., Yoo, S., Watson, J. D., Lee, W. H., Spencer, W. C., Kindt, K. S., et al. (2010). Specific roles for DEG/ENaC and TRP channels in touch and thermosensation in *C. elegans* nociceptors. *Nature Neuroscience, 13*(7), 861–868.

Chen, J., Joshi, S. K., DiDomenico, S., Perner, R. J., Mikusa, J. P., Gauvin, D. M., et al. (2011). Selective blockade of TRPA1 channel attenuates pathological pain without altering noxious cold sensation or body temperature regulation. *Pain, 152*, 1165–1172.

Chen, J., Kang, D., Xu, J., Lake, M., Hogan, J. O., Sun, C., et al. (2013). Species differences and molecular determinant of TRPA1 cold sensitivity. *Nature Communications, 4*, 2501.

Chowdhury, S., Jarecki, B. W., & Chanda, B. (2014). A molecular framework for temperature-dependent gating of ion channels. *Cell, 158*, 1148–1158.

Clapham, D. E., & Miller, C. (2011). A thermodynamic framework for understanding temperature sensing by transient receptor potential (TRP) channels. *Proceedings of the National Academy of Sciences of the United States of America, 108*(49), 19492–19497.

Cordero-Morales, J. F., Gracheva, E. O., & Julius, D. (2011). Cytoplasmic ankyrin repeats of transient receptor potential A1 (TRPA1) dictate sensitivity to thermal and chemical stimuli. *Proceedings of the National Academy of Sciences of the United States of America, 108*(46), E1184–E1191.

Dai, Y., Wang, S., Tominaga, M., Yamamoto, S., Fukuoka, T., Higashi, T., et al. (2007). Sensitization of TRPA1 by PAR2 contributes to the sensation of inflammatory pain. *Journal of Clinical Investigation, 117*(7), 1979–1987.

de Oliveira, C., Garami, A., Lehto, S. G., Pakai, E., Tekus, V., Pohoczky, K., et al. (2014). Transient receptor potential channel ankyrin-1 is not a cold sensor for autonomic thermoregulation in rodents. *Journal of Neuroscience, 34*(13), 4445–4452.

del Camino, D., Murphy, S., Heiry, M., Barrett, L. B., Earley, T. J., Cook, C. A., et al. (2010). TRPA1 contributes to cold hypersensitivity. *Journal of Neuroscience, 30*(45), 15165–15174.

Fajardo, O., Meseguer, V., Belmonte, C., & Viana, F. (2008). TRPA1 channels mediate cold temperature sensing in mammalian vagal sensory neurons: pharmacological and genetic evidence. *Journal of Neuroscience, 28*(31), 7863–7875.

Gawa, N. R., Bannon, A. W., Surapaneni, S., Hovland, D. N., Jr., Lehto, S. G., Gore, A., et al. (2007). The vanilloid receptor TRPV1 is tonically activated in vivo and involved in body temperature regulation. *Journal of Neuroscience, 27*(13), 3366–3374.

Gracheva, E. O., Ingolia, N. T., Kelly, Y. M., Cordero-Morales, J. F., Hollopeter, G., Chesler, A. T., et al. (2010). Molecular basis of infrared detection by snakes. *Nature, 464*(7291), 1006–1011.

Gu, Q., & Lin, R. L. (2010). Heavy metals zinc, cadmium, and copper stimulate pulmonary sensory neurons via direct activation of TRPA1. *Journal of Applied Physiology, 108*(4), 891–897 (1985).

Hinman, A., Chuang, H. H., Bautista, D. M., & Julius, D. (2006). TRP channel activation by reversible covalent modification. *Proceedings of the National Academy of Sciences of the United States of America, 103*(51), 19564–19568.

Hu, H., Bandell, M., Petrus, M. J., Zhu, M. X., & Patapoutian, A. (2009). Zinc activates damage-sensing TRPA1 ion channels. *Nature Chemical Biology, 5*(3), 183–190.

Jabba, S., Goyal, R., Sosa-Pagan, J. O., Moldenhauer, H., Wu, J., Kalmeta, B., et al. (2014). Directionality of temperature activation in mouse TRPA1 ion channel can be inverted by single-point mutations in ankyrin repeat six. *Neuron, 82*, 1–15.

Jaquemar, D., Schenker, T., & Trueb, B. (1999). An ankyrin-like protein with transmembrane domains is specifically lost after oncogenic transformation of human fibroblasts. *Journal of Biological Chemistry, 274*(11), 7325–7333.

Jara-Oseguera, A., & Islas, L. D. (2013). The role of allosteric coupling on thermal activation of thermo-TRP channels. *Biophysical Journal, 104*(10), 2160–2169.

Jordt, S. E., Bautista, D. M., Chuang, H. H., McKemy, D. D., Zygmunt, P. M., Hogestatt, E. D., et al. (2004). Mustard oils and cannabinoids excite sensory nerve fibres through the TRP channel ANKTM1. *Nature, 427*(6971), 260–265.

Julius, D. (2013). TRP channels and pain. *Annual Review of Cell and Developmental Biology, 29*, 355–384.

Kang, K., Panzano, V. C., Chang, E. C., Ni, L., Dainis, A. M., Jenkins, A. M., et al. (2012). Modulation of TRPA1 thermal sensitivity enables sensory discrimination in *Drosophila*. *Nature, 481*(7379), 76–80.

Kang, K., Pulver, S. R., Panzano, V. C., Chang, E. C., Griffith, L. C., Theobald, D. L., et al. (2010). Analysis of *Drosophila* TRPA1 reveals an ancient origin for human chemical nociception. *Nature, 464*(7288), 597–600.

Karashima, Y., Talavera, K., Everaerts, W., Janssens, A., Kwan, K. Y., Vennekens, R., et al. (2009). TRPA1 acts as a cold sensor in vitro and in vivo. *Proceedings of the National Academy of Sciences of the United States of America, 106*(4), 1273–1278.

Kim, D., Cavanaugh, E. J., & Simkin, D. (2008). Inhibition of transient receptor potential A1 channel by phosphatidylinositol-4,5-bisphosphate. *American Journal of Physiology: Cell Physiology, 295*(1), 92–99.

Kindt, K. S., Viswanath, V., Macpherson, L., Quast, K., Hu, H., Patapoutian, A., et al. (2007). *Caenorhabditis elegans* TRPA-1 functions in mechanosensation. *Nature Neuroscience, 10*(5), 568–577.

Knowlton, W. M., Bifolck-Fisher, A., Bautista, D. M., & McKemy, D. D. (2010). TRPM8, but not TRPA1, is required for neural and behavioral responses to acute noxious cold temperatures and cold-mimetics in vivo. *Pain, 150*(2), 340–350.

Kobayashi, K., Fukuoka, T., Obata, K., Yamanaka, H., Dai, Y., Tokunaga, A., et al. (2005). Distinct expression of TRPM8, TRPA1, and TRPV1 mRNAs in rat primary afferent neurons with adelta/c-fibers and colocalization with trk receptors. *Journal of Comparative Neurology, 493*(4), 596–606.

Kremeyer, B., Lopera, F., Cox, J. J., Momin, A., Rugiero, F., Marsh, S., et al. (2010). A gain-of-function mutation in TRPA1 causes familial episodic pain syndrome. *Neuron, 66*(5), 671–680.

Kunkler, P. E., Ballard, C. J., Oxford, G. S., & Hurley, J. H. (2011). TRPA1 receptors mediate environmental irritant-induced meningeal vasodilatation. *Pain, 152*(1), 38–44.

Kwan, K. Y., Allchorne, A. J., Vollrath, M. A., Christensen, A. P., Zhang, D. S., Woolf, C. J., et al. (2006). TRPA1 contributes to cold, mechanical, and chemical nociception but is not essential for hair-cell transduction. *Neuron, 50*(2), 277–289.

Kwan, K. Y., & Corey, D. P. (2009). Burning cold: Involvement of TRPA1 in noxious cold sensation. *Journal of Global Pharma Technology, 133*(3), 251–256.

Kwan, K. Y., Glazer, J. M., Corey, D. P., Rice, F. L., & Stucky, C. L. (2009). TRPA1 modulates mechanotransduction in cutaneous sensory neurons. *Journal of Neuroscience, 29*, 4808–4819.

Macpherson, L. J., Dubin, A. E., Evans, M. J., Marr, F., Schultz, P. G., Cravatt, B. F., et al. (2007). Noxious compounds activate TRPA1 ion channels through covalent modification of cysteines. *Nature, 445*(7127), 541–545.

Macpherson, L. J., Geierstanger, B. H., Viswanath, V., Bandell, M., Eid, S. R., Hwang, S., et al. (2005). The pungency of garlic: activation of TRPA1 and TRPV1 in response to allicin. *Current Biology, 15*(10), 929–934.

Macpherson, L. J., Xiao, B., Kwan, K. Y., Petrus, M. J., Dubin, A. E., Hwang, S., et al. (2007). An ion channel essential for sensing chemical damage. *Journal of Neuroscience, 27*(42), 11412–11415.

Mathie, A., Al-Moubarak, E., & Veale, E. L. (2010). Gating of two pore domain potassium channels. *Journal of Physiology, 588*, 3149–3156.

McKemy, D. D. (2013). The molecular and cellular basis of cold sensation. *ACS Chemical Neuroscience, 4*(2), 238–247.

McNamara, C. R., Mandel-Brehm, J., Bautista, D. M., Siemens, J., Deranian, K. L., Zhao, M., et al. (2007). TRPA1 mediates formalin-induced pain. *Proceedings of the National Academy of Sciences of the United States of America, 104*(33), 13525–13530.

Mishra, S. K., & Hoon, M. A. (2010). Ablation of TrpV1 neurons reveals their selective role in thermal pain sensation. *Molecular and Cellular Neuroscience, 43*(1), 157–163.

Miura, S., Takahashi, K., Imagawa, T., Uchida, K., Saito, S., Tominaga, M., et al. (2013). Involvement of TRPA1 activation in acute pain induced by cadmium in mice. *Molecular Pain, 9*, 7.

Nagata, K., Duggan, A., Kumar, G., & Garcia-Anoveros, J. (2005). Nociceptor and hair cell transducer properties of TRPA1, a channel for pain and hearing. *Journal of Neuroscience, 25*(16), 4052–4061.

Nilius, B., Appendino, G., & Owsianik, G. (2012). The transient receptor potential channel TRPA1: from gene to pathophysiology. *Pflugers Archiv, 464*(5), 425–458.

Nilius, B., Owsianik, G., & Voets, T. (2008). Transient receptor potential channels meet phosphoinositides. *EMBO Journal, 27*, 2809–2816.

Nilius, B., Prenen, J., & Owsianik, G. (2011). Irritating channels: the case of TRPA1. *Journal of Physiology, 589*(Pt 7), 1543–1549.

Oh, M. H., Oh, S. Y., Lu, J., Lou, H., Myers, A. C., Zhu, Z., et al. (2013). TRPA1-dependent pruritus in IL-13-induced chronic atopic dermatitis. *Journal of Immunology, 191*(11), 5371–5382.

Peyrot des Gachons, C., Uchida, K., Bryant, B., Shima, A., Sperry, J. B., Dankulich-Nagrudny, L., et al. (2011). Unusual pungency from extra-virgin olive oil is attributable to restricted spatial expression of the receptor of oleocanthal. *Journal of Neuroscience, 31*(3), 999–1009.

Pogorzala, L. A., Mishra, S. K., & Hoon, M. A. (2013). The cellular code for mammalian thermosensation. *Journal of Neuroscience, 33*(13), 5533–5541.

Prober, D. A., Zimmerman, S., Myers, B. R., McDermott, B. M., Jr., Kim, S. H., Caron, S., et al. (2008). Zebrafish TRPA1 channels are required for chemosensation but not for thermosensation or mechanosensory hair cell function. *Journal of Neuroscience, 28*(40), 10102–10110.

Qin, F. (2007). Regulation of TRP ion channels by phosphatidylinositol-4,5-bisphosphate. *High Energy Physics, 179*, 509–525.

Qin, F. (2013). Demystifying thermal channels: driving a channel both forwards and backwards with a single gear? *Biophysical Journal, 104*(10), 2118–2120.

Rohacs, T., Thyagarajan, B., & Lukacs, V. (2008). Phospholipase C mediated modulation of TRPV1 channels. *Molecular Neurobiology, 37*(2–3), 153–163.

Saito, S., Banzawa, N., Fukuta, N., Saito, C. T., Takahashi, K., Imagawa, T., et al. (2014). Heat and noxious chemical sensor, chicken TRPA1, as a target of bird repellents and identification of its structural determinants by multispecies functional comparison. *Molecular Biology and Evolution, 31*(3), 708–722.

Saito, S., Nakatsuka, K., Takahashi, K., Fukuta, N., Imagawa, T., Ohta, T., et al. (2012). Analysis of transient receptor potential ankyrin 1 (TRPA1) in frogs and lizards illuminates both nociceptive heat and chemical sensitivities and coexpression with TRP vanilloid 1 (TRPV1) in ancestral vertebrates. *Journal of Biological Chemistry, 287*(36), 30743–30754.

Salazar, M., Moldenhauer, H., & Baez-Nieto, D. (2011). Could an allosteric gating model explain the role of TRPA1 in cold hypersensitivity? *Journal of Neuroscience, 31*(15), 5554–5556.

Sawada, Y., Hosokawa, H., Hori, A., Matsumura, K., & Kobayashi, S. (2007). Cold sensitivity of recombinant TRPA1 channels. *Brain Research, 1160*, 39–46.

Schneider, C., Porter, N. A., & Brash, A. R. (2008). Routes to 4-hydroxynonenal: fundamental issues in the mechanisms of lipid peroxidation. *Journal of Biological Chemistry, 283*(23), 15539–15543.

Story, G. M., Peier, A. M., Reeve, A. J., Eid, S. R., Mosbacher, J., Hricik, T. R., et al. (2003). ANKTM1, a TRP-like channel expressed in nociceptive neurons, is activated by cold temperatures. *Cell, 112*(6), 819–829.

Takahashi, N., Kuwaki, T., Kiyonaka, S., Numata, T., Kozai, D., Mizuno, Y., et al. (2011). TRPA1 underlies a sensing mechanism for O_2. *Nature Chemical Biology, 7*(10), 701–711.

Takahashi, N., Mizuno, Y., Kozai, D., Yamamoto, S., Kiyonaka, S., Shibata, T., et al. (2008). Molecular characterization of TRPA1 channel activation by cysteine-reactive inflammatory mediators. *Channels (Austin), 2*(4), 287–298.

Taylor-Clark, T. E., Ghatta, S., Bettner, W., & Undem, B. J. (2009). Nitrooleic acid, an endogenous product of nitrative stress, activates nociceptive sensory nerves via the direct activation of TRPA1. *Molecular Pharmacology, 75*(4), 820–829.

Taylor-Clark, T. E., Kiros, F., Carr, M. J., & McAlexander, M. A. (2009). Transient receptor potential ankyrin 1 mediates toluene diisocyanate-evoked respiratory irritation. *American Journal of Respiratorty Cell and Molecular Biology, 40*(6), 756–762.

Taylor-Clark, T. E., Undem, B. J., Macglashan, D. W., Jr., Ghatta, S., Carr, M. J., McAlexander, M. A., et al. (2008). Prostaglandin-induced activation of nociceptive

neurons via direct interaction with transient receptor potential A1 (TRPA1). *Molecular Pharmacology, 73*(2), 274–281.
Trevisani, M., Siemens, J., Materazzi, S., Bautista, D. M., Nassini, R., Campi, B., et al. (2007). 4-Hydroxynonenal, an endogenous aldehyde, causes pain and neurogenic inflammation through activation of the irritant receptor TRPA1. *Proceedings of the National Academy of Sciences of the United States of America, 104*(33), 13519–13524.
Viswanath, V., Story, G. M., Peier, A. M., Petrus, M. J., Lee, V. M., Hwang, S. W., et al. (2003). Opposite thermosensor in fruitfly and mouse. *Nature, 423*(6942), 822–823.
Voets, T. (2012). Quantifying and modeling the temperature-dependent gating of TRP channels. *Reviews of physiology, biochemistry and pharmacology, 162*, 91–119.
Wang, Y. Y., Chang, R. B., Allgood, S. D., Silver, W. L., & Liman, E. R. (2011). A TRPA1-dependent mechanism for the pungent sensation of weak acids. *Journal of General Physiology, 137*(6), 493–505.
Wang, Y. Y., Chang, R. B., & Liman, E. R. (2010). TRPA1 is a component of the nociceptive response to CO_2. *Journal of Neuroscience, 30*(39), 12958–12963.
Wang, Y. Y., Chang, R. B., Waters, H. N., McKemy, D. D., & Liman, E. R. (2008). The nociceptor ion channel TRPA1 is potentiated and inactivated by permeating calcium ions. *Journal of Biological Chemistry, 283*(47), 32691–32703.
Wang, S., Dai, Y., Fukuoka, T., Yamanaka, H., Kobayashi, K., Obata, K., et al. (2008). Phospholipase C and protein kinase A mediate bradykinin sensitization of TRPA1: a molecular mechanism of inflammatory pain. *Brain, 131*(Pt 5), 1241–1251.
Wang, H., Schupp, M., Zurborg, S., & Heppenstall, P. A. (2013). Residues in the pore region of *Drosophila* transient receptor potential A1 dictate sensitivity to thermal stimuli. *Journal of Physiology, 591*(Pt 1), 185–201.
Wilson, S. R., Gerhold, K. A., Bifolck-Fisher, A., Liu, Q., Patel, K. N., Dong, X., et al. (2011). TRPA1 is required for histamine-independent, mas-related G protein-coupled receptor-mediated itch. *Nature Neuroscience, 14*(5), 595–602.
Xiao, R., Zhang, B., Dong, Y., Gong, J., Xu, T., Liu, J., et al. (2013). A genetic program promotes *C. elegans* longevity at cold temperatures via a thermosensitive TRP channel. *Cell, 152*(4), 806–817.
Xu, H., Delling, M., Jun, J. C., & Clapham, D. E. (2006). Oregano, thyme and clove-derived flavors and skin sensitizers activate specific TRP channels. *Nature Neuroscience, 9*(5), 628–635.
Yonemitsu, T., Kuroki, C., Takahashi, N., Mori, Y., Kanmura, Y., Kashiwadani, H., et al. (2013). TRPA1 detects environmental chemicals and induces avoidance behavior and arousal from sleep. *Scientific Reports, 3*, 3100.
Zakharian, E., Cao, C., & Rohacs, T. (2010). Gating of transient receptor potential melastatin 8 (TRPM8) channels activated by cold and chemical agonists in planar lipid bilayers. *Journal of Neuroscience, 30*, 12526–12534.
Zhong, L., Bellemer, A., Yan, H., Ken, H., Jessica, R., Hwang, R. Y., et al. (2012). Thermosensory and nonthermosensory isoforms of *Drosophila melanogaster* TRPA1 reveal heat-sensor domains of a thermoTRP Channel. *Cell Reports, 1*(1), 43–55.
Zhong, J., Minassi, A., Prenen, J., Taglialatela-Scafati, O., Appendino, G., Nilius, B., et al. (2011). Umbellulone modulates TRP channels. *Pflugers Archiv, 462*(6), 861–870.
Zhou, Y., Suzuki, Y., Uchida, K., & Tominaga, M. (2013). Identification of a splice variant of mouse TRPA1 that regulates TRPA1 activity. *Nature Communications, 4*, 2408.
Zurborg, S., Yurgionas, B., Jira, J. A., Caspani, O., & Heppenstall, P. A. (2007). Direct activation of the ion channel TRPA1 by Ca^{2+}. *Nature Neuroscience, 10*(3), 277–279.

CHAPTER FIVE

Temperature Sensitivity of Two-Pore (K_{2P}) Potassium Channels

Eve R. Schneider[1], Evan O. Anderson[1], Elena O. Gracheva[1,2] and Sviatoslav N. Bagriantsev[1,*]
[1]Department of Cellular & Molecular Physiology, Yale School of Medicine, New Haven, CT, USA
[2]Program in Cellular Neuroscience, Neurodegeneration and Repair, Yale School of Medicine, New Haven, CT, USA
*Corresponding author: E-mail: sviatoslav.bagriantsev@yale.edu

Contents

Abstract

At normal body temperature, the two-pore potassium channels TREK-1 (K_{2P}2.1/*KCNK2*), TREK-2 (K_{2P}10.1/*KCNK10*), and TRAAK (K_{2P}4.1/*KCNK2*) regulate cellular excitability by providing voltage-independent leak of potassium. Heat dramatically potentiates K_{2P} channel activity and further affects excitation. This review focuses on the current understanding of the physiological role of heat-activated K_{2P} current, and discusses the molecular mechanism of temperature gating in TREK-1, TREK-2, and TRAAK.

Current Topics in Membranes, Volume 74
ISSN: 1063-5823
http://dx.doi.org/10.1016/B978-0-12-800181-3.00005-1

1. INTRODUCTION

The two-pore potassium channels (K_{2P}) contribute to the generation of an electric potential on the plasma membrane by providing voltage-independent "leak" of K^+ ions (Enyedi & Czirjak, 2010). The K_{2P}s have a nonconventional topology: a mature channel is formed by two subunits, each containing two nonidentical pore-forming domains arranged in tandem (Figure 5.1). The K_{2P} channels of the TREK/TRAAK group, which includes TREK-1 (K_{2P}2.1, *KCNK2*), TREK-2 (K_{2P}10.1, *KCNK10*), and TRAAK (K_{2P}4.1, *KCNK4*), are expressed in various cell types, including neurons (Acosta et al., 2014; Fink et al., 1996; Kang & Kim, 2006; de la Pena et al., 2012; Talley, Solorzano, Lei, Kim, & Bayliss, 2001), cardiomyocytes (Xian Tao et al., 2006), and arterial myocytes (Bryan et al., 2006; Garry et al., 2007; Heyman et al., 2013). Temperature (Kang, Choe, & Kim, 2005; Maingret et al., 2000), mechanical force (Patel et al., 1998), general anesthetics (Patel et al., 1999), polyunsaturated fatty acids (Patel et al., 1998), and other compounds (Noel, Sandoz, & Lesage, 2011) potentiate the TREK/TRAAK-dependent background potassium efflux and suppress cellular excitability (Acosta et al., 2014; Dey et al., 2014).

TREK-1 is the most well-researched channel in the TREK/TRAAK group. Physiologically, TREK-1 contributes to the perception of temperature, pain, and mechanical force (Alloui et al., 2006; Noel et al., 2009; Plant, 2012); regulation of mood (Dominguez-Lopez, Howell, & Gobbi, 2012; Heurteaux et al., 2006; Kennard et al., 2005), anesthetic responses (Heurteaux et al., 2004); cardiac mechanoelectric feedback (Liu et al.,

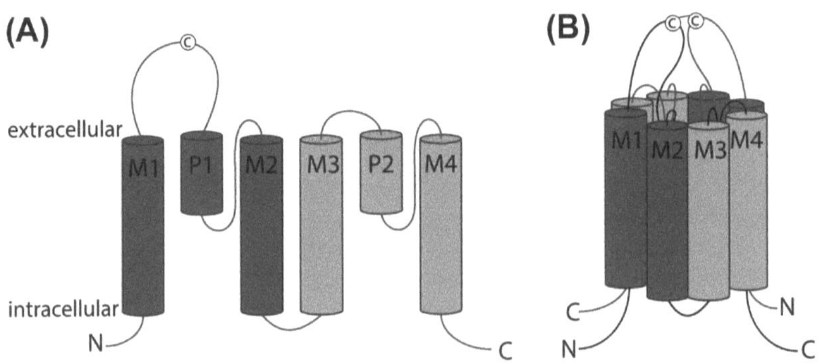

Figure 5.1 Membrane topology of K2P channels. (A) A topology diagram of a single K_{2P} subunit with two pore-forming domains. (B) A mature channel is formed by two subunits covalently linked via the cysteines (C) in the first extracellular loop. M1–M4, transmembrane segment 1–4; P1–2, pore helix 1–2. (See the color plate.)

2008); and vasodilation (Bryan et al., 2006; Garry et al., 2007). Recent studies have suggested unexpected roles for TREK-1 in glutamate conductance (Hwang et al., 2014; Woo et al., 2012) and regulation of blood–brain barrier permeability (Bittner et al., 2013). Channel activity has been linked to several pathological conditions, such as cardiac hypertrophy (Wang et al., 2013), ischemia (Heurteaux et al., 2004; Laigle, Confort-Gouny, Le Fur, Cozzone, & Viola, 2012; Wu et al., 2013), and myocardial infarction (Zhao, Fu, Gao, Xie, & Cao, 2011).

Perhaps one of the most intriguing features of the TREK/TRAAK channels is the robust sensitivity to heat (Kang et al., 2005; Maingret et al., 2000). This feature is not found in other K_{2P}s, which are either insensitive to heat, such as the TASK channels (Bagriantsev, Peyronnet, Clark, Honore, & Minor, 2011), or poorly sensitive to heat, such as THIK-1 (K_{2P}13.1, *KCNK13*) (Kang, Hogan, & Kim, 2013; Rajan et al., 2001). This review will focus on the physiological role and biophysical properties of the heat-evoked activity mediated by the TREK/TRAAK channels.

2. PHYSIOLOGICAL ROLE OF HEAT-ACTIVATED K_{2P} CHANNELS

TREK-1, TREK-2, and TRAAK activate over a broad temperature range: the channels are silent at 14 °C and reach maximum activity above 40 °C (Kang & Kim, 2006; Maingret et al., 2000) (Figure 5.2(A) and (B)). The channels are expressed in the bodies of somatosensory neurons (Alloui et al., 2006; Maingret et al., 2000; Yamamoto, Hatakeyama, & Taniguchi, 2009) where they are thought to control excitation through regulation of temperature-dependent potassium "leak" (Dobler et al., 2007; Kang & Kim, 2006; Kang et al., 2013).

Decreased potassium efflux is expected to cause depolarization and potentiation of excitability. Accordingly, the deletion of *KCNK2* and/or *KCNK4* in mice stimulates firing rate of heat-sensitive somatosensory C-fibers (Alloui et al., 2006) and potentiates heat sensitivity in behavioral tests (Noel et al., 2009). The expression pattern of TREK/TRAAK significantly overlaps with that of TRPV1 (Yamamoto et al., 2009), a heat-activated nonselective cation channel (Caterina et al., 1997) (Figure 5.2(A) and (B)), which is essential for physiological sensitivity to noxious temperatures above 50 °C in behavioral tests (Caterina et al., 2000; Davis et al., 2000; Park et al., 2011; Pogorzala, Mishra, & Hoon, 2013). It was proposed that in the heat-sensing somatosensory neurons, the depolarizing effect of

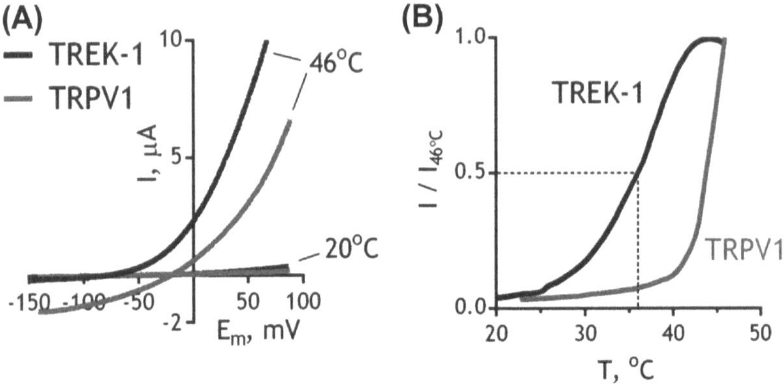

Figure 5.2 A comparison of temperature activation profiles between TREK-1 and TRPV1. (A) Current–voltage plots showing the activity of TREK-1 and TRPV1 recorded by two-electrode voltage clamp in *Xenopus* oocytes at different temperatures. Currents were evoked in a "physiological" solution (2 mM KCl, 96 mM NaCl, 1.8 mM $CaCl_2$, 2 mM $MgCl_2$, 5 mM HEPES pH 7.4) by a 1-s-long voltage ramp from a holding potential of −80 mV. (B) Normalized activity of TREK-1 and TRPV1 at different temperatures, measured at 40 mV.

TRPV1 activation is counterbalanced by the hyperpolarizing activity of TREK/TRAAK. In *KCNK2/4*-knockout mice the balance is shifted toward TRPV1 activity, leading to increased heat sensitivity (Alloui et al., 2006; Noel et al., 2009).

Similarly, TREK/TRAAK expression significantly overlaps with TRPM8 (Yamamoto et al., 2009), a nonselective cold-activated ion channel (McKemy, Neuhausser, & Julius, 2002; Peier et al., 2002) responsible for the detection of mild (nonnoxious) cold (Bautista et al., 2007; Colburn et al., 2007; Dhaka et al., 2007; Knowlton et al., 2013). While the deletion of *KCNK2* alone does not significantly affect cold responses in the sensory periphery (Alloui et al., 2006), the deletion of *KCNK4* or a combined deletion of *KCNK2* and *KCNK4* potentiates cold sensitivity of somatosensory neurons, and facilitates cold avoidance in behavioral tests (Descoeur et al., 2011; Noel et al., 2009). Thus, TREK-1 and TRAAK appear to modify both cold and warm perception, apparently via their effects on excitability of somatosensory neurons. To actively affect temperature sensation, TREK-1 and TRAAK should be expressed in the nerve terminals in the skin. In support of this, it was reported that the channels traffic along peripheral axons (Bearzatto, Lesage, Reyes, Lazdunski, & Laduron, 2000). However, a direct evidence for colocalization of TREK/TRAAK and TRPV1 or TRPM8 in afferent endings is, to our knowledge, missing.

Though explored in the sensory periphery, the molecular basis of thermosensitivity in other types of neurons is unclear. The wide neuronal distribution of TREK/TRAAK suggests that their heat-evoked activity may play a role in various regions of central and peripheral nervous systems (Maingret et al., 2000; Talley et al., 2001). The K_{2P}s were suggested to contribute to temperature-dependent neuronal excitability in the hippocampus (de la Pena et al., 2012), Grueneberg ganglion (Stebe, Schellig, Lesage, Breer, & Fleischer, 2013), and preoptic thermoregulatory area of the hypothalamus (Wechselberger, Wright, Bishop, & Boulant, 2006), a region that determines the set point for body temperature (Kobayashi, Hori, Matsumura, & Hosokawa, 2006; Zhao & Boulant, 2005).

It should be noted, however, that apart from their effects on thermosensitivity, the deletion of *KCNK2* and/or *KCNK4* produces a number of other striking phenotypes, including altered anesthetic (Heurteaux et al., 2004; Vallee, Rostain, & Risso, 2009) and mechanical sensitivity (Noel et al., 2009), increased susceptibility to epilepsy (Heurteaux et al., 2004) and decompression sickness (Vallee, Meckler, Risso, & Blatteau, 2012), and resistance to depression (Heurteaux et al., 2006). Unexpectedly, even though TREK-1 and TRAAK share overall topology, functional properties, and expression pattern (Medhurst et al., 2001; Talley et al., 2001), the deletion of *KCNK2* potentiates ischemia (Heurteaux et al., 2004), while the deletion of *KCNK4* protects against it (Laigle et al., 2012). Thus, the plethora of $KCNK2^{-/-}$ and $KCNK4^{-/-}$ phenotypes strongly suggests that TREK-1 and TRAAK are integral to a number of physiological processes. Therefore, cell-type-specific deletions of the *KCNK* genes will be essential to clarify the exact contribution of these channels in heat sensitivity.

3. MOLECULAR MECHANISM OF TEMPERATURE GATING OF TREK-1, TREK-2, AND TRAAK

3.1 Characteristics of temperature-activated K_{2P} current

Despite their importance for physiology, TREK-1, TREK-2, and TRAAK remain pharmacological orphans (Bagriantsev et al., 2013), which complicates their analysis in native cells. Most of our knowledge about temperature properties of these channels comes from heterologous systems, such as HEK293 and COS7 cells, and *Xenopus* oocytes (Bagriantsev, Clark, & Minor, 2012; Kang et al., 2005; Maingret et al., 2000). At room temperature, TREK-1 exhibits only background potassium leak, which increases with temperature, reaching maximum at ~42 °C (Figure 5.2). Interestingly, TREK-1 has its half-maximal temperature activation point ($T_{1/2}$) at ~37 °C, implying that the midpoint of the channel's dynamic range is centered on the homeostatic thermal set point for most mammals. If this property is maintained in native cells, then TREK-1 activity is set to be maximally sensitive to minute variations in physiological temperature.

The heat activation profile of TREK-1 is notably different from those exhibited by the members of the transient receptor potential family, such as TRPV1 (Caterina et al., 1997; Gracheva et al., 2011). The temperature–activity relationship for TRPV1 has a clearly identifiable inflection point at ~42 °C, after which the rate of change in current per degree Celsius dramatically increases. This point, often referred to as temperature activation threshold, is difficult to identify with regard to TREK-1 (Figure 5.2(B)), because the equilibrium between closed and open channels shifts over a much broader temperature range. An alternative way to define a temperature activation threshold is to define a point at which channel activity begins to significantly exceed background noise. This approach, however, is problematic with regard to TREK-1, because of the "leaky" nature of its current, which increases linearly with the number of channels on the surface. Another difference is in the Q_{10} value, which reports fold change in current amplitude over 10 °C. For TRPV1, Q_{10} estimates vary, depending on the expression system, but in most cases they are around 20 (Caterina et al., 1997; Gracheva et al., 2011), whereas the K_{2P}s exhibit a more modest change of about 10 (Bagriantsev et al., 2012; Kang et al., 2005; Maingret et al., 2000). In summary, even though TREK-1 is heat sensitive, it has a rather modest Q_{10} and does not have a clearly identifiable temperature activation threshold, at least when measured in heterologous expression systems.

3.2 Contribution of the extracellular C-type gate

Like most other potassium channels, K_{2P}s have the canonical Thr-X-Gly-Phe/Tyr-Gly ion selectivity sequence in the structure of the outer pore (Brohawn, del Marmol, & MacKinnon, 2012, 2013; Miller & Long, 2012). In K_{2P}s, the selectivity filter region is a key part of an extracellular gate, which is often referred to as "C-type"-like, because it functions in a way similar to the C-type inactivation gate of voltage-gated potassium channels (Zilberberg, Ilan, & Goldstein, 2001).

During activation, the C-type gate of TREK-1 undergoes structural rearrangements, becoming more potassium selective. Saturation of the selectivity filter with high concentration of extracellular potassium (150 mM) stabilizes the gate in an open conformation and renders it insensitive to gating by extracellular protons (Cohen, Ben-Abu, Hen, & Zilberberg, 2008). Using sensitivity to potassium as readout, it was established that the C-type gate mediates temperature sensitivity of TREK-1 and TREK-2, suggesting that the gating mechanism is conserved among heat-sensitive K_2ps (Bagriantsev et al., 2011).

Another important element of the outer gate is the pore helix, which plays a key role in maintaining proper conformation of the selectivity filter in various ion channel classes (Alagem, Yesylevskyy, & Reuveny, 2003; Cordero-Morales et al., 2006). Crystal structures of TWIK-1 and TRAAK revealed that the pore helix of the K_2ps is located in a typical orientation relative to the selectivity filter (Brohawn et al., 2012, Brohawn, Campbell, & Mackinnon, 2013; Miller & Long, 2012). Mutations in the pore helix dramatically affect gating of various K_2ps, including TWIK-1 (Chatelain et al., 2012), TREK-1, TASK-1, and TASK-3 (Bagriantsev et al., 2012). A G137I mutation in the pore helix 1 of TREK-1 (Figure 5.3) stabilizes the channel in an open, potassium-selective conformation and abrogates gating by temperature. The same effect can be achieved by saturating the selectivity filter with high concentrations of the permeant ion. Thus, temperature responses are significantly attenuated under the conditions of a rigid outer gate, regardless of whether this conformation is achieved by high concentration of extracellular potassium or via a mutation (Bagriantsev et al., 2011, 2012). These data strongly suggest that temperature affects TREK-1 and TREK-2 activity through opening or closing the C-type gate.

The importance of the C-type gate for temperature activation is in accord with the general role of this region in mediating TREK-1 responses to a broad spectrum of gating commands. In addition to temperature, the C-type gate mediates TREK-1 gating by intracellular pH (Piechotta et al., 2011), extracellular pH (Cohen et al., 2008; Ma, Yu et al., 2011; Sandoz, Douguet, Chatelain, Lazdunski, & Lesage, 2009), a small molecule activator ML67-33 (Bagriantsev et al., 2013), phosphorylation (Bagriantsev et al., 2012), and, possibly, membrane stretch (Bagriantsev et al., 2011; Piechotta et al., 2011). Therefore, in TREK-1, the C-type gate is the most crucial, and possibly the sole structural element that controls channel opening in response to different gating modalities, including temperature. In this regard, TREK-1 is similar to cyclic-nucleotide-gated channels, which rely almost exclusively on the extracellular C-type gate for function (Contreras, Srikumar, & Holmgren, 2008; Furini & Domene, 2011).

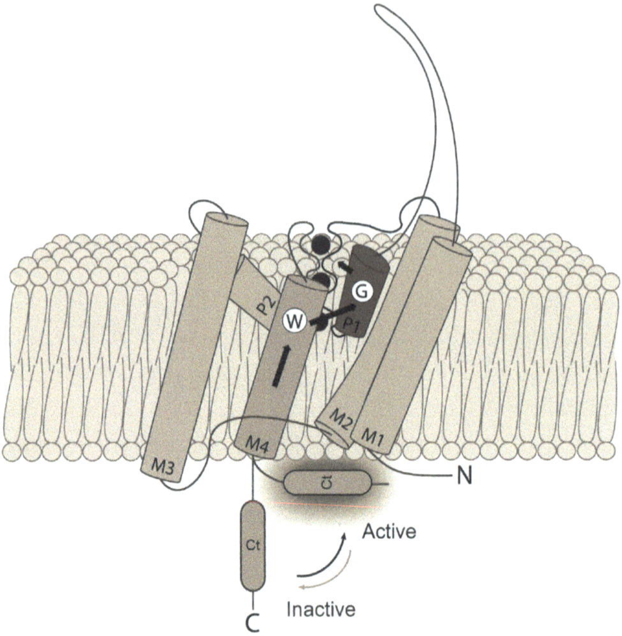

Figure 5.3 A hypothetical cartoon model of how Ct affects TREK-1 activity. A cartoon model of a single TREK-1 subunit showing a hypothetical mechanism of channel activation by temperature. It was proposed that increasing temperature facilitates the transition of Ct from inactive to active conformation, leading to stabilization of an open conformation of the selectivity filter via interaction between Trp275 (W) and Gly137 (G) from a single TREK-1 subunit (Bagriantsev et al., 2012). M1–M4, transmembrane segment 1–4; P1–2, pore helix 1–2; Ct, C-terminal domain. Blue spheres depict potassium ions. (See the color plate.)

3.3 Contribution of the intracellular bundle crossing region

Activity of many potassium channels depends on the movement of the lower activation gate, known as the bundle crossing. The gate is formed by the C-terminal regions of the "inner" transmembrane helixes, topologically corresponding to the M2 and M4 segments of the K_{2P}s. The bundle crossing region may play a role in the gating of TASK-2 (Niemeyer, Cid, Pena-Munzenmayer, & Sepulveda, 2010) and the fly K_{2P} KCNK0 (Ben-Abu, Zhou, Zilberberg, & Yifrach, 2009), but its role in the gating of TREK-1, TREK-2, or TRAAK by temperature or other modalities remains unconfirmed.

Several studies have inquired into the role of bundle crossing in heat-sensitive K_{2P} channel function. Random mutagenesis of TREK-1 identified gain-of-function mutations in various channel regions, except in the C-terminal parts of the M2 and M4 transmembrane helices, suggesting that the potential lower gate exists in a predominantly open conformation (Bagriantsev et al., 2011). An elegant study established that the TREK-1 blocker tetrahexylammonium has unobstructed access to its binding site immediately below the selectivity filter even when the channel is closed (Piechotta et al., 2011). This work demonstrated the absence of a physical barrier between the cytosol and the outer gate, which further supports the idea that the potential bundle crossing region is locked in an open conformation and does not regulate ion flow.

Finally, a crystal structure of TRAAK revealed an opening between the inner helixes measuring ~10 Å (Brohawn et al., 2012), which is a wider opening than in open-state voltage-gated potassium channel structure (Long, Tao, Campbell, & MacKinnon, 2007). Thus, the existence of a functional bundle crossing region in heat-activated K_{2P}s has not yet found experimental support. Though it remains possible that the inner helices of the K_{2P}s may form a gate under certain conditions, the selectivity-filter-based extracellular C-type gate remains the only confirmed physical element that mediates temperature sensitivity of heat-activated K_{2P} channels.

3.4 Contribution of the intracellular C-terminal domain

The C-terminal domain (Ct) is a major intracellular region of TREK-1 that mediates the reception of a number of regulatory commands (Noel et al., 2011). Deletion of Ct produces a striking effect on TREK-1, leading to suppression of basal activity (Patel et al., 1998) and decreased sensitivity to intracellular protons, mechanical force (Maingret, Patel, Lesage, Lazdunski, & Honore, 1999), and temperature (Maingret et al., 2000). These studies have established a key role for Ct in regulation of various modulatory responses, including sensitivity to heat.

However, the plethora of observed effects suggests that a complete deletion of Ct may produce a global impact on the TREK-1 molecule. A more subtle way to probe the importance of Ct for function is to decouple it from the pore-forming domain by introducing a flexible linker at the junction between M4 and Ct. This strategy has been successfully used to decouple cross-talk between functional domains of various classes of ion channels (Findeisen & Minor, 2009; Su, Anishkin, Kung, & Saimi,

2011). With regard to TREK-1, a triple-glycine (3G) or a triple-alanine (3A) linker introduced between M4 and Ct renders the channel insensitive to metabolic stimuli that converge on Ct, suggesting that the mutations obliterate functional communication between Ct and the gating apparatus. Importantly, the C-type gate remains functional in these mutants, as determined by measuring sensitivity to extracellular pH. At the same time, TREK-1 3G and 3A mutants become insensitive to heat (Bagriantsev et al., 2012), suggesting that functional coupling between Ct and the C-type gate is essential for heat sensitivity. These experiments have two major implications. First, they show that Ct is critical for normal heat sensitivity of TREK-1. Second, they clarify the role of the C-type gate: even though the gate mediates channel responses to heat, it lacks robust intrinsic temperature sensitivity that transforms into function.

While temperature almost certainly produces conformational changes throughout the channel, in some regions these changes may have more profound functional implications. The absence of temperature responses in TREK-1 3G and 3A mutants strongly suggest that heat exerts only minimal functional impact on the gate. Instead, current data are most consistent with the idea that temperature affects the intracellular Ct domain, which then potentiates opening of the heat-insensitive C-type gate via an allosteric mechanism (Bagriantsev et al., 2012). This mechanism is in accord with the general gating paradigm established for TREK-1, whereby different sensory elements, such as the extracellular proton sensor His126 (Cohen et al., 2008) or the polymodal intracellular sensor Ct, affect channel function by converging on a common C-type-like extracellular gate (Bagriantsev et al., 2011).

How Ct senses temperature remains unclear. It was proposed (Chemin et al., 2005) that regulatory factors that converge on Ct, such as phospholipids (Lopes et al., 2005), polyunsaturated fatty acids (Patel et al., 1998), intracellular pH (Honore, Maingret, Lazdunski, & Patel, 2002), and phosphorylation (Murbartian, Lei, Sando, & Bayliss, 2005), affect TREK-1 via modulating the affinity of Ct to the phospholipids in the inner leaflet of the plasma membrane. Optical probing of GFP-tagged Ct by total internal reflection fluorescence microscopy provided strong support for this hypothesis by showing that increased association of Ct with the plasma membrane correlates with TREK-1 activation, while dissociation leads to channel inhibition (Sandoz, Bell, & Isacoff, 2011). The dynamic interaction between Ct and the plasma membrane appears to mediate channel responses to a broad range of stimuli, including protons, various metabolites, and the antidepressant drug fluoxetine (Prozac) (Kennard et al., 2005; Sandoz et al., 2011). It is therefore possible, but not confirmed, that heat affects the channel via a similar mechanism, whereby increased temperature facilitates the association between Ct and plasma membrane, leading to channel opening via an allosteric effect on the extracellular C-type gate (Figure 5.3).

3.5 The mechanism connecting the heat-sensing and gating domains of TREK-1

The topological localization of the Ct and the C-type gate of TREK-1 on the opposite sides of the plasma membrane necessitates the existence of a mechanism connecting the domains. Random site-directed mutagenesis identified the N-terminal (near-extracellular) region of M4 as a key element controlling the C-type gate (Bagriantsev et al., 2011). Within this region, the site corresponding to Trp275 of TREK-1 is critical for gating various K_{2P} channels, including TWIK-1 (Chatelain et al., 2012), TREK-2, TASK-1, TASK-2, and TASK-3 (Bagriantsev et al., 2011). Mutagenic analysis showed that substitution of Trp275 with serine, or another amino acid with smaller side chain, stabilizes the selectivity filter of TREK-1 in an open, potassium-selective conformation, causing significant attenuation of temperature responses.

The determination of TRAAK crystal structure showed that the N-terminal portion of M4 is tightly packed against the pore helix 1, and that the side chain of Trp262 (Trp275 in TREK-1) extends to Gly124, a position equivalent to Gly137 in TREK-1 (Brohawn et al., 2012). The close opposition of the two key regulators of the C-type gate provides a mechanistic explanation for how the N-terminal segment of M4 could transmit structural deformations of Ct to the gate. In another high-resolution TRAAK structure, the M4 segment has moved, leading to a rotamer switch of the side chain of Trp262 away from Gly124 and the disappearance of a tight interaction between the N-terminal segment of M4 and pore helix 1 (Brohawn et al., 2013). Thus, functional and structural data suggest that the C-type gate and Ct can be coupled via M4. In this model, a shift in the dynamic equilibrium between membrane-bound and dissociated conformations of Ct can displace M4, causing significant changes around the selectivity filter and thus affecting channel function (Figure 5.3).

A functional TREK-1 molecule contains two Cts. While each Ct is connected to the pore via the M4, it remained unclear whether each Ct affects the gate via the *cis*- or *trans*-M4 segment (or both). This question was resolved using TREK-1 concatamers bearing activating (E306A) and decoupling (3G) mutations in the same or neighboring Ct. The E306A mutation stimulates the association of Ct with the plasma membrane (Chemin et al., 2005; Sandoz et al., 2011), and leads to channel activation via stabilization of the C-type gate (Bagriantsev et al., 2012; Piechotta et al., 2011). When present in only one Ct of a TREK-1 concatamer, E306A causes half-maximal stabilization of the gate. This effect can be eliminated by decoupling the mutated, but not the neighboring wild-type, Ct with a 3G mutation (Bagriantsev et al., 2012). These data showed that Cts act independently, and that the mechanism coupling each Ct and the C-type gate involves M4 segments of the same subunit, i.e., in *cis* configuration.

While selective activation of only one Ct in a TREK-1 by temperature is experimentally challenging, it is possible to compare temperature responses of TREK-1 concatamers with one Ct decoupled from the pore. Experiments showed that decoupling of any one Ct significantly attenuates temperature response, but does not eliminate it. Thus, while both C termini are required to achieve full temperature activation, a partial effect can be achieved with only one Ct (Bagriantsev et al., 2012). The similarity between the effects of E306A and temperature suggest that TREK-1 activation through Ct proceeds through a similar mechanism, regardless of the nature of the activating stimulus.

4. HEAT- AND MECHANOSENSITIVITY OF K_{2P}S: DIFFERENT FACETS OF THE SAME PROCESS?

Heat-sensitive K_{2P}s are truly polymodal ion channels. In addition to temperature, intra- and extracellular protons, fatty acids, phospholipids, and various small molecule compounds, TREK-1, TREK-1, and TRAAK are potently activated by membrane stretch (Bang, Kim, & Kim, 2000; Honore, Patel, Chemin, Suchyna, & Sachs, 2006; Lesage, Terrenoire, Romey, & Lazdunski, 2000; Maingret, Fosset, Lesage, Lazdunski, & Honore, 1999, Maingret, Patel et al., 1999; Patel et al., 1998). Mutations that affect Ct function (Honore et al., 2002; Maingret, Patel et al., 1999; Patel et al., 1998) or stabilize the C-type gate (Bagriantsev et al., 2011) attenuate TREK-1 mechanosensitivity, highlighting a general importance of these regions for TREK-1 gating. Even though the exact role of these domains in mechanosensitivity requires clarification, it is tempting to speculate that temperature and mechanical force activate TREK-1 via similar mechanisms, whereby Ct serves as a mechanosensor and affects channel function via the C-type gate.

The exact structural changes that accompany temperature-evoked gating of the K_{2}ps and other ion channels, are unclear (Clapham & Miller, 2011; Chowdhury et al., 2014). Significant advances have been made with regard to understanding how ion channels are gated by mechanical force, but this largely pertains to prokaryotic channels, such as MscS and MscL. In these channels, membrane stretch invokes drastic structural changes, leading to stabilization of an open conformation with the pore large enough to allow passage of not only ions, but also bulky metabolites (Sukharev & Sachs, 2012). Mammalian mechanogated ion channels, such as the K_{2}ps, Piezo1 and 2 (Coste et al., 2010, Coste et al., 2012), and others (Arnadottir & Chalfie, 2010; Delmas, Hao, & Rodat-Despoix, 2011), exhibit significant ion (or at least charge) preference and presumably require relatively subtle structural perturbations that do not profoundly impact selectivity. This assumption is exemplified in the case of TREK-1 and TRAAK, which remain potassium selective under both low and high membrane tension (Brohawn, Su, & Mackinnon, 2014). The actual gating mechanism(s) that transform membrane tension into channel opening is unclear, but it may, similar to heat-evoked gating, involve stabilization of the C-type gate. It is important to note that such mechanism would not require any additional components other than the lipid membrane, as purified K_{2}ps exhibit robust mechanosensitivity in artificial membranes (Berrier et al., 2013; Brohawn et al., 2014). The emerging similarity in the mechanism of heat- and mechanoactivation of K_{2}ps suggests that these studies may eventually converge to provide a unifying explanation of how the apparently different physical cues affect channel function.

5. FUTURE STUDIES OF K_{2P} CHANNEL THERMAL SENSITIVITY

5.1 Are K_{2P}s intrinsically heat sensitive?

Among key unsolved questions is the requirement of intracellular factors for heat sensitivity of TREK-1 and other K_{2}ps. If such a component (or components) exists, it should be ubiquitous and evolutionarily conserved, as TREK-1 is robustly temperature sensitive in amphibian and mammalian cells (Bagriantsev et al., 2011, 2012; Kang et al., 2005; Maingret et al., 2000). Patch excision was reported to obliterate heat sensitivity of the K_{2}ps in COS7 cells (Kang et al., 2005; Maingret et al., 2000), demonstrating that cell integrity is crucial for temperature activation. Possibly, patch excision leads to dissociation of a cytosolic component (Noel et al., 2011) which interacts with TREK-1 and which is mandatory for heat sensitivity, while the channel itself lacks intrinsic thermal properties. To demonstrate this convincingly would require investigation of thermal properties of purified ion channels in artificial membranes. Recently, this method was used to demonstrate that the nonselective heat-activated ion channel TRPV1 does not require additional proteinaceous components for temperature response (Cao, Cordero-Morales, Liu, Qin, & Julius, 2013).

5.2 What are the physiological roles of the heat-activated K_{2P} current?

In addition to the proposed role in somatosensory heat responses, temperature-activated K_{2P} channels were suggested to contribute to temperature-dependent excitability of the neurons in the hippocampus (de la Pena et al., 2012), Grueneberg ganglion (Stebe et al., 2013), and preoptic area (POA) of the hypothalamus (Wechselberger et al., 2006). The thermoregulatory region of POA contains numerous thermosensitive neurons which are thought to act as sensors for internal temperature. Temperature increases firing rate of a subpopulation of POA neurons (Zhao & Boulant, 2005) through an unknown mechanism. Interestingly, the half-maximal temperature activation point ($T_{1/2}$) of TREK-1 expressed in *Xenopus* oocytes is very close to normal body temperature. While temperature properties of K_{2P} channels in native and heterologous systems may differ, it is worth noting that since temperature changes produce the most significant effect precisely at $T_{1/2}$, the heat-activated K_{2P}s are uniquely positioned to sense minute variations of body temperature. In this regard, it is interesting to consider the contribution of heat-evoked potassium current to the physiology of cells and tissues where TREK-1 or other K_{2P}s have been shown to play functional roles, such as in lung epithelial cells (Davis & Cowley, 2006; Roan, Waters, Teng, Ghosh, & Schwingshackl, 2014; Schwingshackl, Teng, Ghosh, & Waters, 2013), myometrium (Heyman et al., 2013; Wu, Singer, & Buxton, 2012), and endothelium of blood vessels (Bittner et al., 2013; Garry et al., 2007; Namiranian et al., 2010).

While it is well established that K_{2P}s mediate potassium efflux, several reports showed that potassium selectivity of TREK-1 (Bagriantsev et al., 2011; Cohen et al., 2008; Thomas, Plant, Wilkens, McCrossan, & Goldstein, 2008) and other K_{2P}s (Chatelain et al., 2012; Ma, Zhang, & Chen, 2011) can be significantly compromised, leading to increased permeability to sodium. Thus, available evidence suggests that K_{2P}s may lose selectivity to potassium under physiologically relevant conditions. Hypothetically, a near-complete loss of cation selectivity may turn TREK-1/-2/TRAAK into a heat-activated excitatory ion channel similar to TRPV1, i.e., its own functional antipode.

Finally, we note that the striking dependence of TREK-1/-2/TRAAK activity on temperature indicates that this factor must be taken into account in any study investigating K_{2P} channel function in physiological context. As discussed above, multiple modalities that regulate K_{2P} channel function act via the extracellular C-type gate. Therefore, studying the activity of heat-sensitive K_{2P}s at room temperature, i.e., under conditions when the gate does not receive the tonic activating stimulus it receives at 37 °C, will complicate the analysis of the physiological role of these channels.

ACKNOWLEDGMENTS

We thank Willem Laursen for comments on the manuscript. This work was supported by a grant from the American Heart Association (14SDG17880015) to S.N.B., and by fellowships from the Beckman Foundation and Alfred P. Sloan Foundation to E.O.G. Correspondence should be addressed to S.N.B. (sviatoslav.bagriantsev@yale.edu).

REFERENCES

Acosta, C., Djouhri, L., Watkins, R., Berry, C., Bromage, K., & Lawson, S. N. (2014). TREK2 expressed selectively in IB4-binding C-fiber nociceptors hyperpolarizes their membrane potentials and limits spontaneous pain. *Journal of Neuroscience, 34*, 1494–1509.

Alagem, N., Yesylevskyy, S., & Reuveny, E. (2003). The pore helix is involved in stabilizing the open state of inwardly rectifying K^+ channels. *Biophysical Journal, 85*, 300–312.

Alloui, A., Zimmermann, K., Mamet, J., Duprat, F., Noel, J., Chemin, J., et al. (2006). TREK-1, a K+ channel involved in polymodal pain perception. *EMBO Journal, 25*, 2368–2376.

Arnadottir, J., & Chalfie, M. (2010). Eukaryotic mechanosensitive channels. *Annual Review of Biophysics, 39*, 111–137.

Bagriantsev, S. N., Ang, K. H., Gallardo-Godoy, A., Clark, K. A., Arkin, M. R., Renslo, A. R., et al. (2013). A high-throughput functional screen identifies small molecule regulators of temperature- and mechano-sensitive K_{2P} channels. *ACS Chemical Biology, 8*, 1841–1851.

Bagriantsev, S. N., Clark, K. A., & Minor, D. L., Jr., (2012). Metabolic and thermal stimuli control K(2P)2.1 (TREK-1) through modular sensory and gating domains. *EMBO Journal, 31*, 3297–3308.

Bagriantsev, S. N., Peyronnet, R., Clark, K. A., Honore, E., & Minor, D. L., Jr., (2011). Multiple modalities converge on a common gate to control K(2P) channel function. *EMBO Journal, 30*, 3594–3606.

Bang, H., Kim, Y., & Kim, D. (2000). TREK-2, a new member of the mechanosensitive tandem-pore K^+ channel family. *Journal of Biological Chemistry, 275*, 17412–17419.

Bautista, D. M., Siemens, J., Glazer, J. M., Tsuruda, P. R., Basbaum, A. I., Stucky, C. L., et al. (2007). The menthol receptor TRPM8 is the principal detector of environmental cold. *Nature, 448*, 204–208.

Bearzatto, B., Lesage, F., Reyes, R., Lazdunski, M., & Laduron, P. M. (2000). Axonal transport of TREK and TRAAK potassium channels in rat sciatic nerves. *Neuroreport, 11*, 927–930.

Ben-Abu, Y., Zhou, Y., Zilberberg, N., & Yifrach, O. (2009). Inverse coupling in leak and voltage-activated K^+ channel gates underlies distinct roles in electrical signaling. *Nature Structural & Molecular Biology, 16*, 71–79.

Berrier, C., Pozza, A., de Lacroix de Lavalette, A., Chardonnet, S., Mesneau, A., Jaxel, C., et al. (2013). The purified mechanosensitive channel TREK-1 is directly sensitive to membrane tension. *Journal of Biological Chemistry, 288*, 27307–27314.

Bittner, S., Ruck, T., Schuhmann, M. K., Herrmann, A. M., Maati, H. M., Bobak, N., et al. (2013). Endothelial TWIK-related potassium channel-1 (TREK1) regulates immune-cell trafficking into the CNS. *Nature Medicine, 19*, 1161–1165.

Brohawn, S. G., Campbell, E. B., & MacKinnon, R. (2013). Domain-swapped chain connectivity and gated membrane access in a Fab-mediated crystal of the human TRAAK K+ channel. *Proc Natl Acad Sci U S A, 110*, 2129–2134.

Brohawn, S. G., del Marmol, J., & MacKinnon, R. (2012). Crystal structure of the human K_{2P} TRAAK, a lipid- and mechano-sensitive K^+ ion channel. *Science, 335*, 436–441.

Brohawn, S. G., Su, Z., & MacKinnon, R. (2014). Mechanosensitivity is mediated directly by the lipid membrane in TRAAK and TREK1 K+ channels. *Proc Natl Acad Sci U S A, 111*, 3614–3619.

Bryan, R. M., Jr., You, J., Phillips, S. C., Andresen, J. J., Lloyd, E. E., Rogers, P. A., et al. (2006). Evidence for two-pore domain potassium channels in rat cerebral arteries. *American Journal of Physiology - Heart and Circulatory Physiology, 291*, H770–H780.

Cao, E., Cordero-Morales, J. F., Liu, B., Qin, F., & Julius, D. (2013). TRPV1 channels are intrinsically heat sensitive and negatively regulated by phosphoinositide lipids. *Neuron, 77*, 667–679.

Caterina, M. J., Leffler, A., Malmberg, A. B., Martin, W. J., Trafton, J., Petersen-Zeitz, K. R., et al. (2000). Impaired nociception and pain sensation in mice lacking the capsaicin receptor. *Science, 288*, 306–313.

Caterina, M. J., Schumacher, M. A., Tominaga, M., Rosen, T. A., Levine, J. D., & Julius, D. (1997). The capsaicin receptor: a heat-activated ion channel in the pain pathway. *Nature, 389*, 816–824.

Chatelain, F. C., Bichet, D., Douguet, D., Feliciangeli, S., Bendahhou, S., Reichold, M., et al. (2012). TWIK1, a unique background channel with variable ion selectivity. *Proceedings of the National Academy of Sciences of the United States of America, 109*, 5499–5504.

Chemin, J., Patel, A. J., Duprat, F., Lauritzen, I., Lazdunski, M., & Honore, E. (2005). A phospholipid sensor controls mechanogating of the K^+ channel TREK-1. *EMBO Journal, 24*, 44–53.

Chowdhury, S., Jarecki, B. W., & Chanda, B. (2014). A molecular framework for temperature-dependent gating of ion channels. *Cell, 158*, 1148–1158.

Clapham, D. E., & Miller, C. (2011). A thermodynamic framework for understanding temperature sensing by transient receptor potential (TRP) channels. *Proceedings of the National Academy of Sciences of the United States of America, 108*, 19492–19497.

Cohen, A., Ben-Abu, Y., Hen, S., & Zilberberg, N. (2008). A novel mechanism for human K2P2.1 channel gating. Facilitation of C-type gating by protonation of extracellular histidine residues. *Journal of Biological Chemistry, 283*, 19448–19455.

Colburn, R. W., Lubin, M. L., Stone, D. J., Jr., Wang, Y., Lawrence, D., D'Andrea, M. R., et al. (2007). Attenuated cold sensitivity in TRPM8 null mice. *Neuron, 54*, 379–386.

Contreras, J. E., Srikumar, D., & Holmgren, M. (2008). Gating at the selectivity filter in cyclic nucleotide-gated channels. *Proceedings of the National Academy of Sciences of the United States of America, 105*, 3310–3314.

Cordero-Morales, J. F., Cuello, L. G., Zhao, Y., Jogini, V., Cortes, D. M., Roux, B., et al. (2006). Molecular determinants of gating at the potassium-channel selectivity filter. *Nature Structural and Molecular Biology, 13*, 311–318.

Coste, B., Mathur, J., Schmidt, M., Earley, T. J., Ranade, S., Petrus, M. J., et al. (2010). Piezo1 and Piezo2 are essential components of distinct mechanically activated cation channels. *Science, 330*, 55–60.

Coste, B., Xiao, B., Santos, J. S., Syeda, R., Grandl, J., Spencer, K. S., et al. (2012). Piezo proteins are pore-forming subunits of mechanically activated channels. *Nature, 483*, 176–181.

Davis, K. A., & Cowley, E. A. (2006). Two-pore-domain potassium channels support anion secretion from human airway Calu-3 epithelial cells. *Pflugers Archiv, 451*, 631–641.

Davis, J. B., Gray, J., Gunthorpe, M. J., Hatcher, J. P., Davey, P. T., Overend, P., et al. (2000). Vanilloid receptor-1 is essential for inflammatory thermal hyperalgesia. *Nature, 405*, 183–187.

de la Pena, E., Malkia, A., Vara, H., Caires, R., Ballesta, J. J., Belmonte, C., et al. (2012). The influence of cold temperature on cellular excitability of hippocampal networks. *PLoS One, 7*, e52475.

Delmas, P., Hao, J., & Rodat-Despoix, L. (2011). Molecular mechanisms of mechanotransduction in mammalian sensory neurons. *Nature Reviews Neuroscience, 12*, 139–153.

Descoeur, J., Pereira, V., Pizzoccaro, A., Francois, A., Ling, B., Maffre, V., et al. (2011). Oxaliplatin-induced cold hypersensitivity is due to remodelling of ion channel expression in nociceptors. *EMBO Molecular Medicine, 3*, 266–278.

Dey, D., Eckle, V. S., Vitko, I., Sullivan, K. A., Lasiecka, Z. M., Winckler, B., et al. (2014). A potassium leak channel silences hyperactive neurons and ameliorates status epilepticus. *Epilepsia, 55*, 203–213.

Dhaka, A., Murray, A. N., Mathur, J., Earley, T. J., Petrus, M. J., & Patapoutian, A. (2007). TRPM8 is required for cold sensation in mice. *Neuron, 54*, 371–378.

Dobler, T., Springauf, A., Tovornik, S., Weber, M., Schmitt, A., Sedlmeier, R., et al. (2007). TRESK two-pore-domain K^+ channels constitute a significant component of background potassium currents in murine dorsal root ganglion neurons. *Journal of Physiology, 585*, 867–879.

Dominguez-Lopez, S., Howell, R., & Gobbi, G. (2012). Characterization of serotonin neurotransmission in knockout mice: implications for major depression. *Reviews in the Neurosciences, 23*, 429–443.

Enyedi, P., & Czirjak, G. (2010). Molecular background of leak K+ currents: two-pore domain potassium channels. *Physiological Reviews, 90*, 559–605.

Findeisen, F., & Minor, D. L., Jr. (2009). Disruption of the IS6-AID linker affects voltage-gated calcium channel inactivation and facilitation. *Journal of General Physiology, 133*, 327–343.

Fink, M., Duprat, F., Lesage, F., Reyes, R., Romey, G., Heurteaux, C., et al. (1996). Cloning, functional expression and brain localization of a novel unconventional outward rectifier K^+ channel. *EMBO Journal, 15*, 6854–6862.

Furini, S., & Domene, C. (2011). Gating at the selectivity filter of ion channels that conduct Na^+ and K^+ ions. *Biophysical Journal, 101*, 1623–1631.

Garry, A., Fromy, B., Blondeau, N., Henrion, D., Brau, F., Gounon, P., et al. (2007). Altered acetylcholine, bradykinin and cutaneous pressure-induced vasodilation in mice lacking the TREK1 potassium channel: the endothelial link. *EMBO Reports, 8*, 354–359.

Gracheva, E. O., Cordero-Morales, J. F., Gonzalez-Carcacia, J. A., Ingolia, N. T., Manno, C., Aranguren, C. I., et al. (2011). Ganglion-specific splicing of TRPV1 underlies infrared sensation in vampire bats. *Nature, 476*, 88–91.

Heurteaux, C., Guy, N., Laigle, C., Blondeau, N., Duprat, F., Mazzuca, M., et al. (2004). TREK-1, a K^+ channel involved in neuroprotection and general anesthesia. *EMBO Journal, 23*, 2684–2695.

Heurteaux, C., Lucas, G., Guy, N., El Yacoubi, M., Thummler, S., Peng, X. D., et al. (2006). Deletion of the background potassium channel TREK-1 results in a depression-resistant phenotype. *Nature Neuroscience, 9*, 1134–1141.

Heyman, N. S., Cowles, C. L., Barnett, S. D., Wu, Y. Y., Cullison, C., Singer, C. A., Leblanc, N., & Buxton, I. L. (2013). TREK-1 currents in smooth muscle cells from pregnant human myometrium. *Am J Physiol Cell Physiol, 305*, C632–642.

Honore, E., Maingret, F., Lazdunski, M., & Patel, A. J. (2002). An intracellular proton sensor commands lipid- and mechano-gating of the K(+) channel TREK-1. *EMBO Journal, 21*, 2968–2976.

Honore, E., Patel, A. J., Chemin, J., Suchyna, T., & Sachs, F. (2006). Desensitization of mechano-gated K_{2P} channels. *Proceedings of the National Academy of Sciences of the United States of America, 103*, 6859–6864.

Hwang, E. M., Kim, E., Yarishkin, O., Woo, D. H., Han, K. S., Park, N., et al. (2014). A disulphide-linked heterodimer of TWIK-1 and TREK-1 mediates passive conductance in astrocytes. *Nature Communications, 5*, 3227.

Kang, D., Choe, C., & Kim, D. (2005). Thermosensitivity of the two-pore domain K^+ channels TREK-2 and TRAAK. *Journal of Physiology, 564*, 103–116.

Kang, D., Hogan, J. O., & Kim, D. (2014). THIK-1 (K2P13.1) is a small-conductance background K(+) channel in rat trigeminal ganglion neurons. *Pflugers Arch, 466*, 1289–1300.

Kang, D., & Kim, D. (2006). TREK-2 (K2P10.1) and TRESK (K2P18.1) are major background K^+ channels in dorsal root ganglion neurons. *American Journal of Physiology - Cell Physiology, 291*, C138–C146.

Kennard, L. E., Chumbley, J. R., Ranatunga, K. M., Armstrong, S. J., Veale, E. L., & Mathie, A. (2005). Inhibition of the human two-pore domain potassium channel, TREK-1, by fluoxetine and its metabolite norfluoxetine. *British Journal of Pharmacology, 144*, 821–829.

Knowlton, W. M., Palkar, R., Lippoldt, E. K., McCoy, D. D., Baluch, F., Chen, J., et al. (2013). A sensory-labeled line for cold: TRPM8-expressing sensory neurons define the cellular basis for cold, cold pain, and cooling-mediated analgesia. *Journal of Neuroscience, 33*, 2837–2848.

Kobayashi, S., Hori, A., Matsumura, K., & Hosokawa, H. (2006). Point: heat-induced membrane depolarization of hypothalamic neurons: a putative mechanism of central thermosensitivity. *American Journal of Physiology - Regulatory, Integrative and Comparative Physiology, 290*, R1479–R1480. discussion R1484.

Laigle, C., Confort-Gouny, S., Le Fur, Y., Cozzone, P. J., & Viola, A. (2012). Deletion of TRAAK potassium channel affects brain metabolism and protects against ischemia. *PLoS One, 7*, e53266.

Lesage, F., Terrenoire, C., Romey, G., & Lazdunski, M. (2000). Human TREK2, a 2P domain mechano-sensitive K^+ channel with multiple regulations by polyunsaturated fatty acids, lysophospholipids, and Gs, Gi, and Gq protein-coupled receptors. *Journal of Biological Chemistry, 275*, 28398–28405.

Liu, X., Huang, H., Wang, W., Wang, J., Sachs, F., & Niu, W. (2008). Stretch-activated potassium channels in hypotonically induced blebs of atrial myocytes. *Journal of Membrane Biology, 226*, 17–25.

Long, S. B., Tao, X., Campbell, E. B., & MacKinnon, R. (2007). Atomic structure of a voltage-dependent K^+ channel in a lipid membrane-like environment. *Nature, 450*, 376–382.

Lopes, C. M., Rohacs, T., Czirjak, G., Balla, T., Enyedi, P., & Logothetis, D. E. (2005). PIP_2 hydrolysis underlies agonist-induced inhibition and regulates voltage gating of two-pore domain K^+ channels. *Journal of Physiology, 564*, 117–129.

Maingret, F., Fosset, M., Lesage, F., Lazdunski, M., & Honore, E. (1999). TRAAK is a mammalian neuronal mechano-gated K^+ channel. *Journal of Biological Chemistry, 274*, 1381–1387.

Maingret, F., Lauritzen, I., Patel, A. J., Heurteaux, C., Reyes, R., Lesage, F., et al. (2000). TREK-1 is a heat-activated background K(+) channel. *EMBO Journal, 19*, 2483–2491.

Maingret, F., Patel, A. J., Lesage, F., Lazdunski, M., & Honore, E. (1999). Mechano- or acid stimulation, two interactive modes of activation of the TREK-1 potassium channel. *Journal of Biological Chemistry, 274*, 26691--26696.

Ma, X. Y., Yu, J. M., Zhang, S. Z., Liu, X. Y., Wu, B. H., Wei, X. L., et al. (2011). External Ba2+ block of the two-pore domain potassium channel TREK-1 defines conformational transition in its selectivity filter. *Journal of Biological Chemistry, 286*, 39813–39822.

Ma, L., Zhang, X., & Chen, H. (2011). TWIK-1 two-pore domain potassium channels change ion selectivity and conduct inward leak sodium currents in hypokalemia. *Science Signaling, 4*, ra37.

McKemy, D. D., Neuhausser, W. M., & Julius, D. (2002). Identification of a cold receptor reveals a general role for TRP channels in thermosensation. *Nature, 416*, 52–58.

Medhurst, A. D., Rennie, G., Chapman, C. G., Meadows, H., Duckworth, M. D., Kelsell, R. E., et al. (2001). Distribution analysis of human two pore domain potassium channels in tissues of the central nervous system and periphery. *Brain Research Molecular Brain Research, 86*, 101–114.

Miller, A. N., & Long, S. B. (2012). Crystal structure of the human two-pore domain potassium channel K2P1. *Science, 335*, 432–436.

Murbartian, J., Lei, Q., Sando, J. J., & Bayliss, D. A. (2005). Sequential phosphorylation mediates receptor- and kinase-induced inhibition of TREK-1 background potassium channels. *Journal of Biological Chemistry, 280*, 30175–30184.

Namiranian, K., Lloyd, E. E., Crossland, R. F., Marrelli, S. P., Taffet, G. E., Reddy, A. K., et al. (2010). Cerebrovascular responses in mice deficient in the potassium channel, TREK-1. *American Journal of Physiology - Regulatory Integrative and Comparative Physiology, 299*, R461–R469.

Niemeyer, M. I., Cid, L. P., Pena-Munzenmayer, G., & Sepulveda, F. V. (2010). Separate gating mechanisms mediate the regulation of K_{2P} potassium channel TASK-2 by intra- and extracellular pH. *Journal of Biological Chemistry, 285*, 16467–16475.

Noel, J., Sandoz, G., & Lesage, F. (2011). Molecular regulations governing TREK and TRAAK channel functions. *Channels (Austin), 5*, 402–409.

Noel, J., Zimmermann, K., Busserolles, J., Deval, E., Alloui, A., Diochot, S., et al. (2009). The mechano-activated K^+ channels TRAAK and TREK-1 control both warm and cold perception. *EMBO Journal, 28*, 1308–1318.

Park, U., Vastani, N., Guan, Y., Raja, S. N., Koltzenburg, M., & Caterina, M. J. (2011). TRP vanilloid 2 knock-out mice are susceptible to perinatal lethality but display normal thermal and mechanical nociception. *Journal of Neuroscience, 31*, 11425–11436.

Patel, A. J., Honore, E., Lesage, F., Fink, M., Romey, G., & Lazdunski, M. (1999). Inhalational anesthetics activate two-pore-domain background K^+ channels. *Nature Neuroscience, 2*, 422–426.

Patel, A. J., Honore, E., Maingret, F., Lesage, F., Fink, M., Duprat, F., et al. (1998). A mammalian two pore domain mechano-gated S-like K^+ channel. *EMBO Journal, 17*, 4283–4290.

Peier, A. M., Moqrich, A., Hergarden, A. C., Reeve, A. J., Andersson, D. A., Story, G. M., et al. (2002). A TRP channel that senses cold stimuli and menthol. *Cell, 108*, 705–715.

Piechotta, P. L., Rapedius, M., Stansfeld, P. J., Bollepalli, M. K., Ehrlich, G., Andres-Enguix, I., et al. (2011). The pore structure and gating mechanism of K_{2P} channels. *EMBO Journal, 30*, 3607–3619.

Plant, L. D. (2012). A role for K_{2P} channels in the operation of somatosensory nociceptors. *Frontiers in Molecular Neuroscience, 5*, 21.

Pogorzala, L. A., Mishra, S. K., & Hoon, M. A. (2013). The cellular code for Mammalian thermosensation. *Journal of Neuroscience, 33*, 5533–5541.

Rajan, S., Wischmeyer, E., Karschin, C., Preisig-Muller, R., Grzeschik, K. H., Daut, J., et al. (2001). THIK-1 and THIK-2, a novel subfamily of tandem pore domain K^+ channels. *Journal of Biological Chemistry, 276*, 7302–7311.

Roan, E., Waters, C. M., Teng, B., Ghosh, M., & Schwingshackl, A. (2014). The 2-Pore domain potassium channel TREK-1 regulates stretch-induced Detachment of alveolar epithelial cells. *PLoS One, 9*, e89429.

Sandoz, G., Bell, S. C., & Isacoff, E. Y. (2011). Optical probing of a dynamic membrane interaction that regulates the TREK1 channel. *Proceedings of the National Academy of Sciences of the United States of America, 108*, 2605–2610.

Sandoz, G., Douguet, D., Chatelain, F., Lazdunski, M., & Lesage, F. (2009). Extracellular acidification exerts opposite actions on TREK1 and TREK2 potassium channels via a single conserved histidine residue. *Proceedings of the National Academy of Sciences of the United States of America, 106*, 14628–14633.

Schwingshackl, A., Teng, B., Ghosh, M., & Waters, C. M. (2013). Regulation of monocyte chemotactic protein-1 secretion by the two-pore-domain potassium (K2P) channel TREK-1 in human alveolar epithelial cells. *American Journal of Translational Research, 5*, 530–542.

Stebe, S., Schellig, K., Lesage, F., Breer, H., & Fleischer, J. (2014). The thermosensitive potassium channel TREK-1 contributes to coolness-evoked responses of Grueneberg ganglion neurons. *Cellular and molecular neurobiology, 34*, 113–122.

Su, Z., Anishkin, A., Kung, C., & Saimi, Y. (2011). The core domain as the force sensor of the yeast mechanosensitive TRP channel. *Journal of General Physiology, 138*, 627–640.

Sukharev, S., & Sachs, F. (2012). Molecular force transduction by ion channels: diversity and unifying principles. *Journal of Cell Science, 125*, 3075–3083.

Talley, E. M., Solorzano, G., Lei, Q., Kim, D., & Bayliss, D. A. (2001). CNS distribution of members of the two-pore-domain (KCNK) potassium channel family. *Journal of Neuroscience, 21*, 7491–7505.

Thomas, D., Plant, L. D., Wilkens, C. M., McCrossan, Z. A., & Goldstein, S. A. (2008). Alternative translation initiation in rat brain yields K2P2.1 potassium channels permeable to sodium. *Neuron, 58*, 859–870.

Vallee, N., Meckler, C., Risso, J. J., & Blatteau, J. E. (2012). Neuroprotective role of the TREK-1 channel in decompression sickness. *Journal of Applied Physiology, 112*, 1191–1196.

Vallee, N., Rostain, J. C., & Risso, J. J. (2009). Low susceptibility to inert gases and pressure symptoms in TREK-1-deficient mice. *Neuroreport, 20*, 343–347.

Wang, W., Zhang, M., Li, P., Yuan, H., Feng, N., Peng, Y., et al. (2013). An increased TREK-1-like potassium current in ventricular myocytes during rat cardiac hypertrophy. *Journal of Cardiovascular Pharmacology, 61*, 302–310.

Wechselberger, M., Wright, C. L., Bishop, G. A., & Boulant, J. A. (2006). Ionic channels and conductance-based models for hypothalamic neuronal thermosensitivity. *American Journal of Physiology - Regulatory Integrative and Comparative Physiology, 291*, R518–R529.

Woo, D. H., Han, K. S., Shim, J. W., Yoon, B. E., Kim, E., Bae, J. Y., et al. (2012). TREK-1 and Best1 channels mediate fast and slow glutamate release in astrocytes upon GPCR activation. *Cell, 151*, 25–40.

Wu, X., Liu, Y., Chen, X., Sun, Q., Tang, R., Wang, W., et al. (2013). Involvement of TREK-1 activity in astrocyte function and neuroprotection under simulated ischemia conditions. *Journal of Molecular Neuroscience, 49*, 499–506.

Wu, Y. Y., Singer, C. A., & Buxton, I. L. (2012). Variants of stretch-activated two-pore potassium channel TREK-1 associated with preterm labor in humans. *Biology of Reproduction, 87*, 96.

Xian Tao, L., Dyachenko, V., Zuzarte, M., Putzke, C., Preisig-Muller, R., Isenberg, G., et al. (2006). The stretch-activated potassium channel TREK-1 in rat cardiac ventricular muscle. *Cardiovascular Research, 69*, 86–97.

Yamamoto, Y., Hatakeyama, T., & Taniguchi, K. (2009). Immunohistochemical colocalization of TREK-1, TREK-2 and TRAAK with TRP channels in the trigeminal ganglion cells. *Neuroscience Letters, 454*, 129–133.

Zhao, Y., & Boulant, J. A. (2005). Temperature effects on neuronal membrane potentials and inward currents in rat hypothalamic tissue slices. *Journal of Physiology, 564*, 245–257.

Zhao, L. N., Fu, L., Gao, Q. P., Xie, R. S., & Cao, J. X. (2011). Regional differential expression of TREK-1 at left ventricle in myocardial infarction. *Canadian Journal of Cardiology, 27*, 826–833.

Zilberberg, N., Ilan, N., & Goldstein, S. A. (2001). KCNKO: opening and closing the 2-P-domain potassium leak channel entails "C-type" gating of the outer pore. *Neuron, 32*, 635–648.

CHAPTER SIX

Lipid Modulation of Thermal Transient Receptor Potential Channels

Enrique Hernández-García[1] and Tamara Rosenbaum[1]
[1]Departamento de Neurodesarrollo y Fisiología, División de Neurociencias, Instituto de Fisiología Celular, Universidad Nacional Autónoma de México, Distrito Federal, México

Contents

Current Topics in Membranes, Volume 74
ISSN: 1063-5823
http://dx.doi.org/10.1016/B978-0-12-800181-3.00006-3

Abstract

There is a subgroup of transient receptor potential (TRP) ion channels that are responsive to temperature (thermo-TRP channels). These are important to a variety of sensory and physiological phenomena such as pain and taste perception. All thermo-TRP channels known to date are subject to modulation by lipidic molecules of many kinds, from the ubiquitous cholesterol to more specialized molecules such as prostaglandins. Although the mechanisms and sites of binding of lipids on thermo-TRPs are largely unknown, the explosion on research of lipids and ion channels has revealed previously unsuspected roles for them. Diacyl glycerol is a lipid produced by phospholipase C (PLC) and it was discovered to modulate TRP channels in the eye of the fly, and many mammal TRP channels have been found to interact with lipids. While most of the lipids acting on thermo-TRP channels have been found to activate them, there are a few capable of inhibition. Phosphatidylinositol 4,5-bisphosphate is even capable of both inhibition and activation on a couple of thermo-TRPs, depending on the cellular context. More data is required to assess the mechanism through which lipids affect thermo-TRP channel activity and the physiological importance of this interaction.

1. INTRODUCTION

Transient receptor potential (TRP) proteins constitute a family of widely distributed nonselective cation channels found in organisms as diverse as yeasts, flies, worms, tunicates, and humans (Flockerzi, 2007; Nilius & Owsianik, 2011; Venkatachalam & Montell, 2007). The first TRP genes were identified in visually impaired mutants of the fruit fly *Drosophila melanogaster* (Clapham, 2003; Hardie, 2003; Montell, 2012), but have since been implied in several physiological processes other than vision (Latorre, Zaelzer, & Brauchi, 2009; Minke, Cook, 2002). Thermal (Vay, Gu, & McNaughton, 2012) and osmotic sensing (Jin, Berrout, & O'Neil, 2011), Ca^{2+} homeostasis (Lieben & Carmeliet, 2012), and perception of endogenous (Palazzo, Rossi, Novellis, De Maione, & de Novellis, 2013) and exogenous (Calixto,

Kassuya, André, & Ferreira, 2005) stimuli, to name a few, are functions mediated by TRP channels.

Seven TRP subfamilies are identified to date (Nilius & Owsianik, 2011): TRPC (canonical), TRPV (vanilloid), TRPM (melastatin), TRPA (ankyrin), TRPML (mucolipin), TRPP (polycystin), and TRPN (no mechanoreceptor potential C). The first six are present in humans while TRPN is found in *Xenopus laevis*, *Danio rerio*, and *D. melanogaster*, but not in mammals.

Information about the structure of TRP channels was only available as gross topological inferences and functional studies. It was predicted that TRPs were composed of six transmembrane segments with intracellular N and C termini, somewhat resembling cyclic-nucleotide-gated (CNG) and voltage-gated K^+ channels in their membrane topology (Yellen, 2002). In accordance with this propinquity it was proposed that TRP channels assemble and function as tetramers (García-Sanz et al., 2004) (Figure 6.1), just as CNG and voltage-gated K^+ channels, and there is also evidence that at least some of them can form heterotetramers (Lintschinger et al., 2000; Xu, Li, Guggino, & Montell, 1997).

Perhaps one of the greatest advances in this regard is the published structure of the TRPV1 channel in several functional states (Cao, Liao, Cheng, & Julius, 2013; Liao, Cao, Julius, & Cheng, 2013). Using a very insightful

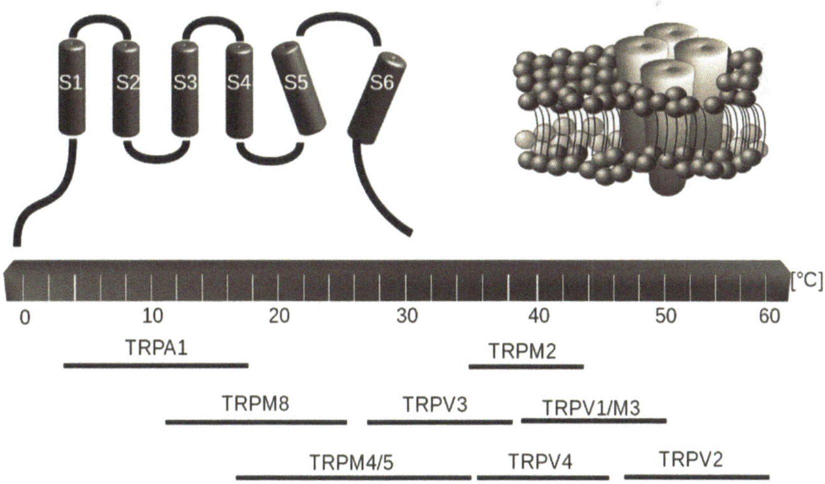

Figure 6.1 General features of thermo-TRP channels. TRP channels are composed of six transmembrane segments (left upper panel), with the S5 and S6 segments and the loop that connects them forming the channel pore. TRP channels are inserted on the membrane (right upper panel) as tetramers. The temperature range to which each thermo-TRP channel responds is depicted as a black line below the thermometer bar. (See the color plate.)

approach of electron cryomicroscopy (Li et al., 2013) instead of the traditional X-ray crystallography, an atomic resolution model of a minimal TRPV1 channel was obtained, thus revealing the tetrameric nature of the protein, the fine details of prominent features such as the vanilloid-binding pocket, and the TRP box domain, and also providing hints for the mechanism of activation of this ion channel.

Several TRP channels have been implicated in physiological/pathophysiological processes associated to inflammation and pain and this is why the discovery of natural and synthetic agonists and antagonists of TRP channels has been of untold importance (Vetter & Lewis, 2011). Indeed, TRP channels are modulated by a miscellanea of molecules and stimuli. From ions and gases (Trevisani et al., 2005) to temperature and mechanical processes of the membranes where they are expressed (Huang, Zhang, & McNaughton, 2006). TRP channels confer a broad array of environmental, local, and internal inputs to guide the cellular behavior in the most efficient fashion. Not surprisingly, for some of the TRPs known to be regulated by molecules, more than one modulator molecule has been described (Bandell et al., 2004; Chianese et al., 2012; Chung, Güler, & Caterina, 2008; Clapham, 2003; Dhaka et al., 2009; Liu & Qin, 2005; Nieto-Posadas et al., 2011), and many of these channels function as signal integrators.

As mentioned above, some TRP channels possess the ability to respond to temperature (Benham, Gunthorpe, & Davis, 2003; Dhaka, Viswanath, & Patapoutian, 2006; Huang et al., 2006). The thermo-TRPs identified to date are (ordered by increasing activation temperature): TRPA1 (Bandell et al., 2004), TRPM8 (Mckemy, Neuhausser, & Julius, 2002; Peier, Moqrich, et al., 2002), TRPM5, TRPM4 (Talavera et al., 2005), TRPM2 (Togashi et al., 2006), TRPV4 (Gu et al., 2002; Watanabe et al., 2002b), TRPV3 (Peier, Reeve, et al., 2002; Smith et al., 2002; Xu et al., 2002), TRPV1 (Tominaga et al., 1998), TRPM3 (Vriens et al., 2011), and TRPV2 (Caterina, Rosen, Tominaga, Brake, & Julius, 1999) (Figure 6.1). Currently, the exact biophysical mechanisms through which thermal stimuli modulate the gating properties of thermo-TRPs are not known and the position of a "temperature sensor," if there is one, is controversial. For instance, Brauchi and colleagues exchanged the C-terminal domains of cold- and hot-activated TRPM8 and TRPV1 and reported an equivalent interchange in their activation temperature (Brauchi, Orta, Salazar, Rosenmann, & Latorre, 2006), whereas Yao and colleagues reported that the temperature sensor of TRPV1 and TRPV2 is located on the N-terminal domain (Yao, Liu, & Qin, 2011). Added to the complexity of these thermo-TRP channels is the fact that many of them respond to a wide array of natural compounds (Morales-Lázaro, Simon, & Rosenbaum, 2013).

Most of the endogenous and many exogenous modulators of TRP channels are molecules of a lipidic nature that exert diverse responses on their targets (Chyb, 1999; Bradshaw, Raboune, & Hollis, 2013; Hardie, 2003). The vast majority of the endogenous lipids that activate TRP channels are produced as second messengers downstream of G protein-coupled receptor (GPCR) signaling pathways (for review, see [Kukkonen, 2011]); and the exogenous ones are mostly compounds found in plants.

The past decade has experienced an explosion of information on the role of lipids as regulators of TRP channel function. Here we provide a detailed description of how thermo-TRP channels are modulated by lipidic molecules (see Table 6.1 for a quick overview).

2. PHOSPHATIDYLINOSITOL 4,5-BISPHOSPHATE (PIP_2)

Phosphatidylinositol 4,5-bisphosphate (PIP_2) is an acidic phospholipid of the inner leaflet of the plasma membrane (Suh & Hille, 2008). It is the prevalent isoform among the doubly phosphorylated inositol lipids, yet it is an order of magnitude less abundant than phosphatidylinositol monophosphate (Vanhaesebroeck et al., 2001).

Many ion channels and transporters are modulated by PIP_2 (Hilgemann, Feng, & Nasuhoglu, 2001; Suh & Hille, 2005), and TRP channels are not an exception (Rohacs, 2007). All thermo-TRPs known to date are modulated by PIP_2 or, at least, are capable of binding it. The effects of PIP_2 on most thermo-TRPs have been studied scarcely, and the role of this molecule on TRPV1 and TRPA1 remains largely debated (Karashima et al., 2008; Kim, Cavanaugh, & Simkin, 2008; Lukacs et al., 2007; Rohacs & Nilius, 2007; Voets & Nilius, 2007). Thermo-TRP channels activated by PIP_2—or that require it for activation—are TRPM2/4/5/8 and TRPV2/4, while TRPV3 is inhibited by PIP_2. This lipid binds to TRPM3, but it is not known whether it exerts a modulatory function on this channel (Holendo'va, Grycova, Jirku, & Teisinger, 2012).

2.1 TRPA1

TRPA1 is present in organs specialized in sensation, namely, nociceptive neurons of peripheral ganglia and the mechanosensory tissue in the inner ear (Kamakura et al., 2013). This ion channel was implicated in

Table 6.1 Lipids that modulate the activity of thermo-TRP channels. At least one example of each group of lipids described in the text is depicted. The thermo-TRP channels affected by the different lipids are shown in the last column with a brief description of the effect of each molecule on the channels. OEA, oleoylethanolamide

Lipid group	Members	Examples	Modulatory effect on thermo-TRP channels
PIP_2	PIP_2	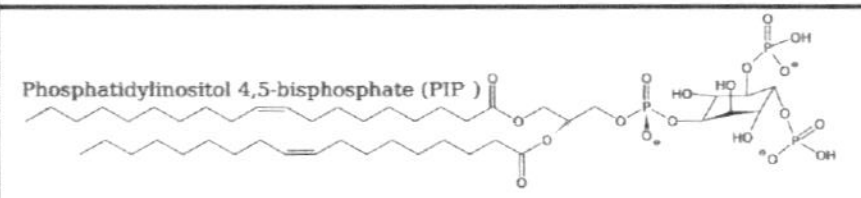 Phosphatidylinositol 4,5-bisphosphate (PIP)	TRPA1/V1: positive or negative TRPM2/4/5/8/V2/V4: positive TRPV3: negative TRPM3: binds to the channel
DAG	DAG	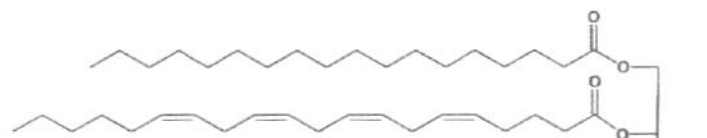 1-stearoyl-2-arachydonyl glycerol (DAG)	TRPA1: positive TRPV1: positive through the same site as capsaicin and cannabinoids
Lipids downstream of PLA_2	AA, EETs, PGs, anandamide, NADA, NATs, HETEs, HPETEs, OEA	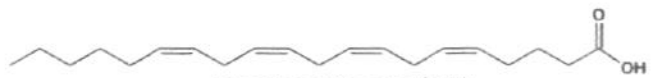 Arachidonic acid (AA)	TRPA1/M5/M8/V1/V3/V4: positive
n-3 PUFAs	EPA, DHA, LNA, OLAMs, resolvins	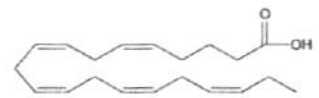 Eicosapentaenoic acid (EPA)	TRPV1: positive modulation by EPA, DHA, LNA and OLAMs; indirectly inhibited by resolvins
Oxidized lipids	HNE, $OA\text{-}NO_2$	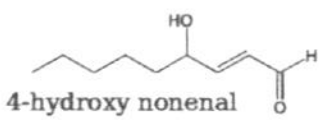 4-hydroxy nonenal 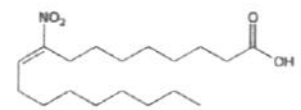 Nitrooleic acid ($OA\text{-}NO_2$)	TRPA1/V1: positive by covalent modification of reactive residues TRPM2: positive
Lysophospholipids	LPC, LPI, LPA	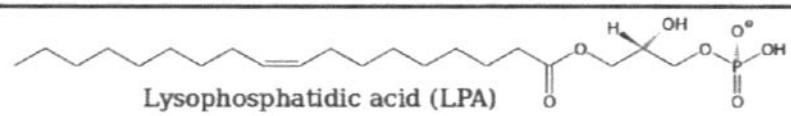 Lysophosphatidic acid (LPA)	TRPM8: positive by modifying membrane properties TRPV1: positive
Cholesterol and steroids	Cholesterol, PS	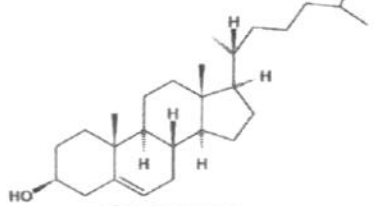 Cholesterol 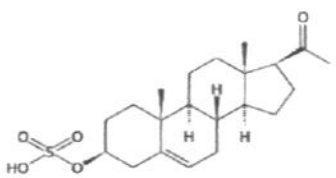 Pregnenolone sulphate (PS)	TRPM3: PS has a positive modulation TRPV1: cholesterol has a negative modulation
Other lipids	SPH, FPP, LA	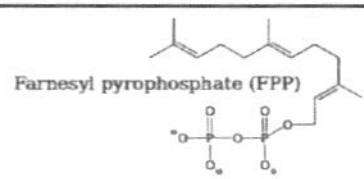 Farnesyl pyrophosphate (FPP)	TRPM3: SPH has a positive modulation TRPV3: FPP and LA have a positive modulation

mechanotransduction in hair cells from ear by Nagata and colleagues (Nagata, Duggan, Kumar, & García-Añoveros, 2005), although later Kwan et al. found that mice lacking functional TRPA1 preserved normal responses to loud noise, had normal balance, and displayed normal transduction currents in vestibular hair cells (Kwan et al., 2006). TRPA1 is responsible for the perception of painful stimuli such as cold temperatures below 18 °C (Dhaka et al., 2006; Sawada, Hosokawa, Hori, Matsumura, & Kobayashi, 2007) and several irritating exogenous molecules such as cinnamaldehyde, and icilin. Other ligands of TRPA1 include dibenzoxazepine, a member of morphanthridines also known as "CR gas" ($EC_{50} = 3$ nM) (Gijsen et al., 2010), and pungent sulfur compounds from mustard, wasabi, and horseradish, collectively known as isothiocyanates (Salazar et al., 2008; Uchida, Miura, Nagai, & Tominaga, 2012). In addition, TRPA1 is inhibited by resveratrol, a compound present in grapes and wine and thought to possess beneficial effects on health (Yu et al., 2013), and camphor, a terpenoid compound obtained from the *Cinnamomum camphora* tree (Macpherson et al., 2006).

When it comes to modulation of TRPA1 (and TRPV1, see below), PIP_2 is a controversial molecule. Dai et al. observed that intracellular PIP_2 abolished the protease-activated receptor 2 (PAR2), a GPCR-mediated TRPA1 potentiation of currents evoked by allyl isothiocyanate. Furthermore, when an anti-PIP_2 antibody was applied to the intracellular side, allyl-isothiocyanate-evoked currents were more robust than those recorded in the absence of the antibody, thus arguing for an inhibitory role of PIP_2 on TRPA1. In 2008, two more reports discussed the regulatory properties of PIP_2 on TRPA1. First, Kim and colleagues showed a decrease in channel activity when PIP_2 was applied intracellularly to inside-out patches of HeLa cells that expressed TRPA1 (Kim et al., 2008a). They found that PIP_2, either synthetic or purified from biological sources, diminished TRPA1 activity in a dose-dependent fashion (interestingly, they found that TRPV1 is activated by the sole application of PIP_2, although the prevalent consensus is that PIP_2 does not activate TRPV1 per se. See, for instance [Klein, Ufret-Vincenty, Hua, & Gordon, 2008; Liu, Zhang, & Qin, 2005]) and their single-channel experiments showed that PIP_2 directly diminished the open probability of the channel by directly interacting with it (Kim et al., 2008a). Later, Karashima and colleagues reported that whole-cell currents from TRPA1-expressing Chinese hamster ovary (CHO) cells were desensitized faster in the presence of the PIP_2 scavenger neomycin (Karashima et al., 2008). They also presented evidence that Ca^{2+}-mediated sensitization was slowed down when PIP_2 was applied intracellularly, in either whole-cell or inside-out configurations, contending for a role of positive modulation of TRPA1 by PIP_2. Thus, as will be discussed for TRPV1, there are still more experiments needed to clarify these contradicting results.

2.2 TRPM2

Thermo-TRPMs span the perception of fresh to noxiously high temperatures. TRPM2 is activated at temperatures above 37 °C (Togashi et al., 2006) and by intracellular Ca^{2+}, adenosine diphosphate (ADP) ribose, the diabetogenic drug alloxan (Inamura et al., 2003), and β-nicotinamide adenine dinucleotide (NAD^+). It is inhibited by ADP, flufenamate, N-(*p*-amylcinnamoyl) anthranilic acid, and the antifungal medication econazole. This channel possesses a Nudix box domain, which catalyzes the hydrolysis of nucleoside diphosphates. TRPM2 has been implicated in the regulation of phagocyte activation, glucose-induced insulin secretion, neuronal apoptosis, and cytokine production (Takahashi, Kozai, Kobayashi, Ebert, & Mori, 2011).

TRPM2 binds PIP_2 with very high affinity. The lipid functions as an inhibitory regulator of the channel. Tóth and Csánady showed that, when using polylysine as a scavenger to mask the head groups of PIP_2 on inside-out patches from *X. laevis* oocytes, Ca^{2+} currents through TRPM2 decreased dramatically (Tóth & Csanády, 2012). Unlike TRPM8, but similar to TRPM4/5, PIP_2 is not capable of opening the channel by itself; it rather increases the apparent affinity of the channel for Ca^{2+}. In a PIP_2-free environment the affinity for Ca^{2+} is >1 mM, but when PIP_2 is applied on the intracellular side, the apparent affinity increases to ~50 μM. It is this further increase in affinity for intracellular Ca^{2+} which ultimately opens the ion channel.

2.3 TRPM3

TRPM3 was the last identified member of the TRPM channels (Oberwinkler & Phillipp, 2007), and the last TRP channel identified as a thermal sensor (Vriens et al., 2011). This channel responds to high noxious temperatures, similar to TRPV1, and it is expressed in kidney and pancreatic β-cells (Grimm, Kraft, Sauerbruch, Schultz, & Harteneck, 2003; Lee et al., 2003; Wagner et al., 2008), as well as in the brain in both neurons and oligodendrocytes (Hoffmann et al., 2010; Zamudio-Bulcock, Everett, Harteneck, & Valenzuela, 2011). It is activated by hypotonicity, d-erythro-sphingosine (SPH), and pregnenolone sulfate (PS).

Holendova and colleagues showed that PIP_2 binds to the N terminus of TRPM3, on the calmodulin (CaM)-binding domains; however, it was not assessed whether this lipid had a regulatory effect on the channel (Holendova et al., 2012) and no further information on how this lipid modulates TRPM3 activity exists as of yet.

2.4 TRPM4/5

TRPM4 and TRPM5 are activated by warm temperatures at a threshold ranging from 15° to 35 °C (Talavera et al., 2005). These channels possess the unique feature of being gated directly by intracellular Ca^{2+} but not exhibiting permeation of this ion (Hofmann, Chubanov, Gudermann, & Montell, 2003; Launay et al., 2002). TRPM5 expression is mostly confined to taste receptor cells (TRCs) (Pérez et al., 2002), while TRPM4 is found in tissues as diverse as skeletal muscle, liver, heart, kidney, spleen, colon, and many others (Launay et al., 2002; Nilius, Droogmans, & Wondergem, 2003; Xu, Moebius, Gill, & Montell, 2001). TRPM4 is activated by decavanadate and inhibited by adenosine triphosphate (ATP) but none of these compounds affect TRPM5 (Nilius, Prenen, Janssens, Voets, & Droogmans, 2004). Both TRPM4 and TRPM5 are inhibited by flufenamate, a nonsteroidal antiinflammatory drug (Ullrich et al., 2005).

The physiological role of TRPM4 has not been yet clearly established (Bödding, 2007); but TRPM5 is thought to participate in the three taste modalities that are mediated by GPCRs: bitter, sweet, and umami (Julius & Nathans, 2012; Kinnamon, 2012). These specialized GPCRs are stimulated in TRCs and their activation, in turn, activates phospholipase C (PLC)-β2 (Huang et al., 1999; Pérez et al., 2002), generating products that cause TRPM5 to be eventually activated (Oike et al., 2006).

In many regards, TRPM4 and TRPM5 are similar (Ullrich et al., 2005) and they both depend on PIP_2 for recovery from rundown (Zhang, Okawa, Wang, & Liman, 2005). In contrast to TRPM8 (where PIP_2 can shift the sensitivity to temperature), PIP_2 alone is not capable of gating either TRPM4 or TRPM5. In experiments performed by Liu and Liman, the patch clamp technique showed that TRPM5 currents activated by intracellular Ca^{2+} suffer rundown that can be recovered with PIP_2 (Liu & Liman, 2003). Moreover, in inside-out recordings, the extent of the rundown depended on how long the patches had been exposed to Ca^{2+} and was partially reversed by adding a short-chain PIP_2.

In a set of experiments similar to those performed by Liu and Liman for TRPM5, Zhang et al. showed that TRPM4 expressed in CHO cells suffers

Ca^{2+}-dependent rundown and that the currents recover from rundown when PIP_2 is applied to the intracellular side (Zhang et al., 2005). Interestingly, the mechanism through which PIP_2 acts on these channels seems to be an increase in the affinity of both TRPM4 and TRPM5 for Ca^{2+} (Nilius et al., 2006).

2.5 TRPM8

TRPM8 is activated below 25 °C (B. Liu & Qin, 2005) and is expressed mainly in sensory neurons. It is found in neurons of the dorsal root, trigeminal, and nodose ganglia (Mckemy et al., 2002; Nealen, Gold, Thut, & Caterina, 2003; Zhang, Jones, Brody, Costa, & Brookes, 2004) and also in smooth muscle (Yang, Lin, McIntosh, & Sham, 2006), liver (Henshall et al., 2003), and prostate (Tsavaler, Shapero, Morkowski, & Laus, 2001). This channel responds to menthol, icilin, and eucalyptol, molecules that produce the sensation of "freshness," and it is inhibited by capsazepine, a synthetic analog of capsaicin.

PIP_2 is a requirement for gating of many TRP channels, but when applied to TRPM8 it is, by itself, enough to make the channel open, as demonstrated by Liu and Qin (Liu & Qin, 2005). By means of electrophysiological assays, these authors showed that in excised membrane patches from *X. laevis* oocytes exposed to menthol, TRPM8 suffered rundown in a short time, but channel activity was restored when exogenous PIP_2 was supplemented. Moreover, when a mixture of phosphatase inhibitors was applied to excised patches, PIP_2 degradation was quenched and rundown was prevented. The most striking finding made by the authors of this study was that application of 20 μM PIP_2 was enough to elicit a current with half the maximal amplitude of that produced by menthol at room temperature. Thus, PIP_2 is a partial agonist of TRPM8 channels and is required for the channel to open in response to other stimuli.

2.6 TRPV1

TRPV1 is expressed mainly in small-diameter neurons from dorsal root, nodose, and trigeminal ganglia (Caterina et al., 1997; Helliwell et al., 1998). This channel is also found in the central nervous system (CNS), smooth muscle, liver, and cells from the immune system. It is activated by temperatures above 42 °C (Caterina et al., 1997), vanilloid compounds, allicin, endocannabinoids, lysophosphatidic acid (LPA), and camphor, among others. This channel is inhibited by capsazepine, a selective TRP channel closed-state stabilizer, and pinosylvin methyl ether, a stilbenoid similar to resveratrol (Yu et al., 2013).

TRPV1 is the most studied member of the TRP ion channel family, and it is well established that PIP_2 exerts some modulation on it (also reviewed in Morales-Lázaro et al., 2013); however, whether this molecule promotes or hinders channel activation has not been established beyond doubt. In whole-cell experiments from *Xenopus* oocytes, PIP_2 was described as an inhibitor of TRPV1 (Prescott & Julius, 2003); additionally, it was found that, when TRPV1 is reconstituted in liposomes of known composition, channel responses to capsaicin, acid, and heat are inhibited in presence of PIP_2 (Cao et al., 2013a). In spite of this, several studies have found that PIP_2 behaves, in most contexts, as a positive modulator of TRPV1 channel activity (Brauchi et al., 2007; Liu et al., 2005; Lukacs, Rives, Sun, Zakharian, & Rohacs, 2013; Senning, Collins, Stratiievska, Ufret-Vincenty, & Gordon, 2014; Sowa, Street, Vihko, & Zylka, 2010; Ufret-Vincenty, Klein, Hua, Angueyra, & Gordon, 2011). For instance, application of PIP_2 to inside-out patches was required for full TRPV1 activation in Dorsal root ganglion (DRG) neurons from mice (Stein, Ufret-Vincenty, Hua, Santana, & Gordon, 2006), and Liu et al. found that synthesis of PIP_2 is required for TRPV1 recovery from desensitization in HEK293 cells (Liu et al., 2005).

Phosphoinositides alter the activity of TRPV1 in complex ways. Lukacs and colleagues argue that PIP_2 is a TRPV1 inhibitor in the cellular context and it is lost/washed off when excised-patch experiments are performed (Lukacs et al., 2007). Lukacs and colleagues also proposed a possible mechanism through which capsaicin-induced Ca^{2+} currents lead to the desensitization of TRPV1 by the activation of PLC which, in turn, leads to depletion of PIP_2 from the membrane (Liu et al., 2005; Yao & Qin, 2009). These authors propose a dual effect of phosphoinositides on TRPV1, similar to the one found in voltage-activated Ca^{2+} channels (Wu, Bauer, Zhen, Xie, & Yang, 2002).

The precise locations of interaction sites through which PIP_2 modulates TRPV1 channel have been equally controversial. The distal (Prescott & Julius, 2003; van Rossum et al., 2005) and proximal (Brauchi et al., 2007; Ufret-Vincenty et al., 2011) C-terminal regions of the TRPV1 channel have been proposed as binding sites for PIP_2. Other studies suggest that modulation occurs indirectly by the action of independent accessory proteins, namely, Pirt (Kim et al., 2008b), and A-kinase anchoring protein 150 (Jeske et al., 2011).

2.7 TRPV2

TRPV2 is the "hottest" thermo-TRP channel, as it activates at temperatures well beyond the noxious threshold (>52 °C). It can be found in dorsal root and trigeminal ganglia (Benham et al., 2003), and is activated by the synthetic ligands 2-aminoethoxydiphenyl borate (2APB) and probenecid.

After activation by temperature or 2APB, this channel goes through a Ca^{2+}-dependent desensitization similar to that of TRPV1. When the PIP_2 concentration in the membrane decreases, TRPV2 is desensitized in the presence of Ca^{2+} in whole-cell recordings from F-11 cells, which endogenously express TRPV2 (Bender et al., 2005). However, when exogenous PIP_2 is applied to the intracellular side of inside-out patches, TRPV2-mediated currents are rescued (Mercado, Gordon-Shaag, Zagotta, & Gordon, 2010). Additionally, it was found that CaM, although capable of binding to TRPV2 in vitro, does not play a role on the regulation of this channel, contrary to the case of TRPV1 (Rosenbaum, Gordon-Shaag, Munari, & Gordon, 2004). These authors suggest that the Ca^{2+}-dependent desensitization of TRPV2 is due to the depletion of PIP_2 from the plasma membrane.

2.8 TRPV3

TRPV3 is expressed in skin, tongue, dorsal root ganglion, trigeminal ganglion, spinal cord, brain (Xu et al., 2002; Xu, Delling, Jun, & Clapham, 2006), and skin keratinocytes (Doerner, Hatt, & Ramsey, 2011). It is activated in response to innocuous temperatures ranging from 31 to 39 °C and becomes sensitized with repeated heat pulses (Dhaka et al., 2006; Grubisha et al., 2013). It is also activated by monoterpenes such as camphor, carvacrol, and thymol (Vogt-Eisele et al., 2007) and by farnesyl pyrophosphate (FPP), an endogenous lipid from the mevalonate pathway. Lipid molecules similar to arachidonic acid (AA) (see Section 3.2), such as 5,8,11-eicosatriynoic acid (ETI), and 5,8,11,14-eicosatetraynoic acid (ETYA) potentiate the channel response to 2APB (Hu et al., 2006). TRPV3 is blocked by some multivalent ions such as Mg^{2+}, but such block can be alleviated by linoleic acid (LA) (Parnas, Peters, & Minke, 2009).

Doerner and colleagues have soundly established that TRPV3 activity is stimulated by PIP_2 depletion (Doerner et al., 2011). These authors observed that in inside-out patches from HM1 cells (a HEK 293 cell clone with a stable translation of the human acetylcholine receptor M1) expressing TRPV3, very few spontaneous channel openings were recorded.

However, when an anti-PIP_2 antibody was applied to the intracellular side, the channel displayed an evident activation. In a key experiment using single-channel recordings, the authors demonstrated that the application of greater concentrations of PIP_2 actually diminished the open probability of TRPV3. Moreover, while two isoforms of PIP_2 hindered channel activity, phosphatidylinositol 3-phosphate (PI(3)P), a product of the hydrolysis of PIP_2, promoted an increase in the open probability of TRPV3 channels.

2.9 TRPV4

TRPV4 is activated at temperatures ranging from 27° to 42 °C (Dhaka et al., 2006). Its messenger RNA can be detected in lung, spleen, testis, and fat, but it is primarily expressed in kidney (Liedtke et al., 2000) and it is believed to play a role in the regulation of the vascular tone (Filosa & Iddings, 2013). In addition to temperature, this channel is activated by changes in osmotic pressure (Strotmann, Harteneck, Nunnenmacher, Schultz, & Plant, 2000), vanillotoxins, and lipidic molecules downstream of PLC activity. Molecules such as 2-phorbol 12,13-didecanoate 20-homovanillate (Cortright & Szallasi, 2009), 5,6- epoxyeicosatrienoic acid (5,6-EET) (Watanabe et al., 2003), phorbol-12-myristate-13-acetate (PMA), and 4α-phorbol-12,13-didecanoate (4a-PDD) (Watanabe et al., 2002a; Xu, Satoh, & Iijima, 2003) also activate TRPV4.

Garcia-Elias and colleagues showed that TRPV4 requires binding of PIP_2 for activation by hypotonicity and heat, and that this binding occurs on the cytosolic N terminus of the channel (Garcia-Elias et al., 2013). PIP_2 requirement for TRPV4 activation by thermal stimulus was elegantly proven by this group using electrophysiological single-channel experiments performed with TRPV4-expressing HEK cells (Garcia-Elias et al., 2013). These authors showed that at 24 °C the channel was not open whether the lipid was present or not. However, at 38 °C the channel opened robustly only in the presence of exogenous PIP_2 applied to the intracellular side of the channel. The authors further investigated the effects of PIP_2 on the arrangement of the intracellular N terminus of this channel using TRPV4-Cyan (CFP)/Yellow (YFP) fluorescent protein chimeric channels and Förster Resonance Energy Transfer. They found that the depletion of PIP_2 by 5-phosphatase IV produced a more compact conformation of the N-terminal domains, while the presence of PIP_2 promoted an "expanded" conformation somewhat similar to PIP_2-regulated K^+ channels (Garcia-Elias et al., 2013).

3. GPCR SIGNALING PATHWAYS

Many endogenous lipidic second messengers are generated from phospholipids of the plasma membrane after activation of signaling pathways that initiate with GPCRs. GPCRs are the largest family of cell surface receptors and they participate in many important physiological processes (Yekkirala, 2013).

In the classical pathways of GPCR signaling, the G-protein-coupled receptors interact with and activate heterotrimeric G-proteins (Gαβγ). These proteins are classified based on the homology of the Gα subunit, and four families are identified: the Gq family activates PLCβ, the Gs family activates adenylyl cyclase (AC), Gi/o inhibits AC, and G12/13 activates guanine nucleotide exchange factors.

Downstream of the Gq family we find PLCβ. This enzyme is of key importance in the genesis of lipidic second messengers, as it catalyzes the degradation of PIP_2 into DAG and inositol 1,4,5-trisphosphate (IP3). Both these molecules are capable of modulating several TRP channels, and DAG is a precursor for the formation of AA. Figure 6.2 shows some of the potential lipid second messengers that can be produced from the activation of PLC.

TRP channel responses are tightly coupled to GPCR signaling. This coupling seems to lie at the heart of the regulation of physiological processes as diverse as mouse pheromone sensing (Kiselyov, van Rossum, & Patterson, 2010), sweet taste perception (Oike et al., 2006; Pérez et al., 2002; Sprous & Palmer, 2010), modulation of cough (Maher, Dubuis, & Belvisi, 2011), and vasoconstriction (Voets & Nilius, 2009). For instance, in 2002, Perez et al. found that the TRPM5 ion channel is expressed in the same mouse circumvallate papillae TRCs as α-gustducin (an α subunit of G-proteins), and PLCβ2. In 2006, Oike and colleagues showed that the equivalent rat TRCs also express cyclooxygenase 2 (COX-2), monoglyceride lipase, and, phospholipase A2 (PLA2). The latter group of enzymes is important for the production of lipid second messengers that activate TRPM5 ion channels, such as AA (See Section 3.2).

3.1 Phospholipase C and diacyl glycerol

Diacyl glycerol (DAG) is a product of PIP_2 hydrolysis by PLC upon GPCR activation, and is known to activate the canonical TRP channel from *Drosophila* during phototransduction (Delgado & Bacigalupo, 2009).

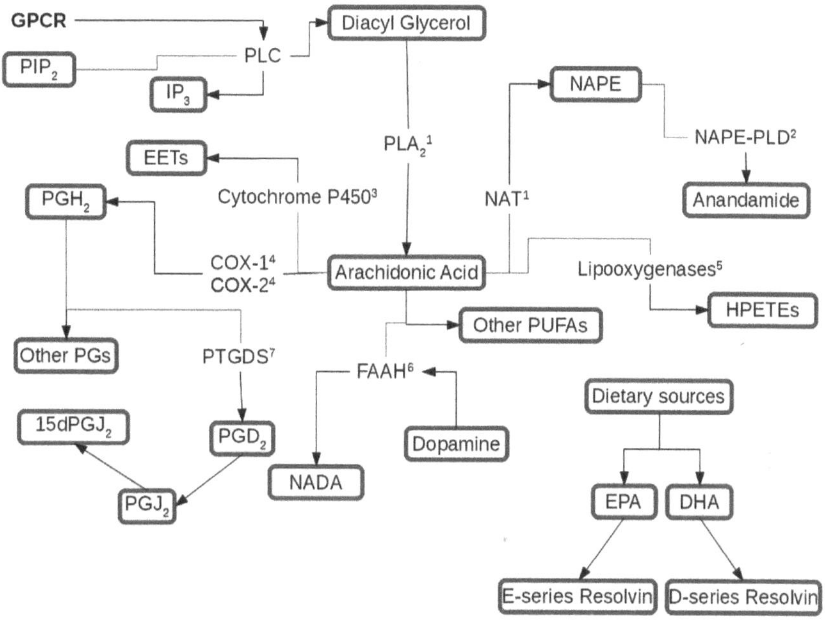

Figure 6.2 Biosynthetic pathways through which lipids relevant to thermo-TRP channel activity are produced. Lipids are shown enclosed in boxes, enzymes are shown in nonenclosed text. Note that the majority of these lipids are products downstream of GPCR activation. [1]*Bingham & Austen, 1999;* [2]*Ellert-Miklaszewska, Ciechomska, & Kaminska, 2013;* [3]*Zeldin, 2001;* [4]*Choi, Aid, Choi, & Bosetti, 2010;* [5]*Pace-Asciak & Asotra, 1989;* [6]*Hu et al., 2009;* [7]*Sun, Chapman, & McGuire, 1977.*

Helliwell and Large published a seminal work in the 1990s in which they demonstrated that a cationic current was activated in portal vein vascular smooth muscle cells (VSMCs) of rabbit in response to DAG, in a protein kinase C (PKC)-independent manner (Helliwell & Large, 1997). Years later, several groups reported findings that shed light on the most likely ion channel candidates for such a current, namely, TRPC3, TRPC6 (Hofmann et al., 1999), and TRPC7 (Okada, 1999). Thus, it comes as no surprise that other TRPs are sensitive to DAG.

3.1.1 TRPA1

In 2004, Bandell et al. showed that CHO cells that express TRPA1 heterologously display an increase in $[Ca^{2+}]_i$ when 1-oleoyl-2-acetyl-sn-glycerol (OAG, a cell-permeable analog of DAG) is applied to them, as measured by ratiometric fluorescence of Fura-2 (Bandell et al., 2004). This result is important in the light of the fact that bradykinin (BK), a peptide released during tissue inflammation and associated with chronic pain (Story et al.,

2003), activates BK receptor 2, a GPCR upstream of a PLC signaling pathway in nociceptive neurons. PLC activation could ultimately lead to the generation of pain and hyperalgesia through TRPA1 (and perhaps TRPV1, see below) in a DAG-dependent way (Palazzo et al., 2013). Notwithstanding this compelling result, it must be noted that the extent to which OAG activates TRPA1 was not fully explored and that the concentrations used to obtain a robust response were rather large (100–200 μM). The mechanism of activation of TRPA1 by DAG, as for many ion channels, is currently unknown.

Interestingly, Dai and colleagues reported that TRPA1-mediated currents were potentiated by PAR2 activation in HEK cells (Dai et al., 2007). Unexpectedly, this latter process was found to be independent of downstream products of PLC, thus contradicting the findings made previously by Bandell and colleagues that DAG activates TRPA1.

3.1.2 TRPV1

DAG was proposed as an endogenous ligand for TRPV1 following the finding that OAG, a synthetic analog of DAG, is capable of producing TRPV1-mediated cation intake in both DRG neurons and heterologously expressing HEK 293 cells (Woo et al., 2008). Using fluorometric calcium assays, Co^{+} uptake, and whole-cell patch clamp experiments, it was demonstrated that TRPV1 responds to the application of OAG in a PLC-independent way, similar to TRPC channels. However, compared to capsaicin, the preferred TRPV1 agonist, OAG produces only a very discrete response of about 20% even at concentrations as large as 100 μM, so the authors of this study suggest that DAG is only a partial agonist (Woo et al., 2008). Remarkably, evidence indicates that OAG, and thus potentially DAG, binds to the same region of the protein as capsaicin (Woo et al., 2008) but to date, no DAG binding site has been found for other TRP channels.

Since the concentration of OAG required to produce TRPV1 activation is somewhat large, further experiments are required to confirm whether DAG is a true physiological activator. An alternate possibility, albeit not yet scrutinized, is that DAG is not to activate but "prime" TRPA1 and TRPV1 for activation. It is known that TRPM8, when in contact with low concentrations of menthol, does not activate but rather shifts the temperature threshold for activation (Mckemy et al., 2002). A similar phenomenon could be taking place with DAG for TRPA1 and TRPV1 in physiological conditions.

3.2 Phospholipase A2, arachidonic acid, and downstream lipids

Phospholipases (PLs) are enzymes activated downstream of GPCR signaling pathways that produce several important bioactive lipids. Perhaps the most important ones for TRP channel modulation are PLCβ2 and PLA2, which release DAG and fatty acids, respectively (Figure 6.2).

One of most common fatty acids found in lipids of the plasma membrane is AA (>10% of the composition of erythrocytes and thrombocytes (Manku, Horrobin, Huang, & Morse, 1983; Simopoulos, 2002); >20% of the composition of the mammalian cell membrane (Contreras et al., 2000; Rapoport, 2008) Figure 6.2). AA is an endogenous n-6 polyunsaturated, 20-carbon essential fatty acid (n-6 polyunsaturated fatty acid (PUFA)) found in many phospholipids, including phosphatidylcholine and phosphatidylethanolamine, and it is released upon the action of PLA2. AA has a short lifetime in neurons as a free lipid since it is quickly incorporated to membrane phospholipids (Horrocks, 1989); nevertheless, it has significant signaling properties and is capable of modulating the activity of several ion channels. Added to its importance, many products of AA metabolism are signaling molecules that regulate the activity of thermo-TRP channels.

3.2.1 TRPA1

Calcium measurements have shown that TRPA1 is activated by AA (Bandell et al., 2004), prostaglandins (PGs) and EETs (Taylor-Clark et al., 2008). AA is a second messenger long known to be important in inflammatory (Vargaftig & Zirinis, 1973) and painful processes (Jaques, 1959) and its nonmetabolizable analog, ETYA, is capable of producing a TRPA1-mediated $[Ca^{2+}]_i$ increase in CHO cells expressing this channel (Bandell et al., 2004).

EETs are products of AA metabolism themselves, by the action of cytochrome P450 epoxygenase (Spector, Fang, Snyder, & Weintraub, 2004). They are also capable of activating thermo-TRP channels in several tissues (Campbell & Fleming, 2010; Earley, 2011; Meves, 1994; Meves, 2008). 5,6-EET activates TRPA1 in both DRG neurons and TRPA1-transfected HEK 293 cells (Sisignano et al., 2012). 5,6-EETs are produced in DRGs and in dorsal spinal cord during nociception, but not inflammation, which argues for a paracrine or autocrine effect of this molecule on sensory neurons with expression of TRPA1 (Sisignano et al., 2012).

PGs are 20-carbon lipid compounds that mediate autocrine and paracrine signaling. These molecules are produced by the sequential oxidation of AA by COX-1 and COX-2 (Figure 6.2). They are responsible for constriction or dilation in VSMCs, and for aggregation or disaggregation of platelets, among other functions. 15-Deoxy-Δ12,14-prostaglandin J2 (15dPGJ2) is a cyclopentenoic PG derived from PGD2. This inflammatory molecule activates TRPA1 directly in trigeminal neurons and HEK293 cells (Taylor-Clark et al., 2008). 15dPGJ2 is a highly electrophilic molecule and its mechanism of action is similar to that proposed for other TRPA1 ligands such as isothiocyanates; namely, it covalently modifies the channel in key cysteine and lysine residues (Macpherson et al., 2007).

3.2.2 TRPM5

Using the whole-cell patch clamp technique, Oike and colleagues showed that AA is capable of activating TRPM5-mediated currents similar to those promoted by release of intracellular Ca^{2+} in HEK 293 cells expressing this channel (Oike et al., 2006). Physiologically TRPM5 is coexpressed, in addition to GPCRs, with PLA2, an enzyme whose products include AA, and the combination of IP3-mediated Ca^{2+} release and PUFA generation could be responsible for the activation of TRPM5 in TRCs. As noted above, the activation of TRPM5 in TRCs appears to be one of the mechanisms through which these sensory cells respond to bitter, sweet, and umami tastants, thus serving as the connection between molecular perception and culinary arts.

3.2.3 TRPM8

TRPM8 is inhibited by AA and other PUFAs, but activated by other phospholipase 2 products (PLA2) in the low micromolar range (Andersson, Nash, & Bevan, 2007; Vanden Abeele et al., 2006). Andersson and colleagues showed that AA and PUFAs such as eicosapentanoic acid (EPA) 20:5, and docosahexanoic acid (DHA) 22:6 inhibit the entrance of $[Ca^{2+}]_i$ through TRPM8 channels stimulated by menthol or cold in CHO cells using Fura-2-AM (Andersson et al., 2007).

De Petrocellis and colleagues have studied the effect of many endocannabinoids on TRPM8 using intracellular calcium measurements (De Petrocellis et al., 2007; De Petrocellis et al., 2008; De Petrocellis et al., 2011). Their findings show, not unexpectedly, that cannabinoids that activate thermo-TRPV channels antagonize TRPM8; especially cannabigerol, cannabidiol, anandamide, and *N*-arachidonoyl dopamine (NADA). The mechanism and the site through which they do so are unknown.

3.2.4 TRPV1

Many molecules downstream of PLA activate TRPV1: anandamide (Zygmunt et al., 1999), NADA (Huang et al., 2002), oleoylethanolamide (Ahern, 2003), etc. They appear to interact directly with the channel by binding to the same site as capsaicin (Sutton et al., 2005). Sutton et al. demonstrated electrophysiologically that mutation of residue S512, located at the N-terminal beginning of the S3 helix of the native human TRPV1 disrupts the gating by both capsaicin and NADA, but not by other stimuli as heat and acidic extracellular pH (Sutton et al., 2005). These authors predict that other vanilloid ligands bind to this same site.

Anandamide is an endocannabinoid synthesized from N-arachidonoyl phosphatidylethanolamine and catabolyzed by the enzyme fatty acid amide hydrolase (FAAH). This molecule has analgesic (Sparling, Giuffrida, Piomelli, Rosskopf, & Dietrich, 2003) and anticarcinogenic effects (De Petrocellis et al., 1998). It acts primarily through cannabinoid receptors, but it is also capable of activating TRPV1. Indeed, anandamide induces vasodilation through the activation of TRPV1 in primary sensory nerves (Zygmunt, Sørgård, Petersson, Johansson, & Högestätt, 2000). Zygmunt and colleagues used patch clamp recordings to test whether anandamide is capable of gating TRPV1. They found that micromolar concentrations of anandamide ($EC_{50} \sim 5\ \mu M$) are capable of opening the TRPV1 ion channel in both inside-out patches from *X. laevis* and in whole-cell recordings from HEK cells expressing the channel (Zygmunt et al., 2000). Anandamide is also capable of producing vasodilation in rat and guinea pig arteries. Although capsaicin produced more robust currents in electrophysiological experiments, anandamide was a more potent vasodilator (Zygmunt et al., 2000).

N-acyl taurines (NATs) are derivatives of AA that are also catabolyzed by FAAH. NATs are proapoptotic, anticarcinogenic cannabinoids (Chatzakos, Slätis, Djureinovic, Helleday, & Hunt, 2012), and it is likely that they play a role in insulin secretion in pancreatic β-cells (Waluk, Vielfort, Derakhshan, Aro, & Hunt, 2013) through TRP channel modulation. As anandamide, NATs are capable of activating TRPV1 (Saghatelian, McKinney, Bandell, Patapoutian, & Cravatt, 2006).

Hydroperoxyeicosatetraenoic acids (HPETEs) are products of lipoxygenase (LOX) activity on AA. They are relevant for the regulation of electrolyte flux, corticosterone synthesis, and release of insulin, histamine, and renin (Pace-Asciak & Asotra, 1989). Arachidonate-12-LOX is an enzyme that converts AA into 12-HPETE, a TRPV1 ligand (Hwang et al., 2000; Sexton, McDonald, Cayla, Thiemermann, & Ahluwalia, 2007). In 2000, Hwang and colleagues showed that TRPV1 activates in response to 12-HPETE in both DRG neurons and transfected HEK cells. Using single-channel recordings, they observed a similar activation in response to a subsaturating concentration of 1 μM capsaicin and 10 μM 12-HPETE (Hwang et al., 2000). The dose–response curve of 12-HPETE attains an EC_{50} of 8 μM and a Hill coefficient of 2.1, suggesting that more than one 12-HPETE binds to the channel. Other metabolites from LOXs, all related to 12-HPETE, were also tested. Among these, 15-HPETE, 5-hydroxyeicosatetraenoic acid (5-HETE), leukotriene B4, and 15-HETE were found to activate TRPV1, although to a lesser extent (see Figure 3(C) from Hwang et al., 2000). Additionally, 20-HETE, another derivative of AA produced by enzymes from the cytochrome P450 family, was found to activate TRPV1 in both native and heterologous systems (Wen et al., 2012).

3.2.5 TRPV3

Hu and colleagues showed that lipidic molecules downstream of PLA are capable of potentiating TRPV3 channel responses to 2APB. Using transfected HEK cells for both inside-out and outside-out single-channel experiments, these authors demonstrated that the amount of opening of TRPV3 increased from 0 to ~0.3 (inside-out) or ~0.1 (outside-out) in 2-s bins in the presence of 2APB (Hu et al., 2006). In contrast, when AA was added concomitantly with 2APB, the number of opening increased by 10- and 5-fold, respectively. 5,8,11-ETI and ETYA had similar potentiating effects (Hu et al., 2006). This result is important since several GPCRs that can lead to PLC activation, such as P2Y and LPA receptors (LPARs), are expressed in keratinocytes (Greig, Linge, Cambrey, & Burnstock, 2003; Ross, Parker, Whitaker, & Reynolds, 2008). P2Y is activated by ATP, which is released during platelet activation due to skin damage (Okazaki, Zhang, Suzuki, & Sakamoto, 1988) and LPARs are activated by LPA, which is also released during inflammation (Barekzi, Roman, Hise, Georas, & Steinke, 2006; Bot et al., 2013; Gierse et al., 2010). The importance of these events can be summarized as follows: generation of AA through PLC and subsequent PLA activation during tissue damage can activate keratinocytes through a TRPV3-dependent mechanism, and keratinocytes, in turn, might produce painful responses through sensory neurons (Shipton, 2013).

3.2.6 TRPV4

Using whole-cell patch clamp experiments, Watanabe and colleagues demonstrated in 2002 that 4a-PDD and PMA activate TRPV4 in transfected HEK cells (Watanabe et al., 2002a). These molecules are similar to DAG and anandamide, which suggested that TRPV4 might be activated by a product downstream of PLC. Later, using a clever combination of AA metabolism blockers and intracellular calcium measurements, this same group identified the endogenous ligand acting in TRPV4 channel, namely, 5,6-EET (Watanabe et al., 2003). This finding was confirmed by single-channel assays from mouse aorta endothelial cells in which application of 3 μM 5,6-EET increased the opening probability of TRPV4 (Watanabe et al., 2003). This ion channel is also activated by the endocannabinoid-like NATs (Saghatelian et al., 2006).

It is perplexing why TRPV4 is upregulated both by PIP_2 and a product downstream of its degradation, namely, 5,6-EET. Fernandes and colleagues showed that IP3 sensitizes TRPV4 for 5,6-EET activation (Fernandes et al., 2008), so it seems plausible that both phosphoinositides, PIP_2 and IP3, bind to TRPV4. Alternatively, it is possible that IP3, but not PIP_2, is the molecule making a synergistic interaction with 5,6-EET in native systems such as ciliate epithelial cells.

4. N-3 PUFAS AND DERIVATIVES

Omega-3 fatty acids are essential lipids to humans and must be obtained through the diet since we are not capable of synthesizing them (Crupi, Marino, & Cuzzocrea, 2013). They possess a double carbon (C=C) bond in the third position counting from the end of the acyl chain. EPA and DHA can be obtained from marine animals, while LA is mainly found in plants.

Oxidized linoleic acid metabolites (OLAMs) are derivatives of LA produced in skin during heat damage and are capable of sensitizing the CNS to allodynia and hyperalgesia.

Resolvins are molecules that promote resolution of cellular inflammation, allowing inflamed tissues to return to homeostasis (Gilroy et al., 2003; Morales-Lázaro et al., 2013). They are derived from DHA and EPA through the COX-2/LOX pathways (Janakiram & Rao, 2009). EPA-derived resolvins are called RvEs and DHA-derived ones are called RvDs. There is an increasing interest in resolvins since they ameliorate chronic pain (Lee, 2012; Van De Ven & Ji, 2013), prevent cardiac disease (Bang, Dyerberg, & Hjøorne, 1976; Frangogiannis, 2012) and mental disorders (Bondi et al., 2013), and improve cognition (Montgomery, Burton, Sewell, Spreckelsen, & Richardson, 2013).

4.1 TRPV1

Omega-3 PUFAs (n-3 PUFAs), DHA, EPA, and linolenic acid (LNA) are all capable of modulating TRPV1 activity (Matta, Miyares, Ahern, & Matta, 2007). These PUFAs possess a double bond (C=C) in the third position from the end of the acyl chain and are essential to physiological functions in humans (Robinson & Mazurak, 2013). Matta and colleagues showed that physiological concentrations of DHA, EPA, and LNA promote activation of TRPV1 with similar potency, but different efficacy. Activation of rat TRPV1 by DHA, EPA, and LNA was augmented when residues S502—located at the beginning of the intracellular N-terminal region of the S3 transmembrane helix—and S800—in the C-terminal portion of the protein—were phosphorylated by PKC. TRPV1 gating by these molecules seems to be through a direct binding to the channel since resiniferatoxin (RTX), a superagonist of TRPV1, was dose dependently displaced by DHA and EPA (Matta et al., 2007). Moreover, the activation by these PUFAs enhances proton-activated currents but inhibits those promoted by vanilloids, suggesting a common binding site with capsaicin (Matta et al., 2007). It was later shown that capsaicin and many PUFAs bind indeed to the same portion of the protein (Sutton et al., 2005).

OLAMs are endogenous ligands of the TRPV1 channel that are generated during spinal cord depolarization. They may play a role in sensitization of the CNS to allodynia and hyperalgesia (Patwardhan, Scotland, Akopian, & Hargreaves, 2009). They also elicit painful responses in vivo through peripheral nociceptors (Patwardhan et al., 2010). Synthetic and endogenous OLAMs like 9-hydroxyoctadecadienoic acid (9-HODE), 13-HODE, 9-odooctadecadienoic acid (9-odoODE), and 13-oxoODE possess the ability to gate TRPV1 channel in both native and heterologous systems, with 9-HODE being the most potent activator as tested in CHO cells (Patwardhan et al., 2009). It is interesting to note that OLAMs are generated in rat and mice skin during exposure to noxious heat. Although TRPV1 is intrinsically sensitive to heat (Cao et al., 2013a), the generation of OLAMs may play a role in the response to high temperatures in vivo.

Resolvins are thought to act through GPCRs rather than directly on TRP channels (Park et al., 2011). They are potent inhibitors of TRPV1 activity, especially RvD2 ($IC_{50} = 0.1$ nM). It has been shown that resolvins have TRPV1 inhibitory activity in vivo (Park et al., 2011; Xu et al., 2010), which renders them as possibly interesting tools for pain control.

5. OXIDIZED LIPIDS

Oxidative stress is a state of imbalance between the production of reactive oxygen species (ROS) and the capacity of a cell to remove them or to repair the damage they produce. One of the effects that oxidative stress produces on cell membranes is lipid peroxidation. Lipid peroxidation results in potentially pathological states such as inflammation and atherosclerosis (Barrera, 2012). Among the oxidized lipids formed during oxidative stress, 4-hydroxynonenal (HNE) is one of the most reactive ones and therefore one of the most studied (Barrera, 2012; Pizzimenti et al., 2010). Nitrooleic acid ($OA\text{-}NO_2$) is another important oxidized lipid produced during inflammation-induced nitrative stress (Radi, 2004).

5.1 TRPA1

TRPA1 is linked to the perception of damage due to oxidative stress. Both HNE and $OA\text{-}NO_2$ are capable of activating TRPA1 on transfected HEK cells, as assessed by calcium measurements (Taylor-Clark, Ghatta, Bettner, & Undem, 2009; Trevisani et al., 2007). The effects of HNE were also tested electrophysiologically in voltage clamp experiments of TRPA1-expressing *X. laevis* oocytes, showing that HNE produces a current similar to that of 2APB, a synthetic compound that activates several thermo-TRPs (Hu, Grandl, Bandell, Petrus, & Patapoutian, 2009).

$OA\text{-}NO_2$ might possess a physiological role in the perception of inflammation since it activates TRPA1 in DRG neurons from adult rats at concentrations in the low micromolar range. (Sculptoreanu, Artim, & de Groat, 2009).

Many electrophilic molecules that activate TRPA1 such as 15dPGJ2, HNE, and $9\text{-}OA\text{-}NO_2$ do so by covalent modification of intracellular cysteines and lysine residues (Kukkonen, 2011), just as allyl isothiocyanate (Macpherson et al., 2007; Salazar et al., 2008) and dibenzoxazepine (Brône et al., 2008). When these residues are mutated, the activation of TRPA1 by electrophilic molecules is lost (Hinman, Chuang, Bautista, & Julius, 2006; Trevisani et al., 2007).

5.2 TRPM2

TRPM2 can function as a redox sensor and it gates in response to molecules generated during oxidative stress (Hara et al., 2002). Using intracellular Ca^{2+} measurements and whole-cell patch clamp experiments, Hara et al. and

Wehage et al. showed that TRPM2 is acti-vated by H_2O_2, and by β-NAD^+ and AA in a dose-dependent manner (Hara et al., 2002; Wehage et al., 2002). It is notable that AA is only capable of producing a response in conditions in which the intracellular structure and composition are maintained (such as the perforated patch variation of the whole-cell mode of patch clamp) thus suggesting that lipid peroxidation alone is not sufficient to activate the channel, as opposed to TRPA1. It is possible that the physiological function of AA is the sensitization of the channel rather than its activation. Physiologically, activation of TRPM2 by ROS might be especially important in cells subject to great oxidative stress, such as those from the immune system like phagocytes and monocytes (Yamamoto, Takahashi, & Mori, 2010).

5.3 TRPV1

TRPV1 responds to OA-NO_2 in DRG neurons from rats. Intracellular calcium measurements with Fura-2 and whole-cell patch clamp experiments by Sculptoreanu and colleagues showed that, in addition to promoting currents through TRPA1, OA-NO_2 is capable of activating TRPV1 in the physiologically relevant, micromolar range (Sculptoreanu et al., 2010).

6. LYSOPHOSPHOLIPIDS

Lysophospholipids (LPLs) are bioactive molecules that possess a large polar or charged head and a single hydrophobic carbon chain. Most of these compounds act through GPCRs, but it has been shown that they are capable of activating TRP channels directly (Nieto-Posadas et al., 2011) or by modifying mechanical properties of biological membranes such as the elastic modulus or the curvature (Andersson et al., 2007). They are usually products of PL activity on sphingolipids or glycerophospholipids. Among the most important molecules of this group are LPA, which accumulates during tissue damage and ischemic conditions; lysophosphatidylcholine (LPC), which promotes the removal of myelin from erythrocytes; lysophosphatidylinositol (LPI), which activates GPR55, a GPCR; and sphingosine-1-phosphate (S1P), which regulates vascular and immune systems.

6.1 TRPM8

LPLs such as LPC and LPI activate TRPM8 in both HEK (Vanden Abeele et al., 2006) and CHO cells (Andersson et al., 2007). Single-channel experiments showed that the open probability of TRPM8 at 37 °C increased with the intracellular application of 3 μM LPC, and previous experiments by Abeele and colleagues showed that LPI is also capable of activating TRPM8-mediated currents in single-channel recordings when applied on the intracellular side. The mechanism of action is not yet fully understood, but it is highly likely that LPC and LPI modify the mechanical properties of the membrane (Lundbaek, Andersen, & Lundbeek, 1994) and that this modification leads to the activation of TRPM8. Indeed, extracellularly applied chlorpromazine 1 μM, which promotes changes in membrane properties such as invaginations, ablated menthol-activated currents, supporting the idea that TRPM8 is modulated by mechanical properties of the membrane (Andersson et al., 2007). This is a significant finding since TRPM8 is known to be expressed on afferent neurons innervating the bladder and the expression has been linked to pathophysiological conditions in the urinary tract (Lashinger et al., 2008; Sun, Li, Dong, & Rong, 2010). TRPM8 might function as a transducer of mechanical stimuli in this organ, thus making it an attractive therapeutic target for overactive bladder and painful bladder syndrome.

It is interesting that two seemingly opposing molecules, PIP_2 and products downstream of PLA2-like LPLs, are capable of eliciting the same effect on TRPM8. Abeele and colleagues found a partial answer for this quandary: while menthol and PIP_2 destabilize the closed state, LPLs act by stabilizing the open state (Vanden Abeele et al., 2006). Perhaps other lipids or products of enzymes other than PLA2 can result in similar physical effects on the membrane, thus generating a synergistic rather than a complementary effect.

6.2 TRPV1

LPA is a small bioactive lipid long implicated in the generation of chronic neuropathic pain (Inoue, Yamaguchi, Kawakami, Chun, & Ueda, 2006; Inoue et al., 2004). This molecule is the product of LPC breakdown by autotaxin, and is capable of direct activation of TRPV1 through a lipid-binding site in the C-terminal domain of the channel (Nieto-Posadas et al., 2011). For TRPV1, a novel scenario was proposed by our laboratory for interaction of this molecule with the channel where LPA does not require the cellular context to exert its action on TRPV1, since excised patches from transfected HEK cells expressing TRPV1 display currents when LPA is applied on the intracellular side. By using a combination of electrophysiological and biochemical techniques, we found that LPA binds to the K710 residue of the proximal C-terminus of TRPV1 (Nieto-Posadas et al., 2011), a result that was later confirmed by another group using TRPV1 reconstituted into liposomes (Cao et al., 2013a). Moreover, the structural requirements of LPA and similar to it have been described by our laboratory. Specifically, chains of at least 18 carbons, a charged head group and the presence of one insaturation are needed for activation of TRPV1 by these lipids (Morales-Lézaro et al., 2014).

7. CHOLESTEROL AND STEROIDS

Cholesterol is a major component of cell membranes and is a precursor of many important compounds, including steroids. The core of these molecules is composed of 20 carbon atoms that form four amalgamated rings: three cyclohexane rings and one cyclopentane ring.

Cholesterol is an important component of membrane microdomains that increases mechanical strength (Brown & London, 1998; Schroeder, Ahmed, Zhu, London, & Brown, 1998). It regulates ion channel function by changing mechanical properties of the membrane (Barrera, Fendos, & Engelman, 2012; Khelashvili & Harries, 2013; Lundbaek, Collingwood, Ingólfsson, Kapoor, & Andersen, 2010), by direct interaction with ion channels (Dopico, Bukiya, & Singh, 2012; Picazo-Juárez et al., 2011), and by maintaining proteins in microdomains (for review, see Levitan, Fang, Rosenhouse-Dantsker, & Romanenko, 2010).

Steroids are hormone products of cholesterol metabolism. They are modulators of many physiological processes such as inflammation, immune functions, and development of sexual characteristics and some of them exert their functions through modulation of ion channels (Bäckström et al., 2013; Ho, 2013).

7.1 Cholesterol and TRPV1

Cholesterol exerts an inhibitory effect on the TRPV1 channel through direct interaction with a binding site located on the S5 segment (Picazo-Juárez et al., 2011). On inside-out membrane patches from both HEK cells and DRG neurons, it was demonstrated by our laboratory that removal of cholesterol by methyl-beta-cyclodextrin (MβCD) produced no effect on currents activated by capsaicin and temperature. However, addition of cholesterol to the patches inactivated TRPV1 channels in a reversible manner (Picazo-Juárez et al., 2011). Cholesterol was found to bind to specific sites along the S5 segment of the channel on cholesterol recognition amino acid consensus motifs (Picazo-Juárez et al., 2011). Humans display polymorphism in these sites, which confer TRPV1 isoforms different physiological sensitivities to cholesterol. Since membrane microdomains are rich in sphingolipids and cholesterol, perhaps TRPV1 channels become inactive when they enter these microdomains (Morales-Lázaro et al., 2013), but further experiments are required to address this issue (Picazo-Juárez et al., 2011).

When exposed to agonists for periods longer than 5 min, TRPV1 undergoes conformational changes that effectively dilate the ion channel pore and large ions, such as N-methyl-D-glucamine, are allowed to permeate (Chung et al., 2008). A work by Jansson and colaborators used whole-cell electrophysiological experiments to show that depletion of cholesterol from cell membranes inhibits this process (Jansson et al., 2013). By gas chromatography–mass spectrometry, Jansson and colleagues showed that the extent of cholesterol depletion they could attain by application of 2.5 mM MβCD:cholesterol (10:1) was ~54%. When the channel was activated with 1 μM capsaicin, the depletion of cholesterol produced a decrease in permeability of NMDG by 70% with respect to sodium, leaving the reversal potential for sodium unaffected (Jansson et al., 2013).

Other steroids modulate the activity of TRPV1 in an indirect fashion. For instance, 17-beta-estradiol increases the expression of this ion channel in the hippocampus of rats (Wu et al., 2010), and it also reduces TRPV1 activation by capsaicin in DRG neurons by a mechanism downstream of estrogen receptor signaling pathways (Xu, Cheng, Keast, & Osborne, 2008).

7.2 Pregnenolone, pregnenolone sulphate, other steroids, and TRPM3

Pregnenolone is a precursor for the synthesis of corticoids and androgens, and it displays neuroactive properties. It enhances learning, memory, and myelination (Marx et al., 2011).

Wagner et al. described the activation of TRPM3 by the neuroactive steroid pregnenolone sulphate (PS) (Wagner et al., 2008). Using calcium measurements, these authors determined that TRPM3-expressing HEK293 cells, but not unstransfected ones, displayed large $[Ca^{2+}]_i$ increases in the presence of PS. Pregnenolone alone produced a small increase in $[Ca^{2+}]_i$, and other steroids such as cholesterol and progesterone (P4) did not promote any effect on TRPM3-mediated currents. The EC_{50} for PS was 12 μM, as determined from whole-cell patch clamp experiments at 80 mV.

On the other hand, P4, another steroid, inhibits TRPM3 by directly binding to the channel (Majeed et al., 2012). Using calcium measurements and whole-cell patch clamp experiments, Majeed and colleagues showed that this inhibition can be detected at physiological concentrations of P4. Furthermore, when a high, nonphysiological concentration of 10 μM of the steroid was used, a robust effect on channel inhibition was seen, and this effect was independent of PS. Direct P4 binding to TRPM3 was assessed by overlay assays (Majeed et al., 2012).

8. OTHER LIPIDS

8.1 Sphingosine and TRPM3

Sphingosine (SPH) is part of the family of sphingolipids similar to sphingomyelin. A phosphorylated version of SPH, sphingosine 1-phosphate, modulates cell proliferation, survival, and migration (Pyne & Pyne, 2011).

Using Ca^{2+} measurements and electrophysiological essays, Grimm and colleagues determined the activation of TRPM3 by extracellular application of SPH (Grimm, Kraft, Schultz, & Harteneck, 2005). This molecule is capable of promoting an intracellular Ca^{2+} increase in cells that express TRPM3. The effect seems to be specific for TRPM3 and independent of PKC since the rise of $[Ca^{2+}]_i$ is seen neither in unstransfected cells nor in cells transfected with other TRP channels (Grimm et al., 2005).

8.2 Farnesyl pyrophosphate and TRPV3

FPP is a lipidic molecule produced by the enzyme FPP synthase (Poulter, Argyle, & Mash, 1978) through the mevalonate pathway. Using patch clamp and calcium measurements, Bang et al. showed that FPP can activate TRPV3 from both the intracellular and extracellular sides (Bang, Yoo, Yang, Cho, & Hwang, 2010). Moreover, they found that FPP is a potent and specific agonist of TRPV3, since it does not activate other TRP channels and the EC_{50} obtained from $[Ca^{2+}]$ measurements is in the high nanomolar range (~130 nM) (Bang et al., 2010). It seems that FPP might be a physiologically relevant molecule since it activates TRPV3 in human keratinocytes and can produce a painful response when injected intradermally into the hind paws of mice after a 3 h pretreatment with carrageenan (Bang et al., 2010). Also, the physiological concentration of FPP in plasma is in the nanomolar range (~15 nM) (Saisho, Morimoto, & Umeda, 1997). Bang and colleagues propose that certain conditions, such as hyperemia and bleeding, can promote TRPV3 activation and subsequent pain (Bang et al., 2010). This mechanism is possible since several groups have shown that keratinocytes, which display rich TRPV3 expression, are capable of transmitting information to sensory neurons (Koizumi et al., 2004; Mandadi et al., 2009; Radtke, Rennekampff, Reimers, Vogt, & Kocsis, 2013).

8.3 Linoleic acid and TRPV3

LA, a precursor in the synthesis of AA, is capable of alleviating open channel block (OCB) by facilitating the passage of multivalent ions that block TRPV3, such as Ca^{2+} and La^{3+} (Parnas, Peters, & Minke, 2009). Parnas et al. used the whole-cell patch clamp technique on transfected S2 cells and showed that extracellular application of 40 μM LA augmented the current activated by 2APB. Moreover, since LA alleviates the OCB, the current–voltage (I–V) relationships of TRPV3 channels are linearized because this channel is not intrinsically voltage dependent. On the contrary, application of LA does not linearize the I–V curves of TRPV1 and TRPM8 because these channels are intrinsically voltage sensitive (Cao et al., 2013a; Hui, Guo, & Feng, 2005).

9. CONCLUDING REMARKS

Many endogenous and exogenous molecules activate thermo-TRP channels, and many of them are of a lipidic nature. These include a large number of metabolites downstream of GPCR signaling, notably PIP_2, which is essential for the function of several TRP channels. Many of these lipid molecules have functions in vivo, for instance, anandamide produces vasodilation, while highly electrophilic molecules, such as oxidized lipids, activate painful responses via TRPV1 and TRPA1 through a mechanism involving covalent modification of reactive cysteine or lysine residues of these channels.

DAG modulates the activity of several TRP channels, including those first discovered in the fly, but the site of action of DAG has been only proposed for TRPV1. Indeed, a wide variety of compounds such as capsaicin and endocannabinoids seem to exert their effects through an intracellular site at the N-terminal beginning of the S3 helix on this channel.

Some thermo-TRP channels have been suggested to act as sensors of the mechanical properties of the membrane (Harrington et al., 2011; Petrus et al., 2007; Walder et al., 2012). Among these, TRPV1 (LPA, cholesterol), TRPM3 (steroids), and TRPM8 (LPLs) are sensitive to the presence of lipidic molecules that modify mechanical properties of membranes and

TRPM8 has been shown to respond directly to the curvature of the membrane (Andersson et al., 2007).

Efforts have been made to reveal the mechanism of action of lipids on TRP channels, but this information is very limited and the sites on which they bind are broadly unknown. Although structural determinants for some lipid interactions for a couple of thermo-TRPs are known, notably TRPV1 and TRPA1 (Bradshaw et al., 2013), many issues remain to be addressed, for instance, the conformational changes that lipids produce on these ion channels (Garcia-Elias et al., 2013), the binding site for these molecules on their targets, and the mechanisms of action through which lipids modulate thermo-TRPs.

ACKNOWLEDGMENTS

This work was supported by grants IN204111 and IN200314 from Dirección General de Asuntos del Personal Académico (DGAPA)-Universidad Nacional Autónoma de México (UNAM); grant 129474 from CONACyT; a scholarship from the Marcos Moshinsky foundation to T.R.; and a CONACyT graduate scholarship to E.H.-G.

REFERENCES

Ahern, G. P. (2003). Activation of TRPV1 by the satiety factor oleoylethanolamide. *Journal of Biological Chemistry, 278*(33), 30429–30434. http://dx.doi.org/10.1074/jbc.M305051200.

Andersson, D. A., Nash, M., & Bevan, S. (2007). Modulation of the cold-activated channel TRPM8 by lysophospholipids and polyunsaturated fatty acids. *Journal of Neuroscience: The Official Journal of the Society for Neuroscience, 27*(12), 3347–3355. http://dx.doi.org/10.1523/JNEUROSCI.4846-06.2007.

Bäckström, T., Bixo, M., Johansson, M., Nyberg, S., Ossewaarde, L., Ragagnin, G., et al. (2013). Allopregnanolone and mood disorders. *Progress in Neurobiology*, 88–94. http://dx.doi.org/10.1016/j.pneurobio.2013.07.005.

Bandell, M., Story, G. M., Hwang, S. W., Viswanath, V., Eid, S. R., Petrus, M. J., et al. (2004). Noxious cold ion channel TRPA1 is activated by pungent compounds and bradykinin. *Neuron, 41*(6), 849–857.

Bang, H. O., Dyerberg, J., & Hjøorne, N. (1976). The composition of food consumed by Greenland Eskimos. *Acta Medica Scandinavica, 200*(1–2), 69–73.

Bang, S., Yoo, S., Yang, T.-J., Cho, H., & Hwang, S. W. (2010). Farnesyl pyrophosphate is a novel pain-producing molecule via specific activation of TRPV3. *Journal of Biological Chemistry, 285*(25), 19362–19371. http://dx.doi.org/10.1074/jbc.M109.087742.

Barekzi, E., Roman, J., Hise, K., Georas, S., & Steinke, J. W. (2006). Lysophosphatidic acid stimulates inflammatory cascade in airway epithelial cells. *Prostaglandins, Leukotrienes, and Essential Fatty Acids, 74*(6), 357–363. http://dx.doi.org/10.1016/j.plefa.2006.03.004.

Barrera, G. (2012). Oxidative stress and lipid peroxidation products in cancer progression and therapy. *ISRN Oncology, 2012*, 137289. http://dx.doi.org/10.5402/2012/137289.

Barrera, F. N., Fendos, J., & Engelman, D. M. (2012). Membrane physical properties influence transmembrane helix formation. *Proceedings of the National Academy of Sciences of the*

United States of America, 109(36), 14422–14427. http://dx.doi.org/10.1073/pnas.1212665109.

Bender, F. L. P., Mederos Y Schnitzler, M., Li, Y., Ji, A., Weihe, E., Gudermann, T., et al. (2005). The temperature-sensitive ion channel TRPV2 is endogenously expressed and functional in the primary sensory cell line F-11. *Cellular Physiology and Biochemistry: International Journal of Experimental Cellular Physiology, Biochemistry, and Pharmacology, 15*(1–4), 183–194. http://dx.doi.org/10.1159/000083651.

Benham, C. D., Gunthorpe, M. J., & Davis, J. B. (2003). TRPV channels as temperature sensors. *Cell Calcium, 33*(5–6), 479–487.

Bingham, C. O., & Austen, K. F. (1999). Phospholipase A2 enzymes in eicosanoid generation. *Proceedings of the Association of American Physicians, 111*(6), 516–524.

Bödding, M. (2007). TRPM6: a Janus-like protein. In *Handbook of experimental pharmacology* (pp. 299–311). http://dx.doi.org/10.1007/978-3-540-34891-7_18.

Bondi, C. O., Taha, A. Y., Tock, J. L., Totah, N. K. B., Cheon, Y., Torres, G. E., et al. (2013). Adolescent behavior and dopamine availability are uniquely sensitive to dietary omega-3 fatty acid deficiency. *Biological Psychiatry, 75*(1). http://dx.doi.org/10.1016/j.biopsych.2013.06.007.

Bot, M., de Jager, S. C. A., MacAleese, L., Lagraauw, H. M., van Berkel, T. J. C., Quax, P. H. A., et al. (2013). Lysophosphatidic acid triggers mast cell-driven atherosclerotic plaque destabilization by increasing vascular inflammation. *Journal of Lipid Research, 54*(5), 1265–1274. http://dx.doi.org/10.1194/jlr.M032862.

Bradshaw, H. B., Raboune, S., & Hollis, J. L. (2013). Opportunistic activation of TRP receptors by endogenous lipids: exploiting lipidomics to understand TRP receptor cellular communication. *Life Sciences, 92*(8–9), 404–409. http://dx.doi.org/10.1016/j.lfs.2012.11.008.

Brauchi, S., Orta, G., Mascayano, C., Salazar, M., Raddatz, N., Urbina, H., et al. (2007). Dissection of the components for PIP_2 activation and thermosensation in TRP channels. *Proceedings of the National Academy of Sciences of the United States of America, 104*(24), 10246–10251. http://dx.doi.org/10.1073/pnas.0703420104.

Brauchi, S., Orta, G., Salazar, M., Rosenmann, E., & Latorre, R. (2006). A hot-sensing cold receptor: C-terminal domain determines thermosensation in transient receptor potential channels. *Journal of Neuroscience: The Official Journal of the Society for Neuroscience, 26*(18), 4835–4840. http://dx.doi.org/10.1523/JNEUROSCI.5080-05.2006.

Brône, B., Peeters, P. J., Marrannes, R., Mercken, M., Nuydens, R., Meert, T., et al. (2008). Tear gasses CN, CR, and CS are potent activators of the human TRPA1 receptor. *Toxicology and Applied Pharmacology, 231*(2), 150–156. http://dx.doi.org/10.1016/j.taap.2008.04.005.

Brown, D. A., & London, E. (1998). Structure and origin of ordered lipid domains in biological membranes. *Journal of Membrane Biology, 164*(2), 103–114.

Calixto, J. B., Kassuya, C. A. L., André, E., & Ferreira, J. (2005). Contribution of natural products to the discovery of the transient receptor potential (TRP) channels family and their functions. *Pharmacology & Therapeutics, 106*(2), 179–208. http://dx.doi.org/10.1016/j.pharmthera.2004.11.008.

Campbell, W. B., & Fleming, I. (2010). Epoxyeicosatrienoic acids and endothelium-dependent responses. *Pflügers Archiv: European Journal of Physiology, 459*(6), 881–895. http://dx.doi.org/10.1007/s00424-010-0804-6.

Cao, E., Cordero-Morales, J. F., Liu, B., Qin, F., & Julius, D. (2013). TRPV1 channels are intrinsically heat sensitive and negatively regulated by phosphoinositide lipids. *Neuron,* 77(4), 667–679. http://dx.doi.org/10.1016/j.neuron.2012.12.016.

Cao, E., Liao, M., Cheng, Y., & Julius, D. (2013). TRPV1 structures in distinct conformations reveal activation mechanisms. *Nature, 504*(7478), 113–118. http://dx.doi.org/10.1038/nature12823.

Caterina, M. J., Rosen, T. A., Tominaga, M., Brake, A. J., & Julius, D. (1999). A capsaicin-receptor homologue with a high threshold for noxious heat. *Nature, 398*(6726), 436–441. http://dx.doi.org/10.1038/18906.

Caterina, M. J., Schumacher, M. a, Tominaga, M., Rosen, T. a, Levine, J. D., & Julius, D. (1997). The capsaicin receptor: a heat-activated ion channel in the pain pathway. *Nature, 389*(6653), 816–824. http://dx.doi.org/10.1038/39807.

Chatzakos, V., Slätis, K., Djureinovic, T., Helleday, T., & Hunt, M. C. (2012). N-acyl taurines are anti-proliferative in prostate cancer cells. *Lipids, 47*(4), 355–361. http://dx.doi.org/10.1007/s11745-011-3639-9.

Chianese, G., Fattorusso, E., Putra, M. Y., Calcinai, B., Bavestrello, G., Moriello, A. S., et al. (2012). Leucettamols, bifunctionalized marine sphingoids, act as modulators of TRPA1 and TRPM8 channels. *Marine Drugs, 10*(11), 2435–2447. http://dx.doi.org/10.3390/md10112435.

Choi, S.-H., Aid, S., Choi, U., & Bosetti, F. (2010). Cyclooxygenases-1 and -2 differentially modulate leukocyte recruitment into the inflamed brain. *The Pharmacogenomics Journal, 10*(5), 448–457. http://dx.doi.org/10.1038/tpj.2009.68.

Chung, M.-K., Güler, A. D., & Caterina, M. J. (2008). TRPV1 shows dynamic ionic selectivity during agonist stimulation. *Nature Neuroscience, 11*(5), 555–564. http://dx.doi.org/10.1038/nn.2102.

Chyb, S., Raghu, P., & Hardie, R. C. (1999). Polyunsaturated fatty acids activate the Drosophila light-sensitive channels TRP and TRPL. *Nature, 397*(6716), 255–259. http://dx.doi.org/10.1038/16703.

Clapham, D. E. (2003). TRP channels as cellular sensors. *Nature, 426*(6966), 517–524. http://dx.doi.org/10.1038/nature02196.

Contreras, M. A., Greiner, R. S., Chang, M. C., Myers, C. S., Salem, N., & Rapoport, S. I. (2000). Nutritional deprivation of alpha-linolenic acid decreases but does not abolish turnover and availability of unacylated docosahexaenoic acid and docosahexaenoyl-CoA in rat brain. *Journal of Neurochemistry, 75*(6), 2392–2400.

Cortright, D. N., & Szallasi, A. (2009). TRP channels and pain. *Current Pharmaceutical Design, 15*(15), 1736–1749.

Crupi, R., Marino, A., & Cuzzocrea, S. (2013). n-3 fatty acids: role in neurogenesis and neuroplasticity. *Current Medicinal Chemistry, 20*(24), 2953–2963.

Dai, Y., Wang, S., Tominaga, M., Yamamoto, S., Fukuoka, T., Higashi, T., et al. (2007). Sensitization of TRPA1 by PAR2 contributes to the sensation of inflammatory pain. *Journal of Clinical Investigation, 117*(7), 1979–1987. http://dx.doi.org/10.1172/JCI30951.

De Petrocellis, L., Ligresti, A., Moriello, A. S., Allarà, M., Bisogno, T., Petrosino, S., et al. (2011). Effects of cannabinoids and cannabinoid-enriched Cannabis extracts on TRP channels and endocannabinoid metabolic enzymes. *British Journal of Pharmacology, 163*(7), 1479–1494. http://dx.doi.org/10.1111/j.1476-5381.2010.01166.x.

De Petrocellis, L., Melck, D., Palmisano, A., Bisogno, T., Laezza, C., Bifulco, M., et al. (1998). The endogenous cannabinoid anandamide inhibits human breast cancer cell proliferation. *Proceedings of the National Academy of Sciences of the United States of America, 95*(14), 8375–8380.

De Petrocellis, L., Starowicz, K., Moriello, A. S., Vivese, M., Orlando, P., & Di Marzo, V. (2007). Regulation of transient receptor potential channels of melastatin type 8 (TRPM8): effect of cAMP, cannabinoid CB(1) receptors and endovanilloids. *Experimental Cell Research, 313*(9), 1911–1920. http://dx.doi.org/10.1016/j.yexcr.2007.01.008.

De Petrocellis, L., Vellani, V., Schiano-Moriello, A., Marini, P., Magherini, P. C., Orlando, P., et al. (2008). Plant-derived cannabinoids modulate the activity of transient receptor potential channels of ankyrin type-1 and melastatin type-8. *Journal of*

Pharmacology and Experimental Therapeutics, 325(3), 1007–1015. http://dx.doi.org/10.1124/jpet.107.134809.

Delgado, R., & Bacigalupo, J. (2009). Unitary recordings of TRP and TRPL channels from isolated *Drosophila* retinal photoreceptor rhabdomeres: activation by light and lipids. *Journal of Neurophysiology, 101*(5), 2372–2379. http://dx.doi.org/10.1152/jn.90578.2008.

Dhaka, A., Uzzell, V., Dubin, A. E., Mathur, J., Petrus, M., Bandell, M., et al. (2009). TRPV1 is activated by both acidic and basic pH. *Journal of Neuroscience: The Official Journal of the Society for Neuroscience, 29*(1), 153–158. http://dx.doi.org/10.1523/JNEUROSCI.4901-08.2009.

Dhaka, A., Viswanath, V., & Patapoutian, A. (2006). TRP ion channels and temperature sensation. *Annual Review of Neuroscience, 29*, 135–161. http://dx.doi.org/10.1146/annurev.neuro.29.051605.112958.

Doerner, J. F., Hatt, H., & Ramsey, I. S. (2011). Voltage- and temperature-dependent activation of TRPV3 channels is potentiated by receptor-mediated $PI(4,5)P_2$ hydrolysis. *Journal of General Physiology, 137*(3), 271–288. http://dx.doi.org/10.1085/jgp.200910388.

Dopico, A. M., Bukiya, A. N., & Singh, A. K. (2012). Large conductance, calcium- and voltage-gated potassium (BK) channels: regulation by cholesterol. *Pharmacology & Therapeutics, 135*(2), 133–150. http://dx.doi.org/10.1016/j.pharmthera.2012.05.002.

Earley, S. (2011). Endothelium-dependent cerebral artery dilation mediated by transient receptor potential and Ca^{2+}-activated K^+ channels. *Journal of Cardiovascular Pharmacology, 57*(2), 148–153. http://dx.doi.org/10.1097/FJC.0b013e3181f580d9.

Ellert-Miklaszewska, A., Ciechomska, I., & Kaminska, B. (2013). Cannabinoid signaling in glioma cells. *Advances in Experimental Medicine and Biology, 986*, 209–220. http://dx.doi.org/10.1007/978-94-007-4719-7_11.

Fernandes, J., Lorenzo, I. M., Andrade, Y. N., Garcia-Elias, A., Serra, S. A., Fernández-Fernández, J. M., et al. (2008). IP3 sensitizes TRPV4 channel to the mechano- and osmotransducing messenger 5'-6'-epoxyeicosatrienoic acid. *Journal of Cell Biology, 181*(1), 143–155. http://dx.doi.org/10.1083/jcb.200712058.

Filosa, J. A., & Iddings, J. A. (2013). Astrocyte regulation of cerebral vascular tone. *American Journal of Physiology. Heart and Circulatory Physiology, 305*(5), H609–H619. http://dx.doi.org/10.1152/ajpheart.00359.2013.

Flockerzi, V. (2007). An introduction on TRP channels. In *Handbook of experimental pharmacology* (pp. 1–19). http://dx.doi.org/10.1007/978-3-540-34891-7_1.

Frangogiannis, N. G. (2012). Regulation of the inflammatory response in cardiac repair. *Circulation Research, 110*(1), 159–173. http://dx.doi.org/10.1161/CIRCRESAHA.111.243162.

Garcia-Elias, A., Mrkonjic, S., Pardo-Pastor, C., Inada, H., Hellmich, U. a, Rubio-Moscardó, F., et al. (2013). Phosphatidylinositol-4,5-biphosphate-dependent rearrangement of TRPV4 cytosolic tails enables channel activation by physiological stimuli. *Proceedings of the National Academy of Sciences of the United States of America, 110*(23), 9553–9558. http://dx.doi.org/10.1073/pnas.1220231110.

García-Sanz, N., Fernández-Carvajal, A., Morenilla-Palao, C., Planells-Cases, R., Fajardo-Sánchez, E., Fernández-Ballester, G., et al. (2004). Identification of a tetramerization domain in the C terminus of the vanilloid receptor. *Journal of Neuroscience: The Official Journal of the Society for Neuroscience, 24*(23), 5307–5314. http://dx.doi.org/10.1523/JNEUROSCI.0202-04.2004.

Gierse, J., Thorarensen, A., Beltey, K., Bradshaw-Pierce, E., Cortes-Burgos, L., Hall, T., et al. (2010). A novel autotaxin inhibitor reduces lysophosphatidic acid levels in plasma and the site of inflammation. *Journal of Pharmacology and Experimental Therapeutics, 334*(1), 310–317. http://dx.doi.org/10.1124/jpet.110.165845.

Gijsen, H. J. M., Berthelot, D., Zaja, M., Brône, B., Geuens, I., & Mercken, M. (2010). Analogues of morphanthridine and the tear gas dibenz[b,f][1,4]oxazepine (CR) as extremely potent activators of the human transient receptor potential ankyrin 1 (TRPA1) channel. *Journal of Medicinal Chemistry, 53*(19), 7011–7020. http://dx.doi.org/10.1021/jm100477n.

Gilroy, D. W., Colville-Nash, P. R., McMaster, S., Sawatzky, D. A., Willoughby, D. A., & Lawrence, T. (2003). Inducible cyclooxygenase-derived 15-deoxy(Delta)12-14PGJ2 brings about acute inflammatory resolution in rat pleurisy by inducing neutrophil and macrophage apoptosis. *FASEB Journal: Official Publication of the Federation of American Societies for Experimental Biology, 17*(15), 2269–2271. http://dx.doi.org/10.1096/fj.02-1162fje.

Greig, A. V. H., Linge, C., Cambrey, A., & Burnstock, G. (2003). Purinergic receptors are part of a signaling system for keratinocyte proliferation, differentiation, and apoptosis in human fetal epidermis. *Journal of Investigative Dermatology, 121*(5), 1145–1149. http://dx.doi.org/10.1046/j.1523-1747.2003.12567.x.

Grimm, C., Kraft, R., Sauerbruch, S., Schultz, G., & Harteneck, C. (2003). Molecular and functional characterization of the melastatin-related cation channel TRPM3. *Journal of Biological Chemistry, 278*(24), 21493–21501. http://dx.doi.org/10.1074/jbc.M300945200.

Grimm, C., Kraft, R., Schultz, G., & Harteneck, C. (2005). Activation of the melastatin-related cation channel TRPM3 by D-erythro-sphingosine [corrected]. *Molecular Pharmacology, 67*(3), 798–805. http://dx.doi.org/10.1124/mol.104.006734.

Grubisha, O., Mogg, A. J., Sorge, J. L., Ball, L.-J., Sanger, H., Ruble, C. L. A., et al. (2013). Pharmacological profiling of the TRPV3 channel in recombinant and native assays. *British Journal of Pharmacology, 171*(10), 2631–2644. http://dx.doi.org/10.1111/bph.12303.

Gu, A. D., Lee, H., Iida, T., Shimizu, I., Tominaga, M., Caterina, M., et al. (2002). Heat-evoked activation of the ion channel, TRPV4. *Journal of Neuroscience: The Official Journal of the Society for Neuroscience, 22*(15), 6408–6414, 20026679.

Hara, Y., Wakamori, M., Ishii, M., Maeno, E., Nishida, M., Yoshida, T., et al. (2002). LTRPC2 Ca^{2+}-permeable channel activated by changes in redox status confers susceptibility to cell death. *Molecular Cell, 9*(1), 163–173.

Hardie, R. C. (2003). Regulation of TRP channels via lipid second messengers. *Annual Review of Physiology, 65*, 735–759. http://dx.doi.org/10.1146/annurev.physiol.65.092101.142505.

Harrington, A. M., Hughes, P. A., Martin, C. M., Yang, J., Castro, J., Isaacs, N. J., et al. (2011). A novel role for TRPM8 in visceral afferent function. *Pain, 152*(7), 1459–1468. http://dx.doi.org/10.1016/j.pain.2011.01.027.

Helliwell, R. M., & Large, W. A. (1997). Alpha 1-adrenoceptor activation of a non-selective cation current in rabbit portal vein by 1,2-diacyl-sn-glycerol. *The Journal of Physiology, 499*(Pt 2), 417–428.

Helliwell, R. J., McLatchie, L. M., Clarke, M., Winter, J., Bevan, S., & McIntyre, P. (1998). Capsaicin sensitivity is associated with the expression of the vanilloid (capsaicin) receptor (VR1) mRNA in adult rat sensory ganglia. *Neuroscience Letters, 250*(3), 177–180.

Henshall, S. M., Afar, D. E. H., Hiller, J., Horvath, L. G., Quinn, D. I., Rasiah, K. K., et al. (2003). Survival analysis of genome-wide gene expression profiles of prostate cancers identifies new prognostic targets of disease relapse. *Cancer Research, 63*(14), 4196–4203.

Hilgemann, D. W., Feng, S., & Nasuhoglu, C. (2001). The complex and intriguing lives of PIP_2 with ion channels and transporters. *Science's STKE: Signal Transduction Knowledge Environment, 2001*(111). http://dx.doi.org/10.1126/stke.2001.111.re19. re19–re19.

Hinman, A., Chuang, H.-H., Bautista, D. M., & Julius, D. (2006). TRP channel activation by reversible covalent modification. *Proceedings of the National Academy of Sciences of the United States of America, 103*(51), 19564–19568. http://dx.doi.org/10.1073/pnas.0609598103.

Ho, W.-S. V. (2013). Modulation by 17β-estradiol of anandamide vasorelaxation in normotensive and hypertensive rats: a role for TRPV1 but not fatty acid amide hydrolase. *European Journal of Pharmacology, 701*(1–3), 49–56. http://dx.doi.org/10.1016/j.ejphar.2013.01.002.

Hoffmann, A., Grimm, C., Kraft, R., Goldbaum, O., Wrede, A., Nolte, C., et al. (2010). TRPM3 is expressed in sphingosine-responsive myelinating oligodendrocytes. *Journal of Neurochemistry, 114*(3), 654–665. http://dx.doi.org/10.1111/j.1471-4159.2010.06644.x.

Hofmann, T., Chubanov, V., Gudermann, T., & Montell, C. (2003). TRPM5 is a voltage-modulated and Ca(2+)-activated monovalent selective cation channel. *Current Biology: CB, 13*(13), 1153–1158.

Hofmann, T., Obukhov, A. G., Schaefer, M., Harteneck, C., Gudermann, T., Schultz, G., et al. (1999). Direct activation of human TRPC6 and TRPC3 channels by diacylglycerol. *Nature, 397*(6716), 259–263. http://dx.doi.org/10.1038/16711.

Holendova, B., Grycova, L., Jirku, M., & Teisinger, J. (2012). PtdIns(4,5)P_2 interacts with CaM binding domains on TRPM3 N-terminus. *Channels (Austin, Texas), 6*(6), 479–482. http://dx.doi.org/10.4161/chan.22177.

Horrocks, L. a (1989). Sources for brain arachidonic acid uptake and turnover in glycerophospholipids. *Annals of the New York Academy of Sciences, 559*, 17–24.

Huang, S. M., Bisogno, T., Trevisani, M., Al-Hayani, A., De Petrocellis, L., Fezza, F., et al. (2002). An endogenous capsaicin-like substance with high potency at recombinant and native vanilloid VR1 receptors. *Proceedings of the National Academy of Sciences of the United States of America, 99*(12), 8400–8405. http://dx.doi.org/10.1073/pnas.122196999.

Huang, L., Shanker, Y. G., Dubauskaite, J., Zheng, J. Z., Yan, W., Rosenzweig, S., et al. (1999). Ggamma13 colocalizes with gustducin in taste receptor cells and mediates IP3 responses to bitter denatonium. *Nature Neuroscience, 2*(12), 1055–1062. http://dx.doi.org/10.1038/15981.

Huang, J., Zhang, X., & McNaughton, P. A. (2006). Modulation of temperature-sensitive TRP channels. *Seminars in Cell & Developmental Biology, 17*(6), 638–645. http://dx.doi.org/10.1016/j.semcdb.2006.11.002.

Hu, S. S.-J., Bradshaw, H. B., Benton, V. M., Chen, J. S.-C., Huang, S. M., Minassi, A., et al. (2009). The biosynthesis of N-arachidonoyl dopamine (NADA), a putative endocannabinoid and endovanilloid, via conjugation of arachidonic acid with dopamine. *Prostaglandins, Leukotrienes, and Essential Fatty Acids, 81*(4), 291–301. http://dx.doi.org/10.1016/j.plefa.2009.05.026.

Hu, H., Grandl, J., Bandell, M., Petrus, M., & Patapoutian, A. (2009). Two amino acid residues determine 2-APB sensitivity of the ion channels TRPV3 and TRPV4. *Proceedings of the National Academy of Sciences of the United States of America, 106*(5), 1626–1631. http://dx.doi.org/10.1073/pnas.0812209106.

Hui, K., Guo, Y., & Feng, Z.-P. (2005). Biophysical properties of menthol-activated cold receptor TRPM8 channels. *Biochemical and Biophysical Research Communications, 333*(2), 374–382. http://dx.doi.org/10.1016/j.bbrc.2005.05.123.

Hu, H., Xiao, R. U. I., Wang, C., Gao, N. A., Colton, C. K., Wood, J. D., et al. (2006). Potentiation of TRPV3 channel function by unsaturated fatty acids. *Journal of Cellular Physiology, 212*, 201–212. http://dx.doi.org/10.1002/jcp. September 2005.

Hwang, S. W., Cho, H., Kwak, J., Lee, S. Y., Kang, C. J., Jung, J., et al. (2000). Direct activation of capsaicin receptors by products of lipoxygenases: endogenous capsaicin-like substances. *Proceedings of the National Academy of Sciences of the United States of America, 97*(11), 6155–6160.

Inamura, K., Sano, Y., Mochizuki, S., Yokoi, H., Miyake, A., Nozawa, K., et al. (2003). Response to ADP-ribose by activation of TRPM2 in the CRI-G1 insulinoma cell line. *Journal of Membrane Biology, 191*(3), 201–207. http://dx.doi.org/10.1007/s00232-002-1057-x.

Inoue, M., Rashid, M. H., Fujita, R., Contos, J. J. A., Chun, J., & Ueda, H. (2004). Initiation of neuropathic pain requires lysophosphatidic acid receptor signaling. *Nature Medicine, 10*(7), 712–718. http://dx.doi.org/10.1038/nm1060.

Inoue, M., Yamaguchi, A., Kawakami, M., Chun, J., & Ueda, H. (2006). Loss of spinal substance P pain transmission under the condition of LPA1 receptor-mediated neuropathic pain. *Molecular Pain, 2*, 25. http://dx.doi.org/10.1186/1744-8069-2-25.

Janakiram, N. B., & Rao, C. V. (2009). Role of lipoxins and resolvins as anti-inflammatory and proresolving mediators in colon cancer. *Current Molecular Medicine, 9*(5), 565–579.

Jansson, E. T., Trkulja, C. L., Ahemaiti, A., Millingen, M., Jeffries, G. D., Jardemark, K., et al. (2013). Effect of cholesterol depletion on the pore dilation of TRPV1. *Molecular Pain, 9*, 1. http://dx.doi.org/10.1186/1744-8069-9-1.

Jaques, R. (1959). Arachidonic acid, an unsaturated fatty acid which produces slow contractions of smooth muscle and causes pain. Pharmacological and biochemical characterisation of its mode of action. *Helvetica Physiologica et Pharmacologica Acta, 17*, 255–267.

Jeske, N. A., Por, E. D., Belugin, S., Chaudhury, S., Berg, K. A., Akopian, A. N., et al. (2011). A-kinase anchoring protein 150 mediates transient receptor potential family V type 1 sensitivity to phosphatidylinositol-4,5-bisphosphate. *Journal of Neuroscience: The Official Journal of the Society for Neuroscience, 31*(23), 8681–8688. http://dx.doi.org/10.1523/JNEUROSCI.0020-11.2011.

Jin, M., Berrout, J., & O'Neil, R. (2011). In M. X. Zhu (Ed.), *Regulation of TRP channels by Osmomechanical stress*. CRC Press.

Julius, D., & Nathans, J. (2012). Signaling by sensory receptors. *Cold Spring Harbor Perspectives in Biology, 4*(1), a005991. http://dx.doi.org/10.1101/cshperspect.a005991.

Kamakura, T., Ishida, Y., Nakamura, Y., Yamada, T., Kitahara, T., Takimoto, Y., et al. (2013). Functional expression of TRPV1 and TRPA1 in rat vestibular ganglia. *Neuroscience Letters, 552*, 92–97. http://dx.doi.org/10.1016/j.neulet.2013.07.019.

Karashima, Y., Prenen, J., Meseguer, V., Owsianik, G., Voets, T., & Nilius, B. (2008). Modulation of the transient receptor potential channel TRPA1 by phosphatidylinositol 4,5-biphosphate manipulators. *Pflügers Archiv: European Journal of Physiology, 457*(1), 77–89. http://dx.doi.org/10.1007/s00424-008-0493-6.

Khelashvili, G., & Harries, D. (2013). How cholesterol tilt modulates the mechanical properties of saturated and unsaturated lipid membranes. *Journal of Physical Chemistry. B, 117*(8), 2411–2421. http://dx.doi.org/10.1021/jp3122006.

Kim, D., Cavanaugh, E. J., & Simkin, D. (2008). Inhibition of transient receptor potential A1 channel by phosphatidylinositol-4,5-bisphosphate. *American Journal of Physiology. Cell Physiology, 295*(1), C92–C99. http://dx.doi.org/10.1152/ajpcell.00023.2008.

Kim, A. Y., Tang, Z., Liu, Q., Patel, K. N., Maag, D., Geng, Y., et al. (2008). Pirt, a phosphoinositide-binding protein, functions as a regulatory subunit of TRPV1. *Cell, 133*(3), 475–485. http://dx.doi.org/10.1016/j.cell.2008.02.053.

Kinnamon, S. C. (2012). Taste receptor signalling - from tongues to lungs. *Acta physiologica (Oxford, England), 204*(2), 158–168. http://dx.doi.org/10.1111/j.1748-1716.2011.02308.x.

Kiselyov, K., van Rossum, D. B., & Patterson, R. L. (2010). TRPC channels in pheromone sensing. *Vitamins and Hormones, 83*, 197–213. http://dx.doi.org/10.1016/S0083-6729(10)83008-0.

Klein, R. M., Ufret-Vincenty, C. a, Hua, L., & Gordon, S. E. (2008). Determinants of molecular specificity in phosphoinositide regulation. Phosphatidylinositol (4,5)-bisphosphate ($PI(4,5)P_2$) is the endogenous lipid regulating TRPV1. *Journal of Biological Chemistry, 283*(38), 26208–26216. http://dx.doi.org/10.1074/jbc.M801912200.

Koizumi, S., Fujishita, K., Inoue, K., Shigemoto-Mogami, Y., Tsuda, M., & Inoue, K. (2004). Ca^{2+} waves in keratinocytes are transmitted to sensory neurons: the involvement of extracellular ATP and P2Y2 receptor activation. *Biochemical Journal, 380*(Pt 2), 329–338. http://dx.doi.org/10.1042/BJ20031089.

Kukkonen, J. P. (2011). A ménage à trois made in heaven: G-protein-coupled receptors, lipids and TRP channels. *Cell Calcium, 50*(1), 9–26. http://dx.doi.org/10.1016/j.ceca.2011.04.005.

Kwan, K. Y., Allchorne, A. J., Vollrath, M. A., Christensen, A. P., Zhang, D.-S., Woolf, C. J., et al. (2006). TRPA1 contributes to cold, mechanical, and chemical nociception but is not essential for hair-cell transduction. *Neuron, 50*(2), 277–289. http://dx.doi.org/10.1016/j.neuron.2006.03.042.

Lashinger, E. S. R., Steiginga, M. S., Hieble, J. P., Leon, L. A., Gardner, S. D., Nagilla, R., et al. (2008). AMTB, a TRPM8 channel blocker: evidence in rats for activity in overactive bladder and painful bladder syndrome. *American Journal of Physiology. Renal Physiology, 295*(3), F803–F810. http://dx.doi.org/10.1152/ajprenal.90269.2008.

Latorre, R., Zaelzer, C., & Brauchi, S. (2009). Structure-functional intimacies of transient receptor potential channels. *Quarterly Reviews of Biophysics, 42*(3), 201–246. http://dx.doi.org/10.1017/S0033583509990072.

Launay, P., Fleig, A., Perraud, A. L., Scharenberg, A. M., Penner, R., & Kinet, J. P. (2002). TRPM4 is a Ca^{2+}-activated nonselective cation channel mediating cell membrane depolarization. *Cell, 109*(3), 397–407.

Lee, C. H. (2012). Resolvins as new fascinating drug candidates for inflammatory diseases. *Archives of Pharmacal Research, 35*(1), 3–7. http://dx.doi.org/10.1007/s12272-012-0121-z.

Lee, N., Chen, J., Sun, L., Wu, S., Gray, K. R., Rich, A., et al. (2003). Expression and characterization of human transient receptor potential melastatin 3 (hTRPM3). *Journal of Biological Chemistry, 278*(23), 20890–20897. http://dx.doi.org/10.1074/jbc.M211232200.

Levitan, I., Fang, Y., Rosenhouse-Dantsker, A., & Romanenko, V. (2010). Cholesterol and ion channels. *Sub-Cellular Biochemistry, 51*, 509–549. http://dx.doi.org/10.1007/978-90-481-8622-8_19.

Liao, M., Cao, E., Julius, D., & Cheng, Y. (2013). Structure of the TRPV1 ion channel determined by electron cryo-microscopy. *Nature, 504*(7478), 107–112. http://dx.doi.org/10.1038/nature12822.

Lieben, L., & Carmeliet, G. (2012). The involvement of TRP channels in Bone homeostasis. *Frontiers in Endocrinology, 3*, 99. http://dx.doi.org/10.3389/fendo.2012.00099. August.

Liedtke, W., Choe, Y., Martí-Renom, M. A., Bell, A. M., Denis, C. S., Sali, A., et al. (2000). Vanilloid receptor-related osmotically activated channel (VR-OAC), a candidate vertebrate osmoreceptor. *Cell, 103*(3), 525–535.

Li, X., Mooney, P., Zheng, S., Booth, C. R., Braunfeld, M. B., Gubbens, S., et al. (2013). Electron counting and beam-induced motion correction enable near-atomic-resolution single-particle cryo-EM. *Nature Methods, 10*(6), 584–590. http://dx.doi.org/10.1038/nmeth.2472.

Lintschinger, B., Balzer-Geldsetzer, M., Baskaran, T., Graier, W. F., Romanin, C., Zhu, M. X., et al. (2000). Coassembly of Trp1 and Trp3 proteins generates diacylglycerol- and Ca^{2+} -sensitive cation channels. *Journal of Biological Chemistry, 275*(36), 27799–27805. http://dx.doi.org/10.1074/jbc.M002705200.

Liu, D., & Liman, E. R. (2003). Intracellular Ca^{2+} and the phospholipid PIP_2 regulate the taste transduction ion channel TRPM5. *Proceedings of the National Academy of Sciences of the United States of America, 100*(25), 15160–15165. http://dx.doi.org/10.1073/pnas.2334159100.

Liu, B., & Qin, F. (2005). Functional control of cold- and menthol-sensitive TRPM8 ion channels by phosphatidylinositol 4,5-bisphosphate. *Journal of Neuroscience: The Official*

Journal of The Society for Neuroscience, 25(7), 1674–1681. http://dx.doi.org/10.1523/JNEUROSCI.3632-04.2005.

Liu, B., Zhang, C., & Qin, F. (2005). Functional recovery from desensitization of vanilloid receptor TRPV1 requires resynthesis of phosphatidylinositol 4,5-bisphosphate. *Journal of Neuroscience: The Official Journal of the Society for Neuroscience, 25*(19), 4835–4843. http://dx.doi.org/10.1523/JNEUROSCI.1296-05.2005.

Lukacs, V., Rives, J.-M., Sun, X., Zakharian, E., & Rohacs, T. (2013). Promiscuous activation of transient receptor potential vanilloid 1 (TRPV1) channels by negatively charged intracellular lipids: the key role of endogenous phosphoinositides in maintaining channel activity. *Journal of Biological Chemistry, 288*(49), 35003–35013. http://dx.doi.org/10.1074/jbc.M113.520288.

Lukacs, V., Thyagarajan, B., Varnai, P., Balla, A., Balla, T., & Rohacs, T. (2007). Dual regulation of TRPV1 by phosphoinositides. *Journal of Neuroscience: The Official Journal of the Society for Neuroscience, 27*(26), 7070–7080. http://dx.doi.org/10.1523/JNEUROSCI.1866-07.2007.

Lundbaek, J. A., Andersen, O. S., & Lundbeek, J. A. (1994). Lysophospholipids modulate channel function by altering the mechanical properties of lipid bilayers. *Journal of General Physiology, 104*(4), 645–673.

Lundbaek, J. A., Collingwood, S. A., Ingólfsson, H. I., Kapoor, R., & Andersen, O. S. (2010). Lipid bilayer regulation of membrane protein function: gramicidin channels as molecular force probes. *Journal of the Royal Society, Interface/the Royal Society,* 7(44), 373–395. http://dx.doi.org/10.1098/rsif.2009.0443.

Macpherson, L. J., Dubin, A. E., Evans, M. J., Marr, F., Schultz, P. G., Cravatt, B. F., et al. (2007). Noxious compounds activate TRPA1 ion channels through covalent modification of cysteines. *Nature, 445*(7127), 541–545. http://dx.doi.org/10.1038/nature05544.

Macpherson, L. J., Hwang, S. W., Miyamoto, T., Dubin, A. E., Patapoutian, A., & Story, G. M. (2006). More than cool: promiscuous relationships of menthol and other sensory compounds. *Molecular and Cellular Neurosciences, 32*(4), 335–343. http://dx.doi.org/10.1016/j.mcn.2006.05.005.

Maher, S. A., Dubuis, E. D., & Belvisi, M. G. (2011). G-protein coupled receptors regulating cough. *Current Opinion in Pharmacology, 11*(3), 248–253. http://dx.doi.org/10.1016/j.coph.2011.06.005.

Majeed, Y., Tumova, S., Green, B. L., Seymour, V. A. L., Woods, D. M., Agarwal, A. K., et al. (2012). Pregnenolone sulphate-independent inhibition of TRPM3 channels by progesterone. *Cell Calcium, 51*(1), 1–11. http://dx.doi.org/10.1016/j.ceca.2011.09.005.

Mandadi, S., Sokabe, T., Shibasaki, K., Katanosaka, K., Mizuno, A., Moqrich, A., et al. (2009). TRPV3 in keratinocytes transmits temperature information to sensory neurons via ATP. *Pflügers Archiv: European Journal of Physiology, 458*(6), 1093–1102. http://dx.doi.org/10.1007/s00424-009-0703-x.

Manku, M. S., Horrobin, D. F., Huang, Y. S., & Morse, N. (1983). Fatty acids in plasma and red cell membranes in normal humans. *Lipids, 18*(12), 906–908. Retrieved from http://www.ncbi.nlm.nih.gov/pubmed/6664259.

Marx, C. E., Bradford, D. W., Hamer, R. M., Naylor, J. C., Allen, T. B., Lieberman, J. A., et al. (2011). Pregnenolone as a novel therapeutic candidate in schizophrenia: emerging preclinical and clinical evidence. *Neuroscience, 191*, 78–90. http://dx.doi.org/10.1016/j.neuroscience.2011.06.076.

Matta, J. A., Miyares, R. L., Ahern, G. P., & Matta, A. (2007). TRPV1 is a novel target for omega-3 polyunsaturated fatty acids. *Journal of Physiology, 2*(Pt 2), 397–411. http://dx.doi.org/10.1113/jphysiol.2006.121988.

Mckemy, D. D., Neuhausser, W. M., & Julius, D. (2002). Identification of a cold receptor reveals a general role for TRP channels in thermosensation. *Nature, 416*(6876), 52–58. http://dx.doi.org/10.1038/nature719.

Mercado, J., Gordon-Shaag, A., Zagotta, W. N., & Gordon, S. E. (2010). Ca^{2+}-dependent desensitization of TRPV2 channels is mediated by hydrolysis of phosphatidylinositol 4,5-bisphosphate. *Journal of neuroscience: The Official Journal of the Society for Neuroscience, 30*(40), 13338–13347. http://dx.doi.org/10.1523/JNEUROSCI.2108-10.2010.

Meves, H. (1994). Modulation of ion channels by arachidonic acid. *Progress in Neurobiology, 43*(2), 175–186.

Meves, H. (2008). Arachidonic acid and ion channels: an update. *British Journal of Pharmacology, 155*(1), 4–16. http://dx.doi.org/10.1038/bjp.2008.216.

Minke, B., & Cook, B. (2002). TRP channel proteins and signal transduction. *Physiological Reviews, 82*(2), 429–472. http://dx.doi.org/10.1152/physrev.00001.2002.

Montell, C. (2012). Drosophila visual transduction. *Trends in Neurosciences, 35*(6), 356–363. http://dx.doi.org/10.1016/j.tins.2012.03.004.

Montgomery, P., Burton, J. R., Sewell, R. P., Spreckelsen, T. F., & Richardson, A. J. (2013). Low blood long chain omega-3 fatty acids in UK children are associated with poor cognitive performance and behavior: a cross-sectional analysis from the DOLAB study. *PloS One, 8*(6), e66697. http://dx.doi.org/10.1371/journal.pone.0066697.

Morales-Lázaro, S. L., Serrano-Flores, B., Llorente, I., Hernández-García, E., González-Ramírez, R., Banerjee, S., et al. (2014). Structural Determinants of the Transient Receptor Potential 1 (TRPV1) Channel Activation by Phospholipid Analogs. *The Journal of Biological Chemistry, 289*(35), 24079–24090. http://dx.doi.org/10.1074/jbc.M114.572503.

Morales-Lázaro, S. L., Simon, S. A., & Rosenbaum, T. (2013). The role of endogenous molecules in modulating pain through transient receptor potential vanilloid 1 (TRPV1). *Journal of Physiology, 591*(Pt 13), 3109–3121. http://dx.doi.org/10.1113/jphysiol.2013.251751.

Nagata, K., Duggan, A., Kumar, G., & García-Añoveros, J. (2005). Nociceptor and hair cell transducer properties of TRPA1, a channel for pain and hearing. *Journal of Neuroscience: The Official Journal of the Society for Neuroscience, 25*(16), 4052–4061. http://dx.doi.org/10.1523/JNEUROSCI.0013-05.2005.

Nealen, M. L., Gold, M. S., Thut, P. D., & Caterina, M. J. (2003). TRPM8 mRNA is expressed in a subset of cold-responsive trigeminal neurons from rat. *Journal of Neurophysiology, 90*(1), 515–520. http://dx.doi.org/10.1152/jn.00843.2002.

Nieto-Posadas, A., Picazo-Juárez, G., Llorente, I., Jara-Oseguera, A., Morales-Lázaro, S., Escalante-Alcalde, D., et al. (2011). Lysophosphatidic acid directly activates trpv1 through a C-terminal binding site. *Nature Chemical Biology, 8*, 1–8. http://dx.doi.org/10.1038/nCHeMBIO.712. November.

Nilius, B., Droogmans, G., & Wondergem, R. (2003). Transient receptor potential channels in endothelium: solving the calcium entry puzzle? *Endothelium: Journal of Endothelial Cell Research, 10*(1), 5–15.

Nilius, B., Mahieu, F., Prenen, J., Janssens, A., Owsianik, G., Vennekens, R., et al. (2006). The Ca^{2+}-activated cation channel TRPM4 is regulated by phosphatidylinositol 4,5-biphosphate. *EMBO Journal, 25*(3), 467–478. http://dx.doi.org/10.1038/sj.emboj.7600963.

Nilius, B., & Owsianik, G. (2011). The transient receptor potential family of ion channels. *Genome Biology, 12*(3), 218. http://dx.doi.org/10.1186/gb-2011-12-3-218.

Nilius, B., Prenen, J., Janssens, A., Voets, T., & Droogmans, G. (2004). Decavanadate modulates gating of TRPM4 cation channels. *Journal of Physiology, 560*(Pt 3), 753–765. http://dx.doi.org/10.1113/jphysiol.2004.070839.

Oberwinkler, J., & Phillipp, S. E. (2007). TRPM3. *Handbook of Experimental Pharmacology, 179*, 253–267.

Oike, H., Wakamori, M., Mori, Y., Nakanishi, H., Taguchi, R., Misaka, T., et al. (2006). Arachidonic acid can function as a signaling modulator by activating the TRPM5 cation channel in taste receptor cells. *Biochimica et Biophysica Acta, 1761*(9), 1078–1084. http://dx.doi.org/10.1016/j.bbalip.2006.07.005.

Okada, T. (1999). Molecular and functional characterization of a novel mouse transient receptor potential protein homologue TRP7. Ca^{2+}-permeable cation channel that is constitutively activated and enhanced by stimulation of G protein-coupled receptor. *Journal of Biological Chemistry, 274*(39), 27359–27370. http://dx.doi.org/10.1074/jbc.274.39.27359.

Okazaki, M., Zhang, L. S., Suzuki, M., & Sakamoto, K. (1988). The measurement of platelet aggregation and ATP-release in mice with liver damage induced by carbon tetrachloride (CCl4) using a whole blood aggregometer. *Japanese Journal of Pharmacology, 48*(4), 407–415.

Pace-Asciak, C. R., & Asotra, S. (1989). Biosynthesis, catabolism, and biological properties of HPETEs, hydroperoxide derivatives of arachidonic acid. *Free Radical Biology & Medicine, 7*(4), 409–433.

Palazzo, E., Rossi, F., Novellis, V., De, Maione, S., & de Novellis, V. (2013). Endogenous modulators of TRP channels. *Current Topics in Medicinal Chemistry, 13*(3), 398–407.

Park, C.-K., Xu, Z.-Z., Liu, T., Lü, N., Serhan, C. N., & Ji, R.-R. (2011). Resolvin D2 is a potent endogenous inhibitor for transient receptor potential subtype V1/A1, inflammatory pain, and spinal cord synaptic plasticity in mice: distinct roles of resolvin D1, D2, and E1. *Journal of Neuroscience: The Official Journal of the Society for Neuroscience, 31*(50), 18433–18438. http://dx.doi.org/10.1523/JNEUROSCI.4192-11.2011.

Parnas, M., Peters, M., & Minke, B. (2009). Linoleic acid inhibits TRP channels with intrinsic voltage sensitivity: Implications on the mechanism of linoleic acid action. *Channels (Austin, Texas), 3*(3), 164–166.

Patwardhan, A. M., Akopian, A. N., Ruparel, N. B., Diogenes, A., Weintraub, S. T., Uhlson, C., et al. (2010). Heat generates oxidized linoleic acid metabolites that activate TRPV1 and produce pain in rodents. *Journal of Clinical Investigation, 120*(5), 1617–1626. http://dx.doi.org/10.1172/JCI41678.

Patwardhan, A. M., Scotland, P. E., Akopian, A. N., & Hargreaves, K. M. (2009). Activation of TRPV1 in the spinal cord by oxidized linoleic acid metabolites contributes to inflammatory hyperalgesia. *Proceedings of the National Academy of Sciences of the United States of America, 106*(44), 18820–18824. http://dx.doi.org/10.1073/pnas.0905415106.

Peier, A. M., Moqrich, A., Hergarden, A. C., Reeve, A. J., Andersson, D. A., Story, G. M., et al. (2002). A TRP channel that senses cold stimuli and menthol. *Cell, 108*(5), 705–715.

Peier, A. M., Reeve, A. J., Andersson, D. A., Moqrich, A., Earley, T. J., Hergarden, A. C., et al. (2002). A heat-sensitive TRP channel expressed in keratinocytes. *Science, 296*(5575), 2046–2049. http://dx.doi.org/10.1126/science.1073140 (New York, N.Y.).

Pérez, C. a, Huang, L., Rong, M., Kozak, J. A., Preuss, A. K., Zhang, H., et al. (2002). A transient receptor potential channel expressed in taste receptor cells. *Nature Neuroscience, 5*(11), 1169–1176. http://dx.doi.org/10.1038/nn952.

Petrus, M., Peier, A. M., Bandell, M., Hwang, S. W., Huynh, T., Olney, N., et al. (2007). A role of TRPA1 in mechanical hyperalgesia is revealed by pharmacological inhibition. *Molecular Pain, 3*, 40. http://dx.doi.org/10.1186/1744-8069-3-40.

Picazo-Juárez, G., Romero-Suárez, S., Nieto-Posadas, A., Llorente, I., Jara-Oseguera, A., Briggs, M., et al. (2011). Identification of a binding motif in the S5 helix that confers cholesterol sensitivity to the TRPV1 ion channel. *Journal of Biological Chemistry, 286*(28), 24966–24976. http://dx.doi.org/10.1074/jbc.M111.237537.

Pizzimenti, S., Menegatti, E., Berardi, D., Toaldo, C., Pettazzoni, P., Minelli, R., et al. (2010). 4-hydroxynonenal, a lipid peroxidation product of dietary polyunsaturated fatty acids, has anticarcinogenic properties in colon carcinoma cell lines through the inhibition of telomerase activity. *Journal of Nutritional Biochemistry, 21*(9), 818–826. http://dx.doi.org/10.1016/j.jnutbio.2009.06.005.

Poulter, C. D., Argyle, J. C., & Mash, E. A. (1978). Farnesyl pyrophosphate synthetase. Mechanistic studies of the 1'-4 coupling reaction with 2-fluorogeranyl pyrophosphate. *Journal of Biological Chemistry, 253*(20), 7227–7233.

Prescott, E. D., & Julius, D. (2003). A modular PIP_2 binding site as a determinant of capsaicin receptor sensitivity. *Science, 300*(5623), 1284–1288. http://dx.doi.org/10.1126/science.1083646 (New York, N.Y.).

Pyne, S., & Pyne, N. J. (2011). Translational aspects of sphingosine 1-phosphate biology. *Trends in Molecular Medicine, 17*(8), 463–472. http://dx.doi.org/10.1016/j.molmed.2011.03.002.

Radi, R. (2004). Nitric oxide, oxidants, and protein tyrosine nitration. *Proceedings of the National Academy of Sciences of the United States of America, 101*(12), 4003–4008. http://dx.doi.org/10.1073/pnas.0307446101.

Radtke, C., Rennekampff, H.-O., Reimers, K., Vogt, P. M., & Kocsis, J. D. (2013). Paracrine loop of keratinocyte proliferation and directed neuritic outgrowth in a neuroepithelial coculture. *Annals of Plastic Surgery, 70*(2), 162–167. http://dx.doi.org/10.1097/SPA.0b013e318276d946.

Rapoport, S. I. (2008). Arachidonic acid and the brain. *Journal of Nutrition, 138*(12), 2515–2520.

Robinson, L. E., & Mazurak, V. C. (2013). N-3 polyunsaturated fatty acids: relationship to inflammation in healthy adults and adults exhibiting features of metabolic syndrome. *Lipids, 48*(4), 319–332. http://dx.doi.org/10.1007/s11745-013-3774-6.

Rohacs, T. (2007). Regulation of TRP channels by PIP(2). *Pflügers Archiv: European Journal of Physiology, 453*(6), 753–762. http://dx.doi.org/10.1007/s00424-006-0153-7.

Rohacs, T., & Nilius, B. (2007). Regulation of transient receptor potential (TRP) channels by phosphoinositides. *Pflügers Archiv: European Journal of Physiology, 455*(1), 157–168. http://dx.doi.org/10.1007/s00424-007-0275-6.

Rosenbaum, T., Gordon-Shaag, A., Munari, M., & Gordon, S. E. (2004). Ca^{2+}/calmodulin modulates TRPV1 activation by capsaicin. *Journal of General Physiology, 123*(1), 53–62. http://dx.doi.org/10.1085/jgp.200308906.

Ross, K., Parker, G., Whitaker, M., & Reynolds, N. J. (2008). Inhibition of calcium-independent phospholipase A impairs agonist-induced calcium entry in keratinocytes. *British Journal of Dermatology, 158*(1), 31–37. http://dx.doi.org/10.1111/j.1365-2133.2007.08298.x.

Saghatelian, A., McKinney, M. K., Bandell, M., Patapoutian, A., & Cravatt, B. F. (2006). A FAAH-regulated class of N-acyl taurines that activates TRP ion channels. *Biochemistry, 45*(30), 9007–9015. http://dx.doi.org/10.1021/bi0608008.

Saisho, Y., Morimoto, A., & Umeda, T. (1997). Determination of farnesyl pyrophosphate in dog and human plasma by high-performance liquid chromatography with fluorescence detection. *Analytical Biochemistry, 252*(1), 89–95. http://dx.doi.org/10.1006/abio.1997.2314.

Salazar, H., Llorente, I., Jara-Oseguera, A., García-Villegas, R., Munari, M., Gordon, S. E., et al. (2008). A single N-terminal cysteine in TRPV1 determines activation by pungent compounds from onion and garlic. *Nature Neuroscience, 11*(3), 255–261. http://dx.doi.org/10.1038/nn2056.

Sawada, Y., Hosokawa, H., Hori, A., Matsumura, K., & Kobayashi, S. (2007). Cold sensitivity of recombinant TRPA1 channels. *Brain Research, 1160*, 39–46. http://dx.doi.org/10.1016/j.brainres.2007.05.047.

Sculptoreanu, A., Artim, D. E., & de Groat, W. C. (2009). Neurokinins inhibit low threshold inactivating K+ currents in capsaicin responsive DRG neurons. *Experimental Neurology, 219*(2), 562–573. http://dx.doi.org/10.1016/j.expneurol.2009.07.016.

Schroeder, R. J., Ahmed, S. N., Zhu, Y., London, E., & Brown, D. A. (1998). Cholesterol and sphingolipid enhance the Triton X-100 insolubility of glycosylphosphatidylinositol-anchored proteins by promoting the formation of detergent-insoluble ordered membrane domains. *Journal of Biological Chemistry, 273*(2), 1150–1157.

Sculptoreanu, A., Kullmann, F. A., Artim, D. E., Bazley, F. A., Schopfer, F., Woodcock, S., et al. (2010). Nitro-oleic acid inhibits firing and activates TRPV1- and TRPA1-mediated inward currents in dorsal root ganglion neurons from adult male rats. *Journal of Pharmacology and Experimental Therapeutics, 333*(3), 883–895. http://dx.doi.org/10.1124/jpet.109.163154.

Senning, E. N., Collins, M. D., Stratiievska, A., Ufret-Vincenty, C. a, & Gordon, S. E. (2014). Regulation of TRPV1 by Phosphoinositide (4,5)-bisphosphate: role of membrane Asymmetry. *Journal of Biological Chemistry*, 1–14. http://dx.doi.org/10.1074/jbc.M114.553180.

Sexton, A., McDonald, M., Cayla, C., Thiemermann, C., & Ahluwalia, A. (2007). 12-Lipoxygenase-derived eicosanoids protect against myocardial ischemia/reperfusion injury via activation of neuronal TRPV1. *FASEB Journal: Official Publication of the Federation of American Societies for Experimental Biology, 21*(11), 2695–2703. http://dx.doi.org/10.1096/fj.06-7828com.

Shipton, E. A. (2013). Skin matters: identifying pain mechanisms and predicting treatment outcomes. *Neurology Research International, 2013*, 329364. http://dx.doi.org/10.1155/2013/329364.

Simopoulos, A. (2002). The importance of the ratio of omega-6/omega-3 essential fatty acids. *Biomedicine & Pharmacotherapy, 56*(8), 365–379. http://dx.doi.org/10.1016/S0753-3322(02)00253-6.

Sisignano, M., Park, C.-K., Angioni, C., Zhang, D. D., von Hehn, C., Cobos, E. J., et al. (2012). 5,6-EET is released upon neuronal activity and induces mechanical pain hypersensitivity via TRPA1 on central afferent terminals. *Journal of Neuroscience: The Official Journal of the Society for Neuroscience, 32*(18), 6364–6372. http://dx.doi.org/10.1523/JNEUROSCI.5793-11.2012.

Smith, G. D., Gunthorpe, M. J., Kelsell, R. E., Hayes, P. D., Reilly, P., Facer, P., et al. (2002). TRPV3 is a temperature-sensitive vanilloid receptor-like protein. *Nature, 418*(6894), 186–190. http://dx.doi.org/10.1038/nature00894.

Sowa, N. A., Street, S. E., Vihko, P., & Zylka, M. J. (2010). Prostatic acid phosphatase reduces thermal sensitivity and chronic pain sensitization by depleting phosphatidylinositol 4,5-bisphosphate. *Journal of Neuroscience: The Official Journal of the Society for Neuroscience, 30*(31), 10282–10293. http://dx.doi.org/10.1523/JNEUROSCI.2162-10.2010.

Sparling, P. B., Giuffrida, A., Piomelli, D., Rosskopf, L., & Dietrich, A. (2003). Exercise activates the endocannabinoid system. *Neuroreport, 14*(17), 2209–2211. http://dx.doi.org/10.1097/01.wnr.0000097048.56589.47.

Spector, A. A., Fang, X., Snyder, G. D., & Weintraub, N. L. (2004). Epoxyeicosatrienoic acids (EETs): metabolism and biochemical function. *Progress in Lipid Research, 43*(1), 55–90.

Sprous, D., & Palmer, K. R. (2010). The T1R2/T1R3 sweet receptor and TRPM5 ion channel taste targets with therapeutic potential. *Progress in Molecular Biology and Translational Science, 91*, 151–208. http://dx.doi.org/10.1016/S1877-1173(10)91006-5.

Stein, A. T., Ufret-Vincenty, C. A., Hua, L., Santana, L. F., & Gordon, S. E. (2006). Phosphoinositide 3-kinase binds to TRPV1 and mediates NGF-stimulated TRPV1 trafficking to the plasma membrane. *Journal of General Physiology, 128*(5), 509–522. http://dx.doi.org/10.1085/jgp.200609576.

Story, G. M., Peier, A. M., Reeve, A. J., Eid, S. R., Mosbacher, J., Hricik, T. R., et al. (2003). ANKTM1, a TRP-like channel expressed in nociceptive neurons, is activated by cold temperatures. *Cell, 112*(6), 819–829.

Strotmann, R., Harteneck, C., Nunnenmacher, K., Schultz, G., & Plant, T. D. (2000). OTRPC4, a nonselective cation channel that confers sensitivity to extracellular osmolarity. *Nature Cell Biology, 2*(10), 695–702. http://dx.doi.org/10.1038/35036318.

Suh, B.-C., & Hille, B. (2005). Regulation of ion channels by phosphatidylinositol 4,5-bisphosphate. *Current Opinion in Neurobiology, 15*(3), 370–378. http://dx.doi.org/10.1016/j.conb.2005.05.005.

Suh, B.-C., & Hille, B. (2008). PIP_2 is a necessary cofactor for ion channel function: how and why? *Annual Review of Biophysics, 37*, 175–195. http://dx.doi.org/10.1146/annurev.biophys.37.032807.125859.

Sun, F. F., Chapman, J. P., & McGuire, J. C. (1977). Metabolism of prostaglandin endoperoxide in animal tissues. *Prostaglandins, 14*(6), 1055–1074.

Sun, B., Li, Q., Dong, L., & Rong, W. (2010). Ion channel and receptor mechanisms of bladder afferent nerve sensitivity. *Autonomic Neuroscience: Basic & Clinical, 153*(1–2), 26–32. http://dx.doi.org/10.1016/j.autneu.2009.07.003.

Sutton, K. G., Garrett, E. M., Rutter, A. R., Bonnert, T. P., Jarolimek, W., & Seabrook, G. R. (2005). Functional characterisation of the S512Y mutant vanilloid human TRPV1 receptor. *British Journal of Pharmacology, 146*(5), 702–711. http://dx.doi.org/10.1038/sj.bjp.0706356.

Takahashi, N., Kozai, D., Kobayashi, R., Ebert, M., & Mori, Y. (2011). Roles of TRPM2 in oxidative stress. *Cell Calcium, 50*(3), 279–287. http://dx.doi.org/10.1016/j.ceca.2011.04.006.

Talavera, K., Yasumatsu, K., Voets, T., Droogmans, G., Shigemura, N., Ninomiya, Y., et al. (2005). Heat activation of TRPM5 underlies thermal sensitivity of sweet taste. *Nature, 438*(7070), 1022–1025. http://dx.doi.org/10.1038/nature04248.

Taylor-Clark, T. E., Ghatta, S., Bettner, W., & Undem, B. J. (2009). Nitrooleic acid, an endogenous product of nitrative stress, activates nociceptive sensory nerves via the direct activation of TRPA1. *Molecular Pharmacology, 75*(4), 820–829. http://dx.doi.org/10.1124/mol.108.054445.

Taylor-Clark, T. E., Undem, B. J., Macglashan, D. W., Ghatta, S., Carr, M. J., & McAlexander, M. A. (2008). Prostaglandin-induced activation of nociceptive neurons via direct interaction with transient receptor potential A1 (TRPA1). *Molecular Pharmacology, 73*(2), 274–281. http://dx.doi.org/10.1124/mol.107.040832.

Togashi, K., Hara, Y., Tominaga, T., Higashi, T., Konishi, Y., Mori, Y., et al. (2006). TRPM2 activation by cyclic ADP-ribose at body temperature is involved in insulin secretion. *EMBO Journal, 25*(9), 1804–1815. http://dx.doi.org/10.1038/sj.emboj.7601083.

Tominaga, M., Caterina, M. J., Malmberg, A. B., Rosen, T. A., Gilbert, H., Skinner, K., et al. (1998). The cloned capsaicin receptor integrates multiple pain-producing stimuli. *Neuron, 21*(3), 531–543.

Tóth, B., & Csanády, L. (2012). Pore collapse underlies irreversible inactivation of TRPM2 cation channel currents. *Proceedings of the National Academy of Sciences of the United States of America, 109*(33), 13440–13445. http://dx.doi.org/10.1073/pnas.1204702109.

Trevisani, M., Patacchini, R., Nicoletti, P., Gatti, R., Gazzieri, D., Lissi, N., et al. (2005). Hydrogen sulfide causes vanilloid receptor 1-mediated neurogenic inflammation in the airways. *British Journal of Pharmacology, 145*(8), 1123–1131. http://dx.doi.org/10.1038/sj.bjp.0706277.

Trevisani, M., Siemens, J., Materazzi, S., Bautista, D. M., Nassini, R., Campi, B., et al. (2007). 4-Hydroxynonenal, an endogenous aldehyde, causes pain and neurogenic inflammation through activation of the irritant receptor TRPA1. *Proceedings of the National Academy of Sciences of the United States of America, 104*(33), 13519–13524. http://dx.doi.org/10.1073/pnas.0705923104.

Tsavaler, L., Shapero, M. H., Morkowski, S., & Laus, R. (2001). Trp-p8, a novel prostate-specific gene, is up-regulated in prostate cancer and other malignancies and shares high homology with transient receptor potential calcium channel proteins. *Cancer Research, 61*(9), 3760–3769.

Uchida, K., Miura, Y., Nagai, M., & Tominaga, M. (2012). Isothiocyanates from *Wasabia japonica* activate transient receptor potential ankyrin 1 channel. *Chemical Senses, 37*(9), 809–818. http://dx.doi.org/10.1093/chemse/bjs065.

Ufret-Vincenty, C. a, Klein, R. M., Hua, L., Angueyra, J., & Gordon, S. E. (2011). Localization of the PIP_2 sensor of TRPV1 ion channels. *Journal of Biological Chemistry, 286*(11), 9688–9698. http://dx.doi.org/10.1074/jbc.M110.192526.

Ullrich, N. D., Voets, T., Prenen, J., Vennekens, R., Talavera, K., Droogmans, G., et al. (2005). Comparison of functional properties of the Ca^{2+}-activated cation channels TRPM4 and TRPM5 from mice. *Cell Calcium, 37*(3), 267–278. http://dx.doi.org/10.1016/j.ceca.2004.11.001.

Van De Ven, T., & Ji, R.-R. (2013). Dietary control of chronic headache: Involvement of pro-resolution lipid mediators. *Pain, 154*(11), 2247–2248. http://dx.doi.org/10.1016/j.pain.2013.08.006.

Van Rossum, D. B., Patterson, R. L., Sharma, S., Barrow, R. K., Kornberg, M., Gill, D. L., et al. (2005). Phospholipase Cgamma1 controls surface expression of TRPC3 through an intermolecular PH domain. *Nature, 434*(7029), 99–104. http://dx.doi.org/10.1038/nature03340.

Vanden Abeele, F., Zholos, A., Bidaux, G., Shuba, Y., Thebault, S., Beck, B., et al. (2006). Ca^{2+}-independent phospholipase A2-dependent gating of TRPM8 by lysophospholipids. *Journal of Biological Chemistry, 281*(52), 40174–40182. http://dx.doi.org/10.1074/jbc.M605779200.

Vanhaesebroeck, B., Leevers, S. J., Ahmadi, K., Timms, J., Katso, R., Driscoll, P. C., et al. (2001). Synthesis and function of 3-phosphorylated inositol lipids. *Annual Review of Biochemistry, 70*, 535–602. http://dx.doi.org/10.1146/annurev.biochem.70.1.535.

Vargaftig, B. B., & Zirinis, P. (1973). Platelet aggregation induced by arachidonic acid is accompanied by release of potential inflammatory mediators distinct from PGE2 and PGF2. *Nature: New biology, 244*(134), 114–116.

Vay, L., Gu, C., & McNaughton, P. a (2012). The thermo-TRP ion channel family: properties and therapeutic implications. *British Journal of Pharmacology, 165*(4), 787–801. http://dx.doi.org/10.1111/j.1476-5381.2011.01601.x.

Venkatachalam, K., & Montell, C. (2007). TRP channels. *Annual Review of Biochemistry, 76*, 387–417. http://dx.doi.org/10.1146/annurev.biochem.75.103004.142819.

Vetter, I., & Lewis, R. J. (2011). Natural product ligands of TRP channels. *Advances in Experimental Medicine and Biology, 704*, 41–85. http://dx.doi.org/10.1007/978-94-007-0265-3_3.

Voets, T., & Nilius, B. (2007). Modulation of TRPs by PIPs. *Journal of Physiology, 582*(Pt 3), 939–944. http://dx.doi.org/10.1113/jphysiol.2007.132522.

Voets, T., & Nilius, B. (2009). TRPCs, GPCRs and the Bayliss effect. *EMBO Journal, 28*(1), 4–5. http://dx.doi.org/10.1038/emboj.2008.261.

Vogt-Eisele, A. K., Weber, K., Sherkheli, M. A., Vielhaber, G., Panten, J., Gisselmann, G., et al. (2007). Monoterpenoid agonists of TRPV3. *British Journal of Pharmacology, 151*(4), 530–540. http://dx.doi.org/10.1038/sj.bjp.0707245.

Vriens, J., Owsianik, G., Hofmann, T., Philipp, S. E., Stab, J., Chen, X., et al. (2011). TRPM3 is a nociceptor channel involved in the detection of noxious heat. *Neuron, 70*(3), 482–494. http://dx.doi.org/10.1016/j.neuron.2011.02.051.

Wagner, T. F. J., Loch, S., Lambert, S., Straub, I., Mannebach, S., Mathar, I., et al. (2008). Transient receptor potential M3 channels are ionotropic steroid receptors in pancreatic beta cells. *Nature Cell Biology, 10*(12), 1421–1430. http://dx.doi.org/10.1038/ncb1801.

Walder, R. Y., Radhakrishnan, R., Loo, L., Rasmussen, L. A., Mohapatra, D. P., Wilson, S. P., et al. (2012). TRPV1 is important for mechanical and heat sensitivity in uninjured animals and development of heat hypersensitivity after muscle inflammation. *Pain, 153*(8), 1664–1672. http://dx.doi.org/10.1016/j.pain.2012.04.034.

Waluk, D. P., Vielfort, K., Derakhshan, S., Aro, H., & Hunt, M. C. (2013). N-Acyl taurines trigger insulin secretion by increasing calcium flux in pancreatic β-cells. *Biochemical and Biophysical Research Communications, 430*(1), 54–59. http://dx.doi.org/10.1016/j.bbrc.2012.11.026.

Watanabe, H., Davis, J. B., Smart, D., Jerman, J. C., Smith, G. D., Hayes, P., et al. (2002). Activation of TRPV4 channels (hVRL-2/mTRP12) by phorbol derivatives. *Journal of Biological Chemistry, 277*(16), 13569–13577. http://dx.doi.org/10.1074/jbc.M200062200.

Watanabe, H., Vriens, J., Prenen, J., Droogmans, G., Voets, T., & Nilius, B. (2003). Anandamide and arachidonic acid use epoxyeicosatrienoic acids to activate TRPV4 channels. *Nature, 424*(6947), 434–438. http://dx.doi.org/10.1038/nature01807.

Watanabe, H., Vriens, J., Suh, S. H., Benham, C. D., Droogmans, G., & Nilius, B. (2002). Heat-evoked activation of TRPV4 channels in a HEK293 cell expression system and in native mouse aorta endothelial cells. *Journal of Biological Chemistry, 277*(49), 47044–47051. http://dx.doi.org/10.1074/jbc.M208277200.

Wehage, E., Eisfeld, J., Heiner, I., Jüngling, E., Zitt, C., & Lückhoff, A. (2002). Activation of the cation channel long transient receptor potential channel 2 (LTRPC2) by hydrogen peroxide. A splice variant reveals a mode of activation independent of ADP-ribose. *Journal of Biological Chemistry, 277*(26), 23150–23156. http://dx.doi.org/10.1074/jbc.M112096200.

Wen, H., Ostman, J., Bubb, K. J., Panayiotou, C., Priestley, J. V., Baker, M. D., et al. (2012). 20-Hydroxyeicosatetraenoic acid (20-HETE) is a novel activator of transient receptor potential vanilloid 1 (TRPV1) channel. *Journal of Biological Chemistry, 287*(17), 13868–13876. http://dx.doi.org/10.1074/jbc.M111.334896.

Woo, D. H., Jung, S. J., Zhu, M. H., Park, C.-K., Kim, Y. H., Oh, S. B., et al. (2008). Direct activation of transient receptor potential vanilloid 1(TRPV1) by diacylglycerol (DAG). *Molecular Pain, 4*, 42. http://dx.doi.org/10.1186/1744-8069-4-42.

Wu, L., Bauer, C. S., Zhen, X., Xie, C., & Yang, J. (2002). Dual regulation of voltage-gated calcium channels by PtdIns(4,5)P2. *Nature, 419*(6910), 947–952. http://dx.doi.org/10.1038/nature01118.

Wu, Y.-W., Bi, Y.-P., Kou, X.-X., Xu, W., Ma, L.-Q., Wang, K.-W., et al. (2010). 17-Beta-estradiol enhanced allodynia of inflammatory temporomandibular joint through upregulation of hippocampal TRPV1 in ovariectomized rats. *Journal of Neuroscience: The Official Journal of the Society for Neuroscience, 30*(26), 8710–8719. http://dx.doi.org/10.1523/JNEUROSCI.6323-09.2010.

Xu, S., Cheng, Y., Keast, J. R., & Osborne, P. B. (2008). 17beta-estradiol activates estrogen receptor beta-signalling and inhibits transient receptor potential vanilloid receptor 1 activation by capsaicin in adult rat nociceptor neurons. *Endocrinology, 149*(11), 5540–5548. http://dx.doi.org/10.1210/en.2008-0278.

Xu, H., Delling, M., Jun, J. C., & Clapham, D. E. (2006). Oregano, thyme and clove-derived flavors and skin sensitizers activate specific TRP channels. *Nature Neuroscience, 9*(5), 628–635. http://dx.doi.org/10.1038/nn1692.

Xu, X. Z., Li, H. S., Guggino, W. B., & Montell, C. (1997). Coassembly of TRP and TRPL produces a distinct store-operated conductance. *Cell, 89*(7), 1155–1164.

Xu, X. Z., Moebius, F., Gill, D. L., & Montell, C. (2001). Regulation of melastatin, a TRP-related protein, through interaction with a cytoplasmic isoform. *Proceedings of the National Academy of Sciences of the United States of America, 98*(19), 10692–10697. http://dx.doi.org/10.1073/pnas.191360198.

Xu, H., Ramsey, I. S., Kotecha, S. A., Moran, M. M., Chong, J. A., Lawson, D., et al. (2002). TRPV3 is a calcium-permeable temperature-sensitive cation channel. *Nature, 418*(6894), 181–186. http://dx.doi.org/10.1038/nature00882.

Xu, F., Satoh, E., & Iijima, T. (2003). Protein kinase C-mediated Ca^{2+} entry in HEK 293 cells transiently expressing human TRPV4. *British Journal of Pharmacology, 140*(2), 413–421. http://dx.doi.org/10.1038/sj.bjp.0705443.

Xu, Z.-Z., Zhang, L., Liu, T., Park, J. Y., Berta, T., Yang, R., et al. (2010). Resolvins RvE1 and RvD1 attenuate inflammatory pain via central and peripheral actions. *Nature Medicine, 16*(5), 592–597. http://dx.doi.org/10.1038/nm.2123, 1p following 597.

Yamamoto, S., Takahashi, N., & Mori, Y. (2010). Chemical physiology of oxidative stress-activated TRPM2 and TRPC5 channels. *Progress in Biophysics and Molecular Biology, 103*(1), 18–27. http://dx.doi.org/10.1016/j.pbiomolbio.2010.05.005.

Yang, X.-R., Lin, M.-J., McIntosh, L. S., & Sham, J. S. K. (2006). Functional expression of transient receptor potential melastatin- and vanilloid-related channels in pulmonary arterial and aortic smooth muscle. *American Journal of Physiology. Lung Cellular and Molecular Physiology, 290*(6), L1267–L1276. http://dx.doi.org/10.1152/ajplung.00515.2005.

Yao, J., Liu, B., & Qin, F. (2011). Modular thermal sensors in temperature-gated transient receptor potential (TRP) channels. *Proceedings of the National Academy of Sciences of the United States of America, 108*(27), 11109–11114. http://dx.doi.org/10.1073/pnas.1105196108.

Yao, J., & Qin, F. (2009). Interaction with phosphoinositides confers adaptation onto the TRPV1 pain receptor. *PLoS Biology,* 7(2), e46. http://dx.doi.org/10.1371/journal.pbio.1000046.

Yekkirala, A. S. (2013). Two to tango: GPCR oligomers and GPCR-TRP channel interactions in nociception. *Life Sciences, 92*(8–9), 438–445. http://dx.doi.org/10.1016/j.lfs.2012.06.021.

Yellen, G. (2002). The voltage-gated potassium channels and their relatives. *Nature, 419*(6902), 35–42. http://dx.doi.org/10.1038/nature00978.

Yu, L., Wang, S., Kogure, Y., Yamamoto, S., Noguchi, K., & Dai, Y. (2013). Modulation of TRP channels by resveratrol and other stilbenoids. *Molecular Pain, 9*(1), 3. http://dx.doi.org/10.1186/1744-8069-9-3.

Zamudio-Bulcock, P. A., Everett, J., Harteneck, C., & Valenzuela, C. F. (2011). Activation of steroid-sensitive TRPM3 channels potentiates glutamatergic transmission at cerebellar Purkinje neurons from developing rats. *Journal of Neurochemistry, 119*(3), 474–485. http://dx.doi.org/10.1111/j.1471-4159.2011.07441.x.

Zeldin, D. C. (2001). Epoxygenase pathways of arachidonic acid metabolism. *Journal of Biological Chemistry, 276*(39), 36059–36062. http://dx.doi.org/10.1074/jbc.R100030200.

Zhang, L., Jones, S., Brody, K., Costa, M., & Brookes, S. J. H. (2004). Thermosensitive transient receptor potential channels in vagal afferent neurons of the mouse. *American Journal of Physiology. Gastrointestinal and Liver Physiology, 286*(6), G983–G991. http://dx.doi.org/10.1152/ajpgi.00441.2003.

Zhang, Z., Okawa, H., Wang, Y., & Liman, E. R. (2005). Phosphatidylinositol 4,5-bisphosphate rescues TRPM4 channels from desensitization. *Journal of Biological Chemistry, 280*(47), 39185–39192. http://dx.doi.org/10.1074/jbc.M506965200.

Zygmunt, P. M., Petersson, J., Andersson, D. A., Chuang, H., Sørgård, M., Di Marzo, V., et al. (1999). Vanilloid receptors on sensory nerves mediate the vasodilator action of anandamide. *Nature, 400*(6743), 452–457. http://dx.doi.org/10.1038/22761.

Zygmunt, P. M., Sørgård, M., Petersson, J., Johansson, R., & Högestätt, E. D. (2000). Differential actions of anandamide, potassium ions and endothelium-derived hyperpolarizing factor in guinea-pig basilar artery. *Naunyn-Schmiedeberg's Archives of Pharmacology, 361*(5), 535–542.

CHAPTER SEVEN

Structure of Thermally Activated TRP Channels

Matthew R. Cohen[1,2] and Vera Y. Moiseenkova-Bell[1,2,*]
[1]Department of Pharmacology, Case Western Reserve University School of Medicine, Cleveland, OH, USA
[2]Department of Physiology & Biophysics, Case Western Reserve University School of Medicine, Cleveland, OH, USA
*Corresponding author: E-mail: vera.moiseenkova-bell@case.edu

Contents

Abstract

Temperature sensation is important for adaptation and survival of organisms. While temperature has the potential to affect all biological macromolecules, organisms have evolved specific thermosensitive molecular detectors that are able to transduce temperature changes into physiologically relevant signals. Among these thermosensors are ion channels from the transient receptor potential (TRP) family. Prime candidates include TRPV1–4, TRPA1, and TRPM8 (the so-called "thermoTRP" channels), which are expressed in sensory neurons and gated at specific temperatures. Electrophysiological and thermodynamic approaches have been employed to determine the nature by which thermoTRPs detect temperature and couple temperature changes to channel gating. To further understand how thermoTRPs sense temperature, high-resolution structures of full-length thermoTRPs channels will be required. Here, we will discuss current progress in unraveling the structures of thermoTRP channels.

Current Topics in Membranes, Volume 74
ISSN: 1063-5823
http://dx.doi.org/10.1016/B978-0-12-800181-3.00007-5

1. INTRODUCTION

All living organisms have the ability to detect temperature changes from the outside environment and convert it into specific biological outputs, allowing them to adapt and survive (Sengupta & Garrity, 2013). Cells employ specific biomolecules that undergo temperature-induced conformational changes, initiating signaling cascades that result in these physiological and behavioral responses (Digel, 2011; Digel, Kayser, & Artmann, 2008). It has been suggested that changes in DNA, RNA, and protein conformation or changes in lipid membrane properties initiate temperature-induced signaling cascades (Digel, 2011; Digel et al., 2008).

In higher organisms, skin forms a protective layer that enables the body to detect changes in the physical, chemical, and thermal environment (Schepers & Ringkamp, 2009). A wide array of specialized sensory neurons that specifically detect and transduce thermal changes over a broad range of temperatures innervate skin (McGlone & Reilly, 2010; Schepers & Ringkamp, 2009). These sensory neurons are activated at distinct temperature thresholds and allow organisms to differentiate between noxious cold (<15 °C) and heat (>43 °C), and pleasant cool (15–25 °C) and warm (30–40 °C) (Figure 7.1) (McKemy, 2013).

The identity of the "molecular devices" that sense and differentiate these temperatures was unraveled by the discovery and characterization of transient receptor potential (TRP) ion channels. The first TRP channel was identified by characterization of a vision-impaired *trp* mutant from *Drosophila* (Cosens & Manning, 1969; Minke, Wu, & Pak, 1975). Presently, the TRP superfamily consists of 28 mammalian members and is subdivided into six major branches: TRPC (canonical), TRPV (vanilloid), TRPM (melastatin), TRPA (ankyrin), TRPP (polycystin), and TRPML (mucolipin). Among them, members of TRPV, TRPA, and TRPM subfamilies have been suggested to play a critical role in temperature sensation (Venkatachalam & Montell, 2007). Specifically, TRPV1 (>43 °C), TRPV2 (>52 °C), TRPV3 (>30–39 °C), and TRPV4 (>25–35 °C) have been implicated in hot and warm sensation, while TRPM8 (<20–28 °C) and TRPA1 (<17 °C) are involved in cool and cold detection, respectively (Belmonte & Viana, 2008).

Thermosensation is likely not limited to TRP channels as the tetrodotoxin-resistant voltage-gated sodium channel Na(v)1.8 has been shown to play a critical role in noxious cold signaling (Abrahamsen et al., 2008; Zimmermann et al., 2007) and two-pore potassium channels

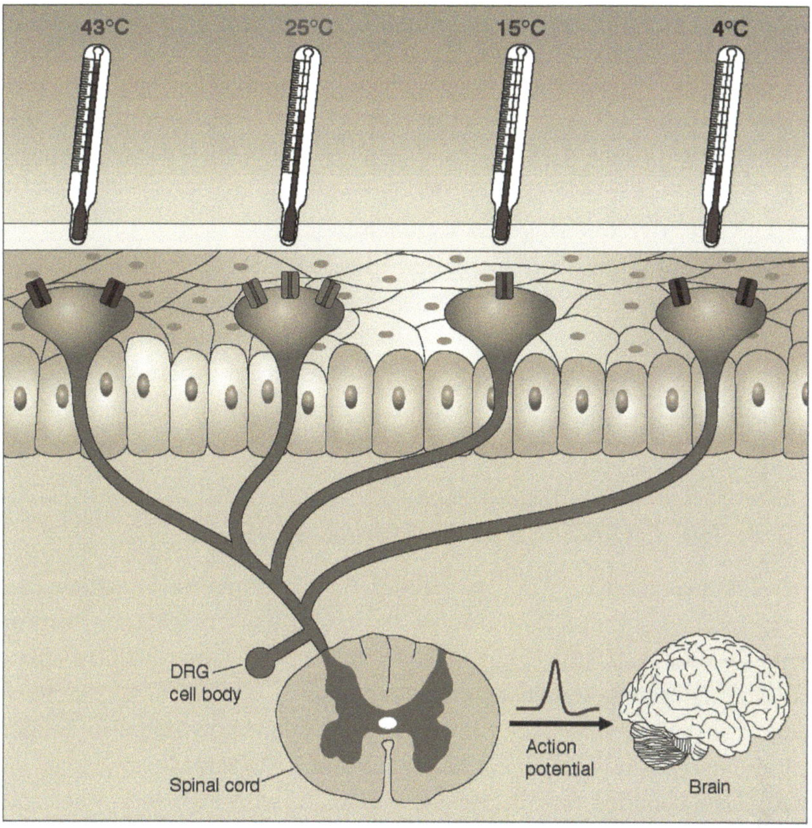

Figure 7.1 Thermosensitive ion channels in sensory neurons. Sensory neurons innervate the skin and contain thermosensitive nonselective cation channels in their terminals that sense a wide range of temperatures. Activation of these channels depolarizes the sensory neuron, leading to propagation of action potentials that are relayed to the spinal cord and eventually reach the brain. *Illustration by Kelly Paralis, Penumbra Design.* (See the color plate.)

TREK-1 and TRAAK have been implicated in cold and warm thermoregulation (Noel et al., 2009). Based on this current knowledge, it is clear that exposure to wide-range temperature changes triggers the generation of Ca^{2+}, K^{+}, and Na^{+} currents, leading to the formation and propagation of action potentials that send signals to the brain (Figure 7.1), thereby modifying behaviors according to the temperature change encountered (Viana, 2011).

Nevertheless, the molecular mechanism of temperature sensation by these ion channels is still unknown. Activity of all proteins is sensitive to the temperature changes; however, only select proteins are considered thermosensors. Temperature sensitivity of proteins is often quantified in terms of

Q_{10}, which represents the ratio of a protein property measured at two temperatures 10 °C apart (Sengupta & Garrity, 2013). Ion channels that exhibit Q_{10} values of ~3 are considered temperature insensitive, while proteins with a Q_{10} value >7 are considered thermosensitive (Sengupta & Garrity, 2013). The Q_{10} of TRPV1 is ~40 and of TRPM8 is ~28 (Maingret et al., 2000; Sengupta & Garrity, 2013), indicating that these channels are especially sensitive to changes in temperature.

These biophysical properties clearly suggest that TRP channels act as cellular thermosensors; however, the structural features of the channels that determine their thermosensitivity are only recently coming to light. In the following sections, we will focus on the recent progress in structure determination of thermosensitive TRP channels and how these structural details could aid in understanding thermosensation at the molecular level.

2. TRP CHANNELS AS THERMAL SENSORS

The mammalian TRP channel superfamily is one of the largest families of cation channels, consisting of 28 mammalian homologues. Among them, TRPV1–4, TRPA1, and TRPM8 have been shown to play critical roles in thermosensation, detecting temperatures from 4 °C to 52 °C. In addition to temperature, thermoTRP channels are polymodal integrators of multiple types of stimuli (e.g., ligand, voltage, stretch).

All TRP channels form functional tetramers, with each monomer consisting of six transmembrane (TM) segments (Figure 7.2). The primary sequence of the TM region is the most conserved between the thermoTRP channels. Interestingly, the intracellular N- and C-termini of these TRP channels contain divergent structural domains. TRPV1–4 are smaller tetramers (340–390 kDa) with 6 N-terminal ankryin repeats and a conserved C-terminal TRP box sequence per subunit (Figure 7.2). TRPA1 and TRPM8 are larger proteins (~530 kDa). TRPA1 contains 14–19 N-terminal ankyrin repeats and no C-terminal TRP box sequence (Figure 7.2). On the other hand, TRPM8 has no ankyrin repeats but contains a conserved TRPM channel specific amino terminal region and a coiled-coil domain and TRP box sequence on the C-terminus (Figure 7.2).

Despite predicted structural diversity amongst thermoTRP channels, they potentially share the common feature of temperature sensation, which has intrigued the field for many years. Extensive and elegant studies have been employed to determine their "molecular temperature switch" (Brauchi, Orta, Salazar, Rosenmann, & Latorre, 2006; Cordero-Morales, Gracheva,

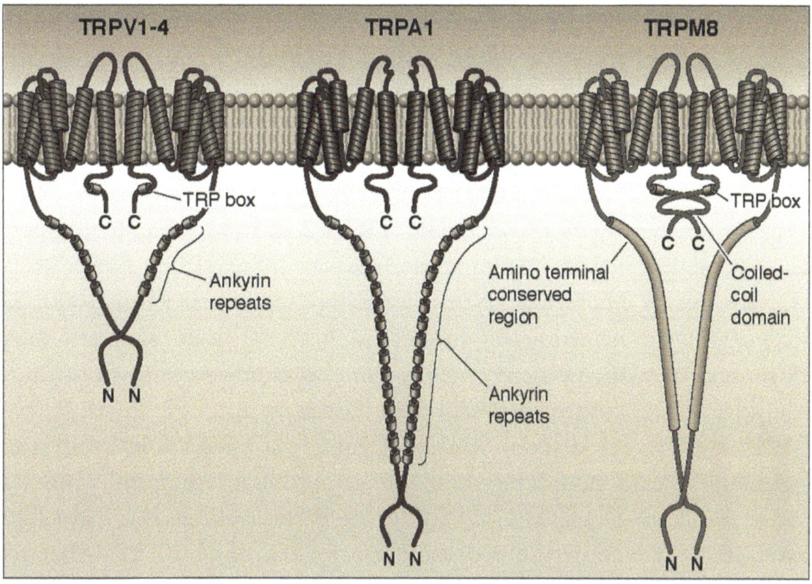

Figure 7.2 Schematic of the domain organization for thermo transient receptor potential (TRP) channels. Representation of the domain organization for TRPV1–4, TRPA1, and TRPM8. For clarity, dimers are shown. A monomer for each channel contains 6 transmembrane segments and large N- and C-termini in the cytoplasm. TRPV1–4 have 6 N-terminal ankryin repeats and a conserved C-terminal TRP box sequence. TRPA1 contains 14–19 N-terminal ankyrin repeats. TRPM8 has an N-terminal conserved region and C-terminal TRP box and coiled-coil domains. *Illustration by Kelly Paralis, Penumbra Design.* (See the color plate.)

& Julius, 2011; Grandl et al., 2008, 2010; Kim, Patapoutian, & Grandl, 2013; Yang, Cui, Wang, & Zheng, 2010; Yao, Liu, & Qin, 2011); however, a temperature-dependent detector may not exist at a specific site since temperature affects the global conformation of these large membrane protein complexes (Clapham & Miller, 2011; Liu, Hui, & Qin, 2003). High-resolution structural information for these channels will greatly accelerate understanding of how the entire protein operates as a thermosensor.

2.1 TRPV subfamily

2.1.1 Thermosensitivity of TRPV channels

TRPV1 was the first thermoTRP channel to be identified and characterized (Caterina et al., 1997). It has been the most extensively studied TRP channel due to its role in pain and temperature sensation. TRPV1 is activated by chemical compounds including capsaicin, the active ingredient in chili peppers, as well as heat, with an activation threshold of 42 °C (Caterina et al.,

1997). TRPV1 shows an intrinsic ability to sense temperature, as purified TRPV1 reconstituted in artificial liposomes shows temperature-sensitive activation (Cao, Cordero-Morales, Liu, Qin, & Julius, 2013). Furthermore, there is evidence that TRPV1 acts as a thermosensor in vivo, as deletion of the TRPV1 gene impairs temperature sensation in mouse and treatment of rodents with TRPV1 antagonists reduces heat sensitivity (Caterina et al., 2000). Other thermoTRPV channels, TRPV2–4, have subsequently been cloned (Caterina, Rosen, Tominaga, Brake, & Julius, 1999; Smith et al., 2002; Strotmann, Harteneck, Nunnenmacher, Schultz, & Plant, 2000; Xu et al., 2002). In heterologous expression systems, these channels show heat sensitivity with different temperature thresholds ranging from 25 to 52 °C (Table 7.1) (Caterina et al., 1999; Guler et al., 2002; Peier, Reeve, et al., 2002). Genetic deletion of these channels in mice has little effect on temperature sensation, leaving their role as in vivo thermosensors unclear (Huang, Li, Yu, Wang, & Caterina, 2011; Park et al., 2011). Therefore, most structure–function studies of temperature sensation by TRPV channels have focused on TRPV1.

2.1.2 In search of the temperature-sensing domain in TRPV channels

TRPV1 functions as a tetramer (Garcia-Sanz et al., 2004; Kedei et al., 2001). The TM region has six α-helical domains. TM5, TM6, and the linker between them form the putative pore of the channel. The cytoplasmic N-terminus of TRPV1 accounts for ~50% of the total molecular weight and contains six ankyrin repeats. Ankyrin repeats consist of 33 amino acids that form an antiparallel helix-turn-helix motif linked by a β-hairpin turn (Gaudet, 2008). Ligands interact with ankyrin repeat domains (ARDs) via the β-turn and the first antiparallel helix from a varying number of repeats (Gaudet, 2008). The C-terminus of TRPV1 contains the conserved TRP box sequence, which may be involved in channel oligomerization (Garcia-Sanz et al., 2004) (Figure 7.3(A)).

In order to identify the temperature-sensitive domains of TRPV1, electrophysiology studies coupled to chimeric and mutagenesis strategies have been employed. Several domains are proposed to act as the temperature sensors of thermoTRPV channels and controversy still exists as to the molecular nature of temperature sensation by thermoTRPVs.

Screens of random mutations in TRPV1 and TRPV3 revealed deficits in temperature sensitivity when residues in the channel pore were mutated, suggesting that the pore domain contributes to thermosensitivity (Grandl

Table 7.1 Biophysical and physiological properties of TRPV1–4 channels

TRPV member	Sequence identity with TRPV1	Temperature threshold (°C)	Q_{10}	KO mouse phenotype	References
TRPV1	–	>43	40	Impaired temperature sensation	Caterina et al. (1997, 2000)
TRPV2	50%	>52	100	No major deficits in thermosensation	Caterina et al. (1999), Park et al. (2011)
TRPV3	43%	>30	22	No major deficits in thermosensation	Huang et al. (2011), Smith et al. (2002), Xu et al. (2002)
TRPV4	40%	>25	20	No major deficits in thermosensation	Huang et al. (2011), Strotmann et al. (2000)

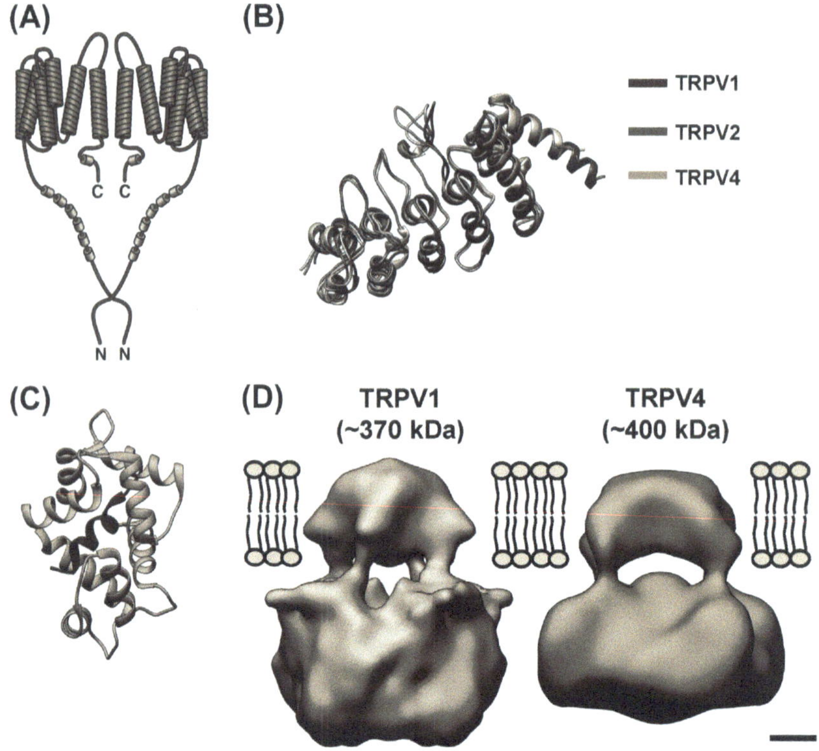

Figure 7.3 Structural analyses of thermoTRPV channels. (A) Domain organization of thermoTRPV channels. (B) Comparison of the crystal structures for the ankyrin repeat domains of TRPV1 (PDB code: 2PNN), TRPV2 (PDB code: 2ETB), and TRPV4 (PDB code: 4DX2) (Phelps, Wang, Choo, & Gaudet, 2010). (C) Crystal structure of a TRPV1 C-terminal peptide (red) in complex with calmodulin (gray) (PDB code: 3SUI) (Lau, Procko, & Gaudet, 2012). (D) Comparison of the cryo-electron microscopy structures for full-length TRPV1 at 19 Å resolution (Moiseenkova-Bell, Stanciu, Serysheva, Tobe, & Wensel, 2008) and TRPV4 at 35 Å resolution (Shigematsu, Sokabe, Danev, Tominaga, & Nagayama, 2010). Scale bar represents 25 Å. (See the color plate.)

et al., 2008, 2010). Later studies showed that the pore turret region of TRPV1, a 24-residue sequence between TM5 and the pore helix, is involved in heat-induced activation of TRPV1 (Yang et al., 2010). Replacement of the pore turret sequence with an artificial linker nearly ablated the response of TRPV1 to heat while maintaining its response to ligands and voltage. However, another group showed that deletion of the pore turret sequence had no effect on temperature-sensitive activation of TRPV1 (Yao, Liu, & Qin, 2010).

Furthermore, both the N- and C-termini of TRPV1 have been implicated in thermosensation. Chimeras of TRPV1 and TRPM8 in which the C-termini were exchanged conferred heat sensitivity to TRPM8 and cold sensitivity to TRPV1, suggesting that the C-terminus is involved in the directionality of temperature sensation (Brauchi et al., 2006). Another study found that the membrane proximal region linking the N-terminal ARD to the TM1 segment dictates temperature sensation together with the last two ankyrin repeats (Yao et al., 2011).

Based on these chimeric and mutagenesis studies, it appears that the pore domain, C-terminus, and N-terminus of TRPV1 all participate in temperature sensation. Controversy remains as the search for the molecular temperature sensor in TRPV1 and other thermoTRPV channels continues.

2.1.3 Crystal structures of isolated TRPV channel domains

Structure determination of full-length TRPV1–4 by X-ray crystallography has been hampered by the lack of an expression system to produce sufficient amounts of protein (Moiseenkova-Bell & Wensel, 2009, 2011). Two different approaches have yielded progress in understanding TRP channel structures: a "divide and conquer" strategy in which structures of smaller soluble domains have been determined by X-ray crystallography (Gaudet, 2009) and cryo-electron microscopy (cryo-EM) analysis of the full-length TRP channels, which allowed for determination of TRP channel structures at moderate and recently at side-chain resolution (Cao, Liao, Cheng, & Julius, 2013; Cvetkov, Huynh, Cohen, & Moiseenkova-Bell, 2011; Huynh et al., 2014; Liao, Cao, Julius, & Cheng, 2013; Moiseenkova-Bell et al., 2008).

The "divide and conquer" approach has yielded structures of the ARDs from TRPV1, TRPV2, TRPV3, TRPV4, and TRPV6 (Inada, Procko, Sotomayor, & Gaudet, 2012; Jin, Touhey, & Gaudet, 2006; Lishko, Procko, Jin, Phelps, & Gaudet, 2007; Phelps, Huang, Lishko, Wang, & Gaudet, 2008; Phelps et al., 2010; Shi, Ye, Cao, Zhang, & Wang, 2013). The ARDs comprise ~60% of the N-terminus and are important in interacting with ligands that regulate channel activity (Binz et al., 2004; Li, Mahajan, & Tsai, 2006; Nakamura et al., 2007). The ARD structures from TRPV channels display overall structural similarity (Figure 7.3(B)). A concave ligand-binding surface is formed by the six antiparallel helices and loops and a large twist is present between repeats 4 and 5 (Gaudet, 2008). Despite the overall similarity of these structures, biochemical studies show that the thermoTRPV ARDs bind different ligands (Phelps et al., 2010). For example, the ARDs of TRPV1, TRPV3, and TRPV4 bind ATP at the concave ligand-binding surface whereas TRPV2, TRPV5, and TRPV6 do not (Phelps et al., 2010). Furthermore, ATP has differential effects on the activity of TRPV1, TRPV3, and TRPV4 (Al-Ansary, Bogeski, Disteldorf, Becherer, & Niemeyer, 2010; Lishko et al., 2007; Phelps et al., 2010). The structures of the thermoTRPV ARDs provided insight into the regulation of these channels by ligands; however, the role of ankyrin repeats in temperature sensation remains unknown.

Recently, the X-ray structure for a 35 amino acid TRPV1 C-terminal peptide (residues 767–801) in complex with calmodulin (CaM) was determined (Figure 7.3(C)) (Lau et al., 2012). TRPV channels do not contain classical CaM-binding motifs, however, it has been proposed that CaM plays a significant role in desensitization of several TRPV channels. In the structure, only 14 out of 35 C-terminal amino acid densities were observed, displaying an overall similarity to previously published structures of CaM-binding peptides (Figure 7.3(C)). Further investigation will be required to gain molecular insights into the role of the C-terminus in temperature sensation and CaM-dependent desensitization of TRPV channels.

2.1.4 EM structures of full-length TRPV channels

Cryo-EM can theoretically provide near atomic resolution structural information for a variety of biological molecules without the need to produce crystals (Henderson, 1995). Recently, this has been confirmed by the determination of near atomic resolution structures for large symmetrical molecules using cryo-EM (Wolf, Garcea, Grigorieff, & Harrison, 2010; Zhang, Jin, Fang, Hui, & Zhou, 2010). A major advantage of cryo-EM versus other structural approaches is the ability to solve protein structures in different functional states, thereby identifying functionally relevant conformational changes (Heymann, Conway, & Steven, 2004). Additionally, cryo-EM requires approximately one hundred times less protein than other structural techniques (Wang & Sigworth, 2006).

Structure determination of full-length eukaryotic ion channels is a challenging task in structural biology. Cryo-EM has provided structures of large ion channels such as the ryanodine receptor and IP_3 receptor at the subnanometer (~10 Å) resolution and revealed precise conformational changes during ligand activation (Ludtke, Serysheva, Hamilton, & Chiu, 2005; Ludtke et al., 2011; Samso, Feng, Pessah, & Allen, 2009; Samso, Wagenknecht, & Allen, 2005; Serysheva et al., 2008). The same methodology was applied toward structural analysis of full-length TRP channels.

The initial cryo-EM structures of functional TRPV1 and TRPV4 were determined at 19 Å and 35 Å, respectively (Figure 7.3(D)) (Moiseenkova-Bell et al., 2008; Shigematsu et al., 2010). Both cryo-EM maps revealed considerable structural similarities, including fourfold symmetry and the presence of two distinct architectural regions likely corresponding to the TM and cytoplasmic domains of the channels (Figure 7.3(D)).

Preparation of stable samples was a major obstacle in structure determination of TRP channels (Cvetkov et al., 2011; Huynh et al., 2014; Liao et al., 2013). The use of traditional detergents such as decyl-β-D-maltoside and dodecyl-β-D-maltoside led to sample heterogeneity (Huynh et al., 2014; Liao et al., 2013); however, utilization of the newly developed maltose-neopentyl glycol (MNG) class of detergents (Huynh et al., 2014) or A8-35 amphipol molecules (Cvetkov et al., 2011) allowed for stabilization of full-length TRPV2 and TRPA1, respectively. Recently, A8-35 amphipols were also used in stabilizing a truncated TRPV1 construct used for cryo-EM analysis (Liao et al., 2013). Enhanced purification and stabilization methods allowed for the determination of TRPV1 and TRPV2 structures by cryo-EM at higher resolutions.

The recent publication of a high-resolution structure of TRPV1 channel concurrent with the determination of the structure of TRPV2 allows for a comparison of their structural features (Huynh et al., 2014; Liao et al., 2013). Despite nearly 50% sequence identity, TRPV1 and TRPV2 diverge in function (Peralvarez-Marin, Donate-Macian, & Gaudet, 2013). While TRPV1 clearly acts as a thermosensor and receptor for endogenous and exogenous vanilloids (Caterina et al., 1997, 2000), the role of TRPV2 in noxious temperature sensation remains in question (Caterina et al., 1999; Park et al., 2011). Also, TRPV2 is not activated by vanilloids such as capsaicin (Caterina et al., 1999). Comparison between structures of TRPV1 and TRPV2 may provide structural insight into functional divergence between these highly homologous proteins.

Cryo-EM studies were performed on full-length TRPV2 and a stable "minimal" TRPV1 construct in which the N- and C-termini were truncated. The 2D class averages for full-length TRPV2 and minimal TRPV1 are similar (Figure 7.4(A)). A TM domain with an MNG detergent or A8-35 amphipol belt is visible, and a cytoplasmic domain, which consists of the ARD, is also apparent in the 2D class averages (Figure 7.4(A)). The structure of TRPV2 was refined to 13.6 Å (Figure 7.4(B)) (Huynh et al., 2014). A homologous potassium channel structure fits well into the TM domain density of the TRPV2 EM map (Figure 7.4(B) and (C)) and the

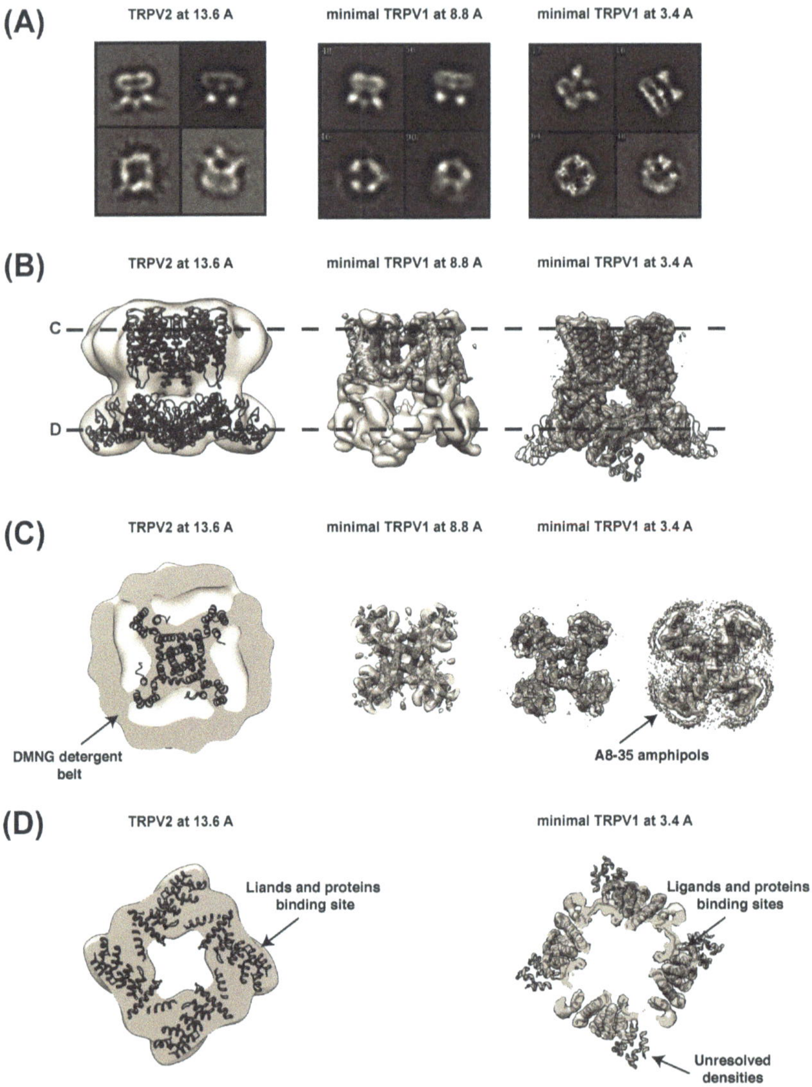

Figure 7.4 Comparison of cryo-electron microscopy (EM) structures of full-length TRPV2 and minimal TRPV1. (A) 2D class averages of TRPV2 at 13.6 Å (Huynh et al., 2014) and minimal TRPV1 at 8.8 Å and 3.4 Å resolution *(Adopted and reprinted from Liao et al. (2013))*. (B) 3D reconstruction of TRPV2 at 13.6 Å resolution (Huynh et al., 2014) and minimal TRPV1 at 8.8 Å and 3.4 Å resolution *(Adopted and reprinted from Liao et al. (2013))*. The TRPV2 EM map is fitted with crystal structures of the MlotiK1 transmembrane (TM) domain (PDB code: 3BEH) and the TRPV2 ankyrin repeat domain (PDB code: 2ETB). Dashed lines correspond to cross-sections represented in (C) and (D). (C) Cross-sectional view through the TM domains of the TRPV2 (Huynh et al., 2014) and minimal TRPV1 EM maps *(Adopted and reprinted from Liao et al. (2013))*. The unfilled densities correspond to the maltose-neopentyl glycol (MNG) detergent belt for TRPV2 and the A8-35 amphipol belt for minimal TRPV1. (D) Cross-sectional view through the cytoplasmic density of the TRPV2 (Huynh et al., 2014) and minimal TRPV1 EM maps. *Liao et al. (2013)*. (See the color plate.)

MNG detergent can be clearly resolved around the TM domain protein density as an uneven belt of ~15–20 Å as previously observed for other membrane proteins (Figure 7.4(C)) (Vahedi-Faridi, Jastrzebska, Palczewski, & Engel, 2013; Westfield et al., 2011).

Remarkably, by employing a new direct electron counting camera and new image processing algorithms, the structure of minimal TRPV1 was refined first to 8.8 Å and then 3.4 Å resolution (Figure 7.4(B)) (Cao, Liao, et al., 2013). The TM helices of minimal TRPV1 are clearly resolved (Figure 7.4(B) and (C)) and a negligible amount of A8-35 amphipol molecules, which were used to stabilize the TM domain of TRPV1, are apparent at 8.8 Å and 3.4 Å resolution (Figure 7.4(C)). The A8-35 amphipol belt, however, can be visualized at the higher isosurface levels, where it is symmetrically diffuse and not directly interacting with the TM helices of the protein density (Figure 7.4(C)). Intriguingly, the TM region of the truncated TRPV1 displays the highest resolution within the structure (Henderson, 2013), while information for the rest of the molecule is at a lower resolution and not as clearly resolved (Figure 7.4(B)). Therefore, the capability of A8-35 amphipols to stabilize the TM region of membrane proteins makes them a very useful tool for studying the structure of membrane proteins at high resolution by cryo-EM (Popot et al., 2011).

The minimal TRPV1 structure at 3.4 Å resolution revealed that the TM region of TRPV channels is nearly identical to that of Na^+ and K^+ channels (Long, Tao, Campbell, & MacKinnon, 2007; Payandeh, Scheuer, Zheng, & Catterall, 2011). Under ligand-free conditions, the TM domain of TRPV1 aligns well with the known structures of Na^+ and K^+ channels (Long et al., 2007; Payandeh et al., 2011). Unlike Na^+ and K^+ channels, the cryo-EM structures of TRPV1 in the presence of activators show that the TM1–TM4 segment of TRPV1 does not undergo conformational changes during channel activation (Cao, Liao, et al., 2013). Conformational changes upon activation occurred in the pore region (Cao, Liao, et al., 2013). Comparison of TRPV1 structures in the absence and presence of activators revealed that the pore expands at two sites: a selectivity filter in the S5–S6 loop (Gly643) and a lower gate at the intracellular end of S6 (Ile679) (Figure 7.5(A)).

Previous electrophysiological studies using a scanning mutagenesis strategy also revealed insight into the pore architecture of TRPV1 (Islas et al., 2009; Salazar et al., 2009; Susankova, Ettrich, Vyklicky, Teisinger, & Vlachova, 2007). These studies showed a dual gate sensitive to capsaicin and heat (Figure 7.5(B)) (Salazar et al., 2009). Consistent with the cryo-EM structure of TRPV1, a gate for larger cations was proposed at the intracellular end of S6 (Leu681) (Figure 7.5(B)). However, these studies also proposed a gate for smaller cations at a more proximal region of the S6 segment (Leu 671)

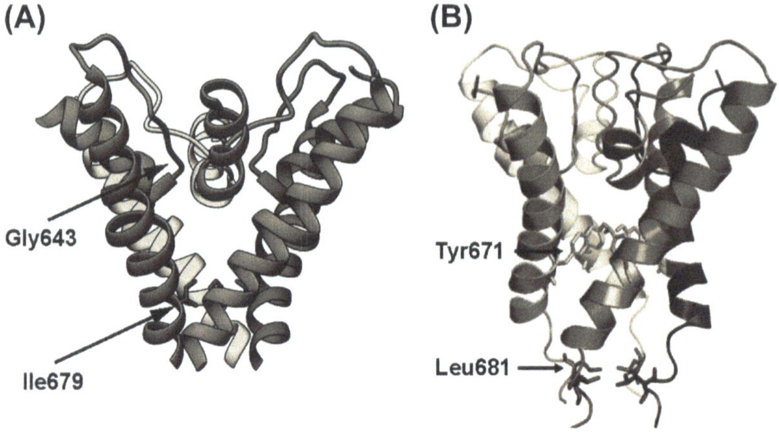

Figure 7.5 Structural analysis of the TRPV1 pore region. (A) Architecture of the TRPV1 pore as determined by cryo-electron microscopy at 3.4 Å (Liao et al., 2013). Residues important for forming the dual gate are indicated by arrows. (B) Predicted pore architecture based on homology modeling and electrophysiological studies using a cysteine accessibility strategy *(Adopted and reprinted from Salazar et al. (2009).)*. A gate for passage of small cations is predicted to begin from the intracellular side at Tyr671. A second gate for larger molecules is predicted to begin from the intracellular side at Leu681 (both residues indicated by arrows). (See the color plate.)

between the selectively filter and lower gate (Figure 7.5(B)) (Salazar et al., 2009), which was not apparent in the cryo-EM structure (Figure 7.5(A)) (Cao, Liao, et al., 2013). Further high-resolution structural studies of TRPV channels in the presence of activators are needed to fully understand the conformational changes and pore dynamics upon channel activation.

The cryo-EM structures of the minimal TRPV1 and full-length TRPV2 diverge most in the cytoplasmic region (Figure 7.4(D)). The ARDs, which comprise ~50% of the total mass of the protein, are involved in protein–ligand and possibly protein–protein interactions (Gaudet, 2008). The crystal structure of the TRPV2 ARDs fit unambiguously into the cytoplasmic density of the TRPV2 EM map (Figure 7.4(B) and (D)). The ARDs were fit into the TRPV2 EM map using computational methods (Chacon & Wriggers, 2002) and were oriented such that the concave ligand-binding surface faces outward toward the cytoplasm. This would provide an accessible binding surface for proteins and ligands to interact with the ARDs without clashing with other portions of the channel protein (Figure 7.4(D)).

The ARDs of the minimal TRPV1 structure were not fully resolved; the density for the first two repeats is absent in the structure (Figure 7.4(B) and (D)). The authors explain that the ARDs of TRPV1 may be especially flexible, which may have prevented them from resolving the first two repeats (Liao et al., 2013). Additionally, the ARDs of the minimal TRPV1 are arranged such that the concave ligand-binding surface, important for protein–protein interactions, is facing inward toward the TM domain and C-terminus of the channel (Figure 7.4(D)). Moreover, a β-strand from the N-terminal membrane proximal linker region is observed interacting with the concave surface of the ARDs, which according to the authors is involved in the stabilization of the channel assembly (Figure 7.4(D)). The authors propose that the rest of the ARD surface is available for interactions between TRPV1 and intracellular partners.

Nevertheless, the concave region of the ARDs is the essential interaction surface for a diverse range of binding partners (Gaudet, 2008). A crystal structure of the complex between Gankyrin and the C-terminal portion of the S6 ATPase of the 26S proteosome shows that the Gankryin interacts with the S6 ATPase via its concave surface (Nakamura et al., 2007). We performed an analysis where the model of the Gankryin-S6 ATPase complex was superimposed with the TRPV channel ARDs as they are arranged in the TRPV cryo-EM maps. The superimposition of the complex is displayed as the S6-ATPase interacting with the TRPV ARDs as it would with Gankryin (Figure 7.6). It is clear that the β-strand from the N-terminal membrane proximal linker of TRPV1 interferes with the interaction of the S6-ATPase protein and the ARD concave ligand-binding surface (Figure 7.6(A)). A

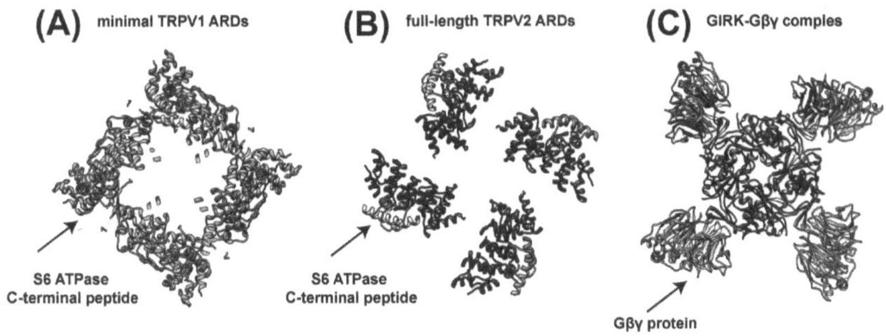

Figure 7.6 Comparison of the ankyrin repeat domain (ARD) orientation in the TRPV2 and minimal TRPV1 electron microscopy (EM) maps. (A) Model of the TRPV1 ARDs as they are arranged in the minimal TRPV1 cryo-EM map and aligned with the crystal structure of the Gankyrin-S6 ATPase C-terminus complex (PDB code: 2DVW). The TRPV1 ARDs are shown in complex with the S6-ATPase protein. (B) Crystal structures of the TRPV2 ARDs (PDB code: 2ETB) as fitted into the TRPV2 cryo-EM map and aligned with the Gankyrin-S6 ATPase C-terminus complex (PDB code: 2DVW). The TRPV2 ARDs are shown in complex with the S6-ATPase protein. (C) Crystal structure of GIRK2 in complex with Gβγ (PDB code: 4KFM). (See the color plate.)

similar analysis performed for the TRPV2 ARDs as they are fitted into the TRPV2 cryo-EM map showed the ARDs have an extensive binding surface with which the S6-ATPase protein and other potential binding proteins can interact (Figure 7.6(B)). This is consistent with the observation from the GIRK2 channel structure in complex with a G protein (Whorton & MacKinnon, 2013), where the channel shows an extensive binding surface exposed to the cytoplasm (Figure 7.6(C)). Divergence in how the ARDs are arranged in the minimal TRPV1 and full-length TRPV2 structures may be due to intrinsic differences in channel architecture, differences in the accuracy with which particle orientation was determined in cryo-EM analysis, resolution limits, or different strategies by which the initial models were generated for each structure (Henderson, 2013). More biochemical and structural studies of TRPV channels are needed to further dissect the different arrangements of the ARDs within the structures of minimal TRPV1 and full-length TRPV2.

A recent FRET-based study provided insight into the domain organization and conformational changes of thermoTRPV channels, specifically TRPV1 (De-la-Rosa, Rangel-Yescas, Ladron-de-Guevara, Rosenbaum, & Islas, 2013). This showed that the C-terminus of TRPV1 is surrounded by the N-terminus, and that the C-terminus is closer to the membrane than the N-terminus. Based on fitting of known structures into the TRPV2 EM map (Figure 7.4(B)), we proposed that the TRPV2 TM domain (Figure 7.4(B)) and the N-terminal ARD (Figure 7.4(B)) are separated by a density that contains C-terminus of the channel (Figure 7.4(B)). Furthermore, this arrangement also suggests that the TRPV2 C-terminus is surrounded by the N-terminal domain (Figure 7.4(B)). A similar overall domain arrangement is also present in the structure of minimal TRPV1 (Figure 7.4(B)). Therefore, the overall architectures of the minimal TRPV1 and full-length TRPV2 EM maps matches well with the in situ FRET data and lends further evidence that these structures represent the general architecture of TRPV channels.

Based on these structural studies, however, it remains difficult to predict how heat activates TRPV1. Utilization of cryo-EM to generate a high-resolution structure of TRPV1 opens new avenues to determine how heat gates thermoTRPV channels (Cao, Liao, et al., 2013).

2.2 TRPA subfamily

2.2.1 Is TRPA1 a cold sensor?

TRPA1 is the only mammalian member of the TRPA subfamily of proteins. It was first discovered as p120 protein in cultured fibroblasts (Jaquemar,

Schenker, & Trueb, 1999) and later identified as ANKTM1 in a bioinformatics screen as a novel noxious cold-sensing channel (Story et al., 2003). TRPA1 is a 525 kDa homotetrameric channel with cytoplasmic N- and C-termini. The N-terminus of each monomer is predicted to contain 14–19 ankyrin repeats followed by a linker region connected to the first TM segment (Figure 7.7(A)).

Evidence that TRPA1 is activated by cold came from patch-clamp experiments demonstrating that temperatures <17 °C and the cooling agent icilin activate currents in Chinese hamster ovary cells heterologously expressing TRPA1 (Story et al., 2003). Since temperatures <17 °C evoke pain in humans, it was concluded that TRPA1 was a sensor for noxious cold; however, this conclusion is controversial (Caspani & Heppenstall, 2009) as it is still unclear if the channel is directly gated by cold or whether such activation is an indirect effect.

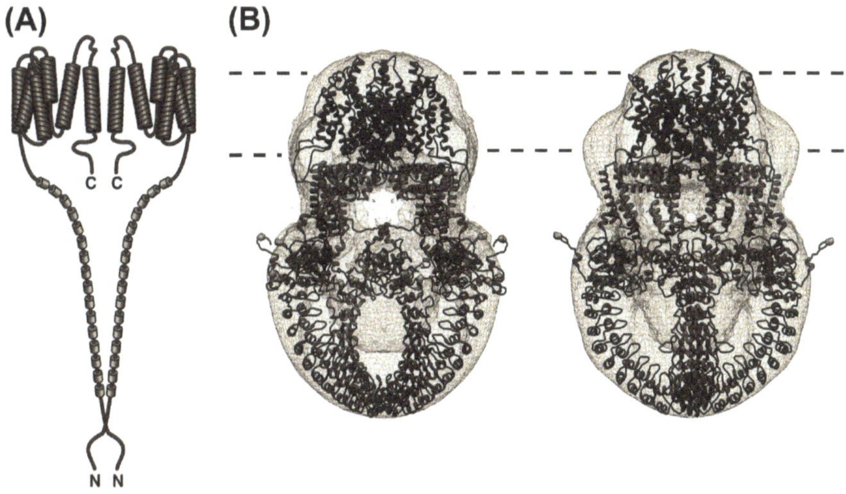

Figure 7.7 Topology model of TRPA1 and TRPA1 electron microscopy (EM) structure at 16 Å resolution. (A) Domain organization of TRPA1. (B) Two side views of the TRPA1 EM structure (Cvetkov et al., 2011). The transmembrane (TM) domain was fitted with a molecular model of the TRPV1 TM domain (purple) (Fernandez-Ballester & Ferrer-Montiel, 2008). A model of the TRPA1 N-terminus was fitted within the cytoplasmic domain of the density map (blue). The cysteine residues critical for electrophilic ligand activation are represented as yellow spheres. A model of the TRPA1 C-terminus was fitted into the 3D density map immediately below the predicted TM domain within the region linking the TM domain to the cytoplasmic domain (light purple). (See the color plate.)

Two lines of TRPA1 knockout mice were generated but did not conclusively show that TRPA1 functions as a cold sensor in vivo (Bautista et al., 2006; Kwan et al., 2006). To follow-up on those original experiments, some laboratories were able to demonstrate that TRPA1 deletion disrupted noxious cold sensation (del Camino et al., 2010; Karashima et al., 2009; Sawada, Hosokawa, Hori, Matsumura, & Kobayashi, 2007), while others found that TRPA1 deletion had little effect on cold sensation (Jordt et al., 2004; Zurborg, Yurgionas, Jira, Caspani, & Heppenstall, 2007). This discrepancy could be due to the different strategies in creating these mice (Bautista et al., 2006; Kwan et al., 2006).

The cold sensitivity of the TRPA1 was also shown to be species dependent. While mouse and rat TRPA1 are cold sensitive, human and monkey TRPA1 are insensitive to cold (Chen et al., 2013). A single glycine residue on the mouse and rat TM segment 5 (Gly878) was suggested to be responsible for cold sensing by the channel. Mutation of Gly878 to valine, the equivalent residue in human and monkey TRPA1 (Val875), abolished cold sensitivity of rat TRPA1 (Chen et al., 2013).

Intriguingly, TRPA1 was identified as a sensor responsible for the infrared heat detection in snakes (Gracheva et al., 2010; Moiseenkova, Bell, Motamedi, Wozniak, & Christensen, 2003; Panzano, Kang, & Garrity, 2010; Pappas, Motamedi, & Christensen, 2004). Moreover, TRPA1 also acts as a heat sensor in fly and mosquitos (Kang et al., 2012; Viswanath et al., 2003), suggesting that TRPA1 evolutionally changed from heat- to cold-sensing channel in vertebrates, and became insensitive to temperature in humans (Chen et al., 2013). The molecular mechanisms by which TRPA1 detects temperature from warm to cold remain to be determined. However, a recent chimeric analysis showed that the TRPA1 ankyrin repeats and N-terminal flexible linker are important for detecting and transducing different stimuli (Cordero-Morales et al., 2011).

In contrast to sensing cold, the role of TRPA1 in pain transduction and inflammatory sensitization is well established (Jordt & Ehrlich, 2007). TRPA1 is best characterized as a chemical nociceptor (Stucky et al., 2009). TRPA1 is activated by numerous exogenous electrophilic compounds and by endogenous reactive species generated during oxidative stress. Recent mutagenesis and chimeric studies have suggested that N-terminal cysteine residues (C622, C642, C666) located on the flexible linker region of the channel (N616-A720) are involved in channel activation and desensitization (Hinman, Chuang, Bautista, & Julius, 2006; Kang et al., 2010; Macpherson et al., 2007; Takahashi et al., 2008, 2011). In addition, non-electrophilic agonists such as 2-amino-phenyl borate (2-APB), δ-9-tetrahydrocannabinol (THC), nicotine, and propofol can activate the channel (Talavera et al., 2009).

2.2.2 EM structures of full-length TRPA1 channel

Structural analysis of TRPA1 can provide a valuable starting point for understanding how TRPA1 homologs respond to such a wide range of temperatures and chemical compounds. Our laboratory determined the structure of mouse TRPA1 at 16 Å resolution using negatively stained, A8-35 amphipol stabilized TRPA1 complexes by EM (Figure 7.7(B)).

The structure revealed a tetrameric assembly with a compact TM domain and a basket-like cytoplasmic domain structure (Cvetkov et al., 2011). The size of our TRPA1 EM map is in agreement with its molecular mass. However, some additional volumetric mass was observed in the TRPA1 EM density due to the presence of the A8-35 amphipol molecules that were used for stabilization of the protein.

Interpretation of the TRPA1 EM map was achieved though molecular modeling of the TRPA1 N- and C-terminal domains using the I-TASSER protein structure prediction server (Roy, Kucukural, & Zhang, 2010) and manually docking these models into the EM density (Cvetkov et al., 2011). The model of the N-terminus (residues 1–721) predicted an ordered region of 12 ankyrin repeats (residues 31–437) and a more flexible downstream region (residues 438–721). This model was placed within the cytoplasmic domain of the density map (Figure 7.7(B), blue). The putative TM domain was fitted with a previously published molecular model of the TRPV1 TM domain (Figure 7.7(B), purple) (Fernandez-Ballester & Ferrer-Montiel, 2008). The model of the TRPA1 C-terminus (residues 964–1125) was placed into the 3D density map immediately below the predicted TM domain within the region linking the TM domain to the cytoplasmic domain (Figure 7.7(B), light purple).

This TRPA1 map reconstruction together with protein modeling provided the first insight into the overall organization of TRPA1 and allowed us to propose a mechanism of TRPA1 activation by electrophilic compounds. In the TRPA1 model fitted into the EM map, we observed that cysteines important for electrophilic activation structurally cluster in the pocket on the flexible region between TM domain and the ankyrin repeats (Figure 7.7(B), yellow spheres). The proximity of these cysteines to one another led us to hypothesize that crosstalk or disufide interactions could be occurring as a part of the channel activation mechanism (Cvetkov et al., 2011).

To test our hypothesis, we performed a mass spectrometry analysis of the in vivo thiol status of TRPA1 and identified that when treated with a thiol-reactive agonist, TRPA1 cysteine residues involved in channel activation form disulfide bonds with each other (Wang, Cvetkov, Chance, & Moiseenkova-Bell, 2012). This disulfide rearrangement could lead to channel activation and the rapid desensitization observed in electrophysiological studies.

We intend to use cryo-EM to achieve a higher resolution structure of fly, mouse, and human TRPA1, which will allow us to resolve structural elements in the channel and identify the root of divergence in temperature sensation between species.

2.3 TRPM subfamily

2.3.1 Thermosensitivity of TRPM channels

The TRPM subfamily of ion channels has eight members, which are subdivided based on their structural similarity into four groups: TRPM1/3, TRPM6/7, TRPM4/5, and TRPM2/8 (Zholos, Johnson, Burdyga, & Melanaphy, 2011). In the TRPM subfamily, TRPM6/7 are Mg^{2+}-permeable and important regulators of intracellular Mg^{2+} homeostasis, while other TRPM channels are nonselective cation channels (Farooqi et al., 2011). Furthermore, TRPM6/7 contain an atypical kinase domain, allowing them to function as "chanzymes," capable of phosphorylating and activating the channel (Montell, 2003). TRPM2 also has catalytic ADP-ribose hydrolase activity (Farooqi et al., 2011). TRPM channels are regulated by oxidative stress, Ca^{2+}, ATP, stretch, and temperature (Zholos et al., 2011).

TRPM8 is the only temperature-sensitive channel of this subfamily (Liu & Qin, 2011). TRPM8 was originally cloned from prostate cancer cells (Tsavaler, Shapero, Morkowski, & Laus, 2001) but was later identified in dorsal root ganglion (DRG) sensory neurons and trigeminal neurons, where it was shown to act as a cold sensor. TRPM8 is activated by cool temperatures (<20–28 °C) as well as cooling agents menthol and icilin (Chuang, Neuhausser, & Julius, 2004; McKemy, Neuhausser, & Julius, 2002; Peier, Moqrich, et al., 2002). TRPM8 acts as a cold sensor in vivo, as deletion of TRPM8 from mice disrupts responses to non-noxious cold temperatures (Bautista et al., 2007; Colburn et al., 2007; Dhaka et al., 2007). Additionally, dissociated sensory neurons from TRPM8 null mice show decreased responses to cold and menthol (Liu & Qin, 2011).

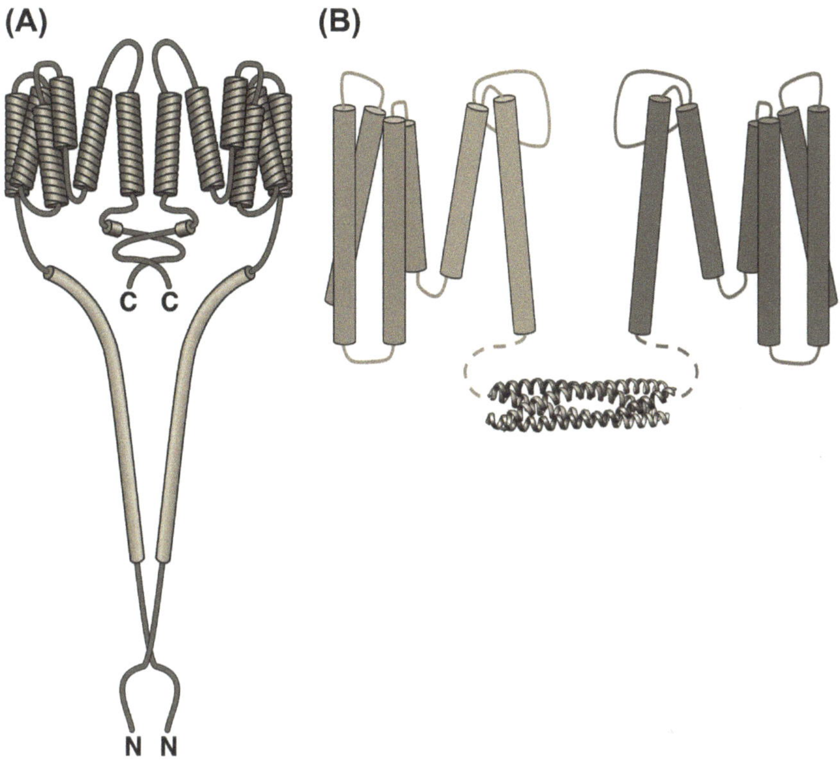

Figure 7.8 Topology model of TRPM channels and crystal structure of TRPM7 coiled-coil domain. (A) Domain organization of TRPM channels. (B) Crystal structure of the TRPM7 coiled-coil domain (PDB code: 3E7K) incorporated into a cartoon model of the TRPM7 transmembrane (TM) domain, showing the antiparallel coiled-coil oriented perpendicular to the TM6 segment. *Fujiwara & Minor (2008).*

TRPM8 is a nonselective, Ca^{2+}-permeable cation channel that functions as a homotetramer (Latorre, Brauchi, Madrid, & Orio, 2011). Its molecular mass is ~510 kDa and it possesses long cytoplasmic N- and C-terminal domains that are important for cold- and ligand-induced channel activity (Figure 7.8(A)) (Latorre et al., 2011). The N-terminus contains the TRPM homology domain, which is conserved among all TRPM family members (Figure 7.8(A)). Its role in channel function remains unclear; however, deletion of this domain disrupts channel activity. The C-terminus contains the conserved TRP box sequence, which is important for activation and trafficking of TRPM8 to the plasma membrane as well as modulating its response to phosphoinositides (Latorre et al., 2011). The C-terminus also harbors a conserved coiled-coil domain, which may be required for

TRPM8 channel tetramerization and temperature- and ligand-induced activation (Tsuruda, Julius, & Minor, 2006). The C-terminus is thought to be directly involved in cold sensation by TRPM8, as TRPV1 chimeras with the TRPM8 C-terminus become cold sensitive, while TRPM8 chimeras with the TRPV1 C-terminus are sensitive to heat (Brauchi et al., 2006).

2.3.2 Structural analysis of TRPM channels

TRPM channels are large membrane proteins, ranging from 500 to 950 kDa in molecular size, with dynamic intracellular regions, which may hamper structural analyses of full-length TRPM channels. To date, the structural and biochemical information for the TRPM subfamily comes from studies of the coiled-coil domain. A crystal structure of the 5.5 kDa TRPM7 coiled-coil complex, which represents ~2.5% of the 850 kDa tetramer, was solved recently and provided insight into its role in tetramer assembly (Fujiwara & Minor, 2008). C-terminal coiled-coil domains for several ion channels, including Kv7, Eag, CNG, TRPC, and TRPV channels have been identified as oligomerization domains (Fujiwara & Minor, 2008; Howard, Clark, Holton, & Minor, 2007).

The isolated coiled-coil domain of TRPM8 self-assembles in solution, and point mutations to key residues involved in coiled-coil domain assembly in full-length TRPM8 disrupted oligomerization and translocation of TRPM8 in cells (Tsuruda et al., 2006). Crystallization of the TRPM7 coiled-coil complex further revealed the diversity among the TRPM subfamily members (Fujiwara & Minor, 2008). The TRPM7 coiled-coil consists of antiparallel helices that assemble by core packing and interactions between polar and charged residues across strands (Figure 7.8(B)). Based on sequence homology, the coiled-coil domains of TRPM channels can be categorized into two groups: those like TRPM7 (TRPM1, TRPM3, and TRPM6) and those unlike TRPM7 (TRPM2, TRPM4, TRPM5, and TRPM8) (Fujiwara & Minor, 2008). Those from the TRPM8 group show poor sequence conservation with the TRPM7 coiled-coil domain. They are shorter, and it is unclear whether they form parallel or antiparallel interactions. TRPM8-like coiled-coil domains would likely form incompatible electrostatic interactions with domains from the TRPM7 group. This provides an explanation as to why TRPM8 functions as a homotetramer.

Since 100 residues separate the coiled-coil domain from the TM6 segment, it has been proposed that the coiled-coil domain of TRPM7 fits beneath the TM domain and arranges in an antiparallel manner parallel to the plasma membrane (Figure 7.8(B)). This would allow for the coiled-coil domain to act as a spacer, allowing symmetrical assembly of kinase domain dimers (Fujiwara & Minor, 2008).

This crystal structure provided a detailed analysis of TRPM channel oligomerization. However, analysis of the structure of full-length TRPM8 will likely be necessary to determine the conformational changes and molecular mechanisms underlying cold sensitivity.

3. OUTLOOK AND PROSPECTIVE

Higher organisms have developed specific mechanisms to detect changes in temperature. Their sensory neurons express an array of polymodal receptors that are activated over a wide range of temperatures from painful cold to noxious heat. Most of these receptors come from the TRP channel family, which when activated, conduct cations that activate the sensory neurons and relay a signal to the brain. Since the discovery of thermosensitive TRP channels, the search has been on for the molecular nature of the temperature sensor within these channels; however, the structural features that convey temperature sensitivity to thermoTRP channels remain elusive.

Structure–function analysis of thermoTRP channels has lent insight into the thermodynamic and molecular mechanisms by which these polymodal receptors sense temperature changes. While mutagenesis and chimeric approaches have identified several channel domains responsible for thermosensation, controversy still remains as to how these channels are gated by temperature. It is likely that the domains identified by structure–function studies are all involved in thermosensation. Higher resolution structural data for all full-length thermoTRP channels will be necessary for understanding how these protein complexes respond to a wide range of temperatures and the conformational changes these channels undergo during activation by temperature changes at atomic levels.

The TRPV subfamily is the most extensively characterized of all TRP channels due to the role of its members in pain and heat sensation. TRPV1–4 are heat sensitive in heterologous expression systems while TRPV1 clearly acts as an intrinsic heat sensor in vivo. Moderate resolution structures of TRPV1, TRPV2, and TRPV4 have revealed the general architecture of thermoTRPV channels. Together with high-resolution structures for the soluble ARDs for thermoTRPV channels, we can start to understand how the domains are organized and how the channels are regulated. The recent high-resolution structure of minimal TRPV1 lends the potential to understand the conformation changes of thermoTRPV channels upon activation by heat at a new detailed level.

Both TRPA1 and TRPM8 are cold sensors that form large dynamic complexes, making structural studies more difficult. Progress is being made in studying TRPA1, as a moderate resolution structure allowed us to understand how TRPA1 is modulated by electrophilic ligands. Obtaining higher resolution information for TRPA1 will allow us to understand how cold temperature gates the channel. Structural understanding of TRPM8 is in its infancy. Better methodologies for stabilizing and purifying TRPM8 will allow for more biophysical biochemical and structural characterization of the channel.

REFERENCES

Abrahamsen, B., Zhao, J., Asante, C. O., Cendan, C. M., Marsh, S., Martinez-Barbera, J. P., et al. (2008). The cell and molecular basis of mechanical, cold, and inflammatory pain. *Science, 321*(5889), 702–705. http://dx.doi.org/10.1126/science.1156916.

Al-Ansary, D., Bogeski, I., Disteldorf, B. M., Becherer, U., & Niemeyer, B. A. (2010). ATP modulates Ca^{2+} uptake by TRPV6 and is counteracted by isoform-specific phosphorylation. *FASEB Journal, 24*(2), 425–435. doi:fj.09-141481, pii:10.1096/fj.09-141481.

Bautista, D. M., Jordt, S. E., Nikai, T., Tsuruda, P. R., Read, A. J., Poblete, J., et al. (2006). TRPA1 mediates the inflammatory actions of environmental irritants and proalgesic agents. *Cell, 124*(6), 1269–1282. doi:S0092-8674(06)00240-6, pii:10.1016/j.cell.2006.02.023.

Bautista, D. M., Siemens, J., Glazer, J. M., Tsuruda, P. R., Basbaum, A. I., Stucky, C. L., et al. (2007). The menthol receptor TRPM8 is the principal detector of environmental cold. *Nature, 448*(7150), 204–208. http://dx.doi.org/10.1038/nature05910.

Belmonte, C., & Viana, F. (2008). Molecular and cellular limits to somatosensory specificity. *Molecular Pain, 4*, 14. http://dx.doi.org/10.1186/1744-8069-4-14.

Binz, H. K., Amstutz, P., Kohl, A., Stumpp, M. T., Briand, C., Forrer, P., et al. (2004). High-affinity binders selected from designed ankyrin repeat protein libraries. *Nature Biotechnology, 22*(5), 575–582. doi:10.1038/nbt962, pii:nbt962.

Brauchi, S., Orta, G., Salazar, M., Rosenmann, E., & Latorre, R. (2006). A hot-sensing cold receptor: C-terminal domain determines thermosensation in transient receptor potential channels. *The Journal of Neuroscience, 26*(18), 4835–4840. http://dx.doi.org/10.1523/JNEUROSCI.5080-05.2006.

del Camino, D., Murphy, S., Heiry, M., Barrett, L. B., Earley, T. J., Cook, C. A., et al. (2010). TRPA1 contributes to cold hypersensitivity. *The Journal of Neuroscience, 30*(45), 15165–15174. http://dx.doi.org/10.1523/JNEUROSCI.2580-10.2010.

Cao, E., Cordero-Morales, J. F., Liu, B., Qin, F., & Julius, D. (2013). TRPV1 channels are intrinsically heat sensitive and negatively regulated by phosphoinositide lipids. *Neuron*, 77(4), 667–679. http://dx.doi.org/10.1016/j.neuron.2012.12.016.

Cao, E., Liao, M., Cheng, Y., & Julius, D. (2013). TRPV1 structures in distinct conformations reveal activation mechanisms. *Nature, 504*(7478), 113–118. http://dx.doi.org/10.1038/nature12823.

Caspani, O., & Heppenstall, P. A. (2009). TRPA1 and cold transduction: an unresolved issue? *The Journal of General Physiology, 133*(3), 245–249. doi:jgp.200810136, pii: 10.1085/jgp.200810136.
Caterina, M. J., Leffler, A., Malmberg, A. B., Martin, W. J., Trafton, J., Petersen-Zeitz, K. R., et al. (2000). Impaired nociception and pain sensation in mice lacking the capsaicin receptor. *Science, 288*(5464), 306–313.
Caterina, M. J., Rosen, T. A., Tominaga, M., Brake, A. J., & Julius, D. (1999). A capsaicin-receptor homologue with a high threshold for noxious heat. *Nature, 398*(6726), 436–441. http://dx.doi.org/10.1038/18906.
Caterina, M. J., Schumacher, M. A., Tominaga, M., Rosen, T. A., Levine, J. D., & Julius, D. (1997). The capsaicin receptor: a heat-activated ion channel in the pain pathway. *Nature, 389*(6653), 816–824. http://dx.doi.org/10.1038/39807.
Chacon, P., & Wriggers, W. (2002). Multi-resolution contour-based fitting of macromolecular structures. *Journal of Molecular Biology, 317*(3), 375–384. http://dx.doi.org/10.1006/jmbi.2002.5438.
Chen, J., Kang, D., Xu, J., Lake, M., Hogan, J. O., Sun, C., et al. (2013). Species differences and molecular determinant of TRPA1 cold sensitivity. *Nature Communications, 4*, 2501. http://dx.doi.org/10.1038/ncomms3501.
Chuang, H. H., Neuhausser, W. M., & Julius, D. (2004). The super-cooling agent icilin reveals a mechanism of coincidence detection by a temperature-sensitive TRP channel. *Neuron, 43*(6), 859–869. http://dx.doi.org/10.1016/j.neuron.2004.08.038.
Clapham, D. E., & Miller, C. (2011). A thermodynamic framework for understanding temperature sensing by transient receptor potential (TRP) channels. *Proceedings of the National Academy of Sciences USA, 108*(49), 19492–19497. http://dx.doi.org/10.1073/pnas.1117485108.
Colburn, R. W., Lubin, M. L., Stone, D. J., Jr., Wang, Y., Lawrence, D., D'Andrea, M. R., et al. (2007). Attenuated cold sensitivity in TRPM8 null mice. *Neuron, 54*(3), 379–386. http://dx.doi.org/10.1016/j.neuron.2007.04.017.
Cordero-Morales, J. F., Gracheva, E. O., & Julius, D. (2011). Cytoplasmic ankyrin repeats of transient receptor potential A1 (TRPA1) dictate sensitivity to thermal and chemical stimuli. *Proceedings of the National Academy of Sciences USA, 108*(46), E1184–E1191. http://dx.doi.org/10.1073/pnas.1114124108.
Cosens, D. J., & Manning, A. (1969). Abnormal electroretinogram from a *Drosophila mutant. Nature, 224*(5216), 285–287.
Cvetkov, T. L., Huynh, K. W., Cohen, M. R., & Moiseenkova-Bell, V. Y. (2011). Molecular architecture and subunit organization of TRPA1 ion channel revealed by electron microscopy. *The Journal of Biological Chemistry, 286*(44), 38168–38176. http://dx.doi.org/10.1074/jbc.M111.288993.
De-la-Rosa, V., Rangel-Yescas, G. E., Ladron-de-Guevara, E., Rosenbaum, T., & Islas, L. D. (2013). Coarse architecture of the transient receptor potential vanilloid 1 (TRPV1) ion channel determined by fluorescence resonance energy transfer. *The Journal of Biological Chemistry, 288*(41), 29506–29517. http://dx.doi.org/10.1074/jbc.M113.479618.
Dhaka, A., Murray, A. N., Mathur, J., Earley, T. J., Petrus, M. J., & Patapoutian, A. (2007). TRPM8 is required for cold sensation in mice. *Neuron, 54*(3), 371–378. http://dx.doi.org/10.1016/j.neuron.2007.02.024.
Digel, I. (2011). Primary thermosensory events in cells. *Advances in Experimental Medicine and Biology, 704*, 451–468. http://dx.doi.org/10.1007/978-94-007-0265-3_25.
Digel, I., Kayser, P., & Artmann, G. M. (2008). Molecular processes in biological thermosensation. *Journal of Biophysics, 2008*, 602870. http://dx.doi.org/10.1155/2008/602870.

Farooqi, A. A., Javeed, M. K., Javed, Z., Riaz, A. M., Mukhtar, S., Minhaj, S., et al. (2011). TRPM channels: same ballpark, different players, and different rules in immunogenetics. *Immunogenetics, 63*(12), 773–787. http://dx.doi.org/10.1007/s00251-011-0570-4.

Fernandez-Ballester, G., & Ferrer-Montiel, A. (2008). Molecular modeling of the full-length human TRPV1 channel in closed and desensitized states. *The Journal of Membrane Biology, 223*(3), 161–172. http://dx.doi.org/10.1007/s00232-008-9123-7.

Fujiwara, Y., & Minor, D. L., Jr. (2008). X-ray crystal structure of a TRPM assembly domain reveals an antiparallel four-stranded coiled-coil. *Journal of Molecular Biology, 383*(4), 854–870. http://dx.doi.org/10.1016/j.jmb.2008.08.059.

Garcia-Sanz, N., Fernandez-Carvajal, A., Morenilla-Palao, C., Planells-Cases, R., Fajardo-Sanchez, E., Fernandez-Ballester, G., et al. (2004). Identification of a tetramerization domain in the C terminus of the vanilloid receptor. *The Journal of Neuroscience, 24*(23), 5307–5314. http://dx.doi.org/10.1523/JNEUROSCI.0202-04.2004.

Gaudet, R. (2008). A primer on ankyrin repeat function in TRP channels and beyond. *Molecular BioSystems, 4*(5), 372–379. http://dx.doi.org/10.1039/b801481g.

Gaudet, R. (2009). Divide and conquer: high resolution structural information on TRP channel fragments. *The Journal of General Physiology, 133*(3), 231–237. http://dx.doi.org/10.1085/jgp.200810137.

Gracheva, E. O., Ingolia, N. T., Kelly, Y. M., Cordero-Morales, J. F., Hollopeter, G., Chesler, A. T., et al. (2010). Molecular basis of infrared detection by snakes. *Nature, 464*(7291), 1006–1011. doi:nature08943, pii:10.1038/nature08943.

Grandl, J., Hu, H., Bandell, M., Bursulaya, B., Schmidt, M., Petrus, M., et al. (2008). Pore region of TRPV3 ion channel is specifically required for heat activation. *Nature Neuroscience, 11*(9), 1007–1013. http://dx.doi.org/10.1038/nn.2169.

Grandl, J., Kim, S. E., Uzzell, V., Bursulaya, B., Petrus, M., Bandell, M., et al. (2010). Temperature-induced opening of TRPV1 ion channel is stabilized by the pore domain. *Nature Neuroscience, 13*(6), 708–714. http://dx.doi.org/10.1038/nn.2552.

Guler, A. D., Lee, H., Iida, T., Shimizu, I., Tominaga, M., & Caterina, M. (2002). Heat-evoked activation of the ion channel, TRPV4. *The Journal of Neuroscience, 22*(15), 6408–6414. doi:20026679.

Henderson, R. (1995). The potential and limitations of neutrons, electrons and X-rays for atomic resolution microscopy of unstained biological molecules. *Quarterly Reviews of Biophysics, 28*(2), 171–193.

Henderson, R. (2013). Structural biology: Ion channel seen by electron microscopy. *Nature, 504*(7478), 93–94. http://dx.doi.org/10.1038/504093a.

Heymann, J. B., Conway, J. F., & Steven, A. C. (2004). Molecular dynamics of protein complexes from four-dimensional cryo-electron microscopy. *Journal of Structural Biology, 147*(3), 291–301. doi:10.1016/j.jsb.2004.02.006, pii:S1047847704000425.

Hinman, A., Chuang, H. H., Bautista, D. M., & Julius, D. (2006). TRP channel activation by reversible covalent modification. *Proceedings of the National Academy of Sciences USA, 103*(51), 19564–19568. doi:0609598103, pii:10.1073/pnas.0609598103.

Howard, R. J., Clark, K. A., Holton, J. M., & Minor, D. L., Jr. (2007). Structural insight into KCNQ (Kv7) channel assembly and channelopathy. *Neuron, 53*(5), 663–675. http://dx.doi.org/10.1016/j.neuron.2007.02.010.

Huang, S. M., Li, X., Yu, Y., Wang, J., & Caterina, M. J. (2011). TRPV3 and TRPV4 ion channels are not major contributors to mouse heat sensation. *Molecular Pain,* 7, 37. http://dx.doi.org/10.1186/1744-8069-7-37.

Huynh, K. W., Cohen, M. R., Chakrapani, S., Holdaway, H. A., Stewart, P. L., & Moiseenkova-Bell, V. Y. (2014). Structural insight into the assembly of TRPV channels. *Structure, 22*(2), 260–268. http://dx.doi.org/10.1016/j.str.2013.11.008.

Inada, H., Procko, E., Sotomayor, M., & Gaudet, R. (2012). Structural and biochemical consequences of disease-causing mutations in the ankyrin repeat domain of the human TRPV4 channel. *Biochemistry, 51*(31), 6195–6206. http://dx.doi.org/10.1021/bi300279b.

Islas, L. D., Salazar, H., Jara-Oseguera, A., Nieto-Posadas, A., Llorente, I., Rangel-Yescas, G., et al. (2009). The helical character of the S6 segment of TRPV1 channels. *Channels (Austin, Tex.), 3*(5), 311–313.

Jaquemar, D., Schenker, T., & Trueb, B. (1999). An ankyrin-like protein with transmembrane domains is specifically lost after oncogenic transformation of human fibroblasts. *The Journal of Biological Chemistry, 274*(11), 7325–7333.

Jin, X., Touhey, J., & Gaudet, R. (2006). Structure of the N-terminal ankyrin repeat domain of the TRPV2 ion channel. *The Journal of Biological Chemistry, 281*(35), 25006–25010. http://dx.doi.org/10.1074/jbc.C600153200.

Jordt, S. E., Bautista, D. M., Chuang, H. H., McKemy, D. D., Zygmunt, P. M., Hogestatt, E. D., et al. (2004). Mustard oils and cannabinoids excite sensory nerve fibres through the TRP channel ANKTM1. *Nature, 427*(6971), 260–265. http://dx.doi.org/10.1038/nature02282.

Jordt, S. E., & Ehrlich, B. E. (2007). TRP channels in disease. *Subcellular Biochemistry, 45*, 253–271.

Kang, K., Panzano, V. C., Chang, E. C., Ni, L., Dainis, A. M., Jenkins, A. M., et al. (2012). Modulation of TRPA1 thermal sensitivity enables sensory discrimination in *Drosophila*. *Nature, 481*(7379), 76–80. http://dx.doi.org/10.1038/nature10715.

Kang, K., Pulver, S. R., Panzano, V. C., Chang, E. C., Griffith, L. C., Theobald, D. L., et al. (2010). Analysis of *Drosophila* TRPA1 reveals an ancient origin for human chemical nociception. *Nature, 464*(7288), 597–600. doi:nature08848, pii:10.1038/nature08848.

Karashima, Y., Talavera, K., Everaerts, W., Janssens, A., Kwan, K. Y., Vennekens, R., et al. (2009). TRPA1 acts as a cold sensor in vitro and in vivo. *Proceedings of the National Academy of Sciences USA, 106*(4), 1273–1278. http://dx.doi.org/10.1073/pnas.0808487106.

Kedei, N., Szabo, T., Lile, J. D., Treanor, J. J., Olah, Z., Iadarola, M. J., et al. (2001). Analysis of the native quaternary structure of vanilloid receptor 1. *The Journal of Biological Chemistry, 276*(30), 28613–28619. http://dx.doi.org/10.1074/jbc.M103272200.

Kim, S. E., Patapoutian, A., & Grandl, J. (2013). Single residues in the outer pore of TRPV1 and TRPV3 have temperature-dependent conformations. *PLoS One, 8*(3), e59593. http://dx.doi.org/10.1371/journal.pone.0059593.

Kwan, K. Y., Allchorne, A. J., Vollrath, M. A., Christensen, A. P., Zhang, D. S., Woolf, C. J., et al. (2006). TRPA1 contributes to cold, mechanical, and chemical nociception but is not essential for hair-cell transduction. *Neuron, 50*(2), 277–289. doi:S0896-6273(06)00269-8, pii:10.1016/j.neuron.2006.03.042.

Latorre, R., Brauchi, S., Madrid, R., & Orio, P. (2011). A cool channel in cold transduction. *Physiology (Bethesda, Md.), 26*(4), 273–285. http://dx.doi.org/10.1152/physiol.00004.2011.

Lau, S. Y., Procko, E., & Gaudet, R. (2012). Distinct properties of Ca^{2+}-calmodulin binding to N- and C-terminal regulatory regions of the TRPV1 channel. *The Journal of General Physiology, 140*(5), 541–555. http://dx.doi.org/10.1085/jgp.201210810.

Liao, M., Cao, E., Julius, D., & Cheng, Y. (2013). Structure of the TRPV1 ion channel determined by electron cryo-microscopy. *Nature, 504*(7478), 107–112. http://dx.doi.org/10.1038/nature12822.

Li, J., Mahajan, A., & Tsai, M. D. (2006). Ankyrin repeat: a unique motif mediating protein-protein interactions. *Biochemistry, 45*(51), 15168–15178. http://dx.doi.org/10.1021/bi062188q.

Lishko, P. V., Procko, E., Jin, X., Phelps, C. B., & Gaudet, R. (2007). The ankyrin repeats of TRPV1 bind multiple ligands and modulate channel sensitivity. *Neuron, 54*(6), 905–918. http://dx.doi.org/10.1016/j.neuron.2007.05.027.

Liu, B., Hui, K., & Qin, F. (2003). Thermodynamics of heat activation of single capsaicin ion channels VR1. *Biophysical Journal, 85*(5), 2988–3006. http://dx.doi.org/10.1016/S0006-3495(03)74719-5.

Liu, Y., & Qin, N. (2011). TRPM8 in health and disease: cold sensing and beyond. *Advances in Experimental Medicine and Biology, 704*, 185–208. http://dx.doi.org/10.1007/978-94-007-0265-3_10.

Long, S. B., Tao, X., Campbell, E. B., & MacKinnon, R. (2007). Atomic structure of a voltage-dependent K^+ channel in a lipid membrane-like environment. *Nature, 450*(7168), 376–382. http://dx.doi.org/10.1038/nature06265.

Ludtke, S. J., Serysheva, II, Hamilton, S. L., & Chiu, W. (2005). The pore structure of the closed RyR1 channel. *Structure, 13*(8), 1203–1211. doi:S0969-2126(05)00215-7, pii: 10.1016/j.str.2005.06.005.

Ludtke, S. J., Tran, T. P., Ngo, Q. T., Moiseenkova-Bell, V. Y., Chiu, W., & Serysheva, II. (2011). Flexible architecture of IP3R1 by Cryo-EM. *Structure, 19*(8), 1192–1199. doi: S0969-2126(11)00180-8, pii:10.1016/j.str.2011.05.003.

Macpherson, L. J., Dubin, A. E., Evans, M. J., Marr, F., Schultz, P. G., Cravatt, B. F., et al. (2007). Noxious compounds activate TRPA1 ion channels through covalent modification of cysteines. *Nature, 445*(7127), 541–545. doi:nature05544, pii:10.1038/nature05544.

Maingret, F., Lauritzen, I., Patel, A. J., Heurteaux, C., Reyes, R., Lesage, F., et al. (2000). TREK-1 is a heat-activated background K(+) channel. *EMBO Journal, 19*(11), 2483–2491. http://dx.doi.org/10.1093/emboj/19.11.2483.

McGlone, F., & Reilly, D. (2010). The cutaneous sensory system. *Neuroscience and Biobehavioral Reviews, 34*(2), 148–159. http://dx.doi.org/10.1016/j.neubiorev.2009.08.004.

McKemy, D. D. (2013). The molecular and cellular basis of cold sensation. *ACS Chemical Neuroscience, 4*(2), 238–247. http://dx.doi.org/10.1021/cn300193h.

McKemy, D. D., Neuhausser, W. M., & Julius, D. (2002). Identification of a cold receptor reveals a general role for TRP channels in thermosensation. *Nature, 416*(6876), 52–58. http://dx.doi.org/10.1038/nature719.

Minke, B., Wu, C., & Pak, W. L. (1975). Induction of photoreceptor voltage noise in the dark in *Drosophila mutant. Nature, 258*(5530), 84–87.

Moiseenkova-Bell, V. Y., Stanciu, L. A., Serysheva, II, Tobe, B. J., & Wensel, T. G. (2008). Structure of TRPV1 channel revealed by electron cryomicroscopy. *Proceedings of the National Academy of Sciences USA, 105*(21), 7451–7455. http://dx.doi.org/10.1073/pnas.0711835105.

Moiseenkova-Bell, V. Y., & Wensel, T. G. (2009). Hot on the trail of TRP channel structure. *The Journal of General Physiology, 133*(3), 239–244. http://dx.doi.org/10.1085/jgp.200810123.

Moiseenkova-Bell, V., & Wensel, T. G. (2011). Functional and structural studies of TRP channels heterologously expressed in budding yeast. *Advances in Experimental Medicine and Biology, 704*, 25–40. http://dx.doi.org/10.1007/978-94-007-0265-3_2.

Moiseenkova, V., Bell, B., Motamedi, M., Wozniak, E., & Christensen, B. (2003). Wide-band spectral tuning of heat receptors in the pit organ of the copperhead snake (Crotalinae). *American Journal of Physiology. Regulatory, Integrative and Comparative Physiology, 284*(2), R598–R606.

Montell, C. (2003). Mg^{2+} homeostasis: the Mg^{2+}nificent TRPM chanzymes. *Current Biology, 13*(20), R799–R801.

Nakamura, Y., Nakano, K., Umehara, T., Kimura, M., Hayashizaki, Y., Tanaka, A., et al. (2007). Structure of the oncoprotein gankyrin in complex with S6 ATPase of the

26S proteasome. *Structure, 15*(2), 179–189. http://dx.doi.org/10.1016/j.str.2006.11.015.

Noel, J., Zimmermann, K., Busserolles, J., Deval, E., Alloui, A., Diochot, S., et al. (2009). The mechano-activated K^+ channels TRAAK and TREK-1 control both warm and cold perception. *EMBO Journal, 28*(9), 1308–1318. http://dx.doi.org/10.1038/emboj.2009.57.

Panzano, V. C., Kang, K., & Garrity, P. A. (2010). Infrared snake eyes: TRPA1 and the thermal sensitivity of the snake pit organ. *Science Signaling, 3*(127), pe22. http://dx.doi.org/10.1126/scisignal.3127pe22.

Pappas, T. C., Motamedi, M., & Christensen, B. N. (2004). Unique temperature-activated neurons from pit viper thermosensors. *American Journal of Physiology. Cell Physiology, 287*(5), C1219–C1228. doi:10.1152/ajpcell.00040.2004, pii:00040.2004.

Park, U., Vastani, N., Guan, Y., Raja, S. N., Koltzenburg, M., & Caterina, M. J. (2011). TRP vanilloid 2 knock-out mice are susceptible to perinatal lethality but display normal thermal and mechanical nociception. *The Journal of Neuroscience, 31*(32), 11425–11436. http://dx.doi.org/10.1523/JNEUROSCI.1384-09.2011.

Payandeh, J., Scheuer, T., Zheng, N., & Catterall, W. A. (2011). The crystal structure of a voltage-gated sodium channel. *Nature, 475*(7356), 353–358. http://dx.doi.org/10.1038/nature10238.

Peier, A. M., Moqrich, A., Hergarden, A. C., Reeve, A. J., Andersson, D. A., Story, G. M., et al. (2002). A TRP channel that senses cold stimuli and menthol. *Cell, 108*(5), 705–715.

Peier, A. M., Reeve, A. J., Andersson, D. A., Moqrich, A., Earley, T. J., Hergarden, A. C., et al. (2002). A heat-sensitive TRP channel expressed in keratinocytes. *Science, 296*(5575), 2046–2049. http://dx.doi.org/10.1126/science.1073140.

Peralvarez-Marin, A., Donate-Macian, P., & Gaudet, R. (2013). What do we know about the transient receptor potential vanilloid 2 (TRPV2) ion channel? *FEBS Journal, 280*(21), 5471–5487. http://dx.doi.org/10.1111/febs.12302.

Phelps, C. B., Huang, R. J., Lishko, P. V., Wang, R. R., & Gaudet, R. (2008). Structural analyses of the ankyrin repeat domain of TRPV6 and related TRPV ion channels. *Biochemistry, 47*(8), 2476–2484. http://dx.doi.org/10.1021/bi702109w.

Phelps, C. B., Wang, R. R., Choo, S. S., & Gaudet, R. (2010). Differential regulation of TRPV1, TRPV3, and TRPV4 sensitivity through a conserved binding site on the ankyrin repeat domain. *The Journal of Biological Chemistry, 285*(1), 731–740. doi: M109.052548, pii:10.1074/jbc.M109.052548.

Popot, J. L., Althoff, T., Bagnard, D., Baneres, J. L., Bazzacco, P., Billon-Denis, E., et al. (2011). Amphipols from A to Z. *Annual Review of Biophysics, 40*, 379–408. http://dx.doi.org/10.1146/annurev-biophys-042910-155219.

Roy, A., Kucukural, A., & Zhang, Y. (2010). I-TASSER: a unified platform for automated protein structure and function prediction. *Nature Protocols, 5*(4), 725–738. doi: nprot.2010.5, pii:10.1038/nprot.2010.5.

Salazar, H., Jara-Oseguera, A., Hernandez-Garcia, E., Llorente, I., Arias, Olguin, II, Soriano-Garcia, M., et al. (2009). Structural determinants of gating in the TRPV1 channel. *Nature Structural and Molecular Biology, 16*(7), 704–710. http://dx.doi.org/10.1038/nsmb.1633.

Samso, M., Feng, W., Pessah, I. N., & Allen, P. D. (2009). Coordinated movement of cytoplasmic and transmembrane domains of RyR1 upon gating. *PLoS Biology, 7*(4), e85. doi: 08-PLBI-RA-5391, pii:10.1371/journal.pbio.1000085.

Samso, M., Wagenknecht, T., & Allen, P. D. (2005). Internal structure and visualization of transmembrane domains of the RyR1 calcium release channel by cryo-EM. *Nature Structural and Molecular Biology, 12*(6), 539–544. doi:nsmb938, pii:10.1038/nsmb938.

Sawada, Y., Hosokawa, H., Hori, A., Matsumura, K., & Kobayashi, S. (2007). Cold sensitivity of recombinant TRPA1 channels. *Brain Research, 1160*, 39–46. http://dx.doi.org/10.1016/j.brainres.2007.05.047.

Schepers, R. J., & Ringkamp, M. (2009). Thermoreceptors and thermosensitive afferents. *Neuroscience and Biobehavioral Reviews, 33*(3), 205–212. http://dx.doi.org/10.1016/j.neubiorev.2008.07.009.

Sengupta, P., & Garrity, P. (2013). Sensing temperature. *Current Biology, 23*(8), R304–R307. http://dx.doi.org/10.1016/j.cub.2013.03.009.

Serysheva, II, Ludtke, S. J., Baker, M. L., Cong, Y., Topf, M., Eramian, D., et al. (2008). Subnanometer-resolution electron cryomicroscopy-based domain models for the cytoplasmic region of skeletal muscle RyR channel. *Proceedings of the National Academy of Sciences USA, 105*(28), 9610–9615. doi:0803189105, pii:10.1073/pnas.0803189105.

Shigematsu, H., Sokabe, T., Danev, R., Tominaga, M., & Nagayama, K. (2010). A 3.5-nm structure of rat TRPV4 cation channel revealed by Zernike phase-contrast cryoelectron microscopy. *The Journal of Biological Chemistry, 285*(15), 11210–11218. http://dx.doi.org/10.1074/jbc.M109.090712.

Shi, D. J., Ye, S., Cao, X., Zhang, R., & Wang, K. (2013). Crystal structure of the N-terminal ankyrin repeat domain of TRPV3 reveals unique conformation of finger 3 loop critical for channel function. *Protein and Cell, 4*(12), 942–950. http://dx.doi.org/10.1007/s13238-013-3091-0.

Smith, G. D., Gunthorpe, M. J., Kelsell, R. E., Hayes, P. D., Reilly, P., Facer, P., et al. (2002). TRPV3 is a temperature-sensitive vanilloid receptor-like protein. *Nature, 418*(6894), 186–190. http://dx.doi.org/10.1038/nature00894.

Story, G. M., Peier, A. M., Reeve, A. J., Eid, S. R., Mosbacher, J., Hricik, T. R., et al. (2003). ANKTM1, a TRP-like channel expressed in nociceptive neurons, is activated by cold temperatures. *Cell, 112*(6), 819–829. doi:S0092867403001582.

Strotmann, R., Harteneck, C., Nunnenmacher, K., Schultz, G., & Plant, T. D. (2000). OTRPC4, a nonselective cation channel that confers sensitivity to extracellular osmolarity. *Nature Cell Biology, 2*(10), 695–702. http://dx.doi.org/10.1038/35036318.

Stucky, C. L., Dubin, A. E., Jeske, N. A., Malin, S. A., McKemy, D. D., & Story, G. M. (2009). Roles of transient receptor potential channels in pain. *Brain Research Reviews, 60*(1), 2–23. doi:S0165-0173(08)00157-4, pii:10.1016/j.brainresrev.2008.12.018.

Susankova, K., Ettrich, R., Vyklicky, L., Teisinger, J., & Vlachova, V. (2007). Contribution of the putative inner-pore region to the gating of the transient receptor potential vanilloid subtype 1 channel (TRPV1). *The Journal of Neuroscience, 27*(28), 7578–7585. http://dx.doi.org/10.1523/JNEUROSCI.1956-07.2007.

Takahashi, N., Kuwaki, T., Kiyonaka, S., Numata, T., Kozai, D., Mizuno, Y., et al. (2011). TRPA1 underlies a sensing mechanism for O(2). *Nature Chemical Biology, 7*(10), 701–711. doi:nchembio.640, pii:10.1038/nchembio.640.

Takahashi, N., Mizuno, Y., Kozai, D., Yamamoto, S., Kiyonaka, S., Shibata, T., et al. (2008). Molecular characterization of TRPA1 channel activation by cysteine-reactive inflammatory mediators. *Channels (Austin, Tex.), 2*(4), 287–298. doi:6745.

Talavera, K., Gees, M., Karashima, Y., Meseguer, V. M., Vanoirbeek, J. A., Damann, N., et al. (2009). Nicotine activates the chemosensory cation channel TRPA1. *Nature Neuroscience, 12*(10), 1293–1299. http://dx.doi.org/10.1038/nn.2379.

Tsavaler, L., Shapero, M. H., Morkowski, S., & Laus, R. (2001). Trp-p8, a novel prostate-specific gene, is up-regulated in prostate cancer and other malignancies and shares high homology with transient receptor potential calcium channel proteins. *Cancer Research, 61*(9), 3760–3769.

Tsuruda, P. R., Julius, D., & Minor, D. L., Jr. (2006). Coiled coils direct assembly of a cold-activated TRP channel. *Neuron, 51*(2), 201–212. http://dx.doi.org/10.1016/j.neuron.2006.06.023.

Vahedi-Faridi, A., Jastrzebska, B., Palczewski, K., & Engel, A. (2013). 3D imaging and quantitative analysis of small solubilized membrane proteins and their complexes by transmission electron microscopy. *Microscopy (Oxford, England), 62*(1), 95–107. http://dx.doi.org/10.1093/jmicro/dfs091.

Venkatachalam, K., & Montell, C. (2007). TRP channels. *Annual Review of Biochemistry, 76*, 387–417. http://dx.doi.org/10.1146/annurev.biochem.75.103004.142819.

Viana, F. (2011). Chemosensory properties of the trigeminal system. *ACS Chemical Neuroscience, 2*(1), 38–50. http://dx.doi.org/10.1021/cn100102c.

Viswanath, V., Story, G. M., Peier, A. M., Petrus, M. J., Lee, V. M., Hwang, S. W., et al. (2003). Opposite thermosensor in fruitfly and mouse. *Nature, 423*(6942), 822–823. http://dx.doi.org/10.1038/423822a.

Wang, L., Cvetkov, T. L., Chance, M. R., & Moiseenkova-Bell, V. Y. (2012). Identification of in vivo disulfide conformation of TRPA1 ion channel. *The Journal of Biological Chemistry, 287*(9), 6169–6176. http://dx.doi.org/10.1074/jbc.M111.329748.

Wang, L., & Sigworth, F. J. (2006). Cryo-EM and single particles. *Physiology (Bethesda, Md.), 21*, 13–18. doi:21/1/13, pii:10.1152/physiol.00045.2005.

Westfield, G. H., Rasmussen, S. G., Su, M., Dutta, S., DeVree, B. T., Chung, K. Y., et al. (2011). Structural flexibility of the G alpha s alpha-helical domain in the beta2-adrenoceptor Gs complex. *Proceedings of the National Academy of Sciences USA, 108*(38), 16086–16091. http://dx.doi.org/10.1073/pnas.1113645108.

Whorton, M. R., & MacKinnon, R. (2013). X-ray structure of the mammalian GIRK2-betagamma G-protein complex. *Nature, 498*(7453), 190–197. http://dx.doi.org/10.1038/nature12241.

Wolf, M., Garcea, R. L., Grigorieff, N., & Harrison, S. C. (2010). Subunit interactions in bovine papillomavirus. *Proceedings of the National Academy of Sciences USA, 107*(14), 6298–6303. http://dx.doi.org/10.1073/pnas.0914604107.

Xu, H., Ramsey, I. S., Kotecha, S. A., Moran, M. M., Chong, J. A., Lawson, D., et al. (2002). TRPV3 is a calcium-permeable temperature-sensitive cation channel. *Nature, 418*(6894), 181–186. http://dx.doi.org/10.1038/nature00882.

Yang, F., Cui, Y., Wang, K., & Zheng, J. (2010). Thermosensitive TRP channel pore turret is part of the temperature activation pathway. *Proceedings of the National Academy of Sciences USA, 107*(15), 7083–7088. http://dx.doi.org/10.1073/pnas.1000357107.

Yao, J., Liu, B., & Qin, F. (2010). Pore turret of thermal TRP channels is not essential for temperature sensing. *Proceedings of the National Academy of Sciences USA, 107*(32), E126–E127. http://dx.doi.org/10.1073/pnas.1008272107. E125; author reply.

Yao, J., Liu, B., & Qin, F. (2011). Modular thermal sensors in temperature-gated transient receptor potential (TRP) channels. *Proceedings of the National Academy of Sciences USA, 108*(27), 11109–11114. http://dx.doi.org/10.1073/pnas.1105196108.

Zhang, X., Jin, L., Fang, Q., Hui, W. H., & Zhou, Z. H. (2010). 3.3 A cryo-EM structure of a nonenveloped virus reveals a priming mechanism for cell entry. *Cell, 141*(3), 472–482. http://dx.doi.org/10.1016/j.cell.2010.03.041.

Zholos, A., Johnson, C., Burdyga, T., & Melanaphy, D. (2011). TRPM channels in the vasculature. *Advances in Experimental Medicine and Biology, 704*, 707–729. http://dx.doi.org/10.1007/978-94-007-0265-3_37.

Zimmermann, K., Leffler, A., Babes, A., Cendan, C. M., Carr, R. W., Kobayashi, J., et al. (2007). Sensory neuron sodium channel Nav1.8 is essential for pain at low temperatures. *Nature, 447*(7146), 855–858. http://dx.doi.org/10.1038/nature05880.

Zurborg, S., Yurgionas, B., Jira, J. A., Caspani, O., & Heppenstall, P. A. (2007). Direct activation of the ion channel TRPA1 by Ca^{2+}. *Nature Neuroscience, 10*(3), 277–279. http://dx.doi.org/10.1038/nn1843.

CHAPTER EIGHT

Thermal Sensitivity of CLC and TMEM16 Chloride Channels and Transporters

Michael Pusch and Giovanni Zifarelli*
Istituto di Biofisica, CNR, Genova, Italy
*Corresponding author: E-mail: zifarelli@ge.ibf.cnr.it

Contents

Abstract

Cl^- transport is of fundamental importance in the most diverse physiological contexts and it is mediated by a variety of ion channels and transporters belonging to different protein families. In particular, the recently identified TMEM16 protein family comprises the long sought Ca^{2+}-activated Cl^- channel (CaCC) and the activity of one of its members, TMEM16A, is highly dependent on temperature and is involved in thermal nociception. Among the other protein families mediating Cl^- transport, CLC proteins are also regulated by temperature although so far the physiological implications of this dependence are unknown.

1. INTRODUCTION

Cl^- transport is of fundamental importance in the most diverse physiological contexts and it is mediated by a variety of ion channels and transporters belonging to different protein families like CLC, GABA and

Current Topics in Membranes, Volume 74
ISSN: 1063-5823
http://dx.doi.org/10.1016/B978-0-12-800181-3.00008-7

glycine receptors, CFTR, and others (Jentsch, Stein, Weinreich, & Zdebik, 2002). The latest protein family discovered is the recently identified TMEM16, comprising the long-sought Ca^{2+}-activated Cl^- channel (CaCC) (Caputo et al., 2008; Schroeder, Cheng, Jan, & Jan, 2008; Yang et al., 2008). Interestingly, a recent study found that TMEM16A activity is highly dependent on temperature and is involved in thermal nociception (Cho et al., 2012). Among the other protein families mediating Cl^- transport, CLC proteins are also regulated by temperature although so far, the physiological implications of this dependence are unknown. In this chapter, we will deal with the thermal sensitivity of CLC and TMEM16 proteins.

2. THERMAL NOCICEPTION

Nociception is the process of sensing potentially damaging stimuli that in general can be of chemical, mechanical, or thermal nature. Nociceptors are sensory neurons that project axons from the dorsal root ganglia (DRG) and the trigeminal ganglia (TG) to peripheral tissues where the noxious stimuli are detected and transduced to electrical signals that are transmitted to the central nervous system from the DRG and TG. Noxious stimuli are detected at the molecular level by ion channels. In particular, for thermal sensitivity it is established that TRP channels have a central role, with the vanilloid subfamily involved in heat response (in particular TRPV1) and the melastatin subfamily (in particular TRPM8) involved in cold response (Julius, 2013). They are nonselective ion channels that are extremely sensitive to changes in temperature to which they respond with a modification of their open probability, P_{open}. Activation of these channels leads to neuron depolarization and firing, conferring thermal sensitivity to the cells in which they are expressed. However, the mechanism of thermal sensitivity is complex and involves also other types of ion channels (Julius, 2013). In particular, while it is well established that TRPV1 is fundamental in heat sensitivity (Caterina et al., 1997), mouse knockout models with deletions of TRPV1 alone (Caterina et al., 2000) or in combination with other TRP channels (Lee, Iida, Mizuno, Suzuki, & Caterina, 2005; Miyamoto, Petrus, Dubin, & Patapoutian, 2011; Park et al., 2011) showed only partial deficits in thermosensation, indicating that a critical component of the heat sensory system was independent of TRP channels and had remained unidentified. Very recently, a new player has entered the stage: The recently identified CaCC TMEM16A (also called Ano1) functions as a thermal sensor involved in nociception (Cho & Oh, 2013; Cho et al., 2012).

3. TMEM16A

After the original description in salamander photoreceptors (Bader, Bertrand, & Schwartz, 1982) and *Xenopus* oocytes (Barish, 1983; Miledi, 1982), currents mediated by CaCC have been characterized in several cell types with fundamental physiological roles that range from epithelial secretion to smooth muscle contraction, sensory transduction, neuronal and cardiac excitability, cell proliferation, and migration and nociception (Duran & Hartzell, 2011; Hartzell, Putzier, & Arreola, 2005). However, the proteins responsible for those currents have been identified only very recently as TMEM16A and TMEM16B (Caputo et al., 2008; Schroeder et al., 2008; Yang et al., 2008). They belong to a protein family with 10 genes in mammals where it is also named Anoctamin or ANO family because the identified members are anion-selective ion channels and putatively have eight transmembrane segments, even though the role as Cl^- channels has been firmly established only for TMEM16A (ANO1) and TMEM16B (ANO2) and the transmembrane topology has been only indirectly inferred from hydropathy analysis.

Chimeric constructs of TMEM16A and 16B in which the putative pore-loop between transmembrane segments five and six were switched indicated a role of this region in regulating the functional expression of these channels at the plasma membrane of HEK cells (Adomaviciene, Smith, Garnett, & Tammaro, 2013). TMEM16A is so far the best-characterized member of the family. At least four splice variants have been identified (Ferrera et al., 2009) that differ in Ca^{2+} sensitivity and voltage dependence (Ferrera et al., 2009; Xiao et al., 2011) which possibly account for the different features of CaCC currents characterized in different cell types (Hartzell et al., 2005). TMEM16A subunits associate in dimers (Fallah et al., 2011; Sheridan et al., 2011) although it is not clear whether this is the functional oligomeric state at the plasma membrane and if each monomer contains a separate Cl^- conduction pathway as in CLC proteins. TMEM16A is gated by both voltage and Ca^{2+}_{int} but the elements of the channel mediating these regulatory mechanisms are still unknown. Similar to CLCs, TMEM16A lacks a transmembrane helix with charged residues that could serve as voltage sensor as in classical voltage-gated ion channels and there is still dispute about possible Ca^{2+}-sensing elements. Recently, from the functional characterization of TMEM16A purified and reconstituted in liposomes, it has been concluded that calmodulin is not required to confer Ca^{2+}-sensitivity (Terashima, Picollo, & Accardi,

2013). Steady-state currents display pronounced outward rectification with a relatively slow activation and a faster deactivation at $Ca^{2+} < 2\ \mu M$ whereas at higher Ca^{2+} concentrations both outward rectification and time dependence of the currents are strongly reduced (Caputo et al., 2008; Schroeder et al., 2008; Xiao et al., 2011; Yang et al., 2008). Voltage and Ca^{2+} gating are synergistic as an increase of Ca^{2+} shifts the voltage dependence of the conductance toward hyperpolarized voltages whereas an increase in voltage decreases the EC_{50} for Ca^{2+} (Xiao et al., 2011). Importantly, at high concentrations, Ca^{2+} alone is sufficient to open the channel even at very hyperpolarized voltages and conversely the channel can be opened in the absence of Ca^{2+} by strong depolarization (Xiao et al., 2011). On the basis of the effect of $[Cl^-]_{ext}$ and Cl^- substitution with NO_3^- on rectification and conductance it has been suggested that TMEM16A voltage gating is dependent on the permeant anion, similar to CLC channels (Xiao et al., 2011) as it has been also recently found for TMEM16B (Betto et al., 2014).

Regarding the other members of the TMEM16 family, their role is still controversial. The debate is particularly animated for TMEM16F with at least three different roles suggested: as an anion channel (Martins et al., 2011), a cation channel (Yang et al., 2012), and a phospholipid scramblase (Suzuki, Umeda, Sims, & Nagata, 2010). For all the other TMEM16 members, the evidence regarding their ability to produce currents at the plasma membrane is still disputed and their biophysical identity as ion channels and possible intracellular roles are yet to be determined (Duran & Hartzell, 2011; Schreiber et al., 2010).

3.1 TMEM16A and nociception

CaCC currents in rat DRG neurons were identified already three decades ago (Mayer, 1985), and the first direct functional role of TMEM16A in nociception was established by the finding that the channel mediates the noxious signals triggered by bradykinin which is released at sites of tissue damage and inflammation (Liu et al., 2010). However, very recently, the study of Cho et al. (2012) indicated that TMEM16A is essential in another aspect of nociception, namely thermal sensitivity.

First of all, the biophysical properties of TMEM16A are extremely sensitive to heat. Temperatures higher than 44 °C robustly activated TMEM16A expressed in HEK cells even in the absence of Ca^{2+}_{int} (Cho et al., 2012). Currents were Cl^--selective, outwardly rectifying and showed time dependence of both activation and deactivation, similar to measurements at room temperature (Caputo et al., 2008; Schroeder et al., 2008; Xiao et al., 2011; Yang et al., 2008). Ca^{2+} and temperature have a synergistic effect on TMEM16A activation (Figure 8.1). When temperature and Ca^{2+} concentration were simultaneously increased, currents were higher compared to experiments where only one of these parameters was increased (Cho et al., 2012). When $[Ca^{2+}]_{int}$ was raised, the apparent temperature threshold for activation decreased (Cho et al., 2012) arguing for a combined effect on channel gating. Notably, the potentiating effect of temperature was present also at saturating Ca^{2+}_{int} concentrations (10 μM) (Cho et al., 2012; Yang et al., 2008). At present, it is not clear whether this is due to an effect of temperature on gating or on single-channel conductance. The thermal sensitivity of TMEM16A was biphasic with a steep activation phase characterized by an apparent temperature coefficient, Q_{10}, of about 20 (Cho et al., 2012).

The temperature dependence of the activation of TMEM16A currents potentially suggests a role in thermal nociception. To definitely prove this point, the authors provided direct physiological evidence. Immunofluorescence of mouse DRG neurons showed that TMEM16A is mostly expressed in small diameter nociceptive neurons. Experiments on cultured DRG neurons from wild type (WT) and TRPV1 knockout showed that heat stimulates Cl^--mediated inward currents of similar amplitude in the two cases, indicating that TMEM16A currents could lead to neuronal depolarization and induce firing of action potentials (Cho et al., 2012).

Behavioral tests on mice further confirmed a role in nociception. Inhibition of TMEM16A with the blocker mefloquine or with small interfering

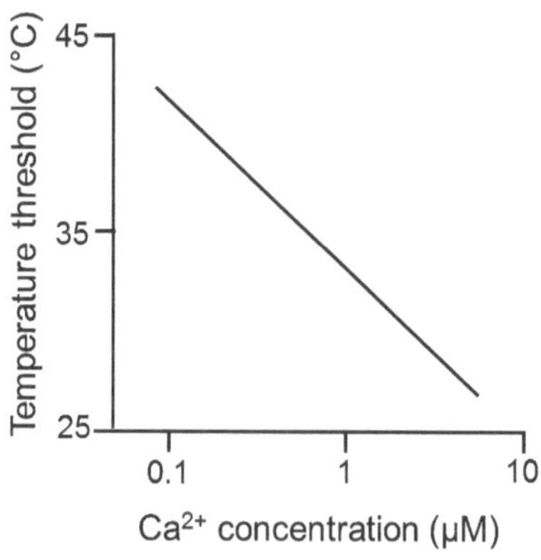

Figure 8.1 Schematic illustration of the Ca^{2+} dependence of the temperature threshold for activation of TMEM16A currents. *Adapted from Cho et al. (2012).*

RNA (siRNA) impaired the normal response to heat in tail-withdrawal tests. Consistent with a role of TMEM16A in nociception, heat ramps applied to neurons dissociated from siRNA-treated mice evoked smaller currents compared to controls (Cho et al., 2012).

In addition, the in vivo role of TMEM16A was further inspected by conditional, tissue-specific knockout mice in DRG neurons (complete deletion of TMEM16A results in death in early neonatal period due to respiratory problems). In DRG neurons from KO mice, heat evoked a much smaller current compared to controls and consistent with this and similar to behavioral experiments after siRNA administration, the KO mice showed higher tolerance to heat (Cho et al., 2012).

The mechanism proposed to account for the depolarizing effect of TMEM16A activation is based on the relatively high $[Cl^-]_{int}$ in DRG neurons compared to other neurons of the CNS due to the elevated expression and activity of the sodium–potassium-chloride co-transporter 1 (NKCC1) in these cells (Alvarez-Leefmans, Gamino, Giraldez, & Nogueron, 1988; Deschenes, Feltz, & Lamour, 1976). This establishes a Cl^- equilibrium potential around −30 mV, more positive than the resting membrane potential (about −60 mV) and implies that upon TMEM16A opening, neurons are depolarized, and can fire action potentials. A similar mechanism has been suggested to explain the excitatory effect of GABA-receptor activation in DRG neurons (Staley, Soldo, & Proctor, 1995).

Among other TMEM16 channels, TMEM16B, but not TMEM16D and TMEM16E, could be activated by heat in a similar temperature range compared to TMEM16A (Cho et al., 2012). However, a critical role of TMEM16B in thermal sensitivity was ruled out as its transcript level in DRG cells were about tenfold lower compared to TMEM16A (Cho et al., 2012).

No mechanism has been proposed to explain the large temperature dependence of TMEM16A (Cho et al., 2012). However, the molecular mechanism underlying the sensitivity to temperature is still debated also for the much more extensively studied TRP channels, especially TRPV1. Different studies proposed that different protein domains could serve as a thermal sensor, in some cases with conflicting results (Voets, 2012). Clapham and Miller (2011) in an elegant study based on fundamental thermodynamics arguments presented a drastically different model whereby temperature sensitivity is not necessarily conferred by specialized protein domains but rather is due to the difference in heat capacity between open and closed conformations. Importantly, this model predicts that the underlying mechanism for cold and heat sensitivity is the same (Clapham & Miller, 2011).

4. CLC PROTEINS

CLC proteins are present in all phyla, with nine distinct CLC genes in humans (Jentsch, 2008; Mindell & Maduke, 2001) where they are fundamental for different physiological functions like muscle excitability and ionic homeostasis (Zifarelli & Pusch, 2007). Among the human homologues, CLC channels (CLC-1, CLC-2, CLC-Ka, CLC-Kb) are localized in the plasma membrane, whereas CLC-3, CLC-4, CLC-5, CLC-6, and CLC-7 are predominantly found in endosomal and lysosomal membranes and are probably all Cl^-/H^+ antiporters with a 2 Cl^-: 1 H^+ stoichiometry, although direct evidence is still lacking for CLC-3 (Stauber & Jentsch, 2013). Some CLC proteins associate with ancillary, β subunits: CLC-K channels (CLC-Ka and CLC-Kb) necessitate the two transmembrane domain protein barttin for proper targeting and function (Estevez et al., 2001); CLC-7 requires the highly glycosylated Ostm1 protein for proper functioning (Lange, Wartosch, Jentsch, & Fuhrmann, 2006; Leisle, Ludwig, Wagner, Jentsch, & Stauber, 2011); and the association of the rather ubiquitous CLC-2 channel with the cell-adhesion molecule GlialCAM in glia modifies both its localization and function (Jeworutzki et al., 2012).

CLCs comprise to date the only protein family in which ion transport is carried out by opposite thermodynamic mechanisms. Passive diffusion (in ion channels) dissipates ionic gradients reducing free energy whereas secondary active antiporters that strictly couple the exchange of two substrates conserve free energy. Surprisingly, these intrinsically different transport modes are carried out by remarkably similar structures. The crystal structures of the bacterial and eukaryotic homologues CLC-ec1 (Dutzler, Campbell, Cadene, Chait, & MacKinnon, 2002; Dutzler, Campbell, & MacKinnon, 2003) and CLC-Cm (Feng, Campbell, Hsiung, & MacKinnon, 2010) show that the protein assembles as a homodimer with two subunits harboring a narrow permeation pathway containing three anion-binding sites defined, from the extracellular to the intracellular side, as S_{ext}, S_{cen}, and S_{in} (Figure 8.2). These structures also explain the critical role of a conserved glutamate residue for the gating of CLC proteins; indeed, it is also called "gating glutamate" (E148 in ClC-ec1). In the structure of WT CLC-ec1, the side chain of E148 occupies S_{ext} and isolates the anion at S_{cen} from the extracellular space (Dutzler et al., 2002). In the structure

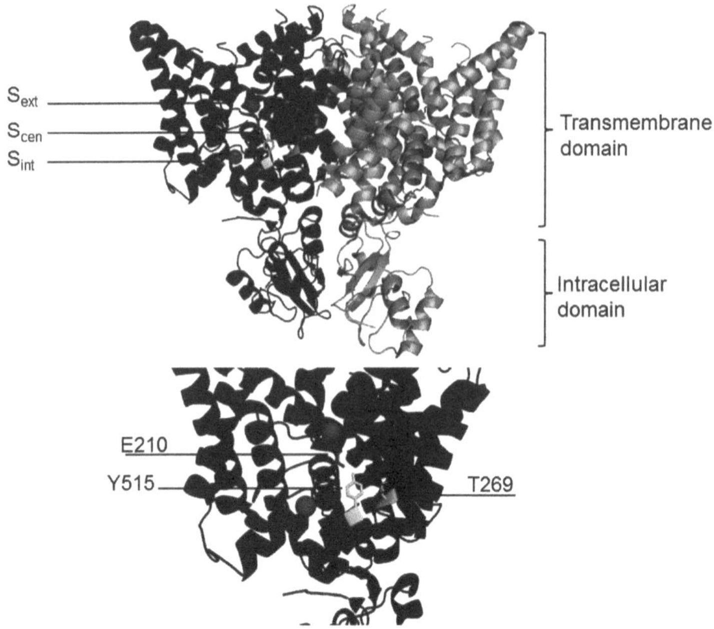

Figure 8.2 Upper panel, dimeric structure of the eukaryotic antiporter CLC-Cm (PDB entry: 3ORG) viewed from the membrane plane (extracellular side above and cytoplasmic side below). The two subunits are shown in green and blue. The residue E210 (also called "gating glutamate") is colored red, T269 is shown in pink, and Y510 in yellow. Cl^- anions bound to S_{ext} and S_{int} are shown in magenta. The position of the three binding sites, S_{ext}, S_{cen}, and S_{int}, is also indicated by horizontal lines. The side chain of E210 occupies S_{cen}. The cytoplasmic C-terminus contains CBS domains. Lower panel, expanded representation of the anion permeation pathway for one of the subunits. In both panels some transmembrane helices were removed for clarity. (See the color plate.)

of the E148Q mutant, an anion occupies S_{ext}, whereas the glutamine side chain is displaced from the pore (Dutzler et al., 2003). Recently, the structure of CLC-Cm revealed that the glutamate side chain can also be oriented toward the intracellular side and occupy S_{cen} (Feng et al., 2010) (Figure 8.2). These structural data rationalize functional evidence showing that mutations of the gating glutamate abrogate gating in CLC channels (Dutzler et al., 2003; Estevez, Schroeder, Accardi, Jentsch, & Pusch, 2003; Fahlke, Yu, Beck, Rhodes, & George, 1997; Traverso, Elia, & Pusch, 2003) and render the transporters CLC-ec1, CLC-5, CLC-6, and CLC-7 passive Cl^- conductances (Accardi, Kolmakova-Partensky, Williams, & Miller, 2004; Friedrich, Breiderhoff, & Jentsch, 1999; Leisle et al., 2011; Neagoe, Stauber, Fidzinski, Bergsdorf, & Jentsch, 2010). Another critical residue for CLC transporters is an off-pore glutamate (E203 in ClC-ec1, E268 in

ClC-5), also called "proton glutamate," because it was thought to be an internal proton acceptor. Mutation of the proton glutamate to glutamine in CLC-ec1 abolishes H^+ transport and renders the transporter a passive Cl^- conductance, similar to mutations of the gating glutamate (Accardi et al., 2005). In CLC-5, neutralization of the proton glutamate completely stalls transport activity, a feature that permitted the resolution of transient capacitive currents (Smith & Lippiat, 2010; Zifarelli, De Stefano, Zanardi, & Pusch, 2012). Interestingly, it has recently emerged that a glutamate residue at this position is not strictly required to mediate Cl^-/H^+-coupled transport as some CLC antiporters have a threonine (Feng et al., 2010) (Figure 8.2) or an isoleucine (Phillips et al., 2012) at the corresponding position. CLC channels invariably have a valine residue at this position.

4.1 Fast and slow gating in muscle-type CLC channels

In the CLC channels CLC-0 and CLC-1, there are two different gating mechanisms defined as slow or common gate and fast or protopore gate. The slow gate acts simultaneously on both protopores of the dimeric channel whereas the fast gate acts on the single protopores independently. The attributes slow and fast are due to the fact that in CLC-0 they operate on timescales that differ by orders of magnitude. In CLC-0, the presence of these two gating mechanism is particularly evident in single-channel measurements where opening of the slow gate leads to bursts of channel activity revealing fast gate transitions (Figure 8.3). The bursts are characterized by three different conductance levels corresponding to the situation where the protopores are both open (higher conductance), both closed (lower conductance), or one is open and one is closed (intermediate

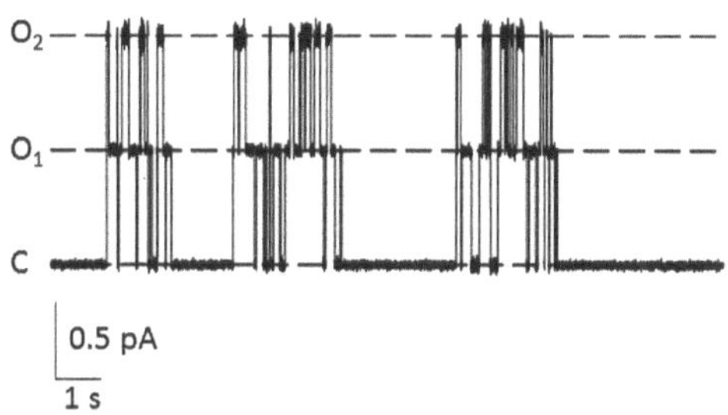

Figure 8.3 Schematic (simulated) single-channel trace of the Torpedo channel CLC-0. Channel activity occurs in bursts that are separated by long closed periods due to closure of the slow gate. Within each burst two open conductance levels (O1 and O2) are seen, where O2 has exactly twice the conductance of O1.

conductance). The skeletal muscle CLC-1 is highly homologous to CLC-0 and in general has similar functional properties, even though the fast and slow gates have comparable kinetics in CLC-1. Fast and common gating of CLC channels depend on different factors, like voltage, anion concentration, temperature, and pH. Most notably, the dependence on the proton concentration shows that CLC channels and transporters might have strong mechanistic and evolutionary links.

The mechanistic relationship of the complex gating observed in other CLC channels like CLC-2 and CLC-Ka and -Kb with the fast and slow gate of CLC-0 and CLC-1 is still unclear.

4.1.1 The fast gate

Gating and permeation are strictly coupled in CLC channels and it has been proposed early on that part of the voltage dependence of fast gating arises indirectly from a voltage-dependent binding and/or translocation of a Cl^- ion entering the channel from the outside (Chen & Miller, 1996; Pusch, Ludewig, Rehfeldt, & Jentsch, 1995). This hypothesis is in agreement with the fact that fast gate opening is favored by increasing $[Cl^-]_{ext}$, and positive voltages, with an apparent "gating valence" of ~1. However, this view has been recently revised as protons from the intracellular side also contribute to voltage-dependent gating (Lisal & Maduke, 2008; Traverso, Zifarelli, Aiello, & Pusch, 2006; Zifarelli, Murgia, Soliani, & Pusch, 2008). In any case, the mechanism of voltage sensing in CLCs bears no relation with the typical "voltage sensor" domains found in cation channels.

Fast gating is mostly attributed to conformational changes of the conserved "gating glutamate" (Glu-166 in CLC-0). In CLC-0, neutralizing Glu-166 completely abolishes the normal voltage-dependent gating relaxations and locks the channel in an open state (Dutzler et al., 2003; Traverso et al., 2003).

Extracellular and intracellular acidification both favor opening of the fast gate (Chen & Chen, 2001; Dutzler et al., 2003) indicating that the protonated side chain of Glu-166 is stabilized in a conformation that allows permeation (Dutzler et al., 2003). However, the molecular mechanisms underlying the effect of pH_{ext} and pH_{int} are different (Traverso et al., 2006) and not completely understood yet. An interesting possibility is that extracellular acidification leads directly to protonation of the gating glutamate whereas protonation from the intracellular side is mediated by a proton originating from the dissociation of a water molecule present in the pore (Zifarelli et al., 2008).

4.1.2 The slow gate

The slow gate controls opening of the two pores of the dimer simultaneously. In CLC-0, it is activated by hyperpolarized potentials (Richard & Miller, 1990) and does not deactivate completely at depolarized voltages (Pusch, Ludewig, & Jentsch, 1997). In CLC-1, the slow gate has an opposite voltage dependence to that of CLC-0, being activated by depolarization and its kinetics are orders of magnitude faster (Accardi & Pusch, 2000; Bennetts, Roberts, Bretag, & Rychkov, 2001; Saviane, Conti, & Pusch, 1999). Slow gate opening is favored by an increase in both $[Cl^-_{ext}]$ and $[Cl^-_{int}]$ (Chen & Miller, 1996; Pusch, Jordt, Stein, & Jentsch, 1999).

The mechanism responsible for the slow gating has not yet been identified. The fact that the slow gate acts on both pores simultaneously suggests, on the structural level, that it relies on subunit interactions (Estevez & Jentsch, 2002), in agreement with the finding that concatamers comprising subunits of different CLC members led invariably to loss of slow gating transitions (Lorenz, Pusch, & Jentsch, 1996; Traverso et al., 2006; Weinreich & Jentsch, 2001) and with the fact that mutations that affect slow gate transitions, while scattered in different protein regions, tend to cluster at the dimer interface, both at the level of the transmembrane domains (Duffield, Rychkov, Bretag, & Roberts, 2003) and of the cytoplasmic domains (Estevez, Pusch, Ferrer-Costa, Orozco, & Jentsch, 2004; Fong, Rehfeldt, & Jentsch, 1998). These mutations include most of the ones that in CLC-1 are responsible for the dominant form of myotonia congenita (Pusch, 2002). Conformational changes associated with slow gate are large as indicated by the analysis of the temperature dependence (see below) and FRET experiments suggesting that slow gate transitions are accompanied by a robust relative movement of the cytoplasmic termini in both CLC-0 and CLC-1 (Bykova, Zhang, Chen, & Zheng, 2006; Ma, Rychkov, Bykova, Zheng, & Bretag, 2011). In particular, for CLC-1 it has been suggested that fast and slow gate are strongly coupled as both depend on the gating glutamate (E232 in CLC-1) (Ma et al., 2011; Traverso et al., 2006). Very recently, this view received further support from the study of Bennetts and Parker (2013) suggesting that a critical determinant of slow gating in CLC-1 is the interaction between the gating glutamate and a highly conserved tyrosine (Y578 in CLC-1) that forms part of S_{cen}.

4.2 Temperature dependence of fast and slow gate

In CLC-0, the different time scale of fast and slow gating transitions allow to easily investigate the two gating mechanisms in isolation. Pusch et al. (1997)

examined the temperature dependence of the deactivation kinetics of the fast gate and found a Q_{10} value of 2.2 that is significantly larger than expected for a diffusion limited process (1.3–1.6). On the other hand, the steady-state voltage dependence of the fast gate was only slightly dependent on temperature suggesting that open and closed states do not differ in terms of structural order. The closing kinetics of the slow gate shows a dramatic temperature dependence with a Q_{10} of about 40, one of the highest ever found in biological processes and is a straightforward indication that slow gating consists of a large conformational change (Pusch et al., 1997). Interestingly, steady-state and single-channel analysis showed that the temperature affects both the degree of maximal activation and the fraction of channel that do not inactivate at positive voltages (i.e., P_{min}) with higher temperatures favoring the closed state of the slow gate (Pusch et al., 1997). On the other hand, the voltage of half-maximal activation was relatively insensitive to temperature (Pusch et al., 1997).

Regarding the temperature dependence, CLC-0 and CLC-1 present a number of intriguing differences. In striking contrast with CLC-0 (Pusch et al., 1997), the slow gate is much less temperature dependent in CLC-1 and the Q_{10} of fast and slow gate is similar (3.1 and 4.1, respectively) (Bennetts et al., 2001). A further difference with CLC-0 is that for CLC-1 opening and closing kinetics of the fast gate appear to differ in terms of thermal sensitivity (Bennetts et al., 2001). In addition, the effect of the temperature on the P_{open} of fast and slow gate of CLC-1 is a shift of the voltage dependence that can be entirely explained by a two-state model in which temperature reduces the free energy of the opening transitions (Bennetts et al., 2001) whereas for CLC-0 the effect of the temperature is complex affecting both P_{max} and P_{min} and can be accounted for only by a complex model that include at least two open and two closed states with temperature-dependent closed–open equilibrium (Pusch et al., 1997). Importantly, in spite of the thermal sensitivity of the gating properties of CLC-0 and CLC-1, the temperature dependence does not appear to be of critical physiological relevance.

4.3 Gating and temperature dependence of CLC exchangers

The term gating describes the process that controls the open probability of ion channels and is normally not associated with transporters. However, the pioneering work of Hilgemann (1996) based on noise analysis indicated that gating exists also for the Na^+/Ca^{2+} exchanger for which transport activity is not continuous as the transporter can be in "active" and "inactive" states. The concept that the activity of CLC transporters could also be gated was originally proposed for CLC-5 on the basis of the characteristic noise spectrum of macroscopic currents (Zdebik et al., 2008). Retrospectively, the very first electrophysiological characterization of CLC-5 already contained a hint of the presence of a gating process in CLC-5 as some mutants displayed a dramatic alteration of activation kinetics compared to WT (Friedrich et al., 1999). However, the first direct demonstration of the presence of a gating process for a CLC transporter has been obtained for the lysosomal antiporter CLC-7, which exhibits very slow activation and deactivation kinetics with time constants in the seconds range (Leisle et al., 2011). Very recently, a mutation of CLC-5 that significantly slowed down current kinetics, D76H, manifested inward tail currents at negative potentials and convincingly proved the presence of a gating mechanism that regulates transport activity also for this transporter (De Stefano, Pusch, & Zifarelli, 2013). However, while the voltage dependence of the gating mechanism is able to completely account for the voltage dependence of macroscopic currents and the rectifying behavior of CLC-7 (Leisle et al., 2011), the situation is less clear for CLC-5 where the rectification may additionally reflect an intrinsic voltage dependence of the transport cycle (De Stefano et al., 2013). Very recently, co-expression of CLC-7 subunits harboring mutations known to affect gating kinetics together with WT subunits and of constructs with C-terminal truncations, led to the conclusion that the gating mechanism of CLC-7 involves both subunits and requires noncovalent binding of the cytoplasmic CBS domains which is reminiscent of the properties of the common gating mechanism of CLC channels (Ludwig, Ullrich, Leisle, Stauber, & Jentsch, 2013).

Voltage-dependent current activation of CLC-7 is quite strongly dependent on temperature with a $Q_{10} \approx 3$ (Leisle et al., 2011). For CLC-5 a comparable analysis performed on the inactivating inward tail currents elicited by the D76H mutant at negative voltages yielded a very similar Q_{10} (Figure 8.4). Collectively these results show that the Q_{10} values associated with the gating of CLC-5 and CLC-7 are compatible with a conformational change that appears qualitatively similar to that observed in CLC channels, but much smaller than the one that characterizes the slow gate in CLC-0. It is not clear whether the temperature dependence of CLC antiporters has any physiological implication.

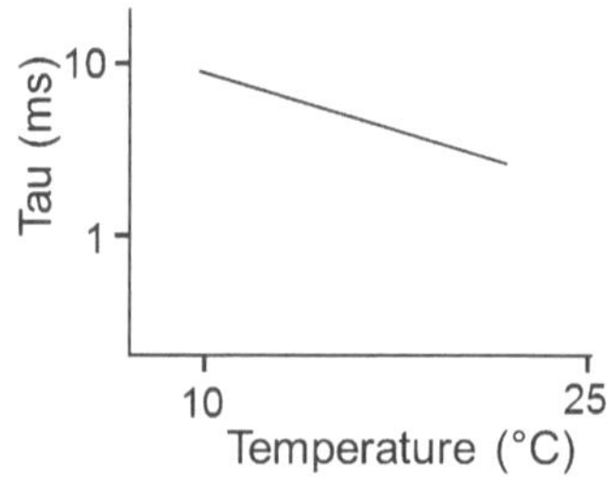

Figure 8.4 Schematic illustration of the temperature dependence of the deactivation kinetics of the currents from the CLC-5 mutant D76H. The vertical scale is logarithmic. *Adapted from De Stefano et al. (2013).*

5. CONCLUSIONS

The recent identification of the TMEM16 protein family comprising the long-sought CaCC has sparked a number of research avenues about the physiological roles of its members. One of the most recent and exciting developments is the contribution of TMEM16A in thermal nociception alongside the well-established role of TRP channels. However, while the conclusion that TMEM16A is involved in thermal sensitivity is based on several lines of evidence, both biophysical and physiological, it should be kept in mind that it also relies on a single study (Cho et al., 2012). In a cautionary approach, we consider that before firmly establishing the role of this protein in thermal nociception, it would be desirable to have other studies converging on the same conclusion. Among the other protein families mediating Cl^- transport, the CLC family comprises channels and transporters with gating properties that are sensitive to temperature even though CLC thermal regulation does not appear to have any physiological impact.

ACKNOWLEDGMENTS

The financial support of Telethon Italy (grants GGP12008 and GEP13101) and Compagnia di San Paolo is gratefully acknowledged.

REFERENCES

Accardi, A., Kolmakova-Partensky, L., Williams, C., & Miller, C. (2004). Ionic currents mediated by a prokaryotic homologue of CLC Cl^- channels. *The Journal of General Physiology, 123*, 109–119.

Accardi, A., & Pusch, M. (2000). Fast and slow gating relaxations in the muscle chloride channel CLC-1. *The Journal of General Physiology, 116*, 433–444.

Accardi, A., Walden, M., Nguitragool, W., Jayaram, H., Williams, C., & Miller, C. (2005). Separate ion pathways in a Cl^-/H^+ exchanger. *The Journal of General Physiology, 126*, 563–570.
Adomaviciene, A., Smith, K. J., Garnett, H., & Tammaro, P. (2013). Putative pore-loops of TMEM16/anoctamin channels affect channel density in cell membranes. *The Journal of Physiology, 591*, 3487–3505.
Alvarez-Leefmans, F. J., Gamino, S. M., Giraldez, F., & Nogueron, I. (1988). Intracellular chloride regulation in amphibian dorsal root ganglion neurones studied with ion-selective microelectrodes. *The Journal of Physiology, 406*, 225–246.
Bader, C. R., Bertrand, D., & Schwartz, E. A. (1982). Voltage-activated and calcium-activated currents studied in solitary rod inner segments from the salamander retina. *The Journal of Physiology, 331*, 253–284.
Barish, M. E. (1983). A transient calcium-dependent chloride current in the immature *Xenopus* oocyte. *The Journal of Physiology, 342*, 309–325.
Bennetts, B., & Parker, M. W. (2013). Molecular determinants of common gating of a ClC chloride channel. *Nature Communications, 4*, 2507.
Bennetts, B., Roberts, M. L., Bretag, A. H., & Rychkov, G. Y. (2001). Temperature dependence of human muscle ClC-1 chloride channel. *The Journal of Physiology, 535*, 83–93.
Betto, G., Cherian, O. L., Pifferi, S., Cenedese, V., Boccaccio, A., & Menini, A. (2014). Interactions between permeation and gating in the TMEM16B/anoctamin2 calcium-activated chloride channel. *The Journal of General Physiology, 143*, 703–718.
Bykova, E. A., Zhang, X. D., Chen, T. Y., & Zheng, J. (2006). Large movement in the C terminus of CLC-0 chloride channel during slow gating. *Nature Structural & Molecular Biology, 13*, 1115–1119.
Caputo, A., Caci, E., Ferrera, L., Pedemonte, N., Barsanti, C., Sondo, E., et al. (2008). TMEM16A, a membrane protein associated with calcium-dependent chloride channel activity. *Science, 322*, 590–594.
Caterina, M. J., Leffler, A., Malmberg, A. B., Martin, W. J., Trafton, J., Petersen-Zeitz, K. R., et al. (2000). Impaired nociception and pain sensation in mice lacking the capsaicin receptor. *Science, 288*, 306–313.
Caterina, M. J., Schumacher, M. A., Tominaga, M., Rosen, T. A., Levine, J. D., & Julius, D. (1997). The capsaicin receptor: a heat-activated ion channel in the pain pathway. *Nature, 389*, 816–824.
Chen, M. F., & Chen, T. Y. (2001). Different fast-gate regulation by external Cl(−) and H(+) of the muscle-type ClC chloride channels. *The Journal of General Physiology, 118*, 23–32.
Chen, T. Y., & Miller, C. (1996). Nonequilibrium gating and voltage dependence of the ClC-0 Cl- channel. *The Journal of General Physiology, 108*, 237–250.
Cho, H., & Oh, U. (2013). Anoctamin 1 mediates thermal pain as a heat sensor. *Current Neuropharmacology, 11*, 641–651.
Cho, H., Yang, Y. D., Lee, J., Lee, B., Kim, T., Jang, Y., et al. (2012). The calcium-activated chloride channel anoctamin 1 acts as a heat sensor in nociceptive neurons. *Nature Neuroscience, 15*, 1015–1021.
Clapham, D. E., & Miller, C. (2011). A thermodynamic framework for understanding temperature sensing by transient receptor potential (TRP) channels. *Proceedings of the National Academy of Sciences of the United States of America, 108*, 19492–19497.
De Stefano, S., Pusch, M., & Zifarelli, G. (2013). A single point mutation reveals gating of the human ClC-5 Cl^-/H^+ antiporter. *The Journal of Physiology, 591*, 5879–5893.
Deschenes, M., Feltz, P., & Lamour, Y. (1976). A model for an estimate in vivo of the ionic basis of presynaptic inhibition: an intracellular analysis of the GABA-induced depolarization in rat dorsal root ganglia. *Brain Research, 118*, 486–493.

Duffield, M., Rychkov, G., Bretag, A., & Roberts, M. (2003). Involvement of helices at the dimer interface in ClC-1 common gating. *The Journal of General Physiology, 121*, 149–161.
Duran, C., & Hartzell, H. C. (2011). Physiological roles and diseases of Tmem16/Anoctamin proteins: are they all chloride channels? *Acta Pharmacologica Sinica, 32*, 685–692.
Dutzler, R., Campbell, E. B., Cadene, M., Chait, B. T., & MacKinnon, R. (2002). X-ray structure of a ClC chloride channel at 3.0 A reveals the molecular basis of anion selectivity. *Nature, 415*, 287–294.
Dutzler, R., Campbell, E. B., & MacKinnon, R. (2003). Gating the selectivity filter in ClC chloride channels. *Science, 300*, 108–112.
Estevez, R., Boettger, T., Stein, V., Birkenhager, R., Otto, E., Hildebrandt, F., et al. (2001). Barttin is a Cl^- channel beta-subunit crucial for renal Cl^- reabsorption and inner ear K^+ secretion. *Nature, 414*, 558–561.
Estevez, R., & Jentsch, T. J. (2002). CLC chloride channels: correlating structure with function. *Current Opinion in Structural Biology, 12*, 531–539.
Estevez, R., Pusch, M., Ferrer-Costa, C., Orozco, M., & Jentsch, T. J. (2004). Functional and structural conservation of CBS domains from CLC chloride channels. *The Journal of Physiology, 557*, 363–378.
Estevez, R., Schroeder, B. C., Accardi, A., Jentsch, T. J., & Pusch, M. (2003). Conservation of chloride channel structure revealed by an inhibitor binding site in ClC-1. *Neuron, 38*, 47–59.
Fahlke, C., Yu, H. T., Beck, C. L., Rhodes, T. H., & George, A. L., Jr. (1997). Pore-forming segments in voltage-gated chloride channels. *Nature, 390*, 529–532.
Fallah, G., Romer, T., Detro-Dassen, S., Braam, U., Markwardt, F., & Schmalzing, G. (2011). TMEM16A(a)/anoctamin-1 shares a homodimeric architecture with CLC chloride channels. *Molecular & Cellular Proteomics: MCP, 10*. M110 004697.
Feng, L., Campbell, E. B., Hsiung, Y., & MacKinnon, R. (2010). Structure of a eukaryotic CLC transporter defines an intermediate state in the transport cycle. *Science, 330*, 635–641.
Ferrera, L., Caputo, A., Ubby, I., Bussani, E., Zegarra-Moran, O., Ravazzolo, R., et al. (2009). Regulation of TMEM16A chloride channel properties by alternative splicing. *The Journal of Biological Chemistry, 284*, 33360–33368.
Fong, P., Rehfeldt, A., & Jentsch, T. J. (1998). Determinants of slow gating in ClC-0, the voltage-gated chloride channel of *Torpedo marmorata*. *The American Journal of Physiology, 274*, C966–973.
Friedrich, T., Breiderhoff, T., & Jentsch, T. J. (1999). Mutational analysis demonstrates that ClC-4 and ClC-5 directly mediate plasma membrane currents. *The Journal of Biological Chemistry, 274*, 896–902.
Hartzell, C., Putzier, I., & Arreola, J. (2005). Calcium-activated chloride channels. *Annual Review of Physiology, 67*, 719–758.
Hilgemann, D. W. (1996). Unitary cardiac Na^+, Ca^{2+} exchange current magnitudes determined from channel-like noise and charge movements of ion transport. *Biophysical Journal, 71*, 759–768.
Jentsch, T. J. (2008). CLC chloride channels and transporters: from genes to protein structure, pathology and physiology. *Critical Reviews in Biochemistry and Molecular Biology, 43*, 3–36.
Jentsch, T. J., Stein, V., Weinreich, F., & Zdebik, A. A. (2002). Molecular structure and physiological function of chloride channels. *Physiological Reviews, 82*, 503–568.
Jeworutzki, E., Lopez-Hernandez, T., Capdevila-Nortes, X., Sirisi, S., Bengtsson, L., Montolio, M., et al. (2012). GlialCAM, a protein defective in a leukodystrophy, serves as a ClC-2 Cl(−) channel auxiliary subunit. *Neuron, 73*, 951–961.
Julius, D. (2013). TRP channels and pain. *Annual Review of Cell and Developmental Biology, 29*, 355–384.

Lange, P. F., Wartosch, L., Jentsch, T. J., & Fuhrmann, J. C. (2006). ClC-7 requires Ostm1 as a beta-subunit to support bone resorption and lysosomal function. *Nature, 440*, 220–223.

Lee, H., Iida, T., Mizuno, A., Suzuki, M., & Caterina, M. J. (2005). Altered thermal selection behavior in mice lacking transient receptor potential vanilloid 4. *The Journal of Neuroscience: The Official Journal of the Society for Neuroscience, 25*, 1304–1310.

Leisle, L., Ludwig, C. F., Wagner, F. A., Jentsch, T. J., & Stauber, T. (2011). ClC-7 is a slowly voltage-gated 2Cl(−)/1H(+)-exchanger and requires Ostm1 for transport activity. *The EMBO Journal, 30*, 2140–2152.

Lisal, J., & Maduke, M. (2008). The ClC-0 chloride channel is a "broken" Cl^-/H^+ antiporter. *Nature Structural & Molecular Biology, 15*, 805–810.

Liu, B., Linley, J. E., Du, X., Zhang, X., Ooi, L., Zhang, H., et al. (2010). The acute nociceptive signals induced by bradykinin in rat sensory neurons are mediated by inhibition of M-type K^+ channels and activation of Ca^{2+}-activated Cl^- channels. *The Journal of Clinical Investigation, 120*, 1240–1252.

Lorenz, C., Pusch, M., & Jentsch, T. J. (1996). Heteromultimeric CLC chloride channels with novel properties. *Proceedings of the National Academy of Sciences of the United States of America, 93*, 13362–13366.

Ludwig, C. F., Ullrich, F., Leisle, L., Stauber, T., & Jentsch, T. J. (2013). Common gating of both CLC transporter subunits underlies voltage-dependent activation of the $2Cl^-/1H^+$ exchanger ClC-7/Ostm1. *The Journal of Biological Chemistry, 288*, 28611–28619.

Martins, J. R., Faria, D., Kongsuphol, P., Reisch, B., Schreiber, R., & Kunzelmann, K. (2011). Anoctamin 6 is an essential component of the outwardly rectifying chloride channel. *Proceedings of the National Academy of Sciences of the United States of America, 108*, 18168–18172.

Ma, L., Rychkov, G. Y., Bykova, E. A., Zheng, J., & Bretag, A. H. (2011). Movement of hClC-1 C-termini during common gating and limits on their cytoplasmic location. *The Biochemical Journal, 436*, 415–428.

Mayer, M. L. (1985). A calcium-activated chloride current generates the after-depolarization of rat sensory neurones in culture. *The Journal of Physiology, 364*, 217–239.

Miledi, R. (1982). A calcium-dependent transient outward current in *Xenopus laevis* oocytes. *Proceedings of the Royal Society of London. Series B, Containing papers of a Biological character. Royal Society, 215*, 491–497.

Mindell, J. A., & Maduke, M. (2001). ClC chloride channels. *Genome Biology, 2*. Reviews 3003.

Miyamoto, T., Petrus, M. J., Dubin, A. E., & Patapoutian, A. (2011). TRPV3 regulates nitric oxide synthase-independent nitric oxide synthesis in the skin. *Nature Communications, 2*, 369.

Neagoe, I., Stauber, T., Fidzinski, P., Bergsdorf, E. Y., & Jentsch, T. J. (2010). The late endosomal ClC-6 mediates proton/chloride countertransport in heterologous plasma membrane expression. *The Journal of Biological Chemistry, 285*, 21689–21697.

Park, U., Vastani, N., Guan, Y., Raja, S. N., Koltzenburg, M., & Caterina, M. J. (2011). TRP vanilloid 2 knock-out mice are susceptible to perinatal lethality but display normal thermal and mechanical nociception. *The Journal of Neuroscience: The Official Journal of the Society for Neuroscience, 31*, 11425–11436.

Phillips, S., Brammer, A. E., Rodriguez, L., Lim, H. H., Stary-Weinzinger, A., & Matulef, K. (2012). Surprises from an unusual CLC homolog. *Biophysical Journal, 103*, L44–46.

Pusch, M. (2002). Myotonia caused by mutations in the muscle chloride channel gene CLCN1. *Human Mutation, 19*, 423–434.

Pusch, M., Jordt, S. E., Stein, V., & Jentsch, T. J. (1999). Chloride dependence of hyperpolarization-activated chloride channel gates. *The Journal of Physiology, 515*(Pt 2), 341–353.

Pusch, M., Ludewig, U., & Jentsch, T. J. (1997). Temperature dependence of fast and slow gating relaxations of ClC-0 chloride channels. *The Journal of General Physiology, 109*, 105–116.

Pusch, M., Ludewig, U., Rehfeldt, A., & Jentsch, T. J. (1995). Gating of the voltage-dependent chloride channel CIC-0 by the permeant anion. *Nature, 373*, 527–531.

Richard, E. A., & Miller, C. (1990). Steady-state coupling of ion-channel conformations to a transmembrane ion gradient. *Science, 247*, 1208–1210.

Saviane, C., Conti, F., & Pusch, M. (1999). The muscle chloride channel ClC-1 has a double-barreled appearance that is differentially affected in dominant and recessive myotonia. *The Journal of General Physiology, 113*, 457–468.

Schreiber, R., Uliyakina, I., Kongsuphol, P., Warth, R., Mirza, M., Martins, J. R., et al. (2010). Expression and function of epithelial anoctamins. *The Journal of Biological Chemistry, 285*, 7838–7845.

Schroeder, B. C., Cheng, T., Jan, Y. N., & Jan, L. Y. (2008). Expression cloning of TMEM16A as a calcium-activated chloride channel subunit. *Cell, 134*, 1019–1029.

Sheridan, J. T., Worthington, E. N., Yu, K., Gabriel, S. E., Hartzell, H. C., & Tarran, R. (2011). Characterization of the oligomeric structure of the Ca(2+)-activated Cl- channel Ano1/TMEM16A. *The Journal of Biological Chemistry, 286*, 1381–1388.

Smith, A. J., & Lippiat, J. D. (2010). Voltage-dependent charge movement associated with activation of the CLC-5 $2Cl^{-}/1H^{+}$ exchanger. *FASEB Journal: Official Publication of the Federation of American Societies for Experimental Biology, 24*, 3696–3705.

Staley, K. J., Soldo, B. L., & Proctor, W. R. (1995). Ionic mechanisms of neuronal excitation by inhibitory GABAA receptors. *Science, 269*, 977–981.

Stauber, T., & Jentsch, T. J. (2013). Chloride in vesicular trafficking and function. *Annual Review of Physiology, 75*, 453–477.

Suzuki, J., Umeda, M., Sims, P. J., & Nagata, S. (2010). Calcium-dependent phospholipid scrambling by TMEM16F. *Nature, 468*, 834–838.

Terashima, H., Picollo, A., & Accardi, A. (2013). Purified TMEM16A is sufficient to form Ca(2+)-activated Cl- channels. *Proceedings of the National Academy of Sciences of the United States of America, 110*, 19354–19359.

Traverso, S., Elia, L., & Pusch, M. (2003). Gating competence of constitutively open CLC-0 mutants revealed by the interaction with a small organic inhibitor. *The Journal of General Physiology, 122*, 295–306.

Traverso, S., Zifarelli, G., Aiello, R., & Pusch, M. (2006). Proton sensing of CLC-0 mutant E166D. *The Journal of General Physiology, 127*, 51–65.

Voets, T. (2012). Quantifying and modeling the temperature-dependent gating of TRP channels. *Reviews of Physiology, Biochemistry and Pharmacology, 162*, 91–119.

Weinreich, F., & Jentsch, T. J. (2001). Pores formed by single subunits in mixed dimers of different CLC chloride channels. *The Journal of Biological Chemistry, 276*, 2347–2353.

Xiao, Q., Yu, K., Perez-Cornejo, P., Cui, Y., Arreola, J., & Hartzell, H. C. (2011). Voltage- and calcium-dependent gating of TMEM16A/Ano1 chloride channels are physically coupled by the first intracellular loop. *Proceedings of the National Academy of Sciences of the United States of America, 108*, 8891–8896.

Yang, Y. D., Cho, H., Koo, J. Y., Tak, M. H., Cho, Y., Shim, W. S., et al. (2008). TMEM16A confers receptor-activated calcium-dependent chloride conductance. *Nature, 455*, 1210–1215.

Yang, H., Kim, A., David, T., Palmer, D., Jin, T., Tien, J., et al. (2012). TMEM16F forms a Ca^{2+}-activated cation channel required for lipid scrambling in platelets during blood coagulation. *Cell, 151*, 111–122.

Zdebik, A. A., Zifarelli, G., Bergsdorf, E. Y., Soliani, P., Scheel, O., Jentsch, T. J., et al. (2008). Determinants of anion–proton coupling in mammalian endosomal CLC proteins. *The Journal of Biological Chemistry, 283*, 4219–4227.

Zifarelli, G., De Stefano, S., Zanardi, I., & Pusch, M. (2012). On the mechanism of gating charge movement of ClC-5, a human Cl(−)/H(+) antiporter. *Biophysical Journal, 102*, 2060–2069.

Zifarelli, G., Murgia, A. R., Soliani, P., & Pusch, M. (2008). Intracellular proton regulation of ClC-0. *The Journal of General Physiology, 132*, 185–198.

Zifarelli, G., & Pusch, M. (2007). CLC chloride channels and transporters: a biophysical and physiological perspective. *Reviews of Physiology, Biochemistry and Pharmacology, 158*, 23–76.

CHAPTER NINE

Structure and Function of the ThermoTRP Channel Pore

Jie Zheng* and Linlin Ma
Department of Physiology and Membrane Biology, University of California School of Medicine, Davis, CA, USA
*Corresponding author: E-mail: jzheng@ucdavis.edu

Contents

Abstract

Temperature-sensitive transient receptor potential (TRP) channels are structurally similar to other tetrameric cation channels, but can be potently activated by heat. Recent studies suggest that the pore-forming region directly participates in activation gating. In this chapter, we summarize major findings from both structural and functional studies concerning the gating role of the pore region, focusing in particular on TRPV1. The emerging picture is that the peripheral S1–S4 region of TRPV1 is rigid and plays a supporting role for the pore to undergo conformational rearrangements. This places the pore region in the center of activation gating.

Current Topics in Membranes, Volume 74
ISSN: 1063-5823
http://dx.doi.org/10.1016/B978-0-12-800181-3.00009-9

1. INTRODUCTION

Ion channels form an aqueduct across the hydrophobic cell membrane to facilitate ion transport in a controlled manner (Hille, 2001). The channel pore, located at the center of symmetrically arranged subunits for most ion channels, plays a central role in carrying out this biological function. Besides determining the ion selectivity profile and conducting speed, the pore is also where gating occurs. Rarely does a channel pore behave as a passive, uncontrolled ion pathway. The activation gate, under the control of specialized sensor modules, opens and closes the pore to allow ion permeation only in the presence of appropriate stimuli. The inactivation gate terminates ion permeation when the activation gate is kept open for an extended time, thus preserving the ion concentration gradient across the cell membrane, permitting rapid signaling kinetics, and helping to ensure unidirectional action potential propagation along the axon. The channel pore also presents high-affinity binding sites for channel blockers (such as pore-blocking toxins) and regulators (such as PIP_2). Similarly, ions and small molecules entering the pore can strongly influence the rate and direction of ion flux (rectification) and even terminate ion flux altogether. The channel pore can present specialized pathways (fenestrations) for regulatory molecules, or harbor modulatory domains or glycan chains. The pores of temperature-sensitive TRP (thermoTRP) channels are known to have many of these features.

Recently a landmark cryo-electron microscopy (EM) study revealed a series of TRPV1 structures at up to 3.4 Å resolution (Cao, Liao, Cheng, & Julius, 2013; Liao, Cao, Julius, & Cheng, 2013). Besides annotating beautifully many previous functional observations, these structures also place the pore in the center of the TRPV1 gating machinery. A comparison of the cryo-EM structures between the unliganded (closed) state and the capsaicin/resiniferatoxin (RTX)/double knot toxin (DkTx) bound (most likely open) state suggests that the peripheral S1–S4 domain remains mostly unchanged while the pore domain exhibits substantial movements. This interpretation is in stark contrast to the view of classic voltage-gated cation channels, for which the S1–S4 domain forms the voltage sensor and its transmembrane movement relocates the equivalent of many elementary charges ($\sim 14\ e_0$ for Shaker potassium channel) across the entire electric field (Sigworth, 1994). It is this substantial voltage-sensor movement that couples depolarization of the membrane potential to the opening of the activation gate. The new TRPV1 structural data thus corroborate a large body of existing functional data in asserting a prominent gating role of the pore. Furthermore, experimental evidence suggests that the conclusions derived from TRPV1 may be applicable to other members of this intriguing family of temperature and pain sensors. Here we summarize structural and functional observations of the thermoTRP channel pore, with a substantial amount of discussions focusing on TRPV1, the archetypal thermoTRP channel for which rich information exists in the literature. In this chapter the channel pore is defined as the integrated structure surrounding the ion permeation pathway. In linear sequence it corresponds to the S5–P loop–S6 region.

2. STRUCTURAL FEATURES OF A THERMOTRP CHANNEL PORE

Despite the low sequence similarity between thermoTRP channels and other cation channels such as the potassium channels, their overall pore architectures resemble each other quite well. This is nicely illustrated by the cryo-EM structures of the truncated rat TRPV1 channel (Cao et al., 2013; Liao et al., 2013). The ion permeation pathway is located in the center of four symmetrically arranged subunits (Figure 9.1). The S5 and S6 transmembrane helical segments wrap around the pore at a tilted angle, forming a tightening at the intracellular end of the pore. The pore helices lay even flatter near the extracellular end of the pore, with their C-terminal ends pointing to the center. The selectivity filter forms the tightest part of the outer pore. Right beneath the selectivity filter there is a water-filled cavity that is surrounded by the S6 segments. The pore domain is surrounded by four S1–S4 domains like flower pedals cradling the stamina. This architectural design is also seen in most tetrameric channels with a known crystal structure (Catterall, 2011, 2014; Gonzalez et al., 2012). It is thus anticipated that at least some of the functional utilities of these structural elements may also be shared.

Interestingly, the TRPV1 channel ion selectivity filter contains the conserved potassium channel signature sequence TX_1GX_2GD, in which X_1 and X_2 are I and M, respectively, in TRPV1, or V and Y for most potassium channels. This is rather intriguing as TRPV1 is known as a non-selective cation channel with substantial Ca^{2+} permeability (Owsianik, Talavera, Voets, & Nilius, 2006). The selectivity filter of potassium channels contains a series of K^+-binding sites formed by the main-chain backbone oxygen atoms that point towards the ion permeation pathway (Doyle

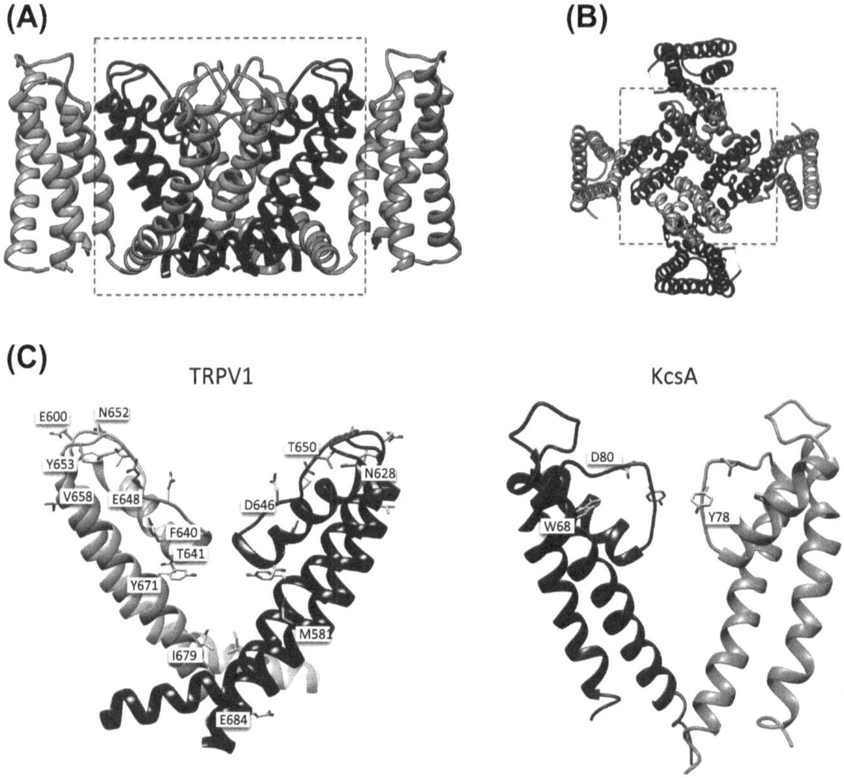

Figure 9.1 Side (A) and top (B) view of the TRPV1 channel transmembrane region in the closed state. Each subunit is shown in a different color. The S1–S4 domains of the red and blue subunits are omitted in (A) to reveal the pore region (highlighted with the box). (C) Comparison of the pore features of TRPV1 (left) and KcsA (right). Some important amino acids discussed in the text are highlighted. Note that 23 amino acids in the turret region are missing in the TRPV1 structure. (See the color plate.)

et al., 1998). The distance between these oxygen atoms is optimized for coordination of K^+ instead of Na^+ or other ions in the physiological solution. A rigid network of aromatic interactions (for example, between W68 and Y78 in KcsA or W434 and Y445 in Shaker) around the selectivity filter is thought to form a scaffold-like framework that helps to ensure the precise positioning of the K^+-coordinating oxygen atoms (Doyle et al., 1998) (Figure 9.1(C), right panel). Perturbing this aromatic scaffold in the Shaker potassium channel by a W434F mutation yields a permanently inactivated channel (Yang, Yan, & Sigworth, 1997). Surprisingly, in the TRPV1 cryo-EM structures, the selectivity filter region is much shorter and likely cannot hold multiple permeant ions. The TIGMGD segments, for the most part, are well separated (Figure 9.1(C), left panel). As a result, the outer pore region takes the shape of a large crater. While a defined extracellular space is also seen in the structures of KcsA (Doyle et al., 1998) and the Kv1.2-Kv2.1 chimera channel (Long, Tao, Campbell, & MacKinnon, 2007), it locates outside of the multi-ion long pore. Here the aspartate residues (D80 in KcsA, Figure 9.1(C), right panel) at the end of the TVGYGD signature sequence from the four subunits help to coordinate a K^+ ion before it enters the selectivity filter, a feature that is thought to contribute to high permeation rate (Zhou, Morais-Cabral, Kaufman, & MacKinnon, 2001). Whether TRPV1 contains a multi-ion long pore has not been determined functionally. Implications of the unique selectivity filter structure remain to be tested.

It should also be mentioned that the TRPV1 structural models were built upon cryo-EM data derived from a truncated rat channel. This minimal functional channel lacks the N-terminal intracellular segment up to the first ankyrin-like repeat domain, a significant portion of the C-terminal intracellular segment, as well as almost the whole pore turret. The deletion of 23 amino acids from the turret region in particular is expected to substantially alter the landscape of the outer pore region. Indeed, in full-length mouse TRPV1 pore models we recently built using the Rosetta method, the pore turret covers a large surface area of the outer pore (Yang, Yarov-Yarovoy, & Zheng, unpublished data). The pre-S6 loop and the S1–S2 linker are both seen occupying a different location compared to those in the cryo-EM structures. Even the selectivity filter segment is seen in most of the conformations to be more "upright" and closer to each other, thus forming a long pore that may accommodate multiple permeant ions. The outer pore of thermoTRP channels has attracted substantial interests as numerous activation stimuli, such as H^+ (Jordt, Tominaga, & Julius, 2000; Tominaga et al., 1998), spider toxin DkTx (Bohlen et al., 2010), heat (Cao, Ma, Yang, Wang, & Zheng, 2014; Cui et al., 2012; Grandl et al., 2008; Grandl et al., 2010; Yang, Cui, Wang, & Zheng, 2010), and divalent cations (Ahern, Brooks, Miyares, & Wang, 2005; Ahern, Wang, & Miyares, 2006; Cao et al., 2014; Luebbert et al., 2010; Ohta, Imagawa, & Ito, 2008; Riera, Vogel, Simon, & le Coutre, 2007; Tousova, Vyklicky, Susankova, Benedikt, & Vlachova, 2005; Yang, Ma, Cao, Wang, & Zheng, 2014) are known to interact with the channel at this region. It is thus important for future studies to further define the structural details and functional roles of the outer pore.

3. FUNCTIONAL TESTS OF PORE STRUCTURES

3.1 Chimera studies

Substantial efforts have been put into generating and testing various chimeras in the pre-cryo-EM-structures era. The design of these constructs was guided by sequence analysis and the anticipated similarities to other tetrameric cation channels with a known structure. Unfortunately, sequence conservation among thermoTRP channels is not very good; similarities between sequences of thermoTRPs and other cation channels are even less. Nonetheless, data from chimeric channels have proven to be quite informative in revealing the location of functional domains and are well supported by other functional studies and the new cryo-EM structures. For example, the identification of capsaicin-binding site in TRPV1 and icilin-binding site in TRPM8 was facilitated by chimeras between rat TRPV1 (capsaicin-sensitive) and chicken TRPV1 (capsaicin-insensitive) (Jordt & Julius, 2002), and between rat and chicken TRPM8 channels that exhibit different icilin sensitivities (Chuang, Neuhausser, & Julius, 2004). Chimeras were also generated for the purpose of locating the heat-sensing domains. Julius et al. generated chimeras between rattlesnake TRPA1 and rat snake TRPA1 as well as between human and snake channels (Cordero-Morales, Gracheva, & Julius, 2011). They identified two portable heat-sensitive modules within the ankyrin-like repeats domain. Interestingly, the regions flanking the ankyrin-like repeats domain can exert regulatory effects on the heat-activation process, making certain fruit fly TRPA1 splice variants insensitive to heat (Zhong et al., 2012). Brauchi et al. made a series of chimeras between the heat-activated TRPV1 and the cold-activated TRPM8 (Brauchi, Orta, Salazar, Rosenmann, & Latorre, 2006). They found that the intracellular C-terminal region can transfer the temperature sensitivity of its parental channel into the chimera channels, a conclusion supporting an earlier study demonstrating that deletion of the last 72 amino acids from the rat TRPV1 substantially affected channel activation by heat (and by other stimuli such as voltage and capsaicin) (Vlachova et al., 2003). In another study the Qin group made chimeras between TRPV1 and TRPV2 to show that the linker region between the ankyrin-like repeats domain and S1 is a modular heat sensor (Yao, Liu, & Qin, 2011). Chimeras between Kv2.1 and TRPV1 or TRPM8 were also generated to test their structural compatibility (Kalia & Swartz, 2013).

Taking advantage of the knowledge that capsaicin is a full agonist for rat TRPV1 but a partial agonist for human TRPV1, the Chuang group generated a series of chimeras between them to search for the region responsible for the potentiation effects of H^+ and Mg^{2+} on capsaicin activation of human TRPV1. The study concluded that the H^+ effect mostly originates from the S1–S4 domain, while the Mg^{2+} effect can be traced to the pore region. Indeed, mutations to the S3–S4 linker residues were found to affect H^+-dependent activation of the rat TRPV1 (Ryu, Liu, Yao, Fu, & Qin, 2007). Our recent study found that the turret is one of the required structures for Mg^{2+} potentiation (Yang et al., 2014). Mutations to two glutamate residues in the outer pore region are known to prevent Mg^{2+} potentiation (Ahern et al., 2005).

Aiming at identification of channel structures involved in heat activation of TRPV3, the Patapoutian group generated a series of chimeras between various TRPV channels of human, rat, mouse, and frog origins (Grandl et al., 2008). These tests focused on the pre-S6 loop and the N-terminal end of S6. They observed that a number of chimeras, containing mostly TRPV3 but the focused region of TRPV1, TRPV2, or TRPV4, exhibited reduced or none heat activation and a shifted agonist sensitivity (to 2-APB and camphor). Noticeably, a chimera containing the pre-S6 loop and the first half of S6 of human TRPV2 channel was found to be heat-insensitive while exhibiting TRPV3 wildtype-like agonist sensitivity. Taken together with results from mutational tests, these observations suggest that the pre-S6 loop may play a special role in the heat activation of TRPV3.

In the past, the design of chimeric constructs was based on sequence alignments. The reliability of such alignments, especially for chimeras between channels of different families, is limited by low sequence similarity. It is likely that the new structural information may help to improve the design of future chimera experiments.

3.2 Structural perturbation by point mutation

A commonly used approach for initial investigation of the structure–function relationship of an ion channel is to generate point mutations and attempt to interpret the functional consequences. This has been done with thermoTRPs in two ways: screening for a particular phenotype resulting from random mutagenesis, and designing site-specific point mutations to test specific predictions.

In one screening of the rat TRPV1 channel, Julius et al. looked for gain-of-function mutations that left the mutant channels constitutively active, killing yeast cells expressing them by cation overload (which could be prevented by a channel blocker, ruthenium red) (Myers, Bohlen, & Julius, 2008). Such gain-of-function mutations were identified throughout the channel protein, including but not limited to the ankyrin-like repeats domain, the S4–S5 linker, the intracellular C-terminal segment, and the pore domain. Mutations exhibiting a "strong growth phenotype" appeared in the pore domain and the immediate post-S6 region at a higher rate. Within the S5–S6 region, they include M581T (which is located in S5 at a position facing a hydrophobic patch of the neighboring S6 segment) and F640L (which is at the C-terminal end of the pore helix) (Figure 9.1(C)). Additional "weak growth phenotype" mutations include N652D in pre-S6 loop and E684 G/V at the C-terminal end of S6, again pointing to the neighboring S6. (In addition, Q560R at the N-terminal end of the S4–S5 linker is also a "weak growth phenotype" mutation.) Functional tests of these mutations confirmed that they render the channel either constitutively open or highly sensitive to activation stimuli. Follow-up targeted mutagenesis tests (with conserved mutations) identified T641S (next to F640) and T650S (in pre-S6 loop) that increase the activity level at the resting condition, as well as N628D (in pore turret) and V658A (the N-terminal end of S6) that significantly elevate H^+-induced activation.

In another random mutagenesis screening of rat TRPV1, Grandl et al. looked for loss-of-function mutations affecting specifically heat activation (Grandl et al., 2010). In their report, they highlighted N628K (in the pore turret), N652T and Y653T (in pre-S6 loop) that individually or in combination strongly inhibit heat-induced channel activation while having negligible or minor effects on capsaicin/2-APB/H^+-induced activation (Figure 9.1(C)). These observations from TRPV1 are consistent with those from a similar screening study of mouse TRPV3 by the same group (Grandl et al., 2008). In that study, Grandl and colleagues also identified a number of mutations in the pre-S6 loop and the extracellular part of S6, such as N643S, I644S, N647Y, L657I, and Y661C, that strongly inhibit heat activation while having minor effects on 2-APB or camphor-induced activation. Additional heat activation-specific point mutations in mouse TRPV3 include S648I, P651A, F654S/K, and L657E. It was further reported that in frog TRPV3 point mutations to each residue in the E638–P643 segment exhibited a heat activation-specific phenotype.

An alanine scanning of the S6 segment of the rat TRPV1 revealed that almost each of these point mutations affected some aspects of channel activation (Susankova, Ettrich, Vyklicky, Teisinger, & Vlachova, 2007). When individual residues of the TRPV1 S6 were mutated to a cysteine (after removing an endogenous cysteine with the C157A mutation), it allowed the detection of state-dependent access to deep pore position and activation-associated conformational changes in the channel pore (Salazar et al., 2009). Many informative studies on individual point mutations have been reported (Winter et al., 2013). Some of them are discussed in the rest of this chapter.

3.3 Fluorescent recording of structural dynamics

Site-directed fluorescence recordings provide information on structure dynamics from functional channels under near-physiological conditions. As fluorescence change is monitored simultaneously with current change using the patch fluorometry method, structural information is obtained in the functional context (Zheng, 2006; Zheng & Zagotta, 2003). Site-directed FRET (fluorescence resonance energy transfer) signals have so far been recorded from chemical fluorophores attached to several extracellular sites of mouse TRPV1, including L461C and N467C (in the S1–S2 linker), K536C (in the S3–S4 linker), C617 and C622 (in the pore turret) (Yang et al., 2010, 2014). FRET between cyan and yellow fluorescent proteins was also used to detect the positioning of intracellular structures of the rat TRPV1 (De-la-Rosa, Rangel-Yescas, Ladron-de-Guevara, Rosenbaum, & Islas, 2013). The detected FRET efficiency values correlated nicely to the distances between fluorophore attachment sites seen in the cryo-EM structures, further confirming the reliability of FRET quantification.

Several general observations from these FRET experiments are of particular interests in term of structure–function relationship of the TRPV1 channel pore. (1) No FRET change was observed from the sites in the S1–S2 linker or the S3–S4 linker, even when the labeled channel clearly underwent activation transitions. In contrast, large FRET changes were observed from the turret. This result is consistent with the notion recently put forward in the cryo-EM study (Liao et al., 2013), that is, the peripheral S1–S4 domains form a static supportive structure surrounding the pore domains that carry out the activation gating motions. (2) No FRET change was observed from the outer pore positions when capsaicin was used to activate the channel. The inability of capsaicin to affect extracellular structures—despite its strong activation effects—is consistent with the generally accepted idea that TRPV1 stimuli are sensed and transmitted by separate pathways/structures. The capsaicin binding site is located near the intracellular side of the transmembrane domains adjacent to the pore domain (Gavva et al., 2004; Jordt & Julius, 2002; Liao et al., 2013). (3) No FRET change was observed when membrane depolarization was used to activate the channel. One possible reason for the lack of an FRET change might come from the fact that depolarization is not a strong activator for TRPV1. Furthermore, voltage-dependent activation of TRPV1 involves the movement of a very small effective gating charge (<1 e_0), which might suggest small and maybe local conformational changes. (4) FRET signal recorded from the pore turret changed significantly when heat or Mg^{2+} was used to activate the channel. The FRET change correlated well with current increase upon channel activation. An increase in FRET was observed when the channel was activated, suggesting that the fluorophore attachment sites on the turret move closer towards the central axis of the channel pore. It is noted that fluorophores at the same pore turret positions did not report any structural change when capsaicin was used to activate the channel, as mentioned above. (5) Extracellular divalent cations (such as Mg^{2+} and Ba^{2+}) could also activate TRPV1 and change FRET between fluorophores attached to the same pore turret sites (Cao et al., 2014; Yang et al., 2014). Intriguingly, the FRET change induced by Mg^{2+} appeared to precede the current increase that signaled channel activation. This is so far the most direct evidence that the pore turret movement may be the cause of channel activation, instead of an accommodating or consequential structural adjustment. Heat-induced activation occurs rapidly so that low temporal resolution fluorescence signal did not show any detectable kinetic difference from the current signal. Undoubtedly, further investigation using site-directed fluorescence recordings will help to reveal detailed structural dynamics associated with TRPV1 activation.

3.4 Toxin binding

A fortunate discovery for the TRPV1 research field is the identification of a peptide toxin from tarantula spider venom that acts as a potent channel agonist (Bohlen et al., 2010; Siemens et al., 2006). The high (submicromolar) apparent affinity of the double-knot toxin, DkTx, is partially a consequence of the divalent design, as binding of one knot substantially enhances the binding of the second knot through enhancement of the effective local concentration (Bohlen et al., 2010). Judging from the elicited currents or intracellular Ca^{2+} change, DkTx is almost as potent as capsaicin in activating rodent TRPV1. Mutagenesis analysis suggested that the outer pore region participates in toxin binding/potentiation. This was confirmed by the cryo-EM study, which revealed that the toxin binds to the outer pore region between subunits (Cao et al., 2013; Liao et al., 2013). The channel/DkTx complex (also contained the potent activator RTX) revealed the conformation of a likely open channel. Comparison with the apo structure suggests that activation conformational changes are mostly limited to the pore domains, involving expansion of both the S6 gate and the selectivity filter gate. It is interesting to note that the binding of capsaicin alone is capable of inducing the S6 gate opening. The binding of DkTx clearly induced substantial rearrangements of the outer pore structures. It remains unclear whether the difference in the S6 gate conformations between the DkTx/RTX-bound structure and the capsaicin-bound structure is due to DkTx or RTX.

4. ACTIVATION GATING OF TRPV1

4.1 Lessons from Shaker and other channels

From the rich literature in the channel gating field, a number of general principles have emerged (each with noticeable exceptions). One widely applicable observation is that ion channel functions are supported by "functional domains," which are well-defined and often transferable structural units. Take the voltage-dependent potassium channels as an example. The main voltage sensor of Shaker and other Kv channels is formed by the S1–S4 domain that remains capable of driving voltage-dependent activation of the bacterial potassium channel KcsA in a chimeric channel construct (Lu, Klem, & Ramu, 2001). The N-type inactivation gate is formed by the amino terminal peptide of either the pore-forming subunits or the auxiliary β-subunits; it can cause inactivation even when applied as a free peptide in solution (Hoshi, Zagotta, & Aldrich, 1990; Zagotta, Hoshi, & Aldrich, 1990). Applying this domain principle to the polymodal thermoTRP channels, it has been assumed that there exists an individual "heat sensor," "voltage sensor," and so on. Indeed, binding sites for capsaicin, H^+ and toxin in TRPV1 and menthol in TRPM8 are found to reside in distinct and even transferable channel structures. Based on the transferable domain assumption, a number of chimeric constructs were made with the goal of locating the heat sensor of thermoTRP channels, as discussed earlier. Important channel structures have been identified this way.

The assumption that thermoTRP channels have the same S1–S4 voltage sensor domain as the Shaker and Kv channels has not found convincing experimental support. Indeed, the S4 segment of many thermoTRP channels contains no charged amino acid. The S4–S5 linker of some thermoTRP channels does contain charged amino acids whose mutations affect the apparent voltage-dependence (Boukalova, Marsakova, Teisinger, & Vlachova, 2010; Voets, Owsianik, Janssens, Talavera, & Nilius, 2007). However, it has been shown in Shaker and related Kv channels that the major gating charges are those in the middle of the S4 segments (Aggarwal & MacKinnon, 1996). Mutations to positive charges towards the two ends of S4 have a much weaker impact on voltage sensitivity. Furthermore, mutations of uncharged residues in S4 (Xu, Ramu, & Lu, 2010) and the S4–S5 linker (e.g., V2, (Schoppa, McCormack, Tanouye, & Sigworth, 1992)), and even the pore domain (Yifrach & MacKinnon, 2002; Zheng & Sigworth, 1997) yielded drastic shifts of the voltage-dependent activation curve that can be attributed to alteration of the voltage-independent free energy change associated with activation gating. In addition, the minimal apparent gating charges observed from the wildtype and mutant thermoTRP channels unavoidably obscure any clear interpretation of experimentally observed changes in voltage dependence.

An important general principle associated with the domain design of ion channels is allosteric coupling among gating domains driven by distinct stimuli. Classic examples of allosteric coupling are found between the voltage sensor and the activation gate in Shaker channels, and between cyclic nucleotide-binding domain and the activation gate of CNG and HCN channels (Gordon & Zagotta, 1995; Hu et al., 2013; Schoppa & Sigworth, 1998; Zagotta, Hoshi, & Aldrich, 1994). Allosteric coupling between different gating stimuli are also convincingly demonstrated, for example, for the interaction between voltage and Ca^{2+} in activating BK channels (Horrigan & Aldrich, 2002). The general formalisms of allosteric coupling have been widely used in the study of thermoTRP channels (Cao et al., 2014; Jara-Oseguera & Islas, 2013; Latorre, Brauchi, Orta, Zaelzer, & Vargas, 2007; Matta & Ahern, 2007; Yao, Liu, & Qin, 2010). These studies have yielded some important observations pertaining to the activation mechanism. The generally accepted finding that polymodal thermoTRP channels contain structurally and functionally distinct gating pathways for activation stimuli makes allosteric coupling an attractive framework to understand channel behaviors.

Interestingly, certain experimental observations may at first sight appear to violate the reciprocal principle of allosteric coupling. But they are not. For example, it was repeatedly observed that mutations to the pore turret exhibited strong disruptive effects on heat activation but spared capsaicin activation (Cui et al., 2012; Yang et al., 2010, 2014). This type of observations, taken as evidence in support of the structural separation of different gating pathways, is in fact fully expected from an allosteric coupling point of view: a perturbation to the structure underlying one gating modal is "sensed" by another allosterically coupled modal ONLY when measurement is made under conditions requiring the participation of the perturbed structure. Specifically, measurements made at temperatures where heat activation is dormant will be insensitive to structural perturbations that favor the resting state of the heat-driven process. To observe an allosteric effect of the turret mutations on capsaicin activation, measurements need to be made under conditions that would activate the heat-driven process. It is anticipated that detailed vigorous analysis of the allosteric coupling will continue to yield valuable information on the activation mechanisms of thermoTRP channels.

4.2 Implications from the cryo-EM structures

The cryo-EM structures of the truncated TRPV1 channel are remarkable in both the structural details and the potential mechanistic insights revealed by the differences (and similarities) between the capsaicin-bound structure, the RTX/DkTx-bound structure and the apo structure. Capsaicin, RTX and DkTx are shown to strongly activate this truncated channel. As pointed out by the authors of this study, the similarity in the S1–S4 domain of these structures, in contrast to major differences in the pore domain, argues for a static, supportive role of the peripheral channel structures (Cao et al., 2013). The structures also underpin the concept of separated activation pathways by identifying the capsaicin/RTX binding site and the DkTx binding site, and by revealing the structural differences in the pore domain between the two assumingly open conformations. Since electron microscopy was performed at a prohibitively low temperature for heat activation and, in our view, the truncated channel lacks the pore turret that is critical for heat activation, no direct evidence on the heat activation pathway is easily available from the existing cryo-EM structures.

Furthermore, the observation that different pore-lining regions move upon binding of capsaicin or RTX/DkTx suggests the existence of two candidate activation gates, one formed by residues from the S6 segment, the other formed by the selectivity filter. The presence of an activation gate at the equivalent positions has been reported in the literature on Shaker and Kv channels, CNG channels, bacterial K channel, and others (Contreras, Srikumar, & Holmgren, 2008; Cordero-Morales, Cuello, & Perozo, 2006; Cordero-Morales, Cuello, Zhao, et al., 2006; Flynn & Zagotta, 2001; Liu, Holmgren, Jurman, & Yellen, 1997; Zheng & Sigworth, 1997). Tightly coupled gating movements at these locations are suggested by functional studies (Flynn & Zagotta, 2001; Uysal et al., 2011; Zheng & Sigworth, 1997). It thus appears that, to a certain extent at least, TRPV1 activation gating may mirror that of the voltage- and ligand-gated channels.

4.3 Functional evidence for the location of TRPV1 activation gate

To search for the activation gate in TRPV1, the Rosenbaum group carried out a detailed accessibility study based on SCAM (substituted cysteine accessibility method) (Salazar et al., 2009). This is the approach that contributed to the identification of the *Shaker* potassium channel S6 activation gate (del Camino & Yellen, 2001; Liu et al., 1997) and demonstration of the absence of an S6 gate in CNG channels (Contreras et al., 2008; Flynn & Zagotta, 2001). For the TRPV1 study, a clear restriction for Ag^{+} to access introduced cysteines in the pore was found at the level of Y671: below this point Ag^{+} reacted with introduced cysteines equally well in the closed and open state, while above it, Ag^{+} reacted with introduced cysteines much faster in the open state than in the closed state (Salazar et al., 2009). This is true no matter how the channel was activated—either by capsaicin or by heat. In strong support of the existence of an S6 helix activation gating movement, the authors demonstrated that S6 exists as a helix and disulfide bond formation between subunits occurs at two positions (L681C and Y671C) only in the closed state but not in the open state, as expected if S6 moves away from the central axis upon activation gating as a rigid body.

Using the pore diameter as a criterion, the cryo-EM structure of the apo state suggests that the gate is formed by I679 (Cao et al., 2013; Liao et al., 2013) (Figure 9.1(C)). This is a position much lower than Y671. There are several possible ways to interpret the apparent discrepancy between the structural and functional data. It is likely that spatial constrains and local environment such as electrostatic properties can all play a role in the rate of reaction between Ag^{+} and introduced cysteines. Functional channels under physiological conditions may exist in a slightly different conformation from that exhibited by channels under the condition for cryoelectron microscopy.

Indeed, Y671 is seen in the cryo-EM structure with its aromatic side-chain pointing towards the central axis of the pore. It is also interesting to note that, in the cryo-EM structure, Y671 is at the same vertical level as the end of the pore helix. It might be possible that the Ag^{+} accessibility data actually reflected not gated access to the pore but rather state-dependent accessibility of this part of the S6 during activation gating. Of course, in order for this scenario to work, one has to assume that TRPV1 functions like the CNG channels in that the activation gate are located at the selectivity filter, with no constrain for permeant ion to enter the pore at the intracellular end, and the movement of S6 serves as a coupling mechanism instead of the opening of the activation gate (Contreras et al., 2008; Flynn & Zagotta, 2001). It is interesting to note that all cysteine mutations along the S6 segment only marginally altered capsaicin activation, with a less than fivefold change in K_D (Salazar et al., 2009). More experimental information will be needed to better match functional data to the structures, and determine the location of the activation gate(s).

4.4 H^{+}-Binding sites in TRPV1 and potential pH gating mechanism

Extracellular H^{+} is an important physiological stimulator for TRPV1. Indeed, TRPV1 proteins are expressed in many internal organs and tissues where the temperature remains stable, and the presence of capsaicin or other exogenous chemical stimuli is unlikely. H^{+}-induced TRPV1 activation may play an important role in mediating pain caused by inflammation and tissue injury (Tominaga et al., 1998). Candidate H^{+}-binding sites have been mapped to a number of acidic residues mostly in the outer pore region. The two major H^{+}-binding sites are thought to be E600 at the junction between S5 and the pore turret, and E648 in the pre-S6 loop (using the rat TRPV1 amino acid numbers) (Jordt et al., 2000) (Figure 9.1(C)). Additional acidic residues whose mutations exhibited an effect on H^{+}-induced activation are located in the turret (D601, E610), pore helix (E636), selectivity filter (D646), pre-S6 loop (E651), as well as the S3–S4 linker (E536). How protonation of these sites strongly promotes channel activation remains unknown. Nonetheless, involvement of a broad region of the outer pore in activation gating is evident. This idea is further supported by findings that mutations to nonprotonatable residues can also substantially hinder H^{+} activation. Examples of this type of mutations include T633A at the pore helix and V538L at the S3–S4 linker (Ryu et al., 2007), as well as F660S at the S6 segment (Aneiros et al., 2011).

As mentioned above, random mutation scanning indicated a number of residues in the outer pore region that affect heat activation. Their close proximity to the H^+ sites is noted. Similarly, DkTx is seen to bind to the outer pore to exert its strong activation effect. The location of the two major H^+ sites is of particular interest. E600 is at the extracellular end of S5 where the transmembrane segment is connected to the pore turret (which is missing in the cryo-EM structures). Mutation E600Q/D/K/H alone is sufficient to substantially alter channel activation by capsaicin, heat and pH (Jordt et al., 2000). Thus, the movement of E600 and its surrounding structure either directly drives activation conformational change or is coupled to the gating motion. The same case can be made for E648, located in the short pre-S6 loop that also carries heat-sensitive residues.

An intriguing observation from multiple studies is that paired mutations to the two major H^+-binding sites, E600Q and E648A, can near completely block the potentiation effect of extracellular divalent and multivalent cations (to be discussed below). One possible explanation for these experimental observations is that E600 and E648 serve as the binding sites for these cations in addition for H^+. Alternatively, E600 and E648 can mediate activation conformational changes induced by extracellular cations. In either case, the prominent gating role of these H^+-binding sites cannot be overlooked.

4.5 Divalent cation-induced TRPV1 activation

As mentioned above, H^+ is not the only extracellular cation that can strongly potentiate TRPV1. Many divalent and multivalent cations exhibit a similar gating effect albeit at much higher concentrations. They either potentiate TRPV1 activation induced by other stimuli or directly induce channel activation. The first example was reported on Mg^{2+} by Ahern and colleagues (Ahern et al., 2005). They demonstrated that millimolar concentrations of Mg^{2+} could induce clearly identifiable currents from TRPV1-expressing cells, which could be enhanced in the presence of another activation stimulus and blocked by mutations E600Q and E648A. Quickly the list of activating cations expanded to include gadolinium (Tousova et al., 2005), polyamines (Ahern et al., 2006), Zn^{2+} and Cu^{2+} (Riera et al., 2007), Ca^{2+} (Ohta et al., 2008), Ni^{2+} (Luebbert et al., 2010), and Ba^{2+} (Cao et al., 2014). These cations appear to potentiate the channel in a similar manner; for example, the same mutations to E600 and E648 could block the potentiation effects of all these cations. It was further shown that the potentiation effect is limited to the extracellular side of the channel. Applying up to 100 mM of intracellular Mg^{2+} failed to exhibit any potentiation of channel activity (Yang et al., 2014).

A recent investigation revealed that extracellular divalent cations potentiate TRPV1 by shifting the heat activation threshold into the room temperature level, so that even in the absence of other stimuli the channel is heat-activated (Cao et al., 2014). A shift in heat activation range can be achieved through either a direct interaction or an indirect, allosteric interaction. It was shown that Mg^{2+} (as well as Ba^{2+}) could directly potentiate heat activation, providing 2.7 kcal/mol activation energy by 10 mM of Mg^{2+}. Even though Mg^{2+} also potentiates voltage and capsaicin activation, it is best explained by an allosteric effect instead of a direct interaction of Mg^{2+} with these processes. Additional evidence in support of a direct interaction of Mg^{2+} and Ba^{2+} with the heat activation process is that channel desensitization induced by these cations only prevents heat activation but not capsaicin activation. The observation suggests that the structure undergoing divalent cation-mediated desensitization is required for heat activation but not for capsaicin activation. The structure responsible for potentiation by divalent and multivalent cations has not been identified, though experiments with mutational perturbations indicated that the pore turret as well as a part of the S1–S2 linker expected to be near the turret are required for Mg^{2+} potentiation, while perturbations to the anticipated peripheral portion of the S1–S2 linker as well as the S3–S4 linker (also away from the pore) did not affect Mg^{2+} activation (Yang et al., 2014).

4.6 Heat activation

The heat activation mechanism is a central question for the study of thermoTRP channels. After nearly 17 years of intensive investigation and great advances in our understanding of the structure and function of these temperature-sensing ion channels, the heat activation mechanism remains a mystery. A recent review summarized existing evidence pertaining to heat activation (Zheng, 2013). A few pieces of new information have been added to the literature and will be updated here. The fascinating new cryo-EM structures, being extremely informative in revealing the structural features and agonist activation mechanism, did not provide direct evidence on how heat might affect the channel. Nonetheless, the interaction between the ankyrin-like repeats domain and three β-strands formed by a pre-S1 region and a post-S6 region is intriguing, because in TRPA1 the ankyrin-like repeats domain (which is much larger than that of TRPV1) and its flanking regions are found to be important for heat activation

(Cordero-Morales et al., 2011; Zhong et al., 2012). The linker between the ankyrin-like repeats domain and S1, which contains two of the three β-strands, was reported to serve as the heat sensor in TRPV1 (Yao et al., 2011). It would be interesting to test whether heat-dependent changes in the interaction between the ankyrin-like repeats domain and the β-strands might occur. Furthermore, the need to remove the pore turret for the cryo-EM study is intriguing. The S1–S2 linker and the S3–S4 linker, having a shorter length (12 amino acids for the S1–S2 linker and two amino acids for the S3–S4 linker versus more than 20 amino acids for the turret), did not affect the protein preparation. The turret sequence is also noticeably less conserved between species. The possibility that the turret might be a flexible protein domain is rather attractive, as it would make the structure compatible for rapid heat-induced structural changes.

As mentioned above, recent investigation of divalent cation potentiation of TRPV1 activation yielded a few interesting observations. First, Mg^{2+} appears to interact specifically with the heat activation process. Second, Mg^{2+} potentiation occurs only from the extracellular side, not from the intracellular side. Third, turret movement induced by Mg^{2+} occurs before channel activation. Taken together with previous findings of the important roles of the turret, the pre-S6 loop, and the pore helix in heat activation, results from Mg^{2+} potentiation of heat activation further suggest that the outer pore region is not only critical for H^+ and toxin induced activation but also critical for heat activation.

Another important idea presented in a recent article is that heat- and cold-induced channel activations might share a common temperature-sensing mechanism (Jara-Oseguera & Islas, 2013). The argument, based on a general allosteric gating framework, is that the opposite temperature sensitivity can theoretically arise from the same temperature-sensing process, as long as the sensor is coupled to the activation gate in opposite ways. A promoting coupling (with a coupling efficiency factor larger than 1) will lead to heat activation, while an inhibiting coupling (with a coupling efficiency factor less than 1) will yield heat deactivation, i.e., cold activation. Finding experimental evidence in support of this idea would substantially advance the field.

5. PORE DILATION

In a study of TRPV1 biophysical properties the Caterina group made an interesting observation, that is, the channel's ion selectivity exhibited time-dependent changes (Chung, Guler, & Caterina, 2008). During this process, permeability to large ions such as $NMDG^+$, $TRIS^+$, and $2\text{-}MAE^+$ increased substantially, as if the size of the channel pore expanded over time. Probing the ion selectivity filter with cysteine-reactive reagents indicated that the pore indeed dilated in a time- and activity-dependent manner. Independent investigation by the Bean group demonstrated that the relative permeability for QX-314, a bulky voltage-gated sodium channel blocker molecule, increased over time, leading to an activity-dependent change in the reversal potential when QX-314 was the only external permeable cation (Puopolo et al., 2013). Earlier the Caterina group also observed activity-dependent permeability change of TRPV3 channels (Chung, Guler, & Caterina, 2005). Similarly, it was reported that TRPA1 activation results in dynamic changes in permeability to $NMDG^+$ (Banke, 2011; Banke, Chaplan, & Wickenden, 2010; Chen et al., 2009). The confirmation of the pore dilation phenomenon in these thermoTRP channels adds them to a short list of ion channels that are known to exhibit a dramatically changing pore structure (Virginio, MacKenzie, Rassendren, North, & Surprenant, 1999).

Interestingly, changes in the ion selectivity filter conformation are also known for a number of well-studied channels. For example, in a mutant voltage-gated Shaker potassium channel, it was shown that depolarization-induced channel activation involves a series of step-wise changes in conductance (subconductances) and associated changes in ion selectivity (Zheng & Sigworth, 1997). It was proposed that the last activation transitions occur at the ion selectivity filter, probably in each of the four subunits. Subconductances occur as the activation transitions in different subunits are not perfectly synchronized (e.g., the last concerted gating transition is not absolutely "concerted"), a phenomenon that was amplified by the T442S mutation at the selectivity filter. The involvement of the selectivity filter in activation gating has been shown in CNG channels and BK channels, among others (Chen & Aldrich, 2011; Chen, Yan, & Aldrich, 2014; Contreras et al., 2008; Cordero-Morales, Cuello, & Perozo, 2006; Cordero-Morales, Cuello, Zhao, et al., 2006; Flynn & Zagotta, 2001). It demonstrates that the selectivity filter is a dynamic instead of a static structure.

Can pore dilation in TRPV1 and TRPV3 be originated from their normal activation process? When capsaicin is used to activate TRPV1, the open probability reaches a level that approaches unity. The activation kinetics is very fast, reaching stable level in milliseconds, comparing to the seconds-to-minutes time scale for the development of pore dilation. It thus seems that pore dilation does not arise from the normal activation transitions, but rather represents a slow migration to a more stable open conformation. In this sense, the pore dilated state of TRPV1 is similar to the C-type inactivated state of the Shaker potassium channel (except that the channel pore becomes conducting larger ions and molecules instead of closing down). This type of slow gating is possible if the stable pore-dilated open state is separated from the normal open state by a large energy barrier, which may come from, for example, a need for the channel structures surrounding the pore to alter their conformations to accommodate the stable open pore conformation. Direct evidence for slow global conformational rearrangements in TRPV1 is not yet available. Nonetheless, the existence of seconds-to-minutes time scale transitions from the normal open state is important for understanding structural studies. As discussed earlier, the conformation of TRPV1 ion selectivity filter seen in the cryo-EM structures is atypical for channels with a homologous sequence. It is noticed that the short filter region with a large extracellular vestibule would be compatible with a dilatable pore to conduct large molecules (Cao et al., 2013). Can the cryo-EM structures represent a stable but hard to access conformation?

While the molecular mechanism underlying pore dilation remains to be investigated, this peculiar gating phenomenon presents an attractive practical usage: the dilated pore can serve as a highly specific pathway to deliver drug molecules into nociceptive neurons expressing high levels of TRPV1. QX-314, for example, is a well-known inhibitor for voltage-gated Na channels whose activation is required for the neuron to initiate action potential firings and the transmission of pain sensation. The QX-314/channel interaction occurs only at the intracellular side of the Na channel protein. Encouraging results suggest that selective inhibition of nociceptive neurons through the TRPV1 portal may be possible (Binshtok, Bean, & Woolf, 2007; Brenneis et al., 2014).

6. CONCLUDING REMARKS

The classical view of ion channels is that the ion permeation pore is a static structure while the surrounding channel structures are responsible for sensing and transmitting activation stimuli. Emerging evidence suggests that for thermoTRP channels the pore domain may be actively involved in sensing activation stimuli. Perhaps to accommodate the active role of the pore domain, the peripheral transmembrane structures of thermoTRP channels appear to be static and supportive. Whether a dynamic pore domain plays a role in temperature sensing remains to be determined.

ACKNOWLEDGMENTS

We are grateful to our lab members for helpful discussions, Fan Yang for assistance with rendering of protein structure images. Research in the Zheng lab is supported by a grant from National Institutes of Health (R01NS072377). Linlin Ma is supported by an Australian National Health and Medical Research Council fellowship.

REFERENCES

Aggarwal, S. K., & MacKinnon, R. (1996). Contribution of the S4 segment to gating charge in the Shaker K^+ channel. *Neuron, 16*(6), 1169–1177.

Ahern, G. P., Brooks, I. M., Miyares, R. L., & Wang, X. B. (2005). Extracellular cations sensitize and gate capsaicin receptor TRPV1 modulating pain signaling. *The Journal of Neuroscience, 25*(21), 5109–5116.

Ahern, G. P., Wang, X., & Miyares, R. L. (2006). Polyamines are potent ligands for the capsaicin receptor TRPV1. *The Journal of Biological Chemistry, 281*(13), 8991–8995.

Aneiros, E., Cao, L., Papakosta, M., Stevens, E. B., Phillips, S., & Grimm, C. (2011). The biophysical and molecular basis of TRPV1 proton gating. *EMBO Journal, 30*(6), 994–1002.

Banke, T. G. (2011). The dilated TRPA1 channel pore state is blocked by amiloride and analogues. *Brain Research, 1381*, 21–30.

Banke, T. G., Chaplan, S. R., & Wickenden, A. D. (2010). Dynamic changes in the TRPA1 selectivity filter lead to progressive but reversible pore dilation. *American Journal of Physiology. Cell Physiology, 298*(6), C1457–C1468.

Binshtok, A. M., Bean, B. P., & Woolf, C. J. (2007). Inhibition of nociceptors by TRPV1-mediated entry of impermeant sodium channel blockers. *Nature, 449*(7162), 607–610.

Bohlen, C. J., Priel, A., Zhou, S., King, D., Siemens, J., & Julius, D. (2010). A bivalent tarantula toxin activates the capsaicin receptor, TRPV1, by targeting the outer pore domain. *Cell, 141*(5), 834–845.

Boukalova, S., Marsakova, L., Teisinger, J., & Vlachova, V. (2010). Conserved residues within the putative S4-S5 region serve distinct functions among thermosensitive vanilloid transient receptor potential (TRPV) channels. *The Journal of Biological Chemistry, 285*(53), 41455–41462.

Brauchi, S., Orta, G., Salazar, M., Rosenmann, E., & Latorre, R. (2006). A hot-sensing cold receptor: C-terminal domain determines thermosensation in transient receptor potential channels. *The Journal of Neuroscience, 26*(18), 4835–4840.

Brenneis, C., Kistner, K., Puopolo, M., Jo, S., Roberson, D., Sisignano, M., et al. (2014). Bupivacaine-induced cellular entry of QX-314 and its contribution to differential nerve block. *British Journal of Pharmacology, 171*(2), 438–451.

del Camino, D., & Yellen, G. (2001). Tight steric closure at the intracellular activation gate of a voltage-gated K(+) channel. *Neuron, 32*(4), 649–656.

Cao, E., Liao, M., Cheng, Y., & Julius, D. (2013). TRPV1 structures in distinct conformations reveal activation mechanisms. *Nature, 504*(7478), 113–118.

Cao, X., Ma, L., Yang, F., Wang, K., & Zheng, J. (2014). Divalent cations potentiate TRPV1 channel by lowering the heat activation threshold. *The Journal of General Physiology, 143*(1), 75–90.

Catterall, W. A. (2011). Voltage-gated calcium channels. *Cold Spring Harbor Perspectives in Biology, 3*(8), a003947.

Catterall, W. A. (2014). Structure and function of voltage-gated sodium channels at atomic resolution. *Experimental Physiology, 99*(1), 35–51.

Chen, X., & Aldrich, R. W. (2011). Charge substitution for a deep-pore residue reveals structural dynamics during BK channel gating. *The Journal of General Physiology, 138*(2), 137–154.

Chen, J., Kim, D., Bianchi, B. R., Cavanaugh, E. J., Faltynek, C. R., Kym, P. R., et al. (2009). Pore dilation occurs in TRPA1 but not in TRPM8 channels. *Molecular Pain, 5*, 3.

Chen, X., Yan, J., & Aldrich, R. W. (2014). BK channel opening involves side-chain reorientation of multiple deep-pore residues. *Proceedings of the National Academy of Sciences USA, 111*(1), E79–E88.

Chuang, H. H., Neuhausser, W. M., & Julius, D. (2004). The super-cooling agent icilin reveals a mechanism of coincidence detection by a temperature-sensitive TRP channel. *Neuron, 43*(6), 859–869.

Chung, M. K., Guler, A. D., & Caterina, M. J. (2005). Biphasic currents evoked by chemical or thermal activation of the heat-gated ion channel, TRPV3. *The Journal of Biological Chemistry, 280*(16), 15928–15941.

Chung, M. K., Guler, A. D., & Caterina, M. J. (2008). TRPV1 shows dynamic ionic selectivity during agonist stimulation. *Nature Neuroscience, 11*(5), 555–564.

Contreras, J. E., Srikumar, D., & Holmgren, M. (2008). Gating at the selectivity filter in cyclic nucleotide-gated channels. *Proceedings of the National Academy of Sciences USA, 105*(9), 3310–3314.

Cordero-Morales, J. F., Cuello, L. G., & Perozo, E. (2006). Voltage-dependent gating at the KcsA selectivity filter. *Nature Structural and Molecular Biology, 13*(4), 319–322.

Cordero-Morales, J. F., Cuello, L. G., Zhao, Y., Jogini, V., Cortes, D. M., Roux, B., et al. (2006). Molecular determinants of gating at the potassium-channel selectivity filter. *Nature Structural and Molecular Biology, 13*(4), 311–318.

Cordero-Morales, J. F., Gracheva, E. O., & Julius, D. (2011). Cytoplasmic ankyrin repeats of transient receptor potential A1 (TRPA1) dictate sensitivity to thermal and chemical stimuli. *Proceedings of the National Academy of Sciences USA, 108*(46), E1184–E1191.

Cui, Y., Yang, F., Cao, X., Yarov-Yarovoy, V., Wang, K., & Zheng, J. (2012). Selective disruption of high sensitivity heat activation but not capsaicin activation of TRPV1 channels by pore turret mutations. *The Journal of General Physiology, 139*(4), 273–283.

De-la-Rosa, V., Rangel-Yescas, G. E., Ladron-de-Guevara, E., Rosenbaum, T., & Islas, L. D. (2013). Coarse architecture of the transient receptor potential vanilloid 1 (TRPV1) ion channel determined by fluorescence resonance energy transfer. *The Journal of Biological Chemistry, 288*(41), 29506–29517.

Doyle, D. A., Morais Cabral, J., Pfuetzner, R. A., Kuo, A., Gulbis, J. M., Cohen, S. L., et al. (1998). The structure of the potassium channel: molecular basis of K^+ conduction and selectivity. *Science, 280*(5360), 69–77.

Flynn, G. E., & Zagotta, W. N. (2001). Conformational changes in S6 coupled to the opening of cyclic nucleotide-gated channels. *Neuron, 30*(3), 689–698.

Gavva, N. R., Klionsky, L., Qu, Y., Shi, L., Tamir, R., Edenson, S., et al. (2004). Molecular determinants of vanilloid sensitivity in TRPV1. *The Journal of Biological Chemistry, 279*(19), 20283–20295.

Gonzalez, C., Baez-Nieto, D., Valencia, I., Oyarzun, I., Rojas, P., Naranjo, D., et al. (2012). K(+) channels: function-structural overview. *Comprehensive Physiology, 2*(3), 2087–2149.

Gordon, S. E., & Zagotta, W. N. (1995). Localization of regions affecting an allosteric transition in cyclic nucleotide-activated channels. *Neuron, 14*(4), 857–864.

Grandl, J., Hu, H., Bandell, M., Bursulaya, B., Schmidt, M., Petrus, M., et al. (2008). Pore region of TRPV3 ion channel is specifically required for heat activation. *Nature Neuroscience, 11*, 1007–1013.

Grandl, J., Kim, S. E., Uzzell, V., Bursulaya, B., Petrus, M., Bandell, M., et al. (2010). Temperature-induced opening of TRPV1 ion channel is stabilized by the pore domain. *Nature Neuroscience, 13*(6), 708–714.

Hille, B. (2001). *Ion channels of excitable membranes*. Sunderland, Massachusetts: Sinauer Associates, Inc.

Horrigan, F. T., & Aldrich, R. W. (2002). Coupling between voltage sensor activation, Ca^{2+} binding and channel opening in large conductance (BK) potassium channels. *The Journal of General Physiology, 120*(3), 267–305.

Hoshi, T., Zagotta, W. N., & Aldrich, R. W. (1990). Biophysical and molecular mechanisms of Shaker potassium channel inactivation. *Science, 250*(4980), 533–538.

Hu, L., Santoro, B., Saponaro, A., Liu, H., Moroni, A., & Siegelbaum, S. (2013). Binding of the auxiliary subunit TRIP8b to HCN channels shifts the mode of action of cAMP. *The Journal of General Physiology, 142*(6), 599–612.

Jara-Oseguera, A., & Islas, L. D. (2013). The role of allosteric coupling on thermal activation of thermo-TRP channels. *Biophysical Journal, 104*(10), 2160–2169.

Jordt, S. E., & Julius, D. (2002). Molecular basis for species-specific sensitivity to "hot" chili peppers. *Cell, 108*(3), 421–430.

Jordt, S. E., Tominaga, M., & Julius, D. (2000). Acid potentiation of the capsaicin receptor determined by a key extracellular site. *Proceedings of the National Academy of Sciences USA, 97*(14), 8134–8139.

Kalia, J., & Swartz, K. J. (2013). Exploring structure-function relationships between TRP and Kv channels. *Scientific Reports, 3*, 1523.

Latorre, R., Brauchi, S., Orta, G., Zaelzer, C., & Vargas, G. (2007). ThermoTRP channels as modular proteins with allosteric gating. *Cell Calcium, 42*(4–5), 427–438.

Liao, M., Cao, E., Julius, D., & Cheng, Y. (2013). Structure of the TRPV1 ion channel determined by electron cryo-microscopy. *Nature, 504*(7478), 107–112.

Liu, Y., Holmgren, M., Jurman, M. E., & Yellen, G. (1997). Gated access to the pore of a voltage-dependent K^+ channel. *Neuron, 19*(1), 175–184.

Long, S. B., Tao, X., Campbell, E. B., & MacKinnon, R. (2007). Atomic structure of a voltage-dependent K^+ channel in a lipid membrane-like environment. *Nature, 450*(7168), 376–382.

Luebbert, M., Radtke, D., Wodarski, R., Damann, N., Hatt, H., & Wetzel, C. H. (2010). Direct activation of transient receptor potential V1 by nickel ions. *Pflugers Archiv: European Journal of Physiology, 459*(5), 737–750.

Lu, Z., Klem, A. M., & Ramu, Y. (2001). Ion conduction pore is conserved among potassium channels. *Nature, 413*(6858), 809–813.

Matta, J. A., & Ahern, G. P. (2007). Voltage is a partial activator of rat thermosensitive TRP channels. *The Journal of Physiology, 585*(Pt 2), 469–482.

Myers, B. R., Bohlen, C. J., & Julius, D. (2008). A yeast genetic screen reveals a critical role for the pore helix domain in TRP channel gating. *Neuron, 58*(3), 362–373.

Ohta, T., Imagawa, T., & Ito, S. (2008). Novel gating and sensitizing mechanism of capsaicin receptor (TRPV1): tonic inhibitory regulation of extracellular sodium through the external protonation sites on TRPV1. *The Journal of Biological Chemistry, 283*(14), 9377–9387.

Owsianik, G., Talavera, K., Voets, T., & Nilius, B. (2006). Permeation and selectivity of TRP channels. *Annual Review of Physiology, 68*, 685–717.

Puopolo, M., Binshtok, A. M., Yao, G. L., Oh, S. B., Woolf, C. J., & Bean, B. P. (2013). Permeation and block of TRPV1 channels by the cationic lidocaine derivative QX-314. *Journal of Neurophysiology, 109*(7), 1704–1712.

Riera, C. E., Vogel, H., Simon, S. A., & le Coutre, J. (2007). Artificial sweeteners and salts producing a metallic taste sensation activate TRPV1 receptors. *American Journal of Physiology. Regulatory, Integrative, Comparative Physiology, 293*(2), R626–R634.

Ryu, S., Liu, B., Yao, J., Fu, Q., & Qin, F. (2007). Uncoupling proton activation of vanilloid receptor TRPV1. *The Journal of Neuroscience, 27*(47), 12797–12807.

Salazar, H., Jara-Oseguera, A., Hernandez-Garcia, E., Llorente, I., Arias, O., II, Soriano-Garcia, M., et al. (2009). Structural determinants of gating in the TRPV1 channel. *Nature Structural and Molecular Biology, 16*(7), 704–710.

Schoppa, N. E., McCormack, K., Tanouye, M. A., & Sigworth, F. J. (1992). The size of gating charge in wild-type and mutant Shaker potassium channels. *Science, 255*(5052), 1712–1715.

Schoppa, N. E., & Sigworth, F. J. (1998). Activation of Shaker potassium channels. III. An activation gating model for wild-type and V2 mutant channels. *The Journal of General Physiology, 111*(2), 313–342.

Siemens, J., Zhou, S., Piskorowski, R., Nikai, T., Lumpkin, E. A., Basbaum, A. I., et al. (2006). Spider toxins activate the capsaicin receptor to produce inflammatory pain. *Nature, 444*(7116), 208–212.

Sigworth, F. J. (1994). Voltage gating of ion channels. *Quarterly Reviews of Biophysics, 27*(1), 1–40.

Susankova, K., Ettrich, R., Vyklicky, L., Teisinger, J., & Vlachova, V. (2007). Contribution of the putative inner-pore region to the gating of the transient receptor potential vanilloid subtype 1 channel (TRPV1). *The Journal of Neuroscience, 27*(28), 7578–7585.

Tominaga, M., Caterina, M. J., Malmberg, A. B., Rosen, T. A., Gilbert, H., Skinner, K., et al. (1998). The cloned capsaicin receptor integrates multiple pain-producing stimuli. *Neuron, 21*(3), 531–543.

Tousova, K., Vyklicky, L., Susankova, K., Benedikt, J., & Vlachova, V. (2005). Gadolinium activates and sensitizes the vanilloid receptor TRPV1 through the external protonation sites. *Molecular and Cellular Neurosciences, 30*(2), 207–217.

Uysal, S., Cuello, L. G., Cortes, D. M., Koide, S., Kossiakoff, A. A., & Perozo, E. (2011). Mechanism of activation gating in the full-length KcsA K^+ channel. *Proceedings of the National Academy of Sciences USA, 108*(29), 11896–11899.

Virginio, C., MacKenzie, A., Rassendren, F. A., North, R. A., & Surprenant, A. (1999). Pore dilation of neuronal P2X receptor channels. *Nature Neuroscience, 2*(4), 315–321.

Vlachova, V., Teisinger, J., Susankova, K., Lyfenko, A., Ettrich, R., & Vyklicky, L. (2003). Functional role of C-terminal cytoplasmic tail of rat vanilloid receptor 1. *The Journal of Neuroscience, 23*(4), 1340–1350.

Voets, T., Owsianik, G., Janssens, A., Talavera, K., & Nilius, B. (2007). TRPM8 voltage sensor mutants reveal a mechanism for integrating thermal and chemical stimuli. *Nature Chemical Biology, 3*(3), 174–182.

Winter, Z., Buhala, A., Otvos, F., Josvay, K., Vizler, C., Dombi, G., et al. (2013). Functionally important amino acid residues in the transient receptor potential vanilloid 1 (TRPV1) ion channel–an overview of the current mutational data. *Molecular Pain, 9*, 30.

Xu, Y., Ramu, Y., & Lu, Z. (2010). A shaker K^+ channel with a miniature engineered voltage sensor. *Cell, 142*(4), 580–589.

Yang, F., Cui, Y., Wang, K., & Zheng, J. (2010). Thermosensitive TRP channel pore turret is part of the temperature activation pathway. *Proceedings of the National Academy of Sciences USA, 107*(15), 7083–7088.

Yang, F., Ma, L., Cao, X., Wang, K., & Zheng, J. (2014). Divalent cations activate TRPV1 through promoting conformational change of the extracellular region. *The Journal of General Physiology, 143*(1), 91–103.

Yang, Y., Yan, Y., & Sigworth, F. J. (1997). How does the W434F mutation block current in Shaker potassium channels? *The Journal of General Physiology, 109*(6), 779–789.

Yao, J., Liu, B., & Qin, F. (2010). Kinetic and energetic analysis of thermally activated TRPV1 channels. *Biophysical Journal, 99*(6), 1743–1753.

Yao, J., Liu, B., & Qin, F. (2011). Modular thermal sensors in temperature-gated transient receptor potential (TRP) channels. *Proceedings of the National Academy of Sciences USA, 108*(27), 11109–11114.

Yifrach, O., & MacKinnon, R. (2002). Energetics of pore opening in a voltage-gated K(+) channel. *Cell, 111*(2), 231–239.

Zagotta, W. N., Hoshi, T., & Aldrich, R. W. (1990). Restoration of inactivation in mutants of Shaker potassium channels by a peptide derived from ShB. *Science, 250*(4980), 568–571.

Zagotta, W. N., Hoshi, T., & Aldrich, R. W. (1994). Shaker potassium channel gating. III: evaluation of kinetic models for activation. *The Journal of General Physiology, 103*(2), 321–362.

Zheng, J. (2006). Patch fluorometry: shedding new light on ion channels. *Physiology (Bethesda, Md.), 21*, 6–12.

Zheng, J. (2013). Molecular mechanism of TRP channels. *Comprehensive Physiology, 3*(1), 221–242.

Zheng, J., & Sigworth, F. J. (1997). Selectivity changes during activation of mutant Shaker potassium channels. *The Journal of General Physiology, 110*(2), 101–117.

Zheng, J., & Zagotta, W. N. (2003). Patch-clamp fluorometry recording of conformational rearrangements of ion channels. *Science's STKE, 2003*(176), PL7.

Zhong, L., Bellemer, A., Yan, H., Ken, H., Jessica, R., Hwang, R. Y., et al. (2012). Thermosensory and nonthermosensory isoforms of *Drosophila melanogaster* TRPA1 reveal heat-sensor domains of a thermoTRP Channel. *Cell Reports, 1*(1), 43–55.

Zhou, Y., Morais-Cabral, J. H., Kaufman, A., & MacKinnon, R. (2001). Chemistry of ion coordination and hydration revealed by a K^+ channel-Fab complex at 2.0 A resolution. *Nature, 414*(6859), 43–48.

CHAPTER TEN

Temperature-Sensitive Gating of Voltage-Gated Proton Channels

Yuichiro Fujiwara and Yasushi Okamura
Laboratory of Integrative Physiology, Graduate School of Medicine, Osaka University, Suita, Japan
*Corresponding author: E-mail: yokamura@phys2.med.osaka-u.ac.jp

Contents

Abstract

The voltage-gated proton channel (Hv) mediates robust proton transport down the proton electrochemical gradient. Hv is mainly expressed in immune cells, including neutrophils and macrophages, the physiological functions of which are temperature sensitive. In those cells, Hv plays key roles in the regulation of reactive oxygen species production and pH homeostasis. Proton transport through Hv is regulated by both the membrane potential and the pH difference across the cell membrane. Earlier studies showed that the properties of Hv, including proton conductance and gating, are highly temperature dependent. Hv consists of a voltage sensor domain involved in both voltage sensing and proton permeation and a C-terminal coiled coil region. Although the channel's activities are innate to the protomers, normally two protomers assemble as a dimer via interaction between C-terminal coiled coils. We recently discovered that

Current Topics in Membranes, Volume 74
ISSN: 1063-5823
http://dx.doi.org/10.1016/B978-0-12-800181-3.00010-5

the coiled-coil region of Hv dissociates at around room temperature, and that subtle changes in the coiled-coil region affect temperature-sensitive gating. In this chapter, we describe the physiological functions and molecular mechanisms of Hv, focusing mainly on the structure and thermosensitive properties of Hv.

1. INTRODUCTION

Cellular homeostasis and intracellular signals depend in part on the pH of their microenvironment, which is regulated by various membrane proteins, including proton pumps and transporters. What's more, different organelles have distinct pH optima that must be maintained; for example, lysosomes require an acidic pH for function (Casey, Grinstein, & Orlowski, 2010). The voltage-gated proton channel (Hv) mediates robust proton transport down the proton electrochemical gradient across biological membranes. The Hv current was first "discovered" in snail neurons (Thomas & Meech, 1982), and was later described in mammalian blood cells and lung epithelial cells (Decoursey, 2003). Hv was also recently shown to be present in cancer cells and phytoplankton (Taylor, Chrachri, Wheeler, Goddard, & Brownlee, 2011), and the biological functions of Hv have been most intensively studied in mammalian phagocytes (Decoursey, 2003; Demaurex & Petheo, 2005). In addition, it is now known that in eosinophils and neutrophils, proton transport through Hv is required to offset the charge imbalance created by the export of electrons for production of superoxide anion (O_2^-) by NADPH oxidase (Decoursey, 2003; Henderson, Chappell, & Jones, 1987).

In 2006, the molecular identity of Hv was established through the cloning of its cDNA from mouse and sea squirt using a bioinformatics approach. The molecule was named "voltage-sensor domain only protein" (VSOP) because its structure consists of only a voltage sensor domain (VSD) similar to that from other voltage-gated ion channels (Sasaki, Takagi, & Okamura, 2006). In addition, a human VSOP ortholog was independently identified and named Hv1 (Ramsey, Moran, Chong, & Clapham, 2006). Hv1/VSOP has a simple molecular architecture consisting of four transmembrane segments and a cytoplasmic coiled-coil domain. The identification of these molecules enabled later study of their in vivo functions through the generation of Hv knockout mice. A gene encoding Hv1/VSOP is named HVCN1 (hydrogen voltage-gated channel). HVCN1-deficient mice exhibit significantly reduced production of reactive oxygen species in neutrophils (Okochi, Sasaki, Iwasaki, & Okamura, 2009; Ramsey et al., 2006), confirming that Hv1 plays an important role in phagocyte function. Further, recent studies using a heterologous expression system have provided details of the molecular properties and structure–function relationships of Hv1/VSOP (DeCoursey, 2013; Okamura, 2012). Given that most of the biophysical properties of Hv are innate to a single polypeptide consisting of only about 140 amino acids (Koch et al., 2008), it is one of the smallest cation channels so far identified from mammalian genomes.

A remarkable feature of Hv is its high temperature dependence (DeCoursey & Cherny, 1998). The Q_{10} for activation gating of Hv is larger than in other voltage-gated ion channels (Decoursey, 2003). On the other hand, phagocytes, which abundantly express Hv, have several temperature-dependent functions related to bacterial killing, including the release of bioactive substances, chemotaxis, apoptosis, and proliferation. Understanding the linkage between the biophysical properties of the protein and its physiological function is one of the main goals of ion channel physiology. In this chapter, we focus on the temperature-dependent mechanisms of Hv in the context of its cellular physiology. With respect to nomenclature, throughout this chapter Hv is used as a general term for the voltage-gated proton channel, Hv1/VSOP is used for the specific Hv protein, and HVCN1 is the gene encoding Hv1/VSOP protein.

2. PROPERTIES AND PHYSIOLOGICAL FUNCTIONS OF HV

Many excellent reviews are available on the properties of Hv (DeCoursey, 2003; DeCoursey, 2013; Demaurex & El Chemaly, 2010; Okamura, 2012). In this section, we will briefly summarize the facts so far established.

2.1 Gating, permeation, and pharmacology

Like other voltage-gated ion channels, Hv exhibits voltage-dependent gating (Figure 10.1); however, its conductance–voltage relationship is shifted by changes in either the intracellular or extracellular pH. Although Hv has a high threshold voltage for activation, its voltage dependence is shifted according to the difference between the intracellular and extracellular pH (Figure 10.2); that is, when the intracellular side is acidic or the extracellular side alkaline, Hv can be activated at membrane potentials within the physiological range. This arrangement gives Hv the unique property of vectorial proton transport without direct utilization of ATP. Thanks to

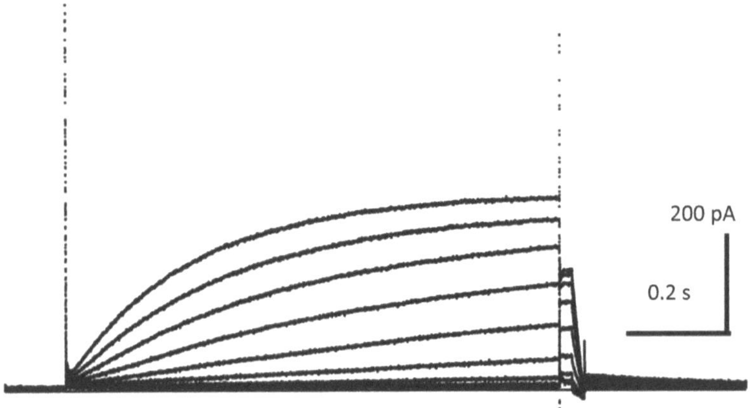

Figure 10.1 Representative traces of currents through Hv in a mouse neutrophil voltage was stepped up to 180 mV from a holding potential of −60 mV. $pH_o/pH_i = 7.0/6.0$. NMDG-methanesulfonate solution was used. *(Reproduced from Okochi et al. (2009).)* Currents were recorded at room temperature.

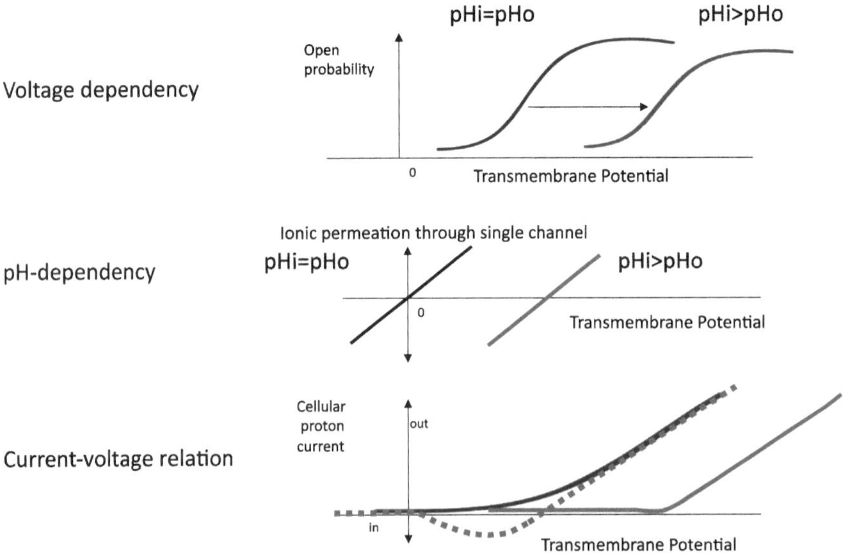

Figure 10.2 Scheme of the pH-dependent shift in the voltage threshold of Hv. The voltage dependency of the channel activation is shifted in manner reflecting the pH difference across the cell membrane (top). This makes the reversal potential always more negative than the activation threshold (middle). Consequently, proton flow is always outward, irrespective of the pH environment (bottom). For simplicity, it is predicted here that single channel amplitude is linearly related to membrane potential, which has not yet been experimentally verified due to small size of single channel currents. It should also be mentioned that voltage dependence is shifted to a level more negative range than the reversal potential for protons, causing inward proton current under some circumstances such as activation of mammalian phagocytes (Banfi et al., 1999). *Reproduced from Okamura, 2007. Copyright by Springer.*

the nature of the pH-sensitivity of its voltage-dependent gating, Hv is opened only when the electrochemical gradient drives an outward proton current. The proton selectivity of Hv is nearly perfect and is the highest among proton transporting proteins. In addition, the proton transfer rate is also the highest among proton transporting proteins.

An important feature of Hv gating is its high heat sensitivity. The Q_{10} of its activation kinetics is about 6, which is significantly larger than the two to four of other voltage-gated ion channels (DeCoursey & Cherny, 1998). Like the gating, Hv conductance is also heat sensitive. The Q_{10} for proton conductance is estimated to be about 2.0 (DeCoursey & Cherny, 1998).

Zinc is the most frequently used Hv inhibitor. Zinc slows Hv gating, most likely by inhibiting the motion of the voltage sensor (Cherny & DeCoursey, 1999; DeCoursey, 2003). Hanatoxin, a toxin derived from tarantula, binds to the VSD of voltage-gated ion channels and is known to suppress Hv activity (Alabi, Bahamonde, Jung, Kim, & Swartz, 2007). The guanidine-related reagent, 2-guanidinobenzimidazole, is an open channel blocker, which binds to the ion conduction pathway of Hv in the open state (Hong, Pathak, Kim, Ta, & Tombola, 2013). Hv activity is enhanced by unsaturated fatty acids, including arachidonic acids (Cherny, Henderson, Xu, Thomas, & DeCoursey, 2001) and endocannabinoids (Lishko, Botchkina, Fedorenko, & Kirichok, 2010). The absence of endocannabinoid receptors in sperm and the enhancement of Hv currents in human sperm by anandamide suggest the action of unsaturated fatty acids on Hv is a direct effect on Hv1/VSOP (Lishko et al., 2010).

2.2 Physiological functions of Hv

Native Hv currents have been described in murine and human blood cells (neutrophils, eosinophils, basophils, mast cells, macrophages, dendritic cells, T-lymphocytes, B-lymphocytes), murine microglia, murine alveolar cells, human airway epithelium, murine osteoclasts, human cardiac fibroblasts, and human sperm. Hv currents have also recently been recorded from human cancer cells (Wang, Li, Wu, Che, & Li, 2012; Wang, Wu, Li, Zhang, & Li, 2013) and, in amphibians, they have been recorded from oocytes (Barish & Baud, 1984) and renal epithelial cells (Gu & Sackin, 1995).

Neutrophils from HVCN1-deficient mice exhibit diminished production of O_2^- (Okochi et al., 2009) and hydrogen peroxide (H_2O_2) (El Chemaly et al., 2010), confirming the previously proposed compensatory function of Hv currents to offset imbalances caused by electron transfer by NADPH oxidase (Figure 10.3). In fact, membrane potentials are more

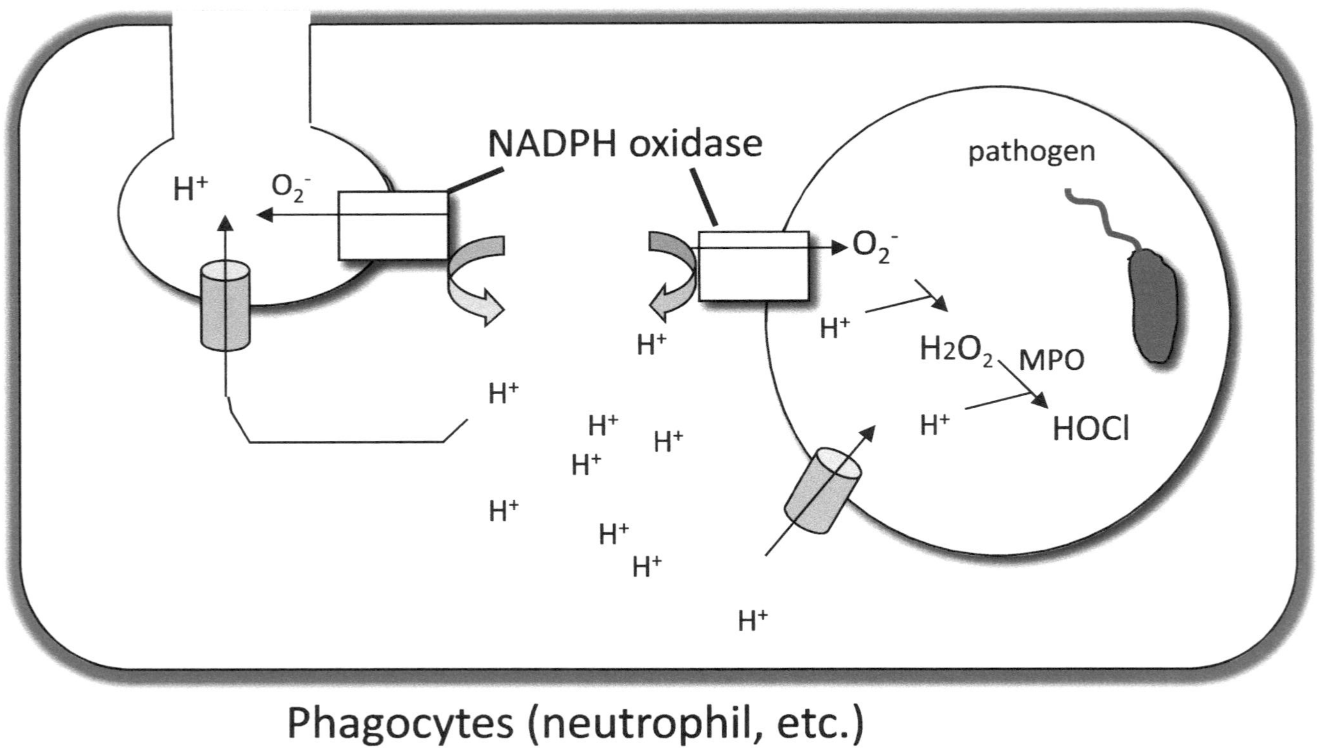

Figure 10.3 Physiological coupling between Hv and NADPH oxidase and NOX2 in the neutrophil membrane. Hv is localized in both the cell membrane and in phagosomes. NADPH oxidase transfers electrons across the cell membrane and phagosomal membranes resulting in the accumulation of protons in the cytoplasm. This leads to charge imbalance that depolarizes the cell membrane and makes the cytoplasm acidic. Hv cancels this charge imbalance and maintains the cytoplasmic pH neutral. Hv also provides protons, which are substrates for myeloperoxidase producing HOCl. Within phagosomes, protons are also required to make H_2O_2 from O_2^-.

depolarized in neutrophils from HVCN1-deficient mice than in those from wild-type mice (El Chemaly et al., 2010). In addition, neutrophils from HVCN1-deficient mice show less motility than normal neutrophils, probably due to a reduction in calcium influx, reflecting a smaller driving force (El Chemaly et al., 2010).

Hv function has also been studied in B-lymphocytes. Studies of HVCN1 knockout mice suggest Hv1/VSOP plays a role in antibody production through the regulation of intracellular signaling via protein phosphorylation (Capasso et al., 2010). Antigen binding to B-cell receptors (BCRs) leads to the phosphorylation of immunoreceptor tyrosine-based activation motifs in the Ig-α/β heterodimer and spleen tyrosine kinase (Syk) Syk binding to BCR, which amplifies the receptor's signaling through recruitment and activation of additional Syk molecules. Syk is negatively regulated by a SH2 domain-containing protein-tyrosine phosphatase, SHP-1, the activity of which is suppressed by oxidation of a Cys residue in the enzyme active site upon exposure to reactive oxygen species (ROS). Hv1/VSOP is situated close to the BCR, where it facilitates the NADPH oxidase activity that suppresses the phosphatase activity of SHP, thereby sustaining the phosphorylated state of Syk (Capasso et al., 2010).

T-lymphocytes also express Hv1/VSOP, and T-lymphocytes from HVCN1-deficient mice exhibit reduced O_2^- production at the late phase following stimulation with phorbol-12-myristate-13-acetate, PMA. HVCN1-deficient mice also frequently exhibit nephritis (Sasaki et al., 2013). In fact, HVCN1 knockout mice often show enhanced production of autoantibodies in serum and immune complex deposition within glomeruli, which is characteristic of autoimmune disorders. Although few details on the mechanism underlying the autoimmune disorder phenotype of HVCN1-deficient mice are available, it appears to result in part from the dysregulated proliferation of activated T-lymphocytes, which may be related to the ROS level. Intriguingly, upon stimulation with PMA, T-lymphocytes show two phases of O_2^- production, the latter of which is completely absent in HVCN1-deficient mice (Sasaki et al., 2013).

Hv1/VSOP is also expressed in other immune cells, including eosinophils, macrophages, and dendritic cells, as well as in osteoclasts (personal communication to Dr. S. Tanaka). Electrophysiological studies of eosinophils have shown that Hv is involved in regulating NADPH oxidase activity (DeCoursey, Cherny, DeCoursey, Xu, & Thomas, 2001; Petheo, Maturana, Spat, & Demaurex, 2003). Hv is also expressed in mast cells, where it is involved in IgE-stimulated degranulation (Musset et al., 2008).

Microglia are macrophage-like cells residing in the brain and are the major cells in brain that express Hv (Okochi et al., 2009). ROS production in microglia is Hv dependent, and is an important contributor to neuronal cell death following brain ischemia. Consequently, HVCN1 knockout mice exhibit milder neuronal cell death following brain ischemia (Wu et al., 2012). On the other hand, the function of microglial Hv in a physiological context remains unknown.

Both mast cells and eosinophils secrete histamine and heparin in response to IgE binding. Although Hv currents have been recorded from these cells (Kuno, Kawawaki, & Nakamura, 1997; Musset et al., 2008), the expression of Hv1/VSOP has not yet been confirmed. In addition, one electrophysiological study (Musset et al., 2008) has provided evidence that Hv is involved in regulating IgE-mediated release of histamine from basophils.

Recent patch clamp recordings from human sperm revealed robust Hv currents across the membrane of the principal piece (Lishko et al., 2010). In has been proposed that Hv regulates the initiation of motility in human sperm. In the epididymis, sperm is exposed to zinc ions at mM concentrations, which inhibits Hv activity, thereby preventing the premature activation of sperm motility (Lishko et al., 2010). When the sperm is released into the female reproductive tract, which contains little zinc, Hv escapes its inhibition, and proton efflux through the Hv channel leads to an increase of intracellular pH. This increase in intracellular pH in turn triggers motility in part by the activating CatSper, an alkalization-activated calcium channel (Lishko et al., 2012). It has also been shown that human sperm express the NADPH oxidase NOX5, which coimmunoprecipitates with Hv1/VSOP, suggesting the two proteins colocalize (Musset et al., 2012). Analogous to the relationship between Hv and NOX2 in immune cells, Hv plays a key role in regulating ROS production in sperm by offsetting the charge imbalance otherwise produced by NOX5 activity. Moreover, given that ROS stimulates sperm motility, it appears that Hv contributes to the regulation of sperm motility through effects on both ROS production and intracellular pH.

3. THERMOSENSITIVE FUNCTIONS OF CELLS EXPRESSING HV

Hv is expressed in cell types whose function is highly temperature sensitive. In the following, we summarize the temperature-sensitive functions of cells expressing Hv.

Like cells in other organs, mammalian blood cell function is optimal at around body temperature, although many physiological and pathological activities of blood cells are often accompanied by an increase of body temperature. Within areas of inflammation, for example, blood flow is increased through vessel dilatation and metabolism is increased, leading to a local temperature increase. In addition, global body temperature may be increased by the presence of fever-inducing substances. Increases in body temperature are also induced by various components of pathogens, such as lipopolysaccharide, which can act at the temperature-regulating center in the CNS. Most blood cell types are exposed to higher temperatures under pathological conditions, such as an infection, allergic response, and inflammation.

The primary function of neutrophils is to attack pathogens, eliminating them through phagocytosis. The effects of temperature changes on several aspects of neutrophil function have been examined. *In vitro*, the bactericidal activity of neutrophils is reportedly optimal at 37 °C; at temperatures higher than 37 °C, killing efficiency increases toward *Pneumococcus* but deceases toward *Escherichia coli* (Sebag, Reed, & Williams, 1977). The production of ROS by neutrophils, which is critical for pathogen elimination, is also temperature dependent and is generally optimal at 37 °C. Some reports have shown that ROS production is reduced at fever-like temperatures (e.g., 42 °C), though the temperature dependence of different ROS species varies; for example, O_2^- production is mildly reduced at 42 °C (by 30%), whereas hypochlorous acid (HOCl) production is more substantially reduced (by 85–87%) (Severns, Collins-Lech, & Sohnle, 1986). HOCl production in neutrophils depends on the activity of myeloperoxidase (MPO), the in vitro activity of which is optimal at body temperature (36–37 °C) and is diminished at lower or higher temperatures (Severns et al., 1986). However, the temperature dependence of MPO activity does not, by itself, explain the marked reduction in HOCl production by neutrophils at 42 °C.

Neutrophils are short lived and often undergo apoptosis in a process that is temperature dependent. At temperatures higher than 37 °C, apoptosis is facilitated by the temperature-dependent translocation of Bax to mitochondria (Pryde, Walker, Rossi, Hannah, & Haslett, 2000) or by abrogating TNF-α-induced NF-κB signaling (Salanova et al., 2005). Neutrophils are recruited to sites of infection or inflammation through "chemotaxis", a process in which cells travel up the concentration gradient of a chemical substance released from a pathogen or responder cells. The chemotactic movement of neutrophils is accompanied by morphological changes, including pseudopod extension, and the chemoattractant formyl–methionyl–leucyl–phenylalanine is known to induce pseudopod extension in a temperature-dependent manner (Chodniewicz, Alteraifi, & Zhelev, 2004). The proliferation of neutrophils is also temperature dependent, so that the recovery of blood neutrophils after X-ray irradiation is more rapid when body temperature is kept higher than 37 °C (Capitano et al., 2012).

Eosinophils are a prominent component of allergic and hypersensitivity disorders, and are thus an important factor in the pathogenesis of these diseases. When the temperature dependence of eosinophil NADPH oxidase activity was quantified based on electron currents measured using the whole cell patch clamp method (Morgan et al., 2003), the enzyme's activity was found to be optimal at 37 °C. Although the temperature dependence of the electron currents (i.e., the oxidase activity) was weak when the temperature was increased abruptly ($Q_{10} = 2.2$), the steady-state magnitude of the electron current was more sharply temperature dependent ($Q_{10} = 4.2$) when the temperature was gradually changed in the range between 20 °C and 30 °C. The sharpness of the voltage dependence on slow changes in temperature is probably due to superimposition of temperature-dependent processes other than the oxidase activity itself, such as protein expression and the intracellular recruitment and/or binding of interacting proteins. Analogous to neutrophil activation, gradual increases in temperature induce Rac-dependent recruitment of NADPH oxidase components to the cell membrane, where they are assembled into the active enzyme complex. This is also consistent with a temperature-priming effect in human neutrophils; that is, ROS production at 37 °C is enhanced when cells are briefly preincubated at 38 °C (Jozefowicz-Okonkwo & Nowak, 2004).

Mast cells release histamine in response to nervous stimulation or the binding of antigen to IgE present on the cell surface. The release of histamine increases linearly with increases in temperature from 15 °C to 27 °C (Lagunoff & Wan, 1974).

Many cancer cells undergo anoxic metabolism with lactic acidosis, and are thus subject to both extracellular and intracellular acidity. Hv1/VSOP was recently shown to be strongly expressed in breast, brain, and colon cancer cells (Wang et al., 2012, 2013). It is well known that cancer cells are less tolerant to higher temperatures than normal cells, which provides a basis for cancer thermotherapy.

4. TEMPERATURE DEPENDENCE OF NATIVE HV

As described above, the activation gating of Hv is highly temperature sensitive. It was recently shown that Hv is recruited from intracellular vesicles or is downregulated through endocytosis in a process dependent on long-term changes in intracellular pH (Sakai et al., 2013). This process is mediated by vesicle turnover, which involves the interaction of numerous enzymes and other proteins. Therefore, in cells expressing Hv, it is possible that the temperature-dependent regulation of the surface expression of the protein may underlie the temperature-sensitive changes in cellular function. In this chapter, we focus on the temperature dependence of two innate molecular properties of Hv, gating and conductance, leaving out the mechanisms underlying metabolism-based Hv turnover.

4.1 Gating kinetics

The temperature dependence of Hv activation kinetics was first described in bone marrow-derived mast cells (Kuno et al., 1997) and then intensively studied in rat alveolar type II cells using whole cell patch clamp (DeCoursey & Cherny, 1998). The time course of Hv currents in mildly depolarized cells is well described by a single exponential rise after a short delay. The Q_{10} of the kinetics of activation (tau act) and deactivation (tau deact) upon step depolarization is estimated to be six to nine at temperatures ranging from 10 °C to 40 °C. Notably, the Q_{10} of the Hv activation kinetics is two or three times larger than in other voltage-gated ion channels (DeCoursey & Cherny, 1998). The Q_{10} values differed slightly in other cell types that endogenously express Hv, including human neutrophils, THP-1 cells (a human macrophage cell line), BV-2 cells (murine microglia cell line). This likely reflects variation in the Hv molecules among these cell types, perhaps due to phosphorylation or a stoichiometric effect. Based on these findings, DeCoursey's group reached the conclusion that Hv channel opening involves multiple subunits, which was subsequently confirmed in studies of heterologous Hv1/VSOP expression (Gonzalez, Koch, Drum, & Larsson, 2010).

4.2 Conductance

Maximum Hv conductance increases about twofold with a 10 °C increase in temperature, which indicates the Q_{10} to be about 2 (DeCoursey & Cherny, 1998). This is consistent with the results of the pioneering study of Hv currents in snail neurons by Byerly and Suen ($Q_{10} = 2.1$) (Byerly & Suen, 1989), and are larger than 1.17, the value for proton conductivity in bulk water (Stokes & Robinson, 1948). The Arrhenius plot for Hv conductance is not linear, and the temperature dependence of the conductance was recently analyzed in detail using a rapid perfusion system with the GM1-R1 rat microglial cell line (Kuno et al., 2009). With this experimental design, temperature jumps could be elicited on a subsecond time scale using a double-barreled tube perfusion system driven by a piezoelectric device. This setup enabled the complete separation of the temperature dependence of the conductance from that of the gating. It also eliminated the possibility of temperature-dependent insertion or endocytosis of Hv proteins. Contributing to the Q_{10} of the conductance were the Q_{10}s of the channel and the access resistance. The access resistance component was estimated by measuring the current in low resistance solution, which contained sorbitol. The estimated Q_{10} of the pure proton permeation was found to be 2.8 at 5 °C and 2.2 at 45 °C. This temperature-dependent change is assumed to occur at the level of the single channel conductance, though measuring single channel currents of Hv is difficult due to their small size. In agreement with an earlier report (DeCoursey & Cherny, 1998), the Q_{10} was not constant over the entire temperature range, which suggests multiple processes underlie the temperature sensitivity of proton conduction. These values indicate a steeper temperature dependence than is seen with other voltage-gated ion channels, but are comparable with the values for ion pumps and transporters (DeCoursey, 2003). This suggests that the permeation mechanism is complex and that proton conduction through Hv does not occur as simple diffusion in bulk solution.

5. MOLECULAR STRUCTURE OF HV1/VSOP

5.1 Structural features of Hv1/VSOP

Hv1/VSOP was first identified using a bioinformatics approach (Sasaki et al., 2006; Ramsey et al., 2006), based on its sequence similarity to voltage sensing phosphatase (VSP) (Murata, Iwasaki, Sasaki, Inaba, & Okamura, 2005). It has four transmembrane segments with high sequence similarity to the VSD of VSP and other voltage-gated ion channels (Figure 10.4, left). The VSD of Hv1/VSOP harbors a signature motif in the fourth transmembrane segment, S4, comprised of a pattern of positively

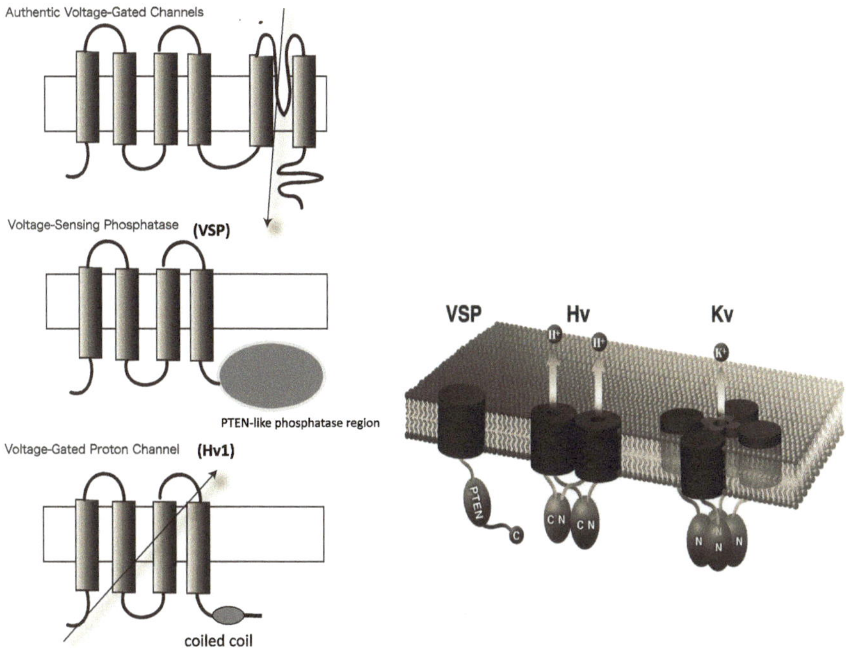

Figure 10.4 The diversity of voltage sensor domain-containing proteins. Authentic voltage-gated ion channels contain a voltage sensor domain (VSD) and a pore domain. Voltage-sensing phosphatase, VSP, is comprised of a VSD and a phosphatase. Hv contains a VSD and a cytoplasmic coiled coil motif, but lacks a pore domain and enzyme region. Voltage-gated ion channels are generally tetramers or pseudo-tetramers. VSP is a monomer. Hv is a dimer.

charged residues (three arginines are highly conserved among Hv orthologs) with two intervening hydrophobic residues. The other three transmembrane segments have acidic residues, some of which serve as counterions stabilizing the positive charges of S4 within the membrane or protein. Asp112 in S1 is known to be critical for proton-selective permeation: mutation of this site renders the channel anion-selective (Musset et al., 2011). A detailed topological profile of the transmembrane helices of mHv1 was recently studied based on the accessibility of maleimide reagent (Kurokawa & Okamura, 2014), which detects amino acid residues that face aqueous environment. Hv1/VSOP also has N-terminal and C-terminal cytoplasmic stretches. The C-terminus contains a coiled coil motif, which is required for dimerization of the channel (Fujiwara et al., 2012; Koch et al., 2008).

5.2 Variations of Hv1/VSOP protein

It appears that HVCN1 uniquely encodes Hv in neutrophils, microglia, B-lymphocytes and T-lymphocytes, as all of these cell types lack Hv currents in HVCN1 knockout mice. There is no known Hv1/VSOP paralog gene.

Hv1/VSOP orthologs have also been identified in various nonmammalian species, including the sea squirt (Sasaki et al., 2006), dinoflagellate (Smith et al., 2011), and sea urchin (unpublished results). Moreover, orthologs from dinoflagellate (Smith et al., 2011), sea urchin (unpublished result), ascidian (*Ciona intestinalis*) (Sasaki et al., 2006), and marine coccolithophorid phytoplankton (Taylor et al., 2011) have proven functional in heterologous expression systems, with the different orthologs exhibiting different activation kinetics and zinc sensitivities. In addition, mammalian Hv1/VSOP exhibits diversity through its phosphorylation and variation in molecular size reflecting different translation start sites (Capasso et al., 2010). The functional significance of the alternative translation start sites remains unknown.

During respiratory bursts by neutrophils and eosinophils, Hv currents increase with a negative shift in the voltage dependence and an acceleration of activation gating; this is known as the "enhanced gating mode." Similarly enhanced Hv currents can be evoked by treating LK35.2 cells, a B cell hybridoma, heterologously expressing hHv1 with PMA. It is also known that Thr29, situated in the N-terminal cytoplasmic region, is phosphorylated by protein kinase C, and that Ser97, in the N-terminal cytoplasmic region, is also reportedly subject to phosphorylation. With mutation of Thr29, there is a shift in the voltage dependence of the channel activation and the enhancement of the current magnitude is somewhat attenuated (Musset, Capasso, et al., 2010), suggesting phosphorylation of these sites may underlie the "enhanced gating mode" of native Hv currents. Single-nucleotide polymorphism of human Hv1 is known to alter Met91 to a Thr (Thr71 in the short form) in the N-terminal cytoplasmic region. This change reduces the pH sensitivity of Hv currents in airway epithelia (Iovannisci, Illek, & Fischer, 2010). These findings suggest the N-terminal cytoplasmic region is important for determining the activation kinetics, though it remains unclear how this region regulates channel gating.

5.3 Dimeric stoichiometry and its molecular basis

Western blot analysis of mouse and human Hv1/VSOP using cross-linkers showed two bands, suggesting Hv1/VSOP is a dimer (Koch et al., 2008;

Lee, Letts, & Mackinnon, 2008), while FRET (Fluorescence Resonance Energy Transfer) experiments demonstrated that sea squirt Hv1/VSOP channels (named Ci-VSOP or CiHv1) are expressed as multimers (Koch et al., 2008). The N- and C-termini of Hv1/VSOP comprise small cytoplasmic domains. The C-terminal cytoplasmic domain, which consists of ~55 residues, is well conserved among species (Figure 10.5(B)). Beginning ~10 residues C-terminal to the last transmembrane domain (S4), there is a distinctive heptad repeat pattern $(abcdefg)_n$, within which the "a" and "d" positions are hydrophobic amino acids. This is typical of a coiled-coil architecture (Figure 10.5(B)). Coimmunoprecipitation assays indicated that the C-terminal domain is required for channel dimerization (Koch et al., 2008). Consistent with that idea, in cross-linking assays using disuccinimidyl suberate (DSS), the dimer fraction is markedly reduced, again indicating the importance of the C-terminal region for dimeric assembly (Fujiwara et al., 2012; Koch et al., 2008).

Currents carried by a truncated form of Hv1/VSOP lacking the entire C-terminus continue to exhibit most of the characteristic properties of Hv

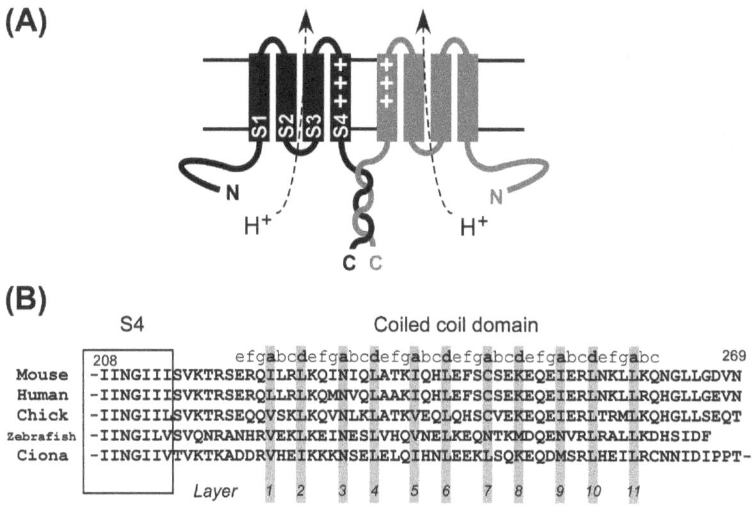

Figure 10.5 The C-terminal region of Hv is the dimer assembly domain. (A) Molecular design of Hv1/VSOP. Hv1/VSOP contains a cytoplasmic coiled-coil domain, which mediates dimerization. (B) Sequence alignment of the C-terminal cytoplasmic coiled-coil domain from various Hv species. The positions of the coiled-coil heptad repeat (abcdefg) are indicated above the alignment. Coiled-coil residues occupying hydrophobic "a" and "d" positions are highlighted. The layer position of each residue in the structure is indicated below the alignment.

currents, indicating the proton permeation pore is innate to the protomer (Koch et al., 2008; Tombola, Ulbrich, & Isacoff, 2008). On the other hand, the activation kinetics are more rapid than in the wild-type channel (see below for more details). This suggests the dimer stoichiometry of Hv is not essential for proton permeation but may be critical for channel gating.

The sedimentation equilibrium analysis of mHv1/VSOP using the purified protein demonstrated the C-terminal domain itself assembles as a dimer. Moreover, data obtained from analytical ultracentrifugation of the coiled-coil domain protein is perfectly fit by a single species model having the molecular mass of a dimer (Fujiwara et al., 2012). Thus, the dimeric nature of the coiled-coil underlies the assembly of the dimeric Hv.

The molecular architecture of the Hv1/VSOP coiled-coil domain was clarified by solving the X-ray crystal structures. The structures of the coiled-coil domains of human and mouse Hv1/VSOP were determined at 2.00 Å (Li et al., 2010) and 1.45 Å resolution (Fujiwara et al., 2012), respectively. Overall, the structure is that of a parallel left-handed, two-stranded coiled coil, ~22 Å wide and ~68 Å long (Figure 10.6(A)). The high-resolution crystal structure of Hv1/VSOP coiled coil reveals the following major features.

1. Intersubunit salt bridges and hydrogen bonds on the outer wall are observed in most coiled coils and reportedly stabilize the coiled-coil assembly (Acharya, Rishi, & Vinson, 2006; Acharya, Ruvinov, Gal, Moll, & Vinson, 2002; Howard, Clark, Holton, & Minor, 2007). They are missing from the coiled coil of Hv1/VSOP, however.
2. The dimer interface shows the typical parallel coiled-coil pattern, with the "a" and "d" positions forming alternate layers along the superhelical core (Figure 10.6(B)). The side chains of the hydrophobic resides positioned at "a"/"d" are packed symmetrically and diagonally in the hydrophobic core (Figure 10.6(B)). This well-ordered architecture of the hydrophobic residues in the core layers is thought to stabilize the coiled-coil assembly (Akey, Malashkevich, & Kim, 2001; Harbury, Zhang, Kim, & Alber, 1993).
3. An I/L core packing pattern, in which Ile/Leu residues are situated at the "a"/"d" position (Figure 10.6(A) and (B)), is periodically present along the length of the coiled coil (layers 1/2, 5/6 and 9/10). Studies of artificially-mutated GCN4 coiled coils suggest that this I/L core alignment prefers dimerization over other stoichiometries (Harbury et al., 1993). Moreover, altering the packing changes the oligomeric state of the full-length channel: channels harboring an I/I or L/L core

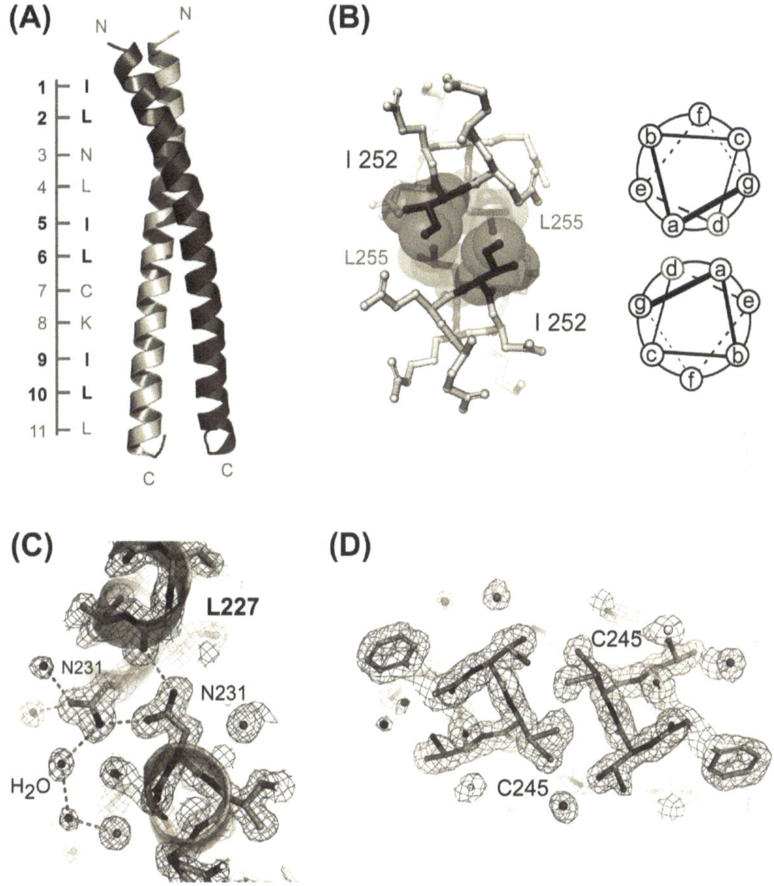

Figure 10.6 Crystal structure of the Hv coiled-coil domain. (A) The coiled-coil domain of mouse Hv1/VSOP (violet, PDB code: 3VMX) (Fujiwara et al., 2012). Layer numbers and the amino acid residues are indicated and Ile and Leu are highlighted. (B) Hydrophobic layers of the coiled-coil core. (B) Representative geometry of the symmetrically packed hydrophobic I/L-type coiled-coil core (left). Van der Waals spheres depicting the side chains of the "a" (red) and "d" (blue) layers on a ribbon backbone (gray) are shown. Highlighted in the hydrophobic core are two Ile residues in layer-9 (red) and two Leu residues in layer-10 (blue). Helical wheel representation of the core-packing arrangements found in the homodimeric coiled coil (right). Heptad repeat positions are labeled a-g. (C) Structure (stick models) and the 2Fo–Fc composite-omit maps in layer-3. Cartoon helix models are also shown to indicate the strand orientation. Red balls depict water oxygen atoms. Maps are contoured at 1.0 σ. Yellow dashed lines depict polar contacts. (D) Structure and 2Fo-Fc maps in layer-7. Maps are contoured at 1.0 σ. Numbers depict occupancies of sulfur atoms in the dual conformations of Cys residues. Neighboring Ser residues also assume dual conformations. *Figures adapted from Fujiwara et al. (2012).* (See the color plate.)

packing pattern are expressed as trimers, whereas channels harboring an L/I pattern are tetramers (Fujiwara, Kurokawa, et al., 2013). The Hv1/VSOP coiled coil is thus optimally designed for assembling dimer units.

4. Hydrophilic Asn residues (N231), not hydrophobic residues, comprise the core at layer-3 (Figure 10.6(C)). These are asymmetrically packed and form hydrogen bonds with water molecules and a carbonyl oxygen in a rational fashion (Figure 10.6(C)). This leads to twisting of the overall structure of the coiled-coil and bending at layer-3 (Figure 10.6(A)). Asn231 at the second "a" position is conserved among species, and its neighboring residues, at positions "g" and "b," are hydrophobic (Ile230 and Ile232) in mHv1/VSOP (Figure 10.5(B)).
5. Cys residues are present in layer-7 (Figure 10.2(D)), where they assume dual conformations, and the distance between each sulfur atom at the interface is 2.7 Å (Figure 10.6(D)). The crystal structures of the

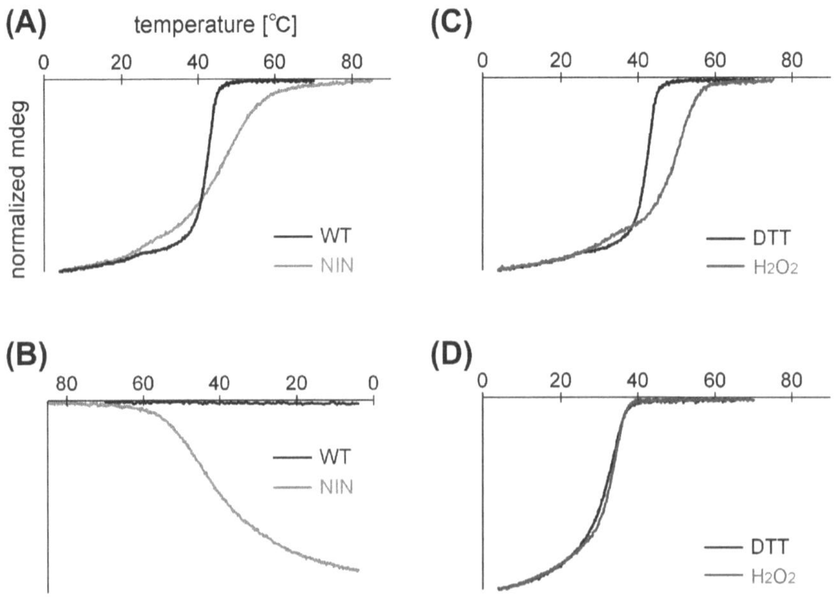

Figure 10.7 Thermal stability of the mHv1/VSOP coiled-coil domain. Melting temperatures of coiled-coil proteins were analyzed using circular dichroism (CD) spectrometry: (A) wild-type (WT), NIN mutant, (C) WT (reduced/oxidized), and (D) C245S mutant (reduced/oxidized). (B) The recoveries after the melting were also analyzed in WT (black) and NIN mutant proteins. *Data are taken from Fujiwara et al. (2012); Fujiwara, Takeshita, et al. (2013).*

coiled-coil domain in reduced, oxidized, and mutated (Cys245Ser, C245S) states show that the structure can change in response to the redox state (Fujiwara, Takeshita, Nakagawa, & Okamura, 2013). The distance between sulfur atoms is 4.4 Å under reducing conditions (Figure 10.3), but is 2.0 Å under oxidizing conditions (Figure 10.7), which is consistent with the length of the disulfide bond. The coiled-coil radius neighborhood of Cys245 also changes in response to the redox state. The apparent redox sensitivity of the coiled coil could be physiologically important for such cellular functions as phagocytosis in neutrophils or antibody production in B-lymphocytes.

There are several differences between the structure of the coiled coil of murine Hv1/VSOP (Fujiwara et al., 2012) from that of human Hv1/VSOP (Li et al., 2010), although their overall structures are similar. Several charged residues reportedly mediate the intersubunit interaction in the human Hv1/VSOP coiled coil, but this is not observed in the murine protein. In addition, the bent structure at the N-terminal end of the coiled-coil region is not clear in the human Hv1/VSOP coiled coil. This is in part due to the length at the N-terminal end of the crystallization construct, which is shorter than in murine Hv1/VSOP.

6. THERMAL STABILITY OF THE COILED-COIL DOMAIN AND THE THERMAL SENSITIVITY OF HV

6.1 Thermal stability of the coiled-coil domain

Circular dichroism (CD) spectroscopy showed that the coiled-coil domain is unfolded at around body temperature. The CD spectrum of the murine Hv1/VSOP coiled coil had minima at 208 and 222 nm, which is characteristic of proteins with a high alpha helix content (Figure 10.7(A)). The temperature dependence of the CD spectrum indicates a cooperative loss of structure, which is consistent with the presence of a well-folded complex with an extensive hydrophobic core within the structure of Hv/VSOP. The melting temperature (for unfolding), T_m, of the Hv1/VSOP coiled coil was 41 °C, and the slope of the melting transition was steeper than those of coiled coils from other proteins, which had T_ms greater than 60 °C (Deng et al., 2006; Hoshino, Yumoto, Yoshikawa, & Goto, 1997; Liu et al., 2007; Tsuruda, Julius, & Minor, 2006). The unfolding of the murine Hv1/VSOP coiled coil was irreversible (Figure 10.7(A) and (B)), indicating that Hv1/VSOP assembles into a dimer based on a thermally labile coiled coil.

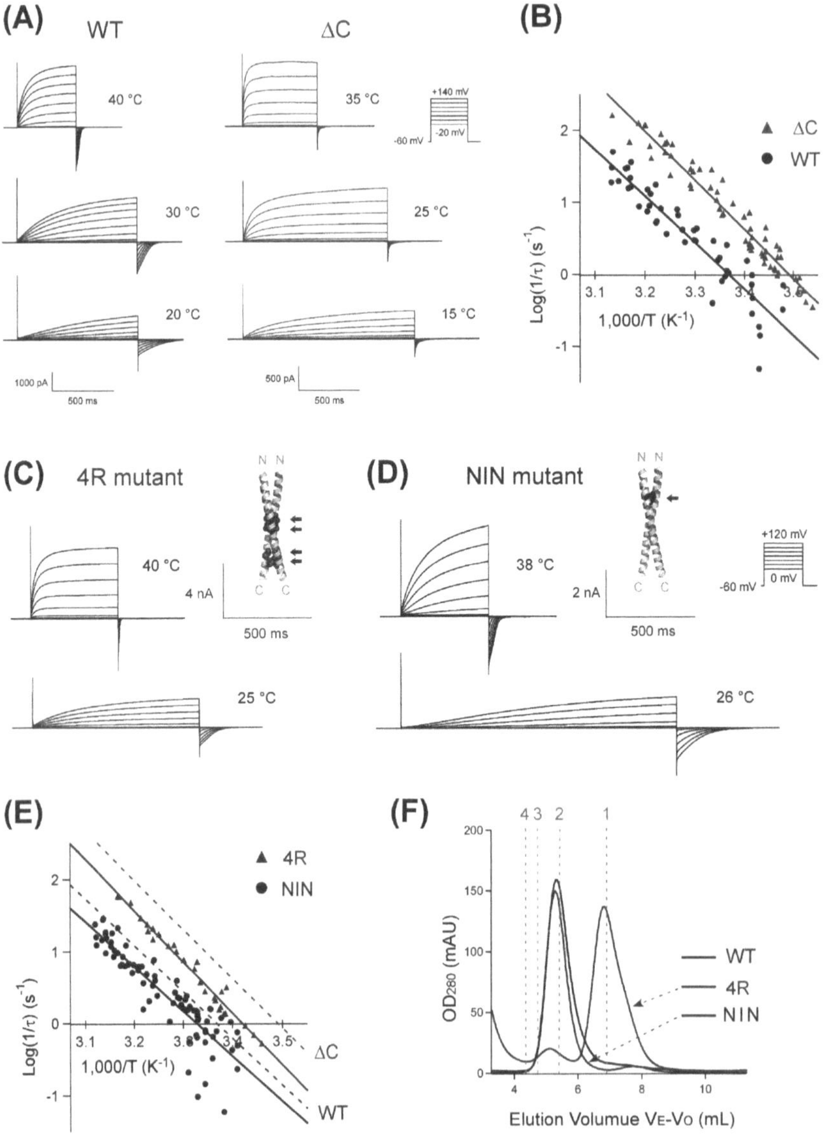

Figure 10.8 Effect of the coiled-coil domain on Hv1/VSOP activation kinetics by (A). Representative currents through WT (black) and ΔC (green) Hv1/VSOP recorded from HEK293T cells at the indicated temperatures (Fujiwara et al., 2012). The holding potential was −60 mV; step pulses from −20 mV to +140 mV were applied in 20-mV increments. The pulse protocol is indicated on the right. (B) Accumulated data from WT (black) and ΔC (green) channels, and linear regression lines showing the correlation of the activation kinetics with temperature. (C,D) Representative currents through 4R (red, in C) and NIN (blue, in D) mHv1/VSOP mutants recorded from HEK293T cells at the indicated temperatures. The pulse protocol is indicated on the right in D. Structure models are shown to indicate the positions of the mutations. (E) Accumulated data from 4R (red) and NIN (blue) mutant channels and their linear regression lines. Symbols used are shown in the figure. Dotted lines depict the linear regression lines. (F) Coiled-coil stoichiometry

The crystal structure showed that the coiled-coil domain twists and bends at layer-3, which is indicative of the weakness of the assembly. This is caused by the presence of the hydrophilic Asn231 residue within the core and the neighboring hydrophobic Ile230 and Ile232 residues (Figure 10.6(B)). In fact, this pattern is uncommon for coiled-coil motifs. Hydrophobic residues are most common at two opposing sites within the coiled coil ("a" and "d" position), and the surrounding residues are usually hydrophilic. Introducing the inverse Asn–Ile–Asn (NIN) sequence in place of the native Ile–Asn–Ile sequence at layer-3 increased the thermal stability of the coiled-coil region: T_m was increased to 45 °C in the NIN mutant, indicating a thermal stabilization of the coiled-coil assembly (Figure 10.7). In addition, the coiled-coil folding was reversible in the NIN mutant (Figure 10.7(A) and (B)), and the steepness of the temperature dependence was shallower than with the wild-type sequence (Figure 10.7(A)).

The thermal stability of the Hv1/VSOP coiled coil is also dependent on the redox conditions (Fujiwara, Takeshita, et al., 2013). Tm is shifted to a higher temperature (apparent Tm = 48–49 °C) by oxidation with H_2O_2 (Figure 10.7(C)). In the presence of DTT, the C245S mutant shows a lower Tm (30–31 °C) than the wild-type protein, and does not show a Tm shift with oxidation (31–32 °C) (Figure 10.7(D)). Thus, the intersubunit disulfide bond formed upon oxidation increases the thermal stability of the coiled-coil assembly. It should be noted that Hv is abundantly expressed in neutrophils and macrophages, in which the intracellular environment can become acutely oxidized due to production of ROS. However, the effect of redox-dependent regulation of the coiled-coil region of Hv on cellular functions remains unclear.

6.2 Relationship between channel activation and the coiled-coil domain

As described above, the Hv is known to exhibit high temperature dependence (DeCoursey & Cherny, 1998; Kuno et al., 1997), and is expressed in phagocytes and contributes to cellular events relevant to immune responses under conditions of local and whole-body fever (Capasso et al., 2010; Morgan et al., 2009; Okochi et al., 2009). The activation kinetics

determined using size exclusion chromatography monitored at 280 nm. For clear size separation, we used HMT-tagged proteins in these experiments. The horizontal axis shows elution volume (ml), and dotted lines indicate the predicted elution volumes for tetrameric, trimeric, dimeric and monomeric proteins. The elution volume VE was corrected for the void elution volume by subtracting the volume of blue dextran (VO). The oligomeric state of the main component was 2.29 for WT and 1.15 for 4R. *Data are taken from Fujiwara et al. (2012).* (See the color plate.)

of Hv1/VSOP evoked by depolarization was highly sensitive to the temperature (Figure 10.8(A)). The activation kinetics was further accelerated by the disruption of the channel dimerization by deletion of the C-terminal coiled-coil domain (Figure 10.8(A)) (Fujiwara et al., 2012; Koch et al., 2008). Activation kinetics reflects the transition between the open and closed states of the channel. Given that the assembly of the Hv1/VSOP coiled coil was thermally labile (Figure 10.7), the relationship between temperature and activation kinetics was studied in detail (Fujiwara et al., 2012) by whole-cell patch clamp recordings. The activation kinetics of the full-length (WT) channel at 40 °C was similar to that of ΔC at 25 °C, while the activation of WT channel at 30 °C was similar to that of ΔC channel at 15 °C (Figure 10.8(A)). This means that the deletion of the coiled-coil domain shifted the thermosensitivity relationship by about 15 °C without changing the slope (Figure 10.8(B)). These findings are distinct from the results of measurements of temperature sensitivity of gating kinetics and conductance of hHv1/VSOP (Musset, Smith, et al., 2010). Q_{10} of activation kinetics, deactivation kinetics and conductance was compared between WT hHv1/VSOP protein (dimer) and ΔC protein (monomer) at two or more temperature points in a range of 9–32 °C. Q_{10} of activation and deactivation kinetics of the ΔC channel was about half of that in dimer channel. The reason for this discrepancy from our results is not clear, but this could be due to species difference between mouse and human proteins, or difference in recording conditions.

Experiments of deleting the whole C-terminal segment cannot completely neglect the possibility whether not just disruption of dimerization but other effects of the deletion of the C-terminus might have altered channel properties. Instead of deleting the whole C-terminal segment, a mutation was introduced into the coiled-coil region so that the dimeric assembly is weakened or lost but the polypeptide of the C-terminal segment was left. Bulky hydrophilic Arg residues were introduced into the well-ordered hydrophobic core layers (layers 5, 6, 9 and 10 in Figure 10.6(A) and (B). Size exclusion chromatography showed that dimeric assembly of the coiled coil was suppressed by this "4R" mutation (Figure 10.8(F)). The 4R mutation accelerated the activation kinetics to some extent (Figure 10.8(C)), and shifted the relationship between activation speed and temperature by ~8 °C in the direction of the ΔC phenotype (Figure 10.8(E)). The NIN mutant was also analyzed as the construct with more stabilized assembly of the coiled coil. The NIN mutant showed a slower activation (Figure 10.8(D)) and shifted the activation temperature relationship by

~5 °C away from the WT phenotype (Figure 10.8(E)), which was consistent with the result of the CD spectrum (Figure 10.7(A)). Thus, the stability of the C-terminal domain determined the channel activation kinetics in a temperature-dependent manner and coiled-coil dimer assembly inhibited channel activation. Interestingly, the NAsn231 is conserved among all species of Hv so far identified (Figure 10.5(A)), suggesting that the pattern with hydrophilic residue in the coiled-coil core has a unique role in optimizing the stability or mobility within the coiled-coil domain for regulation of channel activation (Figure 10.8(D) and (E)).

These experiments clarified two features of the temperature-dependent activation of Hv1/VSOP. (1) The temperature dependence for activation (slope of the relationship between gating activation time constant and the reverse of temperature) of channels with the coiled-coil domain deleted (Figure 10.8(B)) and mutated (Figure 10.8(E)) is virtually identical to those of WT channels. (2) The temperature dependence of the unfolding of the coiled-coil domain is much steeper (Figure 10.7(A)) and occurs over a different temperature range than in the gating of the channel (Figure 10.8(E)). These imply that the coiled-coil domain does not determine the slope of temperature dependence but affects the energy of activation, which is critical for the rate of channel activation over a range of temperature. In other words, the higher energetic impact (the higher temperature) is required for transition to the channel open state in the wild-type channel due to the restriction by the coiled-coil dimeric assembly.

Multimerization of ion channel molecules is critical for their functionality: in the case of voltage-gated K^+ channel, four protomers assemble into a tetramer and the pore is formed in the center of tetramers. Four voltage sensors in one channel operate together via interaction in the pore with steep voltage dependence of channel activation (Bezanilla, 2000; Sigworth, 2003). Further, the assembly of different combinations of subunits forms varieties of heteromultimeric channels yielding functional diversity (Howard et al., 2007; Isacoff, Jan, & Jan, 1990; Nakajo & Kubo, 2008). In Hv, the dimeric coiled-coil assembly regulates the channel activation kinetics in a temperature-dependent manner, such that kinetics of depolarization-induced increase in H^+ conductance is optimized in the range of body temperature (cf. WT and ΔC in Figure 10.8(A) and (B)). This is one of the functional significances of the dimeric molecular design of the Hv. If transition between dimer and monomer occurs in native cells, the thermal stability of the coiled-coil domain may have physiological impact on cellular function. It has been reported that zinc sensitivity is distinct between monomer channels and dimeric channels. Given that native Hv currents are highly sensitive to zinc, the idea of dimer–monomer transition in native cell membranes has been questioned (Musset, Smith, et al., 2010). However, Hvs are also present in endomembranes such as phagosomes (Okochi et al., 2009). It is therefore not yet clear whether dimer–monomer transition occurs in physiological context.

The channel gating occurs in the transmembrane region, and the channel dimerization of Hv1/VSOP depends on the coiled-coil structure of which the dimeric assembly shows thermosensitivity. How is the thermosensor coiled-coil coupled to transmembrane gating? The ΔC channel shows a monomeric pattern of temperature dependent activation, while the WT channel assembles as a dimeric channel and showed a higher thermostability of the gating than ΔC channel (Figure 10.8(A) and (B)). The suppression of the dimerization by 4R mutation destabilized the thermostability of the gating (Figure 10.8(C)–(E)). One possible interpretation for these findings is that the mere spacing of two protomers with close proximity of the channel determines the thermostability of the gating. This idea could be tested by studying concatenated channels, which have a coupled transmembrane domain (Figure 10.9(A)). Two types of concatenated channel harboring/lacking the coiled-coil domain (tandem/ΔC-tandem) were made and their activation was analyzed (Figure 10.9(A)). The "tandem" channel showed similar activation kinetics to WT (Figure 10.9(A)). On the other hand, the "ΔC-tandem" channel showed faster activation kinetics than WT, and the thermosensitivity relationship was indistinguishable from that of the ΔC phenotype (Figure 10.9(A)). These results indicate that the mere positional proximity of two protomers is not sufficient and the structure of coiled-coil domain is necessary for the thermal stability of the WT channels.

The coiled-coil structure begins five residues downstream of the last transmembrane helix, S4. S4 harbors positively charged residues, playing critical role in sensing transmembrane voltage and translocating for channel opening. To address the role of the linker between the transmembrane region and the coiled coil, the linker structure was changed by substituting three amino acid residues, VKT, with Gly–Gly–Gly (GGG) or Ala–Ala–Ala (AAA). The GGG mutant is expected to confer more flexibility to the linker region. AAA is expected to form an α-helix, thus maintaining the stability of the linker region. The thermosensitivity plot (Figure 10.9(B)) of the GGG mutant was similar to that of the ΔC channel (monomer type), while that of the AAA mutant was similar to that of WT channel (dimer type). This suggests that linker

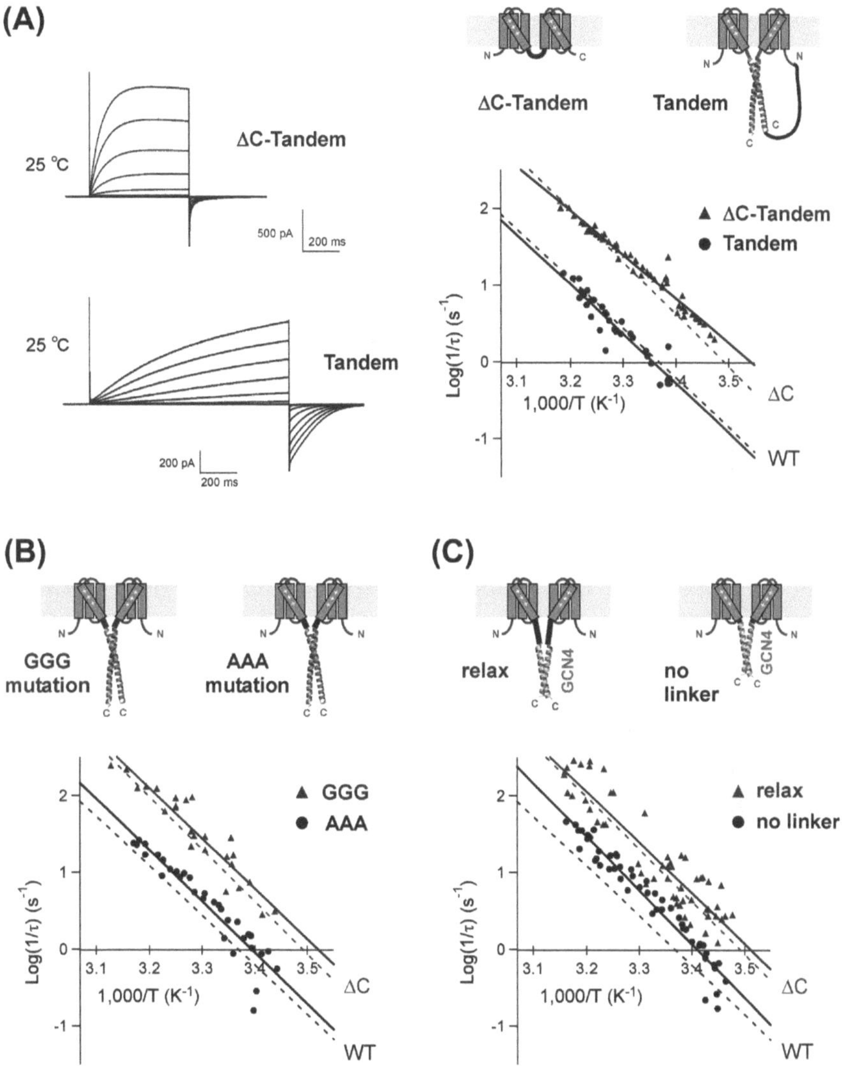

Figure 10.9 The temperature-dependent activation of murine Hv1/VSOP linker mutant channels. (A) Representative currents through murine Hv1/VSOP-Tandem (blue) and ΔC-Tandem (red) channels recorded from HEK293T cells at 25 °C (lower panel). Accumulated data from the Tandem (blue) and ΔC-Tandem (red) channels and the linear regression line (right panel) are shown; the dotted lines depict the linear regression lines for the data from WT (black) and ΔC (green) channels. (B) Accumulated data from the GGG (red) and AAA (blue) mutant channels and their linear regression lines; the dotted lines are the same as in panel (A). (C) Accumulated data from the GCN4 chimeric channel with a long linker (relax) (red) or with no linker (blue) and their linear regression lines; the dotted lines are the same as in panel (A). *Data are taken from Fujiwara et al. (2012).* (See the color plate.)

flexibility/stability plays an important role in the channel gating. This idea was also tested with the chimeric channel harboring the coiled-coil region from the GCN4 protein, a yeast leucine zipper type transcription factor (Fujiwara et al., 2012). The thermosensitivity of a GCN4 chimera with a six amino acid linker (designated "relax") showed the phenotypes of the ΔC mutant channel (Figure 10.9(C)), while that of a GCN4 chimera without intervening linker sequence ("no linker") exhibited the WT phenotype (Figure 10.9(C)). These findings indicate that the mere close positioning of two channel subunits bound by the coiled-coil domain is not sufficient to acquire the thermal stability of channel gating, but that the structural nature, most likely flexibility/stability, of the linker between the transmembrane region and the coiled-coil domain is important. The CD spectroscopy analysis of the isolated coiled-coil polypeptide harboring the amino acids of the WT sequence (Val-Lys-Thr), GGG and AAA support the idea that the linker region folds in an α-helix (Fujiwara et al., 2012). Further, the electrophysiological analysis of mutant channels harboring various lengths of the S4-coiled-coil linker demonstrated that the functional interaction between the coiled coil and the transmembrane domains occurs in the periodicity of α-helix (Fujiwara et al, 2014). These results also support the idea that the rigid structure of the α-helical linker plays a role in coupling between the coiled-coil domain and the transmembrane segment.

6.3 Cooperative channel gating and thermosensitivity

It has been established that gating of the Hv is cooperative: the gating of one protomer channel affects the gating of the other protomer within the dimeric unit (Gonzalez et al., 2010; Musset, Smith, et al., 2010; Tombola, Ulbrich, Kohout, & Isacoff, 2010). This has been proven by quantitative comparison between voltage sensor motion, detected by local fluorescence changes of a fluorophore conjugated to a single amino acid in the lipid–water interface of the voltage sensing transmembrane segment (Gonzalez et al., 2010), as well as the quantitative analysis of channel gating kinetics of mixtures of channels with distinct properties (Tombola et al., 2010). As described above, cooperative gating requires channel dimerization by the coiled-coil assembly, which also affects the thermosensitive gating.

The cooperative channel gating was examined by the limiting slope measurement using whole-cell patch clamp recordings (Figure 10.10). In this analysis, a small Hv current is measured in a voltage range of depolarization that gives only a low open probability to estimate the effective gating

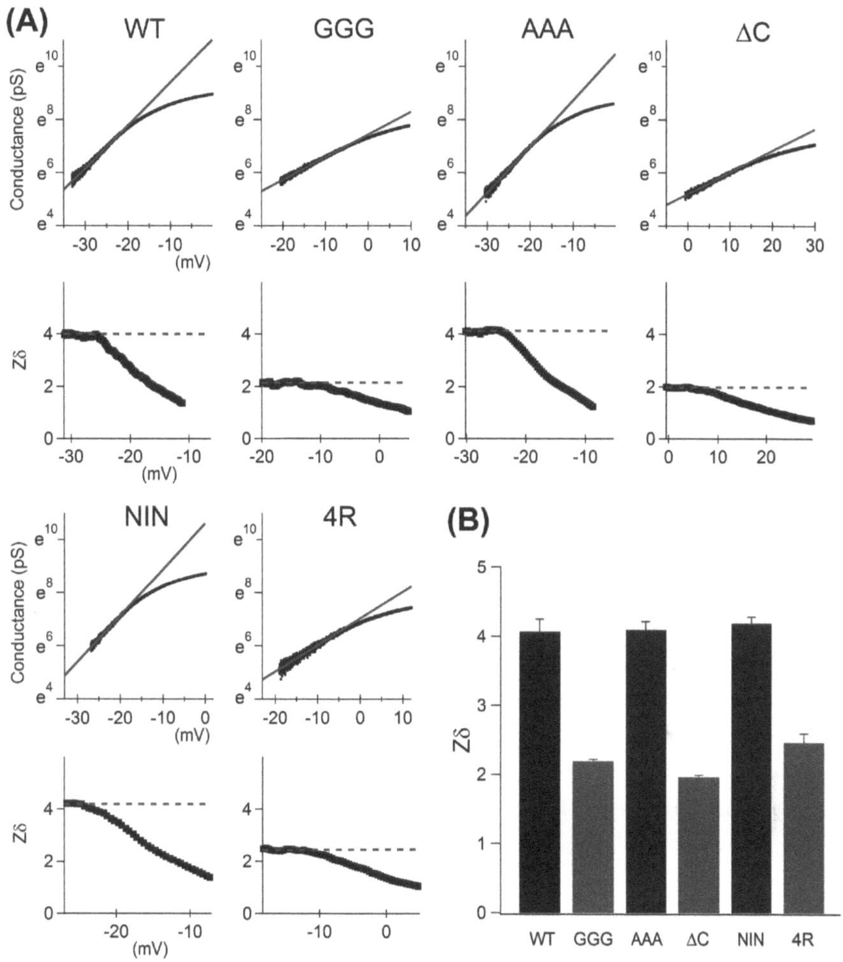

Figure 10.10 Gating cooperativity is dependent on the linker region between S4 and the coiled coil. (A) The representative chord conductance of WT and mutant (top) Hv1/VSOP and fittings of the slope factor Zδ for different voltages (bottom). The chord conductances were calculated based on the prepulse E_{Rev} and the current amplitude evoked by the slow ramp pulse. The slope factors were fitted linearly in the marginal conductance range (top and bottom). Threshold voltage varied among channels (WT = −33 mV, GGG = −20 mV, AAA = −29 mV, ΔC = −3.0 mV, NIN = −28 mV and 4R = −19 mV). (B) Effective gating charge in WT and the mutant channels. Bars depict means and SEM. *Data are taken from Fujiwara et al. (2012).*

charge. Current evoked at low depolarization was carefully used for analysis to avoid effects of H^+ accumulation, since Hv gating is known to be sensitive to pH (DeCoursey, 2003). The values of the effective gating charge (Zδ) were calculated from the slope of the conductance elicited during a ramp pulse. $Z\delta$ in the WT channel was 4.1, and it was 2.0 in ΔC channel (Figure 10.10(A)). This $Z\delta$ value in the WT channel, twofold that with of the monomeric ΔC channel, suggests that two protomers gate in a cooperative manner within dimer. These are consistent with results of limiting slope measurements from CiHv1 by Larsson's group (Gonzalez et al., 2010). Introduction of the GGG mutation into the linker segment decreases $Z\delta$ to half (2.2), while the AAA mutation does not affect the $Z\delta$ value (4.1). This indicates that dimerization is necessary but not sufficient to establish the gating cooperativity and that the linker structure is important for cooperative gating. Effective charge value $Z\delta$ of the GGG mutant and the AAA mutant paralleled the result of thermosensitive activation (Figures 10.9(B) and 10.10(A),(B)). This parallelism was also seen in the 4R mutant of the coiled-coil domain. 4R mutant exhibits unstable thermostability of dimerization and reduced value of $Z\delta$ (2.5), while the NIN mutation showed an intact gating cooperativity ($Z\delta = 4.2$). The results of the effective charge from the GGG/AAA linker mutant and 4R mutant well matched to those in the analysis on the temperature-dependent activation (Figures. 10.9 and 10.10), suggesting that the modulation of temperature-dependent activation by the coiled-coil assembly lies in the same biophysical principle to the cooperative gating of the channel. Thus the gating cooperativity of the Hv1/VSOP channel depends on a structural base consisting of the coiled-coil domain and the linker domain, and the coiled-coil assembly adjusts the range of temperature by elevating the energy of activation for the channel opening.

7. MODEL OF THERMOSENSITIVE CHANNEL GATING

Schematic illustrations of the gating properties of wild-type and mutant channels are shown in Figure 10.11. The stabilization of the temperature-dependent activation via the coiled coil is observed in the wild-type channel (Figure 10.11(A), red arrow) but not in the 4R or GGG channels, where the temperature threshold of the activation kinetics is set lower (Figure 10.11(B) and(C), red broken arrow). Gating cooperativity is also lost in the 4R and GGG mutants, where the 4R mutation destabilizes the coiled-coil assembly and the GGG mutation loosens the connection between S4 and the coiled coil domain (Figure 10.11(B) and (C)). The phenotypes of the temperature-dependent activation and the cooperative gating are the same in the AAA channel as in the wild-type channel. The ΔC channel lacks a coiled-coil domain and gating partner, and therefore regulation of the temperature-dependent activation is

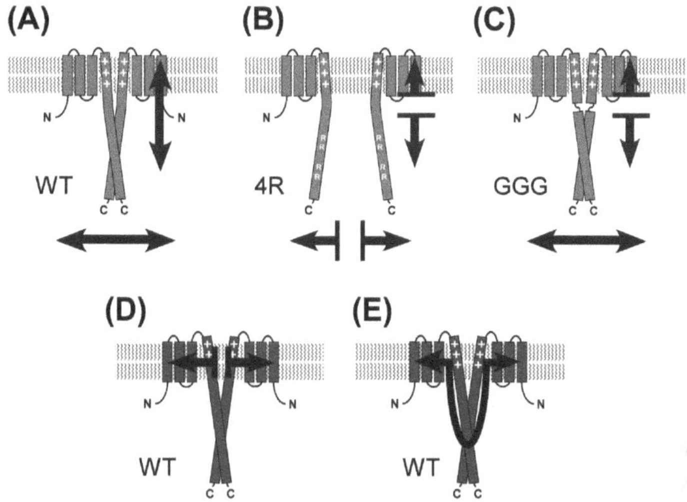

Figure 10.11 Schemes of the thermosensitive gating of Hv. (A) Schematic drawings of the temperature-dependent gating of the WT (or the AAA mutant) channel. Both the interaction between the coiled-coil region and voltage-sensing transmembrane segment via a rigid linker structure (red arrow) and dimerization mediated by the coiled coil (blue arrow) are important for the thermostability of Hv activation and its cooperative gating. (B) Model of the 4R mutant channel. Channel dimerization is loosened due to instability of the coiled-coil assembly (broken blue arrow). This results in instability and shifted temperature dependence of activation gating (red broken arrow). (C) Model of the GGG mutant channel. Although the channel is dimerized (blue arrow), the thermostability of channel activation is not mediated by the coiled-coil domain (red broken arrow) due to the looseness of the connection cause by GGG. (D) Model in which the two channel subunits directly interact in the transmembrane region. Gating movement derived from one subunit is directly transmitted to the other via physical contact in the transmembrane region. (E) Model in which the operations of the two channel subunits interact with one another via the cytoplasmic coiled-coil assembly. In this model, voltage-dependent motion of the transmembrane region is strained due to the nature of the cytoplasmic coiled-coil domain and the gating of the voltage sensor in one subunit affects that of the other through the continuous helical structure from coiled-coil domain. *Figures are adapted from Fujiwara et al. (2012).* (For interpretation of the references to color in this figure legend, the reader is referred to the online version of this book.)

destabilized and cooperative gating disappears. Thus, the rigid structure of the linker between S4 and the coiled coil mediates the stability of the coiled coil in relation to the transmembrane domain and therefore plays a critical role in achieving cooperative gating within the dimer. Physical intersubunit contact between the two VSDs within a dimer has been suggested (Lee et al., 2008; Musset, Smith, et al., 2010). This may lead one to think that two transmembrane regions in close proximity within a dimer may physically interact with one another, and that the primary role of the coiled-coil domain is to facilitate the close positioning of those two transmembrane regions (Figure 10.11(D)). However, that idea is at variance with the observation that GGG mutation of the linker region eliminates the wild-type gating phenotype, though the dimeric assembly is retained (Figure 10.9(B)) (Fujiwara et al., 2012). In an alternative model, cooperative gating of the two channel units is mediated by mechanical interaction via the cytoplasmic coiled coil (Figure 10.11(E)). The linker structure appears to be α-helical, based on the results of the CD spectrometry (Fujiwara et al., 2012) and our recent mutagenesis studies (Fujiwara et al., in press). This suggests the force is transmitted along a single α-helix consisting of the S4 and the coiled-coil strands (Figure 10.11(A) and (E)). In Hv, the coiled coil structure functions as a cytoplasmic machine that accomplishes both subunit assembly and temperature-dependent gating modulation. Future high-resolution structural information will further our understanding of the detailed molecular mechanisms underlying the temperature-sensitive channel properties of Hv.

REFERENCES

Acharya, A., Ruvinov, S. B., Gal, J., Moll, J. R., & Vinson, C. (2002). A heterodimerizing leucine zipper coiled coil system for examining the specificity of a position interactions: amino acids I, V, L, N, A, and K. *Biochemistry, 41*(48), 14122–14131.

Acharya, A., Rishi, V., & Vinson, C. (2006). Stability of 100 homo and heterotypic coiled-coil a-a' pairs for ten amino acids (A, L, I, V, N, K, S, T, E, and R). *Biochemistry, 45*(38), 11324–11332.

Akey, D. L., Malashkevich, V. N., & Kim, P. S. (2001). Buried polar residues in coiled-coil interfaces. *Biochemistry, 40*(21), 6352–6360.

Alabi, A. A., Bahamonde, M. I., Jung, H. J., Kim, J. I., & Swartz, K. J. (2007). Portability of paddle motif function and pharmacology in voltage sensors. *Nature, 450*(7168), 370–375. http://dx.doi.org/10.1038/nature06266.

Banfi, B., Schrenzel, J., Nusse, O., Lew, D. P., Ligeti, E., Krause, K. H., et al. (1999). A novel H(+) conductance in eosinophils: unique characteristics and absence in chronic granulomatous disease. *Journal of Experimental Medicine, 190*(2), 183–194.

Barish, M. E., & Baud, C. (1984). A voltage-gated hydrogen ion current in the oocyte membrane of the axolotl, *Ambystoma. Journal of Physiology, 352*, 243–263.

Bezanilla, F. (2000). The voltage sensor in voltage-dependent ion channels. *Physiological Reviews, 80*(2), 555–592.

Byerly, L., & Suen, Y. (1989). Characterization of proton currents in neurones of the snail, *Lymnaea stagnalis. Journal of Physiology, 413*, 75–89.

Capasso, M., Bhamrah, M. K., Henley, T., Boyd, R. S., Langlais, C., Cain, K., et al. (2010). HVCN1 modulates BCR signal strength via regulation of BCR-dependent generation of reactive oxygen species. *Nature Immunology, 11*(3), 265–272.

Capitano, M. L., Nemeth, M. J., Mace, T. A., Salisbury-Ruf, C., Segal, B. H., McCarthy, P. L., et al. (2012). Elevating body temperature enhances hematopoiesis and neutrophil recovery after total body irradiation in an IL-1-, IL-17-, and G-CSF-dependent manner. *Blood, 120*(13), 2600–2609. http://dx.doi.org/10.1182/blood-2012-02-409805.

Casey, J. R., Grinstein, S., & Orlowski, J. (2010). Sensors and regulators of intracellular pH. *Nature Reviews Molecular Cell Biology, 11*(1), 50–61. http://dx.doi.org/10.1038/nrm2820.

Cherny, V. V., & DeCoursey, T. E. (1999). pH-dependent inhibition of voltage-gated H(+) currents in rat alveolar epithelial cells by Zn(2+) and other divalent cations. *Journal of General Physiology, 114*(6), 819–838.

Cherny, V. V., Henderson, L. M., Xu, W., Thomas, L. L., & DeCoursey, T. E. (2001). Activation of NADPH oxidase-related proton and electron currents in human eosinophils by arachidonic acid. *Journal of Physiology, 535*(Pt 3), 783–794.

Chodniewicz, D., Alteraifi, A. M., & Zhelev, D. V. (2004). Experimental evidence for the limiting role of enzymatic reactions in chemoattractant-induced pseudopod extension in human neutrophils. *Journal of Biological Chemistry, 279*(23), 24460–24466. http://dx.doi.org/10.1074/jbc.M312764200.

DeCoursey, T. E. (2003). Voltage-gated proton channels and other proton transfer pathways. *Physiological Reviews, 83*(2), 475–579.

DeCoursey, T. E. (2013). Voltage-gated proton channels: molecular biology, physiology, and pathophysiology of the H(V) family. *Physiological Reviews, 93*(2), 599–652. http://dx.doi.org/10.1152/physrev.00011.2012.

DeCoursey, T. E., & Cherny, V. V. (1998). Temperature dependence of voltage-gated H^+ currents in human neutrophils, rat alveolar epithelial cells, and mammalian phagocytes. *Journal of General Physiology, 112*(4), 503–522.

DeCoursey, T. E., Cherny, V. V., DeCoursey, A. G., Xu, W., & Thomas, L. L. (2001). Interactions between NADPH oxidase-related proton and electron currents in human eosinophils. *Journal of Physiology, 535*(Pt 3), 767–781. PHY_12304.

Demaurex, N., & El Chemaly, A. (2010). Physiological roles of voltage-gated proton channels in leukocytes. *Journal of Physiology, 588*(Pt 23), 4659–4665. http://dx.doi.org/10.1113/jphysiol.2010.194225.

Demaurex, N., & Petheo, G. L. (2005). Electron and proton transport by NADPH oxidases. *Philosophical Transactions of the Royal Society London B: Biological Sciences, 360*(1464), 2315–2325.

Deng, Y., Liu, J., Zheng, Q., Eliezer, D., Kallenbach, N. R., & Lu, M. (2006). Antiparallel four-stranded coiled coil specified by a 3-3-1 hydrophobic heptad repeat. *Structure, 14*(2), 247–255.

El Chemaly, A., Okochi, Y., Sasaki, M., Arnaudeau, S., Okamura, Y., & Demaurex, N. (2010). VSOP/Hv1 proton channels sustain calcium entry, neutrophil migration, and superoxide production by limiting cell depolarization and acidification. *Journal of Experimental Medicine, 207*(1), 129–139. http://dx.doi.org/10.1084/jem.20091837.

Fujiwara, Y., Kurokawa, T., Takeshita, K., Kobayashi, M., Okochi, Y., Nakagawa, A., et al. (2012). The cytoplasmic coiled-coil mediates cooperative gating temperature sensitivity in the voltage-gated H(+) channel Hv1. *Nature Communications, 3*, 816. http://dx.doi.org/10.1038/ncomms1823.

Fujiwara, Y., Kurokawa, T., Takeshita, K., Nakagawa, A., Larsson, H. P., & Okamura, Y. (2013). Gating of the designed trimeric/tetrameric voltage-gated H^+ channel. *Journal of Physiology, 591*(Pt 3), 627–640. http://dx.doi.org/10.1113/jphysiol.2012.243006.

Fujiwara, Y., Takeshita, K., Nakagawa, A., & Okamura, Y. (2013). Structural characteristics of the redox-sensing coiled coil in the voltage-gated H+ channel. *Journal of Biological Chemistry, 288*(25), 17968–17975. http://dx.doi.org/10.1074/jbc.M113.459024.

Fujiwara, Y., Kurokawa, T., & Okamura, Y. (2014). Long-helices projecting from the membrane as the dimer interface in the voltage-gated H^+ channel. *Journal of General Physiology, 143*(3), 377–386.

Gonzalez, C., Koch, H. P., Drum, B. M., & Larsson, H. P. (2010). Strong cooperativity between subunits in voltage-gated proton channels. *Nature Structural & Molecular Biology, 17*(1), 51–56.

Gu, X., & Sackin, H. (1995). Effect of pH on potassium and proton conductance in renal proximal tubule. *American Journal of Physiology, 269*(3 Pt 2), F289–F308.

Harbury, P. B., Zhang, T., Kim, P. S., & Alber, T. (1993). A switch between two-, three-, and four-stranded coiled coils in GCN4 leucine zipper mutants. *Science, 262*(5138), 1401–1407.

Henderson, L. M., Chappell, J. B., & Jones, O. T. (1987). The superoxide-generating NADPH oxidase of human neutrophils is electrogenic and associated with an H^+ channel. *Biochemistry Journal, 246*(2), 325–329.

Hong, L., Pathak, M. M., Kim, I. H., Ta, D., & Tombola, F. (2013). Voltage-sensing domain of voltage-gated proton channel hv1 shares mechanism of block with pore domains. *Neuron, 77*(2), 274–287. http://dx.doi.org/10.1016/j.neuron.2012.11.013.

Hoshino, M., Yumoto, N., Yoshikawa, S., & Goto, Y. (1997). Design and characterization of the anion-sensitive coiled-coil peptide. *Protein Science, 6*(7), 1396–1404.

Howard, R. J., Clark, K. A., Holton, J. M., & Minor, D. L., Jr. (2007). Structural insight into KCNQ (Kv7) channel assembly and channelopathy. *Neuron, 53*(5), 663–675.

Iovannisci, D., Illek, B., & Fischer, H. (2010). Function of the HVCN1 proton channel in airway epithelia and a naturally occurring mutation, M91T. *J General Physiology, 136*(1), 35–46. http://dx.doi.org/10.1085/jgp.200910379.

Isacoff, E. Y., Jan, Y. N., & Jan, L. Y. (1990). Evidence for the formation of heteromultimeric potassium channels in *Xenopus* oocytes. *Nature, 345*(6275), 530–534.

Jozefowicz-Okonkwo, G., & Nowak, D. (2004). Temperature can prime human peripheral blood neutrophils in a p38MAPKalpha-dependent manner. *Archivum Immunologiae et Therapiae Experimentalis (Warsz), 52*(6), 447–456.

Koch, H. P., Kurokawa, T., Okochi, Y., Sasaki, M., Okamura, Y., & Larsson, H. P. (2008). Multimeric nature of voltage-gated proton channels. *Proceedings of the National Academy of Sciences of the United States of America, 105*(26), 9111–9116.

Kuno, M., Kawawaki, J., & Nakamura, F. (1997). A highly temperature-sensitive proton current in mouse bone marrow-derived mast cells. *Journal of General Physiology, 109*(6), 731–740.

Kuno, M., Ando, H., Morihata, H., Sakai, H., Mori, H., Sawada, M., et al. (2009). Temperature dependence of proton permeation through a voltage-gated proton channel. *Journal of General Physiology, 134*(3), 191–205. http://dx.doi.org/10.1085/jgp.200910213.

Kurokawa, T., & Okamura, Y. (2014). Mapping of sites facing aqueous environment of voltage-gated proton channel at resting state: a study with PEGylation protection. *Biochimica et Biophysica Acta, 1838*(1), 382–387. http://dx.doi.org/10.1016/j.bbamem.2013.10.001.

Lagunoff, D., & Wan, H. (1974). Temperature dependence of mast cell histamine secretion. *Journal of Cell Biology, 61*(3), 809–811.

Lee, S. Y., Letts, J. A., & Mackinnon, R. (2008). Dimeric subunit stoichiometry of the human voltage-dependent proton channel Hv1. *Proceedings of the National Academy of Sciences of the United States of America, 105*(22), 7692–7695.

Li, S. J., Zhao, Q., Zhou, Q., Unno, H., Zhai, Y., & Sun, F. (2010). The role and structure of the carboxyl-terminal domain of the human voltage-gated proton channel Hv1. *Journal of Biological Chemistry, 285*(16), 12047–12054.

Lishko, P. V., Botchkina, I. L., Fedorenko, A., & Kirichok, Y. (2010). Acid extrusion from human spermatozoa is mediated by flagellar voltage-gated proton channel. *Cell, 140*(3), 327–337. http://dx.doi.org/10.1016/j.cell.2009.12.053.

Lishko, P. V., Kirichok, Y., Ren, D., Navarro, B., Chung, J. J., & Clapham, D. E. (2012). The control of male fertility by spermatozoan ion channels. *Annual Review of Physiology, 74*, 453–475. http://dx.doi.org/10.1146/annurev-physiol-020911-153258.

Liu, J., Zheng, Q., Deng, Y., Li, Q., Kallenbach, N. R., & Lu, M. (2007). Conformational specificity of the lac repressor coiled-coil tetramerization domain. *Biochemistry, 46*(51), 14951–14959.

Morgan, D., Cherny, V. V., Murphy, R., Xu, W., Thomas, L. L., & DeCoursey, T. E. (2003). Temperature dependence of NADPH oxidase in human eosinophils. *Journal of Physiology, 550*(Pt. 2), 447–458. http://dx.doi.org/10.1113/jphysiol.2003.041525.

Morgan, D., Capasso, M., Musset, B., Cherny, V. V., Rios, E., Dyer, M. J., et al. (2009). Voltage-gated proton channels maintain pH in human neutrophils during phagocytosis. *Proceedings of the National Academy of Sciences of the United States of America, 106*(42), 18022–18027.

Murata, Y., Iwasaki, H., Sasaki, M., Inaba, K., & Okamura, Y. (2005). Phosphoinositide phosphatase activity coupled to an intrinsic voltage sensor. *Nature, 435*(7046), 1239–1243.

Musset, B., Morgan, D., Cherny, V. V., MacGlashan, D. W., Jr, Thomas, L. L., Rios, E., et al. (2008). A pH-stabilizing role of voltage-gated proton channels in IgE-mediated activation of human basophils. *Proceedings of the National Academy of Sciences of the United States of America, 105*(31), 11020–11025. http://dx.doi.org/10.1073/pnas.0800886105, 0800886105.

Musset, B., Capasso, M., Cherny, V. V., Morgan, D., Bhamrah, M., Dyer, M. J., et al. (2010). Identification of Thr29 as a critical phosphorylation site that activates the human proton channel Hvcn1 in leukocytes. *Journal of Biological Chemistry, 285*(8), 5117–5121. http://dx.doi.org/10.1074/jbc.C109.082727.

Musset, B., Smith, S. M., Rajan, S., Cherny, V. V., Sujai, S., Morgan, D., et al. (2010). Zinc inhibition of monomeric and dimeric proton channels suggests cooperative gating. *Journal of Physiology, 588*(Pt 9), 1435–1449.

Musset, B., Smith, S. M., Rajan, S., Morgan, D., Cherny, V. V., & Decoursey, T. E. (2011). Aspartate 112 is the selectivity filter of the human voltage-gated proton channel. *Nature, 480*(7376), 273–277. http://dx.doi.org/10.1038/nature10557.

Musset, B., Clark, R. A., DeCoursey, T. E., Petheo, G. L., Geiszt, M., Chen, Y., & El Jamali, A. (2012). NOX5 in human spermatozoa: expression, function, and regulation. *Journal of Biological Chemistry, 287*(12), 9376–9388. http://dx.doi.org/10.1074/jbc.M111.314955.

Nakajo, K., & Kubo, Y. (2008). Second coiled-coil domain of KCNQ channel controls current expression and subfamily specific heteromultimerization by salt bridge networks. *Journal of Physiology, 586*(Pt 12), 2827–2840.

Okamura, Y. (2007). Biodiversity of voltage sensor domain proteins. *Pflugers Archiv, 453*(3), 361–371.

Okamura, Y. (2012). Voltage-gated proton channels. *Comprehensive Biophysics, 6.8*, 199–222.

Okochi, Y., Sasaki, M., Iwasaki, H., & Okamura, Y. (2009). Voltage-gated proton channel is expressed on phagosomes. *Biochemical and Biophysical Research Communications, 382*(2), 274–279.

Petheo, G. L., Maturana, A., Spat, A., & Demaurex, N. (2003). Interactions between electron and proton currents in excised patches from human eosinophils. *Journal of General Physiology, 122*(6), 713–726.

Pryde, J. G., Walker, A., Rossi, A. G., Hannah, S., & Haslett, C. (2000). Temperature-dependent arrest of neutrophil apoptosis. Failure of Bax insertion into mitochondria at 15 degrees C prevents the release of cytochrome c. *Journal of Biological Chemistry, 275*(43), 33574–33584. http://dx.doi.org/10.1074/jbc.M001008200.

Ramsey, I. S., Moran, M. M., Chong, J. A., & Clapham, D. E. (2006). A voltage-gated proton-selective channel lacking the pore domain. *Nature, 440*(7088), 1213–1216.

Sakai, H., Li, G., Hino, Y., Moriura, Y., Kawawaki, J., Sawada, M., et al. (2013). Increases in intracellular pH facilitate endocytosis and decrease availability of voltage-gated proton channels in osteoclasts and microglia. *Journal of Physiology, 591*(Pt 23), 5851–5866. http://dx.doi.org/10.1113/jphysiol.2013.263558.

Salanova, B., Choi, M., Rolle, S., Wellner, M., Scheidereit, C., Luft, F. C., et al. (2005). The effect of fever-like temperatures on neutrophil signaling. *FASEB Journal, 19*(7), 816–818. http://dx.doi.org/10.1096/fj.04-2983fje.

Sasaki, M., Takagi, M., & Okamura, Y. (2006). A voltage sensor-domain protein is a voltage-gated proton channel. *Science, 312*(5773), 589–592.

Sasaki, M., Tojo, A., Okochi, Y., Miyawaki, N., Kamimura, D., Yamaguchi, A., et al. (2013). Autoimmune disorder phenotypes in Hvcn1-deficient mice. *Biochemical Journal, 450*(2), 295–301. http://dx.doi.org/10.1042/BJ20121188.

Sebag, J., Reed, W. P., & Williams, R. C., Jr. (1977). Effect of temperature on bacterial killing by serum and by polymorphonuclear leukocytes. *Infection and Immunity, 16*(3), 947–954.

Severns, C., Collins-Lech, C., & Sohnle, P. G. (1986). Effect of temperature on production of hypochlorous acid by stimulated human neutrophils. *Journal of Laboratory and Clinical Medicine, 107*(1), 29–35.

Sigworth, F. J. (2003). Structural biology: life's transistors. *Nature, 423*(6935), 21–22.

Smith, S. M., Morgan, D., Musset, B., Cherny, V. V., Place, A. R., Hastings, J. W., et al. (2011). Voltage-gated proton channel in a dinoflagellate. *Proceedings of the National Academy of Sciences of the United States of America, 108*(44), 18162–18167. http://dx.doi.org/10.1073/pnas.1115405108.

Stokes, R. H., & Robinson, R. A. (1948). Ionic hydration and activity in electrolyte solutions. *Journal of the American Chemical Society, 70*(5), 1870–1878.

Taylor, A. R., Chrachri, A., Wheeler, G., Goddard, H., & Brownlee, C. (2011). A voltage-gated H^+ channel underlying pH homeostasis in calcifying coccolithophores. *PLoS Biology, 9*(6), e1001085. http://dx.doi.org/10.1371/journal.pbio.1001085.

Thomas, R. C., & Meech, R. W. (1982). Hydrogen ion currents and intracellular pH in depolarized voltage-clamped snail neurones. *Nature, 299*(5886), 826–828.

Tombola, F., Ulbrich, M. H., & Isacoff, E. Y. (2008). The voltage-gated proton channel Hv1 has two pores, each controlled by one voltage sensor. *Neuron, 58*(4), 546–556.

Tombola, F., Ulbrich, M. H., Kohout, S. C., & Isacoff, E. Y. (2010). The opening of the two pores of the Hv1 voltage-gated proton channel is tuned by cooperativity. *Nature Structural & Molecular Biology, 17*(1), 44–50.

Tsuruda, P. R., Julius, D., & Minor, D. L., Jr. (2006). Coiled coils direct assembly of a cold-activated TRP channel. *Neuron, 51*(2), 201–212.

Wang, Y., Li, S. J., Wu, X., Che, Y., & Li, Q. (2012). Clinicopathological and biological significance of human voltage-gated proton channel Hv1 protein overexpression in breast cancer. *Journal of Biological Chemistry, 287*(17), 13877–13888. http://dx.doi.org/10.1074/jbc.M112.345280.

Wang, Y., Wu, X., Li, Q., Zhang, S., & Li, S. J. (2013). Human voltage-gated proton channel hv1: a new potential biomarker for diagnosis and prognosis of colorectal cancer. *PLoS One, 8*(8), e70550. http://dx.doi.org/10.1371/journal.pone.0070550.

Wu, L. J., Wu, G., Akhavan Sharif, M. R., Baker, A., Jia, Y., Fahey, F. H., & Clapham, D. E. (2012). The voltage-gated proton channel Hv1 enhances brain damage from ischemic stroke. *Nature Neuroscience, 15*(4), 565–573. http://dx.doi.org/10.1038/nn.3059.

CHAPTER ELEVEN

Intimacies and Physiological Role of the Polymodal Cold-Sensitive Ion Channel TRPM8

Rodolfo Madrid* and María Pertusa*

Departamento de Biología, Facultad de Química y Biología, Universidad de Santiago de Chile, Santiago, Chile

*Corresponding authors: E-mail: rodolfo.madrid@usach.cl; maria.pertusa@usach.cl

Contents

Abstract

The detection of environmental temperature is critical for the survival of the most diverse organisms. Thermosensitive transient receptor potential (thermoTRP) channels have evolved as a class of ion channels activated by a wide range of temperatures. These molecular thermal sensors are spread through the different TRP channel subfamilies. Among the Melastatin subfamily of TRP channels, the eighth member, TRPM8, is a

Current Topics in Membranes, Volume 74
ISSN: 1063-5823
http://dx.doi.org/10.1016/B978-0-12-800181-3.00011-7

calcium-permeable cationic ion channel activated by cold, by substances that evoke cold sensation such as menthol, and by voltage. This channel is considered the main molecular entity responsible for the sensitivity to cold of primary sensory neurons of the somatosensory system. Here we present to the readers a summary of some the most relevant biophysical properties, physiological role, and molecular intimacies of this polymodal thermoTRP channel.

1. INTRODUCTION

Cold is detected by specific subpopulations of primary sensory neurons of the somatosensory system. These neurons express a wide variety of transduction and voltage-dependent channels that act in concert to give shape to their net electrical response to decreases in environmental temperature. In the free nerve endings of these thermosensitive neurons, TRPM8 is the most critical molecular entity responsible for the development of the depolarizing receptor potential in response to cold stimulation.

TRPM8 is a homotetrameric, nonselective Ca^{2+} permeable cation channel activated by cold, by chemical compounds that induce cooling sensations such as menthol, and by voltage. Thermosensitivity of this channel is quite notable, with a Q_{10} temperature coefficient between 10 and 25. Nowadays the key role of TRPM8 in cold thermotransduction is widely accepted. In this chapter, we will review some biophysical properties of TRPM8, its physiological role in cold transduction and some of the intracellular mechanisms and post-translational modifications that influence its contribution as a molecular thermal sensor.

2. TRPM8, A COLD-ACTIVATED POLYMODAL ION CHANNEL

2.1 Molecular cloning of TRPM8

TRPM8 was originally identified as a gene expressed in normal prostate epithelial cells and upregulated in prostate carcinomas by Tsavaler and coworkers (Tsavaler, Shapero, Morkowski, & Laus, 2001). This group described for the first time the full-length human TRPM8 cDNA, formerly called trp-p8. The transcript corresponded to a 1104 amino acid protein, with a marked homology to the TRP channels family. The expression of TRPM8 in sensory neurons was not reported until 2002, when it was cloned by two groups independently and characterized as a cold- and menthol-activated ion channel (McKemy, Neuhausser, & Julius, 2002;

Peier et al., 2002). Two different approaches were used. Peier and colleagues identified TRPM8 by searching TRP-like proteins in genomic DNA databases, and selecting the putative exons with similarity to the four and six transmembrane domains of TRPV1 (Peier et al., 2002). One of these sequences was used to design primers to amplify a 699 nucleotide fragment of TRPM8 from dorsal root ganglion cDNA library. To obtain the full-length sequence of this protein, the authors used a RACE-PCR (**r**apid **a**mplification of **c**DNA **e**nds-PCR), in combination with an exon-prediction software (Peier et al., 2002). McKemy and coworkers on the other hand used a different strategy, where the main goal was to identify the putative cold and menthol receptor expressed in primary somatosensory neurons. To this end, they constructed a cDNA expression library of trigeminal ganglia, and the screening was carried out by transfecting HEK293 cells with pools of this library and using calcium imaging technique. By isolating cells that exhibited responses to 500 μM menthol, the group was able to identify a single cDNA that conferred menthol and cold sensitivity to transfected HEK293 cells (McKemy et al., 2002).

2.2 Polymodal nature and structural properties of TRPM8 channels

TRPM8 behaves as a polymodal ion channel (for recent reviews, see (Almaraz, Manenschijn, de la Pena, & Viana, 2014; Babes, Ciobanu, Neacsu, & Babes, 2011; Latorre, Brauchi, Madrid, & Orio, 2011; McCoy, Knowlton, & McKemy, 2011; Yudin & Rohacs, 2012)). It is activated by several plant-derived and synthetic cooling compounds including menthol (McKemy et al., 2002; Peier et al., 2002), a mint-derived alcohol that activates the channel in a dose-dependent manner. TRPM8 is also gated by voltage, although it exhibits a weak voltage dependence. As a consequence, in the absence of agonists, the activation of this channel requires strongly depolarized membrane potentials (Brauchi, Orio, & Latorre, 2004; Voets et al., 2004). The neutralization of positively charged residues in the S4 domain and the S4–S5 linker of TRPM8 causes a decrease in its voltage dependence, suggesting that this region would be part of the voltage sensor of the channel (Voets, Owsianik, Janssens, Talavera, & Nilius, 2007). The activation of TRPM8 by cold and menthol, and by most of its chemical agonists, is related to a shift in its voltage–activation curve to more negative voltages, which cause the channel to open at physiologically relevant membrane potentials (Brauchi et al., 2004; Voets et al., 2004). Despite the advances in the study of the voltage and temperature dependence of

TRPM8, both the voltage sensor and the molecular determinants of its temperature dependence remain elusive (see below).

TRPM8 is directly activated by cold, with a threshold varying from 33 to 20 °C. A thermodynamic analysis of the ionic currents that result from the opening of TRPM8 channels by cold, studied in heterologous systems, indicates that the high Q_{10} of the temperature sensitivity of this channel is mediated by a large enthalpic change (ΔH) of −150 kcal/mol, between closed and open (Brauchi et al., 2004). The entropic change compensates the enthalpic change ($T\Delta S = -113$ kcal/mol), so that the free energy change is small, and the negative sign of the entropy change implies that the open state is more ordered than the closed state. Nevertheless, temperature gating in thermoTRP channels is poorly understood and the structural basis of their thermosensitivity is still not fully deciphered. Concerning the cold sensitivity of TRPM8, Brauchi and coworkers engineered chimeras in which the entire C-terminal domains were exchanged between TRPM8 and TRPV1 (Brauchi, Orta, Salazar, Rosenmann, & Latorre, 2006). The functional phenotype of these chimeras suggests that the C-terminal domain has a role in the temperature dependence of these thermoTRP channels. After these findings, the same group identified the region located outside the TRP domain comprising the TRPV1 C-terminal amino acids Q727 and W752 as the minimal portion able to turn TRPM8 into a heat-sensitive channel (Brauchi et al., 2007). However, these chimeras displayed lower Q_{10} values compared to the wild-type channels, suggesting the loss of some specific interactions present in these channels (Brauchi et al., 2006). The results obtained from these TRPM8/TRPV1 chimeric channels suggest that the C-terminal domain could participate in the thermal modulation of channel gating, and contribute to the evidence that temperature-, agonist-, and voltage-dependent gating could be independent processes that act in an allosteric fashion (Brauchi et al., 2004, 2007; Matta & Ahern, 2007). Nevertheless, the precise molecular determinants of the TRPM8 activation by cold remain unknown.

In TRPM8 channels, several binding sites of chemical agonists and antagonists are localized in the transmembrane domains. Using high throughput mutagenesis, positions Y745, in S2, and Y1005 and L1009, in the C-terminal domain, have been identified as key residues in the activation of TRPM8 by menthol (Bandell et al., 2006). Mutations in the S4 and in the S4–S5 linker also affect menthol affinity, suggesting that a direct interaction between menthol and the voltage sensor in this channel (Voets et al., 2007).

The activation of TRPM8 by the artificial cooling compound icilin, unlike menthol, requires the presence of intracellular Ca^{2+} and is modulated by intracellular pH (Chuang, Neuhausser, & Julius, 2004). The icilin-sensitive sites are located within the S2–S3 linker and S3 domain, and specifically the residues N799, D802, and G805 are required for icilin response of TRPM8 (Chuang et al., 2004). The function of TRPM8 is also sensitive to external Ca^{2+}, since the responses of the channel are inversely proportional to the extracellular calcium concentration (Mahieu et al., 2010).

On the other hand, regarding the sites that mediate the effects of antagonists, Mälkiä and coworkers have found that tyrosine residue 745, which is crucial for menthol activation, is also critical for the SKF96365-induced inhibition of the channel (Mälkiä, Pertusa, Fernandez-Ballester, Ferrer-Montiel, & Viana, 2009). A single point mutation of TRPM8, where Y745 is replaced by histidine, revealed that the inhibition of the channel by 4-(3-chloro-pyridin-2-yl)-piperazine-1-carboxylic acid (4-tert-butyl-phenyl)-amide (BCTC) is largely unaffected, and the inhibitory effect of clotrimazole, econazole and capsazepine is only partially reduced (Mälkiä et al., 2009). Despite this difference, blockers such as BCTC, 1-[2-(4-methoxyphenyl)-2-[3-(4-methoxyphenyl) propoxy] ethyl-1*H*-imidazole hydrochloride (SKF-96365) and clotrimazole share a similar mechanism of inhibition, shifting the voltage–activation curve toward more positive voltages and reducing the open probability of the channel at negative membrane potentials (Mälkiä et al., 2007; Meseguer et al., 2008).

Functional TRPM8 ion channels are homotetramers (Clapham, 2003; Ramsey, Delling, & Clapham, 2006; Stewart, Egressy, Lim, & Edwardson, 2010). Each subunit has a cytoplasmic N- and C-termini, and six transmembrane domains. The intracellular N-terminal domain, which compromises 693 amino acids, is the largest intracellular part of the protein. It contains four stretches of residues, called MHR (**M**elastatin **H**omology **R**egions), that share some sequence similarity with the rest of the members of the TRPM family (Clapham, 2003; Fleig & Penner, 2004). Although these regions do not represent known structural motifs and their function remains to be established, a quaternary structural model of TRPM8 channels suggests that these MHR domains are implicated in stabilization of the tetramer (Pedretti, Marconi, Bettinelli, & Vistoli, 2009). Outside these MHR, it has been suggested that the region encompassing residues 40 to 86 is required for the proper trafficking of TRPM8 to the plasma membrane (Phelps & Gaudet, 2007). Nevertheless, the role of the large N-terminus domain is still poorly understood.

Recently, we have identified two distinct regions within the initial N-terminus of TRPM8 that contribute differentially to cold and menthol sensitivity and also to proper folding and assembly of this channel. The deletion or substitution of the first 40 residues of TRPM8 yields channels with augmented responses to cold and menthol, and single point mutations at positions S26 or S27 by proline cause a comparable functional effect. On the other hand, different deletions and mutations within the positions 40–60 yield nonfunctional channels that are retained in the endoplasmic reticulum. Altogether, these results suggest a critical contribution of the initial N-terminal domain to thermal and chemical sensitivity, and the proper biogenesis of TRPM8 (Pertusa, González, Hardy, Madrid, & Viana, 2014).

As in other TRP channels, the relatively short C-terminal tail of TRPM8 contains a TRP domain located at the proximal region (Clapham, 2003; Montell, 2005). The neutralization of lysines and arginines in this domain greatly decreases the sensitivity of TRPM8 to phosphatidylinositol-4,5-bisphosphate ($PI(4,5)P_2$) (Rohacs, Lopes, Michailidis, & Logothetis, 2005). The coiled-coil motif of TRPM8, another relevant region within this domain, is formed by the last 40 residues (Erler et al., 2006; Tsuruda, Julius, & Minor, 2006). This domain is capable of self-assembly (Tsuruda, Julius, & Minor, 2006) and it appears to have also an important role in the tetramerization of TRPM8 (Erler et al., 2006; Tsuruda et al., 2006). Nevertheless, other authors have claimed that the deletion of the coiled coil prevents function but not the tetramerization of the channel (Phelps & Gaudet, 2007).

Regarding the transmembrane domains of TRPM8 (S1–S6), it has been shown that charge-neutralizing mutations in the transmembrane segment 4 (S4), and the S4–S5 linker, modify its voltage sensitivity (Voets et al., 2007). The pore module consists of the S5 and S6 domains and the interconnecting loop, where the pore filter is located. TRPM8 shows little discrimination among monovalent cations, and high permeability for calcium ions (McKemy et al., 2002). The permeation characteristics of TRP channels are mostly defined by the central pore region (Owsianik, Talavera, Voets, & Nilius, 2006), but in the case of TRPM8, the distal part of the S6 segment also plays a role in its ion selectivity (Kuhn, Knop, & Luckhoff, 2007). On the other hand, TRPM8 is N-glycosylated in the third extracellular loop, specifically in the Asn residue at position 934, near the pore region (Dragoni, Guida, & McIntyre, 2006; Erler et al., 2006; Pertusa, Madrid, Morenilla-Palao, Belmonte, & Viana, 2012). The N-glycosylation site is flanked by two cysteine residues that form a conserved double Cys motif

(C929 and C940), which is essential for channel function (Dragoni et al., 2006) (see below).

2.3 Tissue distribution of TRPM8

TRPM8 is robustly expressed in a subpopulation of primary sensory neurons of trigeminal and dorsal root ganglia, and is also found in nodose and geniculate ganglia (Katsura, Tsuzuki, Noguchi, & Sakagami, 2006; Zhang, Jones, Brody, Costa, & Brookes, 2004). In the last years, TRPM8 expression has been detected in genitourinary tract (Stein et al., 2004; Thebault et al., 2005), lung epithelial cells (Sabnis, Shadid, Yost, & Reilly, 2008), vascular smooth muscle (Johnson et al., 2009; Yang, Lin, McIntosh, & Sham, 2006), sperm (De Blas et al., 2009; Martinez-Lopez et al., 2011), odontoblasts (El Karim et al., 2010), corneal endothelial cells (Mergler et al., 2013), and salivary glands (Sobhan et al., 2013), among others. TRPM8 expression is also associated with different tumors (Bidaux et al., 2007; Chodon et al., 2010; Tsavaler et al., 2001; Yamamura, Ugawa, Ueda, Morita, & Shimada, 2008; Yee, Zhou, & Lee, 2010), where this channel appears as a pharmacological target with an interesting therapeutic potential (McKemy, 2013). Nevertheless, the physiological role of TRPM8 in tissues that are not exposed to large temperature drops is not fully understood.

Anatomical evidence of the expression pattern of TRPM8 has been provided by the generation of transgenic mice expressing green fluorescent protein under control of the channel promoter. TRPM8-positive primary sensory neurons project to the lamina I and II_0 within the spinal cord (Dhaka, Earley, Watson, & Patapoutian, 2008; Takashima, Ma, & McKemy, 2010; Takashima et al., 2007; Wrigley, Jeong, & Vaughan, 2009). TRPM8-expressing neurons show immunoreactivity to some classical somatosensory and nociceptor markers. Thus, TRPM8 can be found colocalizing with peripherin, a marker of C-fibers, and with intermediate filaments NF200, a marker of Aδ fibers. Primary sensory neurons that bind the plant lectin isolectin-B4, a marker for nonpeptidergic neurons, do not appear to express TRPM8 (Takashima et al., 2007). Instead, this channel appears to coexpress with nociceptor markers such as calcitonin gene-related peptide, substance P, and TRPV1 channels (Axelsson et al., 2009; Babes, Zorzon, & Reid, 2004; Dhaka et al., 2008; Okazawa et al., 2004; Parra et al., 2010; Takashima et al., 2007; Viana, de la Pena, & Belmonte, 2002; Xing, Ling, Chen, & Gu, 2006). The finding that cold- and menthol-induced responses of TRPM8-expressing sensory neurons are diminished by treatment with inflammatory mediators such as bradykinin, prostaglandin

E2, or histamine, both in cultured neurons and intact nerve endings of cold thermoreceptors, suggest a role of this channel in sensory alterations associated to inflammatory states (Linte, Ciobanu, Reid, & Babes, 2007; Premkumar, Raisinghani, Pingle, Long, & Pimentel, 2005; Zhang et al., 2012) (see below).

3. PHYSIOLOGICAL ROLE OF TRPM8

3.1 TRPM8 and cold thermotransduction

Cold is detected by specific thermoreceptor neurons that innervate the surface of the body. Cold-sensitive neurons (CSNs) from trigeminal ganglia (TG) and dorsal root ganglia (DRG) detect environmental temperature decreases with exquisitely temperature-sensitive molecular machinery, concentrated in the free axonal endings that innervate superficial layers of the skin and mucosae. These thermosensitive neurons correspond to subpopulations of unmyelinated primary afferent C-fibers and thinly myelinated Aδ-fibers of the somatosensory system (Hensel & Zotterman, 1951a; Hensel, 1981; Iriuchijima & Zotterman, 1960). Cold thermoreceptor neurons exhibit spontaneous ongoing electrical activity at resting temperature of the skin (33 °C), which increases in response to temperature reductions. The basal action potential firing of these primary somatosensory neurons is sensitized by the TRPM8 channel agonist menthol, and suppressed by mild heating of the receptive field (Brock, Pianova, & Belmonte, 2001; Hensel & Zotterman, 1951b). CSNs can be separated in low- and high-threshold neurons, the latter involved in signaling cold-induced discomfort and pain. Canonical CSNs have a low rheobase current, fire action potentials of shorter duration than those observed in nociceptors, and express a prominent I_h current dependent on HCN1-2 channels (Madrid et al., 2006; Madrid, de la Peña, Donovan-Rodriguez, Belmonte, & Viana, 2009; Orio et al., 2009, 2012; Reid, Babes, & Pluteanu, 2002; Viana et al., 2002). These neurons are sensitive to the TRPM8 agonist menthol, and their cold sensitivity is dampened by the expression of I_{KD}, an inhibitory outward slow-inactivating K^+-current, dependent on Kv1.1 and Kv1.2 Shaker-like potassium channels (Madrid et al., 2009). Differential functional expression of TRPM8 channels and I_{KD} in both populations of CSNs is intimately linked to the thermosensitive phenotype of these neurons, where low-threshold CSNs express higher levels of TRPM8 and lower levels of I_{KD} than high-threshold ones, and vice versa (Madrid et al., 2009). Cold nociceptors on the other hand are silent at resting temperature of the skin

(33 °C), and they are sensitive to low temperatures (below 15 °C); the activation of polymodal nociceptors by intense cold evokes sensations of pain (Campero, Serra, & Ochoa, 1996; Croze, Duclaux, & Kenshalo, 1976; LaMotte & Thalhammer, 1982; Simone & Kajander, 1996).

Nowadays, a long list of evidence supports the idea that TRPM8 channel expression is not only restricted to CSNs signaling innocuous cold, but is also expressed in primary sensory neurons that are part of nociceptive pathways, indicative of a relevant role of TRPM8 in sensing both unpleasant and painful cold (Bautista et al., 2007; Campero, Baumann, Bostock, & Ochoa, 2009; Knowlton, Bifolck-Fisher, Bautista, & McKemy, 2010; Knowlton et al., 2013; Madrid et al., 2009; Takashima et al., 2007); see also (Belmonte, Brock, & Viana, 2009; McKemy, 2013). TRPM8-knockout mice, and also the transgenic mice whose TRPM8-expressing neurons have been selectively ablated, exhibit conspicuous cold-insensitive phenotypes in both the innocuous and noxious range of cold temperatures (see below).

Regarding the thermal threshold of TRPM8 expressing cells, a striking difference in the cooling threshold can be observed between recombinant and native TRPM8 channels, since the mean temperature threshold is notably shifted to lower temperatures in heterologous expression systems compared to primary somatosensory neurons. In 2005, De la Peña and co-workers demonstrated that the overexpression of TRPM8 in hippocampus induces cold sensitivity in these otherwise cold-insensitive neurons, endogenously devoid of cold-activated thermoTRP channels TRPM8 or TRPA1 (de la Peña, Malkia, Cabedo, Belmonte, & Viana, 2005). The active and passive electrical properties of hippocampal neurons are comparable to cultured primary sensory neurons, and despite this similarity, cold sensitivity of the TRPM8-transfected hippocampal neurons showed a mean excitation threshold close to 26 °C, a value similar to the observed in TRPM8-transfected HEK293 cells. This thermal threshold is significantly higher than the one observed in TG and DRG CSNs, which are often already activated at temperatures as high as 32 °C (Madrid et al., 2006, 2009; Mälkiä et al., 2007; Reid et al., 2002; Viana et al., 2002; de la Peña et al., 2005). This discrepancy is not due to differences in the expression levels of the channels, since the current density of TRPM8 in the transfected hippocampal neurons can be even larger than in TG neurons (de la Peña et al., 2005).

While the cold-induced current in CSNs from TG exhibit similar pharmacological properties to TRPM8 at a fixed membrane potential (Madrid et al., 2006), comparison of the potential for half maximal activation ($V_{1/2}$) by cold of native versus recombinant channels show a shift towards hyperpolarizing values resulting in a difference in the thermal excitation threshold of CSNs (Mälkiä et al., 2007). The $V_{1/2}$ of TRPM8 activation by cold is approximately 140 mV more negative in native channels compared to recombinant TRPM8 at the same temperature. Thus, activation of TRPM8 in response to equivalent innocuous cold stimulation results in a large open probability of TRPM8 at physiological membrane potentials in primary somatosensory neurons. This is not a trivial point. Due to the pronounced thermal gradients across the skin, and the powerful thermoregulatory reflexes that control cutaneous irrigation, the temperature at sensory nerve endings in humans probably will not descend beyond few degrees below the resting temperature of the skin, even during sustained surface exposure to noxious cold temperatures (Mauderli, Vierck, Cannon, Rodrigues, & Shen, 2003; Morin & Bushnell, 1998). This means that during slight reductions of the skin temperature, a transduction channel whose activation threshold is far from the resting temperature of the skin, could not be responsible for the extraordinary thermal sensitivity of innocuous cold-sensory endings. Native TRPM8 channels are able to accomplish this task, due to their exquisite thermal sensitivity tuned at the physiologically relevant thermal range in primary sensory neurons. Some clues about the molecular basis of this discrepancy are discussed below.

3.2 TRPM8 knockout mice and selective ablation of TRPM8 expressing neurons

The strongest evidence for the central role of TRPM8 in the detection of innocuous and noxious cold in vivo appeared a few years ago, from behavioral experiments with knockout animals of this thermal sensor. In 2007, mice lacking functional expression of TRPM8 channels were generated by three groups independently (Bautista et al., 2007; Colburn et al., 2007; Dhaka et al., 2007). Survival and general appearance of these TRPM8 knockout mice are normal, although some differences in body weigh but not body size have been reported, possibly related to an altered insulin homeostasis (McCoy et al., 2013). TRPM8 knockout mice do not present significant differences in mean core body temperature compared to wild-type animals, but the systemic blockage of the channel in wild-type animals has a large hypothermic effect (Knowlton, Daniels, Palkar, McCoy, & McKemy, 2011). Although slight variations in thermosensitivity have been described among them in different behavioral tests (see (McCoy et al., 2011) for a comprehensive comparison and discussion), cold sensitivity of all these TRPM8 knockout strains is strongly compromised.

Bautista and coworkers generated TRPM8 knockout mice by deletion of the coding region encompassing residues 594 to 661 within the N-terminal domain of TRPM8, introducing a stop codon within the protein-coding region (Bautista et al., 2007). In these animals, in situ hybridization against nonfunctional truncated transcript shows that the loss of TRPM8 functional protein does not eliminate those neurons that normally express the cold thermoreceptor channel. Behavioral studies using TRPM8 knockout animals revealed a large impairment in discrimination between cold and warm environments. Calcium imaging in cultured primary sensory neurons from these knockout animals also revealed a decrease in the incidence and magnitude of cold-induced responses. Cold-sensitive C-fibers of animals lacking TRPM8 expression present a lower basal action potential firing, but the molecular ablation of TRPM8 does not affect the general excitability of these primary sensory neurons, since both the conduction velocity and the electrical activation threshold of cold sensitive fibers are unaffected. Interestingly, electrophysiological recordings of single sensory fibers from these mice show a strong diminution in the responses to temperature reductions in both low and high-threshold primary afferents.

Dhaka and coworkers developed a TRPM8 knockout mouse targeting a construct to delete a 2–29 region in the N-terminal domain, knocked in using a farnesylated enhanced green fluorescent protein (EGFP-F) followed by an SV40polyA tail in frame with the start codon of the TRPM8 channel, that prevents the transcription of the full TRPM8 mRNA (Dhaka et al., 2007). This strategy was used to create a TRPM8 deficient mouse, where the EGFP allows the identification of sensory neurons that would express the channel under normal conditions. This mouse strain lacking functional expression of TRPM8 exhibits a strongly reduced avoidance to cold in both two-temperature assays and in thermotaxis assays of temperature gradients. Molecular ablation of this channel also reduced the behavioral responses to cooling agents (Dhaka et al., 2007).

A TRPM8-deficient strain was also generated by Colburn and coworkers, by using homologous recombination where a 2 kb fragment of TRPM8 covering the first two exons was replaced by a LacZ/Neo selection cassette (Colburn et al., 2007). TRPM8 gene disruption in these mice reduces cold sensitivity of primary sensory neurons in culture studied by calcium imaging. In behavioral tests, TRPM8 knockout mice fail to respond to systemic stimulation of TRPM8 using icilin, the most potent agonist

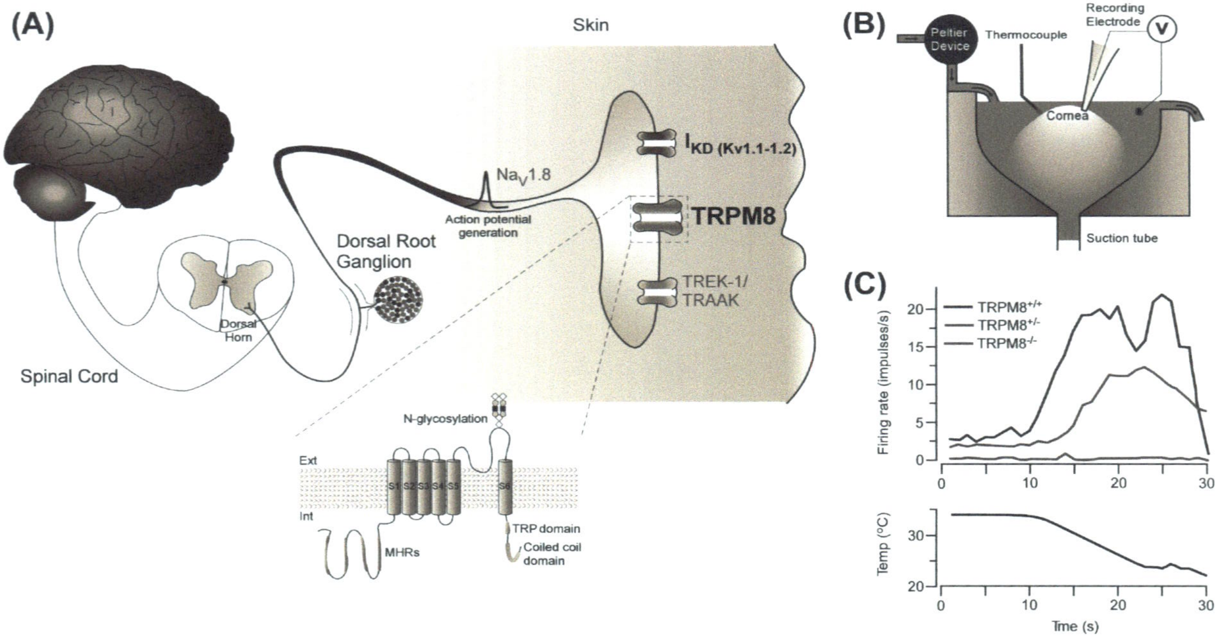

Figure 11.1 TRPM8 is the main molecular determinant of cold sensitivity of primary cold thermoreceptor neurons. (A) The schematic representation of a cold-sensitive primary afferent neuron. The site of cold transduction (enlarged nerve ending at the right) is depicted as separate from the site of action potential generation (axon at left). In this picture, TRPM8 is represented as the most prominent channel. Kv1.1-1.2 (the molecular counterpart of the break current I_{KD}) and TREK-1/TRAAK channels are also important molecular determinants that contribute to the net response to temperature reductions in cold thermoreceptor neurons. (B) Experimental arrangement used to record nerve terminal impulses (NTI) of cold thermoreceptor neurons from isolated mouse eye in vitro. The eye is placed in a small recording chamber and secured by the application of slight suction to a tube connected to the conical bottom. A glass pipette (50 μm in diameter) is

described for this channel. Moreover, in a test to evaluate thermal preference, the knockout mice spend the same fraction of time in cold surfaces than in comfortable warm ones, unlike wild-type animals that exclusively prefer warm surfaces. The authors also explored the development of cold-induced pain that occurs after injury and they observed that TRPM8 knockout mice show a significant reduction of this sensory alteration compared to wild-type animals (Colburn et al., 2007) (see below).

Another strategy to study the contribution of TRPM8 to cold sensing has been the generation of mice lacking TRPM8-expressing neurons. Recently, Knowlton and coworkers have shown that selective elimination of TRPM8-expressing neurons, using diphtheria toxin mediated ablation, induces cold insensitivity in both innocuous and noxious range, generating a more profound phenotype than in TRPM8 knockout animals (Knowlton et al., 2013). Animals depleted of TRPM8 neurons have normal mechanical, propioceptive and heat behaviors, suggesting that TRPM8 expressing neurons could be dispensable for other somatosensory modalities (Knowlton et al., 2013). Besides, Pogorzala and coworkers, using a similar ablation strategy to eliminate TRPM8 expressing neurons in adult mice, showed that this maneuver strongly reduced the responses of these animals to a wide range of cold temperatures (Pogorzala, Mishra, & Hoon, 2013).

3.3 Contribution of TRPM8 channels in sensing cold as pain in response to axonal damage

A large number of transduction and voltage-gated ion channels operate concertedly in primary sensory neurons, giving shape to their excitability (Figure 11.1(A)). Changes in the exquisite functional balance of these channels appear to be intimately linked to the development of diverse forms of untreatable pain induced by peripheral nerve damage, including painful hypersensitivity to cold. In the last years, several molecular entities have been described as key elements with important roles in this sensory alteration.

brought into contact with the cornea to record NTIs. The temperature of the perfusing solution is adjusted with a Peltier device at the entrance of the chamber, controlled by computer, and is measured at the corneal surface using a thermocouple. (C) The mean firing rate (*top*) of a population of corneal cold-sensitive nerve endings from TRPM8$^{+/+}$, TRPM8$^{\pm}$, and TRPM8$^{-/-}$ mice during cold stimulation (cooling ramp at bottom). Basal firing rate (ongoing activity) and cold responses in these free terminals are proportional to the expression level of TRPM8 channels. *((C) adapted from Parra et al., 2010.)* (See the color plate.)

They include the thermoTRP channels TRPM8 and TRPA1, some voltage-gated and background K^+ channels, voltage-gated Na^+ channels, and HCN channels, among others (Abrahamsen et al., 2008; Dray, 2008; Emery, Young, Berrocoso, Chen, & McNaughton, 2011; Patapoutian, Tate, & Woolf, 2009).

The complexity of cold sensing as pain in response to axonal damage is influenced by the type of the injury and by the contribution of the diverse channels with a role in the excitability of primary sensory neurons to the final phenotype. The contribution of TRPM8 to painful hypersensitivity to cold has been explored, and it is now accepted that this channel plays an important role in this often disabling pathology. An increase in the expression level of TRPM8 induced by peripheral nerve damage has been linked to the development and maintenance of painful hypersensitivity to cold (Frederick, Buck, Matson, & Cortright, 2007; Su et al., 2011; Xing, Chen, Ling, Tan, & Gu, 2007). On this regard, Xing and coworkers, using a model of peripheral neuronal damage by chronic constriction nerve injury (CCI) in rats that induce painful cold hypersensitivity, have found that peripheral damage resulted in a significant increase in the percentage of menthol- and cold-sensitive neurons, and a substantial enhancement of the responsiveness of these cells to both menthol and innocuous cold. These changes were observed in capsaicin-sensitive primary sensory neurons, a subpopulation of allegedly nociceptive-like neurons (Xing et al., 2007). Similarly, Su and coworkers described an increase in TRPM8 expression in response to CCI by using TRPM8 protein staining in DRG sections and immunoblotting (Su et al., 2011), supporting the idea that an increase in the expression of TRPM8 is related to the development of cold hypersensitivity. Nevertheless, some interesting complementary approaches that would help an unequivocal interpretation of this kind of results have been less explored. The distribution of threshold temperatures in the different populations of CSNs and an accurate study of their active and passive biophysical membrane properties, would allow to distinguish whether those neurons overexpressing TRPM8 after injury correspond to high-threshold CSNs that coexpress TRPV1 channels or to polymodal nociceptors.

The strongest evidences of the participation of TRPM8 in this sensory alteration induced by nerve injury arise from the use of genetically modified mice and new antagonists of this channel. Thus, the development of cold hypersensitivity induced by axonal damage is unequivocally diminished in TRPM8 knockout animals (Colburn et al., 2007) and pharmacological blockage of TRPM8 channels reduces the nociceptive behavior associated to cold stimulation in mice with peripheral nerve injury (Knowlton et al., 2011). Consistent with these findings, animals that lack TRPM8-expressing neurons also show a significantly attenuated reduction in cold pain associated to nerve injury (Knowlton et al., 2013), indicating that sensory neurons that express this thermal sensor have a relevant role in the phenomenon.

It is important to mention that there are also evidences to suggest that cold hypersensitivity in response to axonal damage may not necessarily be correlated with significant increases in TRPM8 channel expression (Caspani, Zurborg, Labuz, & Heppenstall, 2009; Katsura et al., 2006; Namer, Kleggetveit, Handwerker, Schmelz, & Jorum, 2008; Staaf, Oerther, Lucas, Mattsson, & Ernfors, 2009), and interestingly, it has been also proposed that TRPM8 expressing neurons have an critical role in cold and menthol mediated analgesia (Knowlton et al., 2013; Liu et al., 2013; Proudfoot et al., 2006).

Thus, TRPM8 emerges as the most critical component of the molecular and neural machinery underlying cold sensing in a wide range of temperatures under physiological conditions. This thermal sensor is also involved in altered cold sensitivity in response to axonal damage of primary somatosensory neurons, and almost paradoxically in cold-induced analgesia, which makes the channel an attractive target for the design of future treatments of cold-induced sensory alterations.

3.4 The role of TRPM8-expressing neurons as humidity detectors

Changes in the continuous aqueous fluid secretion by exocrine glands that maintain the wetness of ocular surface can lead to eye dryness syndrome. The neural mechanisms underlying basal tearing that keeps the humidity of the cornea were described few years ago, unveiling a critical role of TRPM8 channel in this phenomenon (Parra et al., 2010).

Among the preparations used to record the electrical activity of primary sensory neurons, the cornea constitutes one of the most useful and powerful tools (Brock, McLachlan, & Belmonte, 1998; Parra et al., 2010) (Figure 11.1(B)). Trigeminal sensory neurons densely innervate this structurally simple tissue, where nerve endings of cold thermoreceptors, mechanoreceptors, and polymodal nociceptors are largely accessible to electrophysiological recording in vitro. Using focal recording of the electrical activity of unitary cold-sensitive nerve endings in the mouse cornea, Parra and coworkers demonstrated that not only the responses to cold but also the ongoing electrical activity of these exquisitely cold-sensitive sensory neurons were proportional to the expression level of TRPM8 channels (Parra et al., 2010) (Figure 11.1(C)). Basal tearing varies with environmental conditions, and blinking rate produces small changes in the temperature of the ocular surface. Parra and coworkers hypothesized that cold thermoreceptor neurons innervating the cornea serve as humidity detectors, sensing the small temperature drops of the corneal surface caused by the evaporation of the tear film during interblink periods. Thus, almost 80% of the cold-sensitive corneal nerve fibers respond to temperature decreases of 2 °C or less, a change equivalent to interblink temperature drops. Basal firing of corneal cold thermoreceptors and their responses to cold and menthol were absent in TRPM8 knockout mice, and proportionally reduced in heterozygous animals. Interestingly, mice lacking TRPM8 showed a largely reduced basal tearing compared to wild-type animals, and a normal nociceptor-mediated irritative tearing in response to activators of TRPV1 and TRPA1 channels. In wild-type animals, basal tearing is strongly reduced by heating the corneal surface, a maneuver that reduces the firing of cold thermoreceptors and also reduces the basal tearing in humans. Thus, TRPM8-expressing neurons innervating the cornea work as humidity detectors, a critical neural component of the physiological machinery that maintains the basal wetness of the ocular surface and, probably, other exposed mucosae.

4. TRAFFICKING, N-GLYCOSYLATION AND MODULATION OF TRPM8 CHANNEL FUNCTION

Function and biophysical properties of TRPM8 channels can be regulated by intracellular cascades of second messengers (Figure 11.2) and post-translational modifications such as phosphorylation and N-glycosylation, which influence trafficking and thermal and chemical sensitivity of this polymodal thermal sensor.

4.1 Trafficking of TRPM8 channels

As we mentioned before, cold thermotransduction in mammals is mainly based on the activation of TRPM8 channels expressed in the nerve endings of the somatosensory neurons. Therefore, the regulation of the amount of functional channels present at the plasma membrane level is an important mechanism to control the cold-induced responses. The number of channels found on the cell surface is determined by the balance of the exocytosis and endocytosis processes. The trafficking of ion channels to the plasma membrane can occur in a nonmodulated pathway (constitutive exocytosis) or signal-dependent manner (regulated-exocytosis). It has been demonstrated that different molecular signals mediate the insertion of thermoTRP channels in the plasma membrane. For example, TRPV1 levels at the cell surface are modulated by NGF stimulation (Zhang, Huang, & McNaughton, 2005), TRPV2 after exposure to IGF-I (Kanzaki et al., 1999), TRPC5 by EGF stimulation (Bezzerides, Ramsey, Kotecha, Greka, & Clapham, 2004), and TRPA1 after treatments with mustard oil (Schmidt, Dubin, Petrus, Earley, & Patapoutian, 2009). Regarding TRPM8, it has been reported that some channels are resident in the plasma membrane, whereas a distinct pool is located in vesicles near the cell surface that are translocated through a kiss-and-linger mechanism (Veliz et al., 2010). The authors suggest that this mobile population represent a reserve of TRPM8 channels that in response to specific stimuli could be incorporated by plasma membrane insertion, increasing the number of available TRPM8 channels in the cell surface and thereby providing a significant amount of signal amplification (Veliz et al., 2010).

Within the plasma membrane, it has been reported that TRPM8 is mainly located in lipid rafts, both in vivo and in heterologous expression systems (Morenilla-Palao, Pertusa, Meseguer, Cabedo, & Viana, 2009). Lipid rafts are plasma membrane microdomains characterized by considerably higher levels of both cholesterol and sphingomyelin than the average membrane composition (Simons & Ikonen, 1997). The association of TRPM8 channels with lipid rafts causes a decrease in both menthol- and cold-mediated activation of TRPM8 channels, suggesting that structural/compositional characteristics of lipid rafts negatively modulate TRPM8 activity (Morenilla-Palao et al., 2009). The precise molecular and/or biophysical mechanisms involved in lipid raft modulation of TRPM8 channel activity need further study. Lipid rafts have been linked to many vital cell processes such as trafficking and membrane protein sorting (reviewed by (Pristera & Okuse, 2012)), and although still unexplored, this association could be critical for the correct targeting and trafficking dynamics of TRPM8 in somatosensory neurons.

4.2 TRPM8 is an N-glycosylated ion channel

One event that plays an important role in the association of TRPM8 with lipid rafts is the N-glycosylation of the channel, which facilitates the segregation of TRPM8 to these membrane microdomains (Morenilla-Palao et al., 2009). The N-glycosylation, *i.e.* modification by the covalent addition

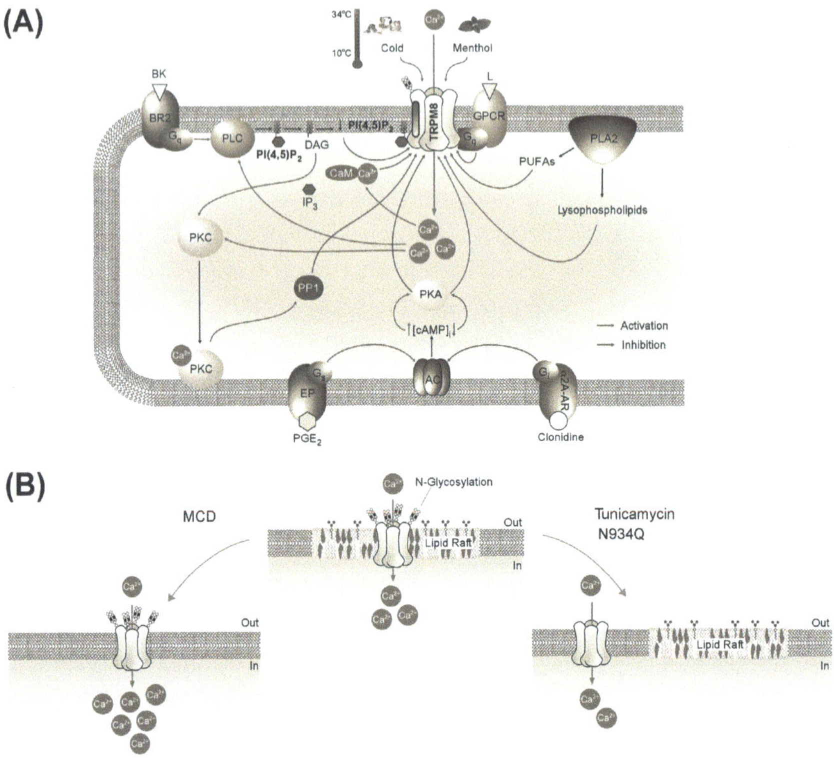

Figure 11.2 The modulation of TRPM8 channel by second messengers, protein–protein interactions and post-translational modifications. TRPM8 is represented as a homotetrameric channel in the scheme. (A) Activation of PLC through the activation of bradykinin (BK) B2 receptor (BR2) induces a reduction in PI(4,5)P_2 levels in the plasma membrane, by the cleavage of PI(4,5)P_2 in diacylglycerol (DAG) and inositol trisphosphate (PI(1,4,5)P_3), which leads to the desensitization of the channel. The activation of prostaglandin receptor (EP) induces an increase in cyclic adenosine monophosphate (cAMP) by the activation of adenylate cyclase (AC). cAMP activates PKA, which can induce a downregulation of the channel by an unknown mechanism. Conversely, the inhibition of α2A adrenoreceptor (α2A-AR) by clonidine induces an inhibition of AC and a reduction of PKA activity that induces a diminution of the phosphorylation level of TRPM8, causing a decrease of channel activity. The activation of TRPM8 channels by cold or menthol can induce an increase in the intracellular Ca^{2+} concentration ($[Ca^{2+}]_i$). This intracellular Ca^{2+} increase can activate PKC, which can modulate protein phosphatase 1 (PP1), leading to the inhibition of the channel by dephosphorylation. The acute Ca^{2+}-dependent desensitization of TRPM8-mediated responses during single sustained application of agonist would be related to Calmodulin (CaM) activation by Ca^{2+}. On the other hand, increases in phospholipase A2 (PLA_2) activity can generate PUFAs and lysophospholipids from glycerophospholipids; PUFAs exert inhibitory effects on TRPM8, and lysophospholipids upregulate the channel function. Gαq subunits bind directly to TRPM8, and through a conformational change of Gαq that follows the activation

of a sugar moiety to an asparagine, takes place in the Asn residue at position 934, in the extracellular loop between transmembrane domains 5 and 6, near the pore region (Dragoni et al., 2006; Erler et al., 2006). Protein extracts from HEK293 cells transfected with TRPM8 show three distinguishable bands, corresponding to different maturation states of the channel with the lowest band representing the nonglycosylated channel. The intermediate band corresponds to the early N-glycosylated form produced in the ER, which results from the transfer of a mannose-rich precursor to the protein. The higher molecular weight band represents the more mature N-glycosylated form, which is generated during the transit of the channel through the Golgi complex and contains terminal sialic acid residues (Morenilla-Palao et al., 2009; Pertusa et al., 2012). All the studies focused on determining the role of this post-translational modification in TRPM8 activity, have concluded that the nonglycosylated TRPM8 mutant (N934Q) displays smaller responses to cold and menthol (Dragoni et al., 2006; Erler et al., 2006; Pertusa et al., 2012). However, some differences arise in the proposed mechanism to explain the decrease in the activity of the protein. Whereas some authors suggest that this diminution is mainly the result of a reduction in the number of TRPM8 channels in the cell surface (Dragoni et al., 2006; Erler et al., 2006), it has been demonstrated that the lack of this post-translational modification has a direct effect on TRPM8 biophysical properties, by shifting the voltage-dependence curve of opening almost 60 mV toward more positive membrane potentials (Pertusa et al., 2012).

In order to understand the physiological relevance of N-glycosylation of TRPM8, it is important to consider that this post-translational modification also occurs in vivo, both in DRG and TG neurons. In cultured primary sensory neurons from trigeminal ganglia, the unglycosylation of native TRPM8 channels dramatically modifies the thermal sensitivity of cold thermoreceptor neurons, shifting the mean threshold of the cold-induced response by ~6 °C to lower temperatures (Pertusa et al., 2012). These

of the G protein-coupled receptors (GPCR) by ligand (L) binding, induces a direct inhibition of the channel. PHB modification and N-glycosylation are represented by a red bar and a schematic sugar moiety in one of the subunits of the channel, respectively. (B) The disruption of lipid rafts by methyl-beta-cyclodextrin (MCD) increases the channel activity. Unglycosylation in response to treatment with Tunicamycin and single point mutation of Asp934 to Gln induce a reduction of TRPM8 sensitivity to cold and menthol, despite the main localization of the channel outside of lipid rafts. (See the color plate.)

results reveal the importance of N-glycosylation in the physiological function of TRPM8, and suggest that in native systems this post-translational modification is a critical molecular determinant in the establishment of cold sensitivity of cold thermoreceptors.

4.3 TRPM8 activation in the endoplasmic reticulum

TRPM8 was originally identified in cells of the human prostate. In cell lines of prostate cancer, TRPM8 has been found to localize not only in the plasma membrane, but also in the endoplasmic reticulum (ER). The role of TRPM8 in the ER is unclear but studies have shown that strong TRPM8 stimulation (cold or 500 μM menthol) activates the channel also in the ER compartment (Valero, Morenilla-Palao, Belmonte, & Viana, 2011; Zhang & Barritt, 2004). As a consequence, in prostate cancer cells the intracellular Ca^{2+} elevation in response to a strong activation of TRPM8 has two components: a main fraction is due to the flow of calcium through the channels in the plasma membrane and a smaller one that is due to Ca^{2+} mobilization from intracellular stores (Valero et al., 2011; Zhang et al., 2004). A similar behavior occurs in transfected HEK293 cells where TRPM8 is overexpressed (Pertusa et al., 2012). Altogether, these results demonstrate that assembled TRPM8 channels can be found in the ER compartment, and that they are functional. However, in both prostate cancer lines and in this recombinant system, a strong nonendogenous stimuli is needed to elicit responses from TRPM8 channels located in the ER (*i.e.* temperature drops from 33 to 18 °C, a high concentration of menthol, or a combination of both stimuli) (Pertusa et al., 2012; Valero et al., 2011; Zhang et al., 2004), suggesting that the TRPM8 channels at the ER are indeed less sensitive to physical and chemical stimulation. The aforesaid has to be kept in mind, in order to consider the activity of these intracellular TRPM8 channels as physiologically relevant.

4.4 PI(4,5)P_2 modulation and Ca^{2+}-dependent desensitization of TRPM8 channels

TRPM8 currents present a strong rundown in excised membranes patches (Liu & Qin, 2005; Rohacs et al., 2005). This rundown, that in some cases can be related to protein or lipid dephosphorylation, provided the first clue to link the TRPM8 activity to PI(4,5)P_2. Two works independently showed that application of PI(4,5)P_2 both restores rundown responses to external agonists and is able to activate the channel in the absence of cold or menthol stimulation (Liu & Qin, 2005; Rohacs et al., 2005). Both groups suggest that $PI(4,5)P_2$ appears to be an obligatory cofactor in TRPM8 activation. The positions that are candidate to mediate the interaction of TRPM8 with $PI(4,5)P_2$ are K995, R998, and R1008, located within the TRP domain (Rohacs et al., 2005). The fact that $PI(4,5)P_2$ can activate TRPM8 in the absence of additional stimulation, and that this lipid is also necessary for its response to exogenous stimuli, suggests a model where cooling agents, temperature, voltage and $PI(4,5)P_2$ regulate the channel in a complex manner, where each one affects the responses to the others (Rohacs et al., 2005). Interestingly, $PI(4,5)P_2$ depletion causes a shift in the $V_{1/2}$ value of voltage activation towards a more positive membrane potential (Daniels, Takashima, & McKemy, 2009).

The role of $PI(4,5)P_2$ in TRPM8 activation suggests that its relative content in the plasma membrane could be a critical factor in the modulation of TRPM8 in a physiological context. Membrane $PI(4,5)P_2$ levels primarily results from the balance between the activity of phosphoinositide kinases, which synthesize $PI(4,5)P_2$ from membrane precursors, and phospholipase C (PLC) that hydrolyzes it generating membrane-bound diacylglycerol (DAG) and cytosolic inositol trisphosphate ($PI(1,4,5)P_3$).

Adaptation of TRPM8 channels depends on extracellular calcium (McKemy et al., 2002). The main molecular mechanism proposed to explain the Ca^{2+}-dependent adaptation of TRPM8 in response to menthol and cold stimulation in the presence of external Ca^{2+} is based on the depletion of $PI(4,5)P_2$ (Daniels et al., 2009; Liu & Qin, 2005; Rohacs et al., 2005). It has been suggested that adaptation is caused by Ca^{2+} influx through the channel, which activates a Ca^{2+}-dependent PLC, diminishing $PI(4,5)P_2$ and therefore reducing the activity of TRPM8 (Daniels et al., 2009; Liu & Qin, 2005; Rohacs et al., 2005) (Figure 11.2). On the other hand, Sarria and coworkers have showed evidences supporting the idea that acute Ca^{2+}-dependent desensitization of TRPM8-mediated responses, which occurs during a single sustained application of agonist, would be related to calmodulin activation and depends on $PI(4,5)P_2$ availability. Tachyphylaxis, which is also Ca^{2+}-dependent and corresponds to the reduction of the responses to repeated agonist applications, would be mediated by hydrolysis of $PI(4,5)P_2$ by PLC and the activation of a PKC/PP1 cascade (Sarria, Ling, Zhu & Gu, 2011).

4.5 Modulation of TRPM8 by protein kinases and G proteins

It has been reported that TRPM8 function can also be modulated by the activity of different protein kinases. Among them, stands out the

Ca^{2+}-dependent protein kinase C (PKC), which can be activated after the PLC activation pathway described in the previous section. It has been shown that the activation of PKC produced a downregulation of TRPM8 activity (Abe, Hosokawa, Sawada, Matsumura, & Kobayashi, 2006; Premkumar et al., 2005). Premkumar and coworkers found that the mechanism of PKC-mediated downregulation of TRPM8 appears to be related to a decrease in the TRPM8 levels of serine-phosphorylation (Premkumar et al., 2005). The authors showed that TRPM8 channels present a basal phosphorylation that is reduced by PKC activation, probably via activation of protein phosphatase-1 (Premkumar et al., 2005) (Figure 11.2). Changes in the phosphorylation state of the channel correlated with the activity displayed by TRPM8. Thus, activators of PKC produce a decrease in phosphorylation levels and also a decrease in channel activity, whereas phosphatase inhibitors and PKC inhibitors reverse these effects (Premkumar et al., 2005). However, the mechanism behind this constitutive phosphorylation, its physiological relevance, and the residues involved in the phenomenon remain as unsolved questions.

The aforementioned examples of intracellular modulation of TRPM8 can explain the desensitization of the channel induced by different inflammatory mediators. For example, acutely applied bradykinin reduces the responses to cold and menthol in TRPM8-expressing neurons, through the activation of Gq protein and PKC mechanisms (Linte et al., 2007; Premkumar et al., 2005). However, other explanations to this observation have recently emerged, indicating that the inhibition of TRPM8 activity is largely independent of cell signaling pathways downstream of $G\alpha q$-coupled receptors, such as PLC and PKC activity (Zhang et al., 2012). Instead, the authors propose a novel mechanism where the $G\alpha q$ subunit binds directly to TRPM8, and through the conformational change of $G\alpha q$ that follows the activation of the G protein coupled receptor, causes a direct *trans*-inhibition of the channel (Zhang et al., 2012) (Figure 11.2). Interestingly, other authors that also reported the interaction of the $G\alpha q$ and TRPM8, proposed that the stimulation of the channel by menthol also can activate the $G\alpha q$ pathway inducing a Ca^{2+} release from intracellular stores independent of the influx of ions through the channel, suggesting that the operational range of TRPM8 may include an additional metabotropic capacity (Klasen et al., 2012).

On the other hand, different studies have also correlated a decreased phosphorylation level of the channel with a reduction in its activity. Bavencoffe and coworkers reported a phosphorylation of the residues S9 and T17 of the channel associated to Protein Kinase A (PKA) activity. The authors proposed that activation of the G_i protein and subsequent inhibition of the adenylate cyclase (AC)/cAMP/PKA pathway results in a decrease in the PKA-dependent phosphorylation of TRPM8, causing a reduction of the channel activity (Bavencoffe et al., 2010) (Figure 11.2). The effects of G protein-PKA activation on TRPM8 channel function are diverse. In this study, the inhibition of PKA via G_i-coupled α2A-adrenoreceptor produces a reduction in the responses of CSNs to temperature decreases (Bavencoffe et al., 2010) (Figure 11.2). Nevertheless, other authors have reported an opposite effect: activators of the AC/cAMP/PKA pathway, such as Forskolin and 8-Br-cAMP, induce a diminution of TRPM8 activity (De Petrocellis et al., 2007; Linte et al., 2007) (Figure 11.2). Moreover, PKA activation is involved in the TRPM8 inhibition caused by prostaglandin E2 (Linte et al., 2007) (Figure 11.2).

To sum up, several lines of evidence suggest that the activity of TRPM8 in CSNs can be finely modulated in sensory neurons and heterologous expression systems, and some of the molecular mechanisms explained here could be relevant in pathologies such as neuropathic pain. However, more efforts are needed to fully understand the real relevance of these processes on the TRPM8 function in vivo under both physiological and physiopathological conditions.

4.6 Modulation of TRPM8 channels by PLA_2

Calcium insensitive phospholipase A_2 ($iPLA_2$) also appears to have a role in the modulation of TRPM8 activity. This enzyme hydrolyzes the sn-2 ester of glycerophospholipids to release a free polyunsaturated fatty acid (PUFA) and a lysophospholipids (LPL). It has been reported that both products are able to differentially modulate TRPM8 activity. Whereas lysophospholipids act as positive modulators, PUFAs inhibit TRPM8 activity (Andersson, Nash, & Bevan, 2007; Vanden Abeele, Zholos, Bidaux, Shuba, Thebault, Beck, Flourakis, Panchin, Skryma, & Prevarskaya, 2006) (Figure 11.2). Cells expressing TRPM8 display reduced responses to cold and icilin after incubation with $iPLA_2$ inhibitors, suggesting that an $iPLA_2$, and the levels of the LPLs generated during its constitutive activity, are important for TRPM8 function. LPLs and PUFAs are both released by $iPLA_2$ activation, suggesting that the stimulatory effect of LPLs may be cancelled out by the simultaneous inhibitory action of the PUFAs. Nevertheless, $iPLA_2$ inhibition causes a reduction of TRPM8 responses to cold, menthol and icilin, suggesting an overall stimulatory effect on the channel of the $iPLA_2$ products. Indeed, in experiments where equimolar concentrations of LPLs and arachidonic acid are used, the LPLs can still activate TRPM8 suggesting that LPLs are more efficient as activators than PUFAs as inhibitors of the channel (Andersson et al., 2007). The authors proposed that endogenous modulators, such as LPLs and the aforementioned PI(4,5)P_2, would shift the temperature threshold of TRPM8 to warmer temperatures, supporting the idea that the composition of the lipid bilayer could have a significant role in the temperature sensitivity of cold thermoreceptors.

4.7 Other modulatory mechanisms in TRPM8

Recently, other partner proteins that are also able to modulate TRPM8 function have been identified. Among them, it has been described that Pirt, a membrane protein with a restricted expression in the peripheral nervous system (Kim et al., 2008), directly interacts with TRPM8 acting as an endogenous regulator (Tang et al., 2013). Calcium imaging and electrophysiological experiments in HEK293 cells expressing TRPM8, with or without Pirt, evidenced that this protein, mainly through the positively charged residues in its C-terminal domain, increases the sensitivity of TRPM8 to cold and menthol (Tang et al., 2013). In agreement with the results obtained using this heterologous expression system, Pirt knockout mice show a reduced cold sensitivity, suggesting a role of this protein in the tuning of cold responses in vivo (Tang et al., 2013). The results of this work suggest that the positive modulation cold and menthol responses by Pirt is due to changes in the biophysical properties of TRPM8, and rule out an effect of Pirt in TRPM8 expression and membrane trafficking (Tang et al., 2013).

Other interactions modulate TRPM8 function regulating trafficking events. It has been described that the μ-opioid receptor 1 (OPRM1) and TRPM8 directly interacts (Shapovalov et al., 2013). In conditions where OPRM1 is activated (e.g. by morphine), a decreased response of TRPM8 to cold and menthol can be observed in both cultured DRG and transfected HEK293 cells (Shapovalov et al., 2013). This reduction is not the result of changes in channel gating properties. Instead, activation of OPRM1 leads to internalization of TRPM8 causing a reduction in the surface TRPM8 expression (Shapovalov et al., 2013).

Interestingly, TRPM8 activity can be also modulated by inorganic polyphosphate (polyP) (Zakharian, Thyagarajan, French, Pavlov, & Rohacs, 2009). The TRPM8 channel form a complex with polyP associated via ionic interactions. PolyP appears to be an important modulator of

TRPM8 channel activity, since enzymatic degradation of this polymer causes an inhibition of channel (Zakharian et al., 2009). The same authors suggested that polyP-TRPM8 complex might be supported by the presence of other homopolymer, poly-(R)-3-hydroxybutyrate (PHB), and demonstrated that TRPM8 is actually conjugated with this polymer (Cao et al., 2013; Zakharian et al., 2009). The PHB molecule is comprised of hydrophobic methyl groups alternating with hydrophilic ester groups, and it has a CoA-ester binding group at its C-terminal. The association with TRPM8 occurs in specific serine residues via formation of ester bonds (Cao et al., 2013). Mass spectrometry analysis showed several PHB modifications in the N-terminal domain of TRPM8. However, only one PHB-modification site was identified on the extracellular side of TRPM8, specifically in the S3–S4 linker. The authors demonstrated that removal of the polyester modification in this extracellular position resulted in a significant reduction of TRPM8 channel activity, suggesting that this post-translational modification is also required for TRPM8 normal function (Cao et al., 2013).

5. CONCLUSIONS

A large list of cellular, molecular, electrophysiological, and behavioral studies demonstrates that TRPM8 is the main molecular determinant of cold sensitivity in primary somatosensory neurons. The uncontested involvement of TRPM8 in innocuous cold thermotransduction has also been extended nowadays in part to noxious cold detection. TRPM8 expressing neurons are fundamental players in the maintenance of the wetness of the surface of the eye, and also have an important role in cold-induced pain resulting from axonal damage. The expression of this channel in different tissues and tumors suggests that its general physiological role can be extended beyond thermosensation. The function and biophysical properties of the channel can be finely regulated by diverse intracellular cascades of second messengers, protein–protein interactions and post-translational modifications, which strongly influence trafficking and thermal and chemical sensitivity of this allosterically gated polymodal ion channel.

ACKNOWLEDGMENTS

Supported by Grants FONDECYT 1131064 (RM), FONDECYT 11130144 (MP) and CONICYT Anillo ACT-1113 (RM and MP). We thank Dr. Annika Mälkiä for comments to the manuscript and A. González for help with the illustrations.

REFERENCES

Abe, J., Hosokawa, H., Sawada, Y., Matsumura, K., & Kobayashi, S. (2006). Ca^{2+}-dependent PKC activation mediates menthol-induced desensitization of transient receptor potential M8. *Neuroscience Letters, 397*, 140–144.

Abrahamsen, B., Zhao, J., Asante, C. O., Cendan, C. M., Marsh, S., Martinez-Barbera, J. P., et al. (2008). The cell and molecular basis of mechanical, cold, and inflammatory pain. *Science, 321*, 702–705.

Almaraz, L., Manenschijn, J. A., de la Pena, E., & Viana, F. (2014). Trpm8. *Handbook of Experimental Pharmacology, 222*, 547–579.

Andersson, D. A., Nash, M., & Bevan, S. (2007). Modulation of the cold-activated channel TRPM8 by lysophospholipids and polyunsaturated fatty acids. *Journal of Neuroscience, 27*, 3347–3355.

Axelsson, H. E., Minde, J. K., Sonesson, A., Toolanen, G., Hogestatt, E. D., & Zygmunt, P. M. (2009). Transient receptor potential vanilloid 1, vanilloid 2 and melastatin 8 immunoreactive nerve fibers in human skin from individuals with and without Norrbottnian congenital insensitivity to pain. *Neuroscience, 162*, 1322–1332.

Babes, A., Ciobanu, A. C., Neacsu, C., & Babes, R. M. (2011). TRPM8, a sensor for mild cooling in mammalian sensory nerve endings. *Current Pharmaceutical Biotechnology, 12*, 78–88.

Babes, A., Zorzon, D., & Reid, G. (2004). Two populations of cold-sensitive neurons in rat dorsal root ganglia and their modulation by nerve growth factor. *European Journal of Neuroscience, 20*, 2276–2282.

Bandell, M., Dubin, A. E., Petrus, M. J., Orth, A., Mathur, J., Hwang, S. W., et al. (2006). High-throughput random mutagenesis screen reveals TRPM8 residues specifically required for activation by menthol. *Nature Neuroscience, 9*, 493–500.

Bautista, D. M., Siemens, J., Glazer, J. M., Tsuruda, P. R., Basbaum, A. I., Stucky, C. L., et al. (2007). The menthol receptor TRPM8 is the principal detector of environmental cold. *Nature, 448*, 204–208.

Bavencoffe, A., Gkika, D., Kondratskyi, A., Beck, B., Borowiec, A. S., Bidaux, G., et al. (2010). The transient receptor potential channel TRPM8 is inhibited via the alpha 2A adrenoreceptor signaling pathway. *Journal of Biological Chemistry, 285*, 9410–9419.

Belmonte, C., Brock, J. A., & Viana, F. (2009). Converting cold into pain. *Experimental Brain Research, 196*, 13–30.

Bezzerides, V. J., Ramsey, I. S., Kotecha, S., Greka, A., & Clapham, D. E. (2004). Rapid vesicular translocation and insertion of TRP channels. *Nature Cell Biology, 6*, 709–720.

Bidaux, G., Flourakis, M., Thebault, S., Zholos, A., Beck, B., Gkika, D., et al. (2007). Prostate cell differentiation status determines transient receptor potential melastatin member 8 channel subcellular localization and function. *Journal of Clinical Investigation, 117*, 1647–1657.

Brauchi, S., Orio, P., & Latorre, R. (2004). Clues to understanding cold sensation: thermodynamics and electrophysiological analysis of the cold receptor TRPM8. *Proceedings of the National Academy of Sciences of the United States of America, 101*, 15494–15499.

Brauchi, S., Orta, G., Mascayano, C., Salazar, M., Raddatz, N., Urbina, H., et al. (2007). Dissection of the components for PIP_2 activation and thermosensation in TRP channels. *Proceedings of the National Academy of the Sciences of the United States of America, 104*, 10246–10251.

Brauchi, S., Orta, G., Salazar, M., Rosenmann, E., & Latorre, R. (2006). A hot-sensing cold receptor: c-terminal domain determines thermosensation in transient receptor potential channels. *Journal of Neuroscience, 26*, 4835–4840.

Brock, J. A., McLachlan, E. M., & Belmonte, C. (1998). Tetrodotoxin-resistant impulses in single nociceptor nerve terminals in guinea-pig cornea. *Journal of Physiology, 512*(Pt. 1), 211–217.

Brock, J. A., Pianova, S., & Belmonte, C. (2001). Differences between nerve terminal impulses of polymodal nociceptors and cold sensory receptors of the guinea-pig cornea. *Journal of Physiology, 533*, 493–501.

Campero, M., Baumann, T. K., Bostock, H., & Ochoa, J. L. (2009). Human cutaneous C fibres activated by cooling, heating and menthol. *Journal of Physiology, 587*, 5633–5652.

Campero, M., Serra, J., & Ochoa, J. L. (1996). C-polymodal nociceptors activated by noxious low temperature in human skin. *Journal of Physiology, 497*(Pt. 2), 565–572.

Cao, C., Yudin, Y., Bikard, Y., Chen, W., Liu, T., Li, H., et al. (2013). Polyester modification of the mammalian TRPM8 channel protein: implications for structure and function. *Cell Reports, 4*, 302–315.

Caspani, O., Zurborg, S., Labuz, D., & Heppenstall, P. A. (2009). The contribution of TRPM8 and TRPA1 channels to cold allodynia and neuropathic pain. *PLoS One, 4*, e7383.

Chodon, D., Guilbert, A., Dhennin-Duthille, I., Gautier, M., Telliez, M. S., Sevestre, H., et al. (2010). Estrogen regulation of TRPM8 expression in breast cancer cells. *BMC Cancer, 10*, 212.

Chuang, H. H., Neuhausser, W. M., & Julius, D. (2004). The super-cooling agent icilin reveals a mechanism of coincidence detection by a temperature-sensitive TRP channel. *Neuron, 43*, 859–869.

Clapham, D. E. (2003). TRP channels as cellular sensors. *Nature, 426*, 517–524.

Colburn, R. W., Lubin, M. L., Stone, D. J., Jr., Wang, Y., Lawrence, D., D'Andrea, M. R., et al. (2007). Attenuated cold sensitivity in TRPM8 null mice. *Neuron, 54*, 379–386.

Croze, S., Duclaux, R., & Kenshalo, D. R. (1976). The thermal sensitivity of the polymodal nociceptors in the monkey. *Journal of Physiology, 263*, 539–562.

Daniels, R. L., Takashima, Y., & McKemy, D. D. (2009). Activity of the neuronal cold sensor TRPM8 is regulated by phospholipase C via the phospholipid phosphoinositol 4,5-bisphosphate. *Journal of Biological Chemistry, 284*, 1570–1582.

De Blas, G. A., Darszon, A., Ocampo, A. Y., Serrano, C. J., Castellano, L. E., Hernandez-Gonzalez, E. O., et al. (2009). TRPM8, a versatile channel in human sperm. *PLoS One, 4*, e6095.

de la Peña, E., Malkia, A., Cabedo, H., Belmonte, C., & Viana, F. (2005). The contribution of TRPM8 channels to cold sensing in mammalian neurones. *Journal of Physiology, 567*, 415–426.

De Petrocellis, L., Starowicz, K., Moriello, A. S., Vivese, M., Orlando, P., Di, M., et al. (2007). Regulation of transient receptor potential channels of melastatin type 8 (TRPM8): effect of cAMP, cannabinoid CB(1) receptors and endovanilloids. *Experimental Cell Research, 313*, 1911–1920.

Dhaka, A., Earley, T. J., Watson, J., & Patapoutian, A. (2008). Visualizing cold spots: TRPM8-expressing sensory neurons and their projections. *Journal of Neuroscience, 28*, 566–575.

Dhaka, A., Murray, A. N., Mathur, J., Earley, T. J., Petrus, M. J., & Patapoutian, A. (2007). TRPM8 is required for cold sensation in mice. *Neuron, 54*, 371–378.

Dragoni, I., Guida, E., & McIntyre, P. (2006). The cold and menthol receptor TRPM8 contains a functionally important double cysteine motif. *Journal of Biological Chemistry, 281*, 37353–37360.

Dray, A. (2008). Neuropathic pain: emerging treatments. *British Journal of Anaesthesia, 101*, 48–58.

El Karim, I., Linden, G. J., Curtis, T. M., About, I., McGahon, M. K., Irwin, C. R., et al. (2010). Human odontoblasts express functional thermo-sensitive TRP channels: Implications for dentin sensitivity. *Pain, 152*, 2211–2223.

Emery, E. C., Young, G. T., Berrocoso, E. M., Chen, L., & McNaughton, P. A. (2011). HCN2 ion channels play a central role in inflammatory and neuropathic pain. *Science, 333*, 1462–1466.

Erler, I., Al-Ansary, D. M., Wissenbach, U., Wagner, T. F., Flockerzi, V., & Niemeyer, B. A. (2006). Trafficking and assembly of the cold-sensitive TRPM8 channel. *Journal of Biological Chemistry, 281*, 38396–38404.

Fleig, A., & Penner, R. (2004). The TRPM ion channel subfamily: molecular, biophysical and functional features. *Trends in Pharmacological Sciences, 25*, 633–639.

Frederick, J., Buck, M. E., Matson, D. J., & Cortright, D. N. (2007). Increased TRPA1, TRPM8, and TRPV2 expression in dorsal root ganglia by nerve injury. *Biochemical and Biophysical Research Communications, 358*, 1058–1064.

Hensel, H. (1981). Thermoreception and temperature regulation. *Monographs of the Physiological Society, 38*, 1–321.

Hensel, H., & Zotterman, Y. (1951). Action potentials of cold fibres and intracutaneous temperature gradient. *Journal of Neurophysiology, 14*, 377–385.

Hensel, H., & Zotterman, Y. (1951). The effect of menthol on the thermoreceptors. *Acta Physiologica Scandinavica, 24*, 27–34.

Iriuchijima, J., & Zotterman, Y. (1960). The specificity of afferent cutaneous C fibres in mammals. *Acta Physiol Scand, 49*, 267–278.

Johnson, C. D., Melanaphy, D., Purse, A., Stokesberry, S. A., Dickson, P., & Zholos, A. V. (2009). Transient receptor potential melastatin 8 channel involvement in the regulation of vascular tone. *American Journal of Physiology - Heart Circulatory Physiology, 296*, H1868–H1877.

Kanzaki, M., Zhang, Y. Q., Mashima, H., Li, L., Shibata, H., & Kojima, I. (1999). Translocation of a calcium-permeable cation channel induced by insulin-like growth factor-I. *Nature Cell Biology, 1*, 165–170.

Katsura, H., Obata, K., Mizushima, T., Yamanaka, H., Kobayashi, K., Dai, Y., et al. (2006). Antisense knock down of TRPA1, but not TRPM8, alleviates cold hyperalgesia after spinal nerve ligation in rats. *Experimental Neurology, 200*, 112–123.

Katsura, H., Tsuzuki, K., Noguchi, K., & Sakagami, M. (2006). Differential expression of capsaicin-, menthol-, and mustard oil-sensitive receptors in naive rat geniculate ganglion neurons. *Chemical Senses, 31*, 681–688.

Kim, A. Y., Tang, Z., Liu, Q., Patel, K. N., Maag, D., Geng, Y., et al. (2008). Pirt, a phosphoinositide-binding protein, functions as a regulatory subunit of TRPV1. *Cell, 133*, 475–485.

Klasen, K., Hollatz, D., Zielke, S., Gisselmann, G., Hatt, H., & Wetzel, C. H. (2012). The TRPM8 ion channel comprises direct Gq protein-activating capacity. *Pflugers Archiv, 463*, 779–797.

Knowlton, W. M., Bifolck-Fisher, A., Bautista, D. M., & McKemy, D. D. (2010). TRPM8, but not TRPA1, is required for neural and behavioral responses to acute noxious cold temperatures and cold-mimetics in vivo. *Pain, 150*, 340–350.

Knowlton, W. M., Daniels, R. L., Palkar, R., McCoy, D. D., & McKemy, D. D. (2011). Pharmacological blockade of TRPM8 ion channels alters cold and cold pain responses in mice. *PLoS One, 6*, e25894.

Knowlton, W. M., Palkar, R., Lippoldt, E. K., McCoy, D. D., Baluch, F., Chen, J., et al. (2013). A sensory-labeled line for cold: TRPM8-expressing sensory neurons define the cellular basis for cold, cold pain, and cooling-mediated analgesia. *Journal of Neuroscience, 33*, 2837–2848.

Kuhn, F. J., Knop, G., & Luckhoff, A. (2007). The transmembrane segment S6 determines cation versus anion selectivity of TRPM2 and TRPM8. *Journal of Biological Chemistry, 282*, 27598–27609.

LaMotte, R. H., & Thalhammer, J. G. (1982). Response properties of high-threshold cutaneous cold receptors in the primate. *Brain Research, 244*, 279–287.

Latorre, R., Brauchi, S., Madrid, R., & Orio, P. (2011). A cool channel in cold transduction. *Physiology (Bethesda), 26*, 273–285.

Linte, R. M., Ciobanu, C., Reid, G., & Babes, A. (2007). Desensitization of cold- and menthol-sensitive rat dorsal root ganglion neurones by inflammatory mediators. *Experimental Brain Research, 178*, 89–98.

Liu, B., Fan, L., Balakrishna, S., Sui, A., Morris, J. B., & Jordt, S. E. (2013). TRPM8 is the principal mediator of menthol-induced analgesia of acute and inflammatory pain. *Pain, 154*, 2169–2177.

Liu, B., & Qin, F. (2005). Functional control of cold- and menthol-sensitive TRPM8 ion channels by phosphatidylinositol 4,5-bisphosphate. *Journal of Neuroscience, 25*, 1674–1681.

Madrid, R., de la Peña, E., Donovan-Rodriguez, T., Belmonte, C., & Viana, F. (2009). Variable threshold of trigeminal cold-thermosensitive neurons is determined by a balance between TRPM8 and Kv1 potassium channels. *Journal of Neuroscience, 29*, 3120–3131.

Madrid, R., Donovan-Rodriguez, T., Meseguer, V., Acosta, M. C., Belmonte, C., & Viana, F. (2006). Contribution of TRPM8 channels to cold transduction in primary sensory neurons and peripheral nerve terminals. *Journal of Neuroscience, 26*, 12512–12525.

Mahieu, F., Janssens, A., Gees, M., Talavera, K., Nilius, B., & Voets, T. (2010). Modulation of the cold-activated cation channel TRPM8 by surface charge screening. *Journal of Physiology, 588*, 315–324.

Mälkiä, A., Madrid, R., Meseguer, V., de la, P. E., Valero, M., Belmonte, C., et al. (2007). Bidirectional shifts of TRPM8 channel gating by temperature and chemical agents modulate the cold sensitivity of mammalian thermoreceptors. *Journal of Physiology, 581*, 155–174.

Mälkiä, A., Pertusa, M., Fernandez-Ballester, G., Ferrer-Montiel, A., & Viana, F. (2009). Differential role of the menthol-binding residue Y745 in the antagonism of thermally gated TRPM8 channels. *Molecular Pain, 5*, 62.

Martinez-Lopez, P., Trevino, C. L., de la Vega-Beltran, J. L., De, B. G., Monroy, E., Beltran, C., et al. (2011). TRPM8 in mouse sperm detects temperature changes and may influence the acrosome reaction. *Journal of Cellular Physiology, 226*, 1620–1631.

Matta, J. A., & Ahern, G. P. (2007). Voltage is a partial activator of rat thermosensitive TRP channels. *Journal of Physiology, 585*, 469–482.

Mauderli, A. P., Vierck, C. J., Jr., Cannon, R. L., Rodrigues, A., & Shen, C. (2003). Relationships between skin temperature and temporal summation of heat and cold pain. *Journal of Neurophysiology, 90*, 100–109.

McCoy, D. D., Knowlton, W. M., & McKemy, D. D. (2011). Scraping through the ice: uncovering the role of TRPM8 in cold transduction. *American Journal of Physiology Regulatory, Integrrative and Comparative Physiology, 300*, R1278–R1287.

McCoy, D. D., Zhou, L., Nguyen, A. K., Watts, A. G., Donovan, C. M., & McKemy, D. D. (2013). Enhanced insulin clearance in mice lacking TRPM8 channels. *American Journal of Physiology - Endocrinology and Metabolism, 305*, E78–E88.

McKemy, D. D. (2013). The molecular and cellular basis of cold sensation. *ACS Chemical Neuroscience, 4*, 238–247.

McKemy, D. D., Neuhausser, W. M., & Julius, D. (2002). Identification of a cold receptor reveals a general role for TRP channels in thermosensation. *Nature, 416*, 52–58.

Mergler, S., Mertens, C., Valtink, M., Reinach, P. S., Castelo, S. V., et al. (2013). Functional significance of thermosensitive transient receptor potential melastatin channel 8 (TRPM8) expression in immortalized human corneal endothelial cells. *Experimental Eye Research, 116*, 337–349.

Meseguer, V., Karashima, Y., Talavera, K., D'Hoedt, D., Donovan-Rodriguez, T., Viana, F., et al. (2008). Transient receptor potential channels in sensory neurons are targets of the antimycotic agent clotrimazole. *Journal of Neuroscience, 28*, 576–586.

Montell, C. (2005). The TRP superfamily of cation channels. *Science's.STKE, 2005*, re3.

Morenilla-Palao, C., Pertusa, M., Meseguer, V., Cabedo, H., & Viana, F. (2009). Lipid raft segregation modulates TRPM8 channel activity. *Journal of Biological Chemistry, 284*, 9215–9224.

Morin, C., & Bushnell, M. C. (1998). Temporal and qualitative properties of cold pain and heat pain: a psychophysical study. *Pain, 74*, 67–73.

Namer, B., Kleggetveit, I. P., Handwerker, H., Schmelz, M., & Jorum, E. (2008). Role of TRPM8 and TRPA1 for cold allodynia in patients with cold injury. *Pain, 139*, 63–72.

Okazawa, M., Inoue, W., Hori, A., Hosokawa, H., Matsumura, K., & Kobayashi, S. (2004). Noxious heat receptors present in cold-sensory cells in rats. *Neuroscience Letters, 359*, 33–36.

Orio, P., Madrid, R., de la, P. E., Parra, A., Meseguer, V., Bayliss, D. A., et al. (2009). Characteristics and physiological role of hyperpolarization activated currents in mouse cold thermoreceptors. *Journal of Physiology, 587*, 1961–1976.

Orio, P., Parra, A., Madrid, R., Gonzalez, O., Belmonte, C., & Viana, F. (2012). Role of I_h in the firing pattern of mammalian cold thermoreceptor endings. *Journal of Neurophysiology, 108*, 3009–3023.

Owsianik, G., Talavera, K., Voets, T., & Nilius, B. (2006). Permeation and selectivity of TRP channels. *Annual Review of Physiology, 68*, 685–717.

Parra, A., Madrid, R., Echevarria, D., Del, O. S., Morenilla-Palao, C., Acosta, M. C., et al. (2010). Ocular surface wetness is regulated by TRPM8-dependent cold thermoreceptors of the cornea. *Nature Medicine, 16*, 1396–1399.

Patapoutian, A., Tate, S., & Woolf, C. J. (2009). Transient receptor potential channels: targeting pain at the source. *Nature Review Drug Discovery, 8*, 55–68.

Pedretti, A., Marconi, C., Bettinelli, I., & Vistoli, G. (2009). Comparative modeling of the quaternary structure for the human TRPM8 channel and analysis of its binding features. *Biochimica et Biophysica Acta, 1788*, 973–982.

Peier, A. M., Moqrich, A., Hergarden, A. C., Reeve, A. J., Andersson, D. A., Story, G. M., et al. (2002). A TRP channel that senses cold stimuli and menthol. *Cell, 108*, 705–715.

Pertusa, M., González, A., Hardy, P., Madrid, R., & Viana, F. (2014). Bidirectional modulation of thermal and chemical sensitivity of TRPM8 channels by the initial region of the N-terminal domain. *J Biol Chem, 289*, 21828–21843.

Pertusa, M., Madrid, R., Morenilla-Palao, C., Belmonte, C., & Viana, F. (2012). N-Glycosylation of TRPM8 ion channels modulates temperature sensitivity of cold thermoreceptor neurons. *Journal of Biological Chemistry, 287*, 18218–18229.

Phelps, C. B., & Gaudet, R. (2007). The role of the N terminus and transmembrane domain of TRPM8 in channel localization and tetramerization. *Journal of Biological Chemistry, 282*, 36474–36480.

Pogorzala, L. A., Mishra, S. K., & Hoon, M. A. (2013). The cellular code for mammalian thermosensation. *Journal of Neuroscience, 33*, 5533–5541.

Premkumar, L. S., Raisinghani, M., Pingle, S. C., Long, C., & Pimentel, F. (2005). Downregulation of transient receptor potential melastatin 8 by protein kinase c-mediated dephosphorylation. *Journal of Neuroscience, 25*, 11322–11329.

Pristera, A., & Okuse, K. (2012). Building excitable membranes: lipid rafts and multiple controls on trafficking of electrogenic molecules. *Neuroscientist, 18*, 70–81.

Proudfoot, C. J., Garry, E. M., Cottrell, D. F., Rosie, R., Anderson, H., Robertson, D. C., et al. (2006). Analgesia mediated by the TRPM8 cold receptor in chronic neuropathic pain. *Current Biology, 16*, 1591–1605.

Ramsey, I. S., Delling, M., & Clapham, D. E. (2006). An introduction to TRP channels. *Annual Review of Physiology, 68*, 619–647.

Reid, G., Babes, A., & Pluteanu, F. (2002). A cold- and menthol-activated current in rat dorsal root ganglion neurones: properties and role in cold transduction. *Journal of Physiology, 545*, 595–614.

Rohacs, T., Lopes, C. M., Michailidis, I., & Logothetis, D. E. (2005). PI(4,5)P_2 regulates the activation and desensitization of TRPM8 channels through the TRP domain. *Nature Neuroscience, 8*, 626–634.

Sabnis, A. S., Shadid, M., Yost, G. S., & Reilly, C. A. (2008). Human lung epithelial cells express a functional cold-sensing TRPM8 variant. *American Journal of Respiratory Cell and Molecular Biology, 39*, 466–474.

Sarria, I., Ling, J., Zhu, M. X., & Gu, J. G. (2011). TRPM8 acute desensitization is mediated by calmodulin and requires PIP(2): distinction from tachyphylaxis. *J Neurophysiol, 106*, 3056–3066.

Schmidt, M., Dubin, A. E., Petrus, M. J., Earley, T. J., & Patapoutian, A. (2009). Nociceptive signals induce trafficking of TRPA1 to the plasma membrane. *Neuron, 64*, 498–509.

Shapovalov, G., Gkika, D., Devilliers, M., Kondratskyi, A., Gordienko, D., Busserolles, J., et al. (2013). Opiates modulate thermosensation by internalizing cold receptor TRPM8. *Cell Reports, 4*, 504–515.

Simone, D. A., & Kajander, K. C. (1996). Excitation of rat cutaneous nociceptors by noxious cold. *Neuroscience Letters, 213*, 53–56.

Simons, K., & Ikonen, E. (1997). Functional rafts in cell membranes. *Nature, 387*, 569–572.

Sobhan, U., Sato, M., Shinomiya, T., Okubo, M., Tsumura, M., Muramatsu, T., et al. (2013). Immunolocalization and distribution of functional temperature-sensitive TRP channels in salivary glands. *Cell and Tissue Research, 354*, 507–519.

Staaf, S., Oerther, S., Lucas, G., Mattsson, J. P., & Ernfors, P. (2009). Differential regulation of TRP channels in a rat model of neuropathic pain. *Pain, 144*, 187–199.

Stein, R. J., Santos, S., Nagatomi, J., Hayashi, Y., Minnery, B. S., Xavier, M., et al. (2004). Cool (TRPM8) and hot (TRPV1) receptors in the bladder and male genital tract. *Journal of Urology, 172*, 1175–1178.

Stewart, A. P., Egressy, K., Lim, A., & Edwardson, J. M. (2010). AFM imaging reveals the tetrameric structure of the TRPM8 channel. *Biochemical and Biophysical Research Communications, 394*, 383–386.

Su, L., Wang, C., Yu, Y. H., Ren, Y. Y., Xie, K. L., & Wang, G. L. (2011). Role of TRPM8 in dorsal root ganglion in nerve injury-induced chronic pain. *BMC Neuroscience, 12*, 120.

Takashima, Y., Daniels, R. L., Knowlton, W., Teng, J., Liman, E. R., & McKemy, D. D. (2007). Diversity in the neural circuitry of cold sensing revealed by genetic axonal labeling of transient receptor potential melastatin 8 neurons. *Journal of Neuroscience, 27*, 14147–14157.

Takashima, Y., Ma, L., & McKemy, D. D. (2010). The development of peripheral cold neural circuits based on TRPM8 expression. *Neuroscience, 169*, 828–842.

Tang, Z., Kim, A., Masuch, T., Park, K., Weng, H., Wetzel, C., et al. (2013). Pirt functions as an endogenous regulator of TRPM8. *Nature Communications, 4*, 2179.

Thebault, S., Lemonnier, L., Bidaux, G., Flourakis, M., Bavencoffe, A., Gordienko, D., et al. (2005). Novel role of cold/menthol-sensitive transient receptor potential melastatine family member 8 (TRPM8) in the activation of store-operated channels in LNCaP human prostate cancer epithelial cells. *Journal of Biological Chemistry, 280*, 39423–39435.

Tsavaler, L., Shapero, M. H., Morkowski, S., & Laus, R. (2001). Trp-p8, a novel prostate-specific gene, is up-regulated in prostate cancer and other malignancies and shares high homology with transient receptor potential calcium channel proteins. *Cancer Research, 61*, 3760–3769.

Tsuruda, P. R., Julius, D., & Minor, D. L., Jr. (2006). Coiled coils direct assembly of a cold-activated TRP channel. *Neuron, 51*, 201–212.

Valero, M., Morenilla-Palao, C., Belmonte, C., & Viana, F. (2011). Pharmacological and functional properties of TRPM8 channels in prostate tumor cells. *Pflugers Archiv, 461*, 99–114.

Vanden Abeele, F., Zholos, A., Bidaux, G., Shuba, Y., Thebault, S., Beck, B., et al. (2006). Ca^{2+}-independent phospholipase A2-dependent gating of TRPM8 by lysophospholipids. *J Biol Chem, 281*, 40174–40182.
Veliz, L. A., Toro, C. A., Vivar, J. P., Arias, L. A., Villegas, J., Castro, M. A., et al. (2010). Near-membrane dynamics and capture of TRPM8 channels within transient confinement domains. *PLoS One, 5*, e13290.
Viana, F., de la Pena, E., & Belmonte, C. (2002). Specificity of cold thermotransduction is determined by differential ionic channel expression. *Nature Neuroscience, 5*, 254–260.
Voets, T., Droogmans, G., Wissenbach, U., Janssens, A., Flockerzi, V., & Nilius, B. (2004). The principle of temperature-dependent gating in cold- and heat-sensitive TRP channels. *Nature, 430*, 748–754.
Voets, T., Owsianik, G., Janssens, A., Talavera, K., & Nilius, B. (2007). TRPM8 voltage sensor mutants reveal a mechanism for integrating thermal and chemical stimuli. *Nature Chemical Biology, 3*, 174–182.
Wrigley, P. J., Jeong, H. J., & Vaughan, C. W. (2009). Primary afferents with TRPM8 and TRPA1 profiles target distinct subpopulations of rat superficial dorsal horn neurones. *British Journal of Pharmacology, 157*, 371–380.
Xing, H., Chen, M., Ling, J., Tan, W., & Gu, J. G. (2007). TRPM8 mechanism of cold allodynia after chronic nerve injury. *Journal of Neuroscience, 27*, 13680–13690.
Xing, H., Ling, J., Chen, M., & Gu, J. G. (2006). Chemical and cold sensitivity of two distinct populations of TRPM8-expressing somatosensory neurons. *Journal of Neurophysiology, 95*, 1221–1230.
Yamamura, H., Ugawa, S., Ueda, T., Morita, A., & Shimada, S. (2008). TRPM8 activation suppresses cellular viability in human melanoma. *American Journal of Physiology - Cell Physiology, 295*, C296–C301.
Yang, X. R., Lin, M. J., McIntosh, L. S., & Sham, J. S. (2006). Functional expression of transient receptor potential melastatin- and vanilloid-related channels in pulmonary arterial and aortic smooth muscle. *American Journal of Physiology - Lung Cellular and Molecular.-Physiology, 290*, L1267–L1276.
Yee, N. S., Zhou, W., & Lee, M. (2010). Transient receptor potential channel TRPM8 is over-expressed and required for cellular proliferation in pancreatic adenocarcinoma. *Cancer Letters, 297*, 49–55.
Yudin, Y., & Rohacs, T. (2012). Regulation of TRPM8 channel activity. *Molecular and Cellular Endocrinology, 353*, 68–74.
Zakharian, E., Thyagarajan, B., French, R. J., Pavlov, E., & Rohacs, T. (2009). Inorganic polyphosphate modulates TRPM8 channels. *PLoS One, 4*, e5404.
Zhang, L., & Barritt, G. J. (2004). Evidence that TRPM8 is an androgen-dependent Ca^{2+} channel required for the survival of prostate cancer cells. *Cancer Research, 64*, 8365–8373.
Zhang, X., Huang, J., & McNaughton, P. A. (2005). NGF rapidly increases membrane expression of TRPV1 heat-gated ion channels. *EMBO J, 24*, 4211–4223.
Zhang, L., Jones, S., Brody, K., Costa, M., & Brookes, S. J. (2004). Thermosensitive transient receptor potential channels in vagal afferent neurons of the mouse. *American Journal of Physiology - Gastrointestinal Liver Physiology, 286*, G983–G991.
Zhang, X., Mak, S., Li, L., Parra, A., Denlinger, B., Belmonte, C., et al. (2012). Direct inhibition of the cold-activated TRPM8 ion channel by Galphaq. *Nature Cell Biology, 14*, 851–858.

CHAPTER TWELVE

Thermally Activated TRPV3 Channels

Jialie Luo and Hongzhen Hu*
The Center for the Study of Itch, Washington University Pain Center, The Department of Anesthesiology, Washington University School of Medicine in St. Louis, St. Louis, MO, USA
*Corresponding author: E-mail: huh@anest.wustl.edu

Contents

Current Topics in Membranes, Volume 74
ISSN: 1063-5823
http://dx.doi.org/10.1016/B978-0-12-800181-3.00012-9

Abstract

TRPV3 is a temperature-sensitive transient receptor potential (TRP) ion channel. The TRPV3 protein functions as a Ca^{2+}-permeable nonselective cation channel with six transmembrane domains forming a tetrameric complex. TRPV3 is known to be activated by warm temperatures, synthetic small-molecule chemicals, and natural compounds from plants. Its function is regulated by a variety of physiological factors including extracellular divalent cations and acidic pH, intracellular adenosine triphosphate, membrane voltage, and arachidonic acid. TRPV3 shows a broad expression pattern in both neuronal and non-neuronal tissues including epidermal keratinocytes, epithelial cells in the gut, endothelial cells in blood vessels, and neurons in dorsal root ganglia and CNS. TRPV3 null mice exhibit abnormal hair morphogenesis and compromised skin barrier function. Recent advances suggest that TRPV3 may play critical roles in inflammatory skin disorders, itch, and pain sensation. Thus, identification of selective TRPV3 activators and inhibitors could potentially lead to beneficial pharmacological interventions in several diseases. The intent of this review is to summarize our current knowledge of the tissue expression, structure, function, and mechanisms of activation of TRPV3.

1. INTRODUCTION

Transient receptor potential (TRP) channels comprise a large family of nonselective cation channels, containing nearly 30 mammalian members. The vanilloid TRP family comprises TRPV1 to TRPV6 and constitutes a subgroup in the TRP superfamily. TRPV3 appears to be evolutionally conserved among amphibians (shown in western clawed frog), reptiles (green anole), birds (chicken), and mammals but is not present in bony fishes or invertebrates. Like other TRP channels, TRPV3 is a tetramer, and each subunit is composed of six transmembrane (TM) domains and two relatively large cytoplasmic amino- (N-) and carboxy- (C-) termini on which additional structural motifs are present, notably the ankyrin repeats on the N terminus. The putative pore loop is located between TM5 and TM6, which forms the pore for ion flux (Xu et al., 2002). Although the central TM domains are well conserved, the amino acid sequences of the N- and C-termini of the TRPV3 channels are highly diversified among different species (Saito, Fukuta, Shingai, & Tominaga, 2011; Saito & Shingai, 2006) (Figure 12.1). TRPV3 is highly selective for Ca^{2+} (Ca^{2+}/Na^{+} ratio $\approx$ 12:1) and has the largest unitary conductance among all TRPV channels ($\approx$170 pS) (Xu et al., 2002).

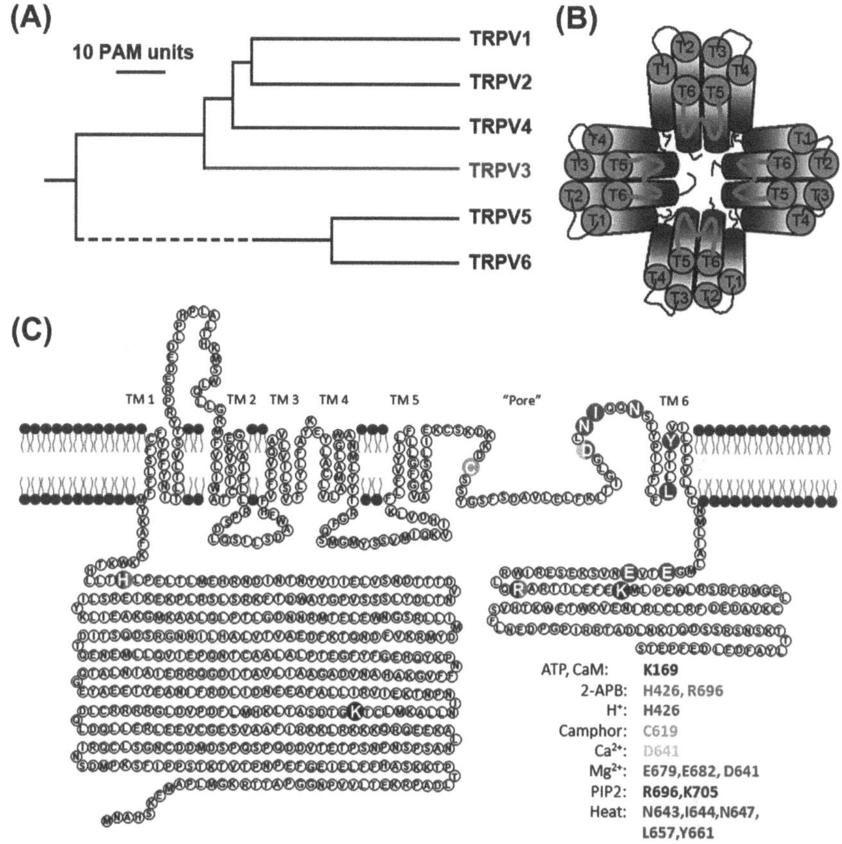

Figure 12.1 (A) The phylogenetic tree of TRPV subfamily. (B) Schematic diagram illustrating the putative tetrameric structure of TRPV3 composed of four identical subunits. The putative pore loops lie between TM5 and TM6 (yellow). (C) Predicted structural topology of TRPV3 with all the amino acid residues indicated. Residues involved in activation/modulation by 2-APB, camphor, heat, H^+, Ca^{2+}, Mg^{2+}, adenosine triphosphate (ATP), CaM, and PIP2 are color coded. (See the color plate.)

TRPV3 is predominantly expressed in skin, where it is activated by warm temperatures and chemical activators. Selective genetic ablation of TRPV3 from keratinocytes in mice causes defects in hair morphogenesis and skin barrier function, whereas in rodents and humans TRPV3 gain-of-function mutations are associated with skin inflammation and pruritus (Asakawa et al., 2006; Cheng et al., 2010; Danso-Abeam et al., 2013; Imura et al., 2007; Imura, Yoshioka, Hirasawa, & Sakata, 2009; Lai-Cheong et al., 2012; Lin et al., 2012; Steinhoff & Biro, 2009; Sulk et al., 2012; Yamamoto-Kasai et al., 2012; Yoshioka et al., 2009). Activation of epidermal TRPV3 has been shown to increase the release of diffusible molecules that are capable of activating or sensitizing primary sensory neurons, leading to sensory hypersensitivity (Danso-Abeam et al., 2013; Huang et al., 2008; Mandadi et al., 2009). Besides keratinocytes, TRPV3 is also expressed by many neuronal and non-neuronal tissues, indicating that TRPV3 might play important roles in many unknown cellular and physiological functions.

As the founding member in the TRPV subfamily, TRPV1 has been intensively studied since its cloning in 1997 (Caterina et al., 1997) and more than 3000 papers have been published to date. Although TRPV3 is a close relative and shares approximately 40% homology in amino acid sequence with TRPV1 (Hu, Grandl, Bandell, Petrus, & Patapoutian, 2009), the study of TRPV3 has been hampered, largely because of the absence of efficient research tools, including highly selective TRPV3 agonists and antagonists as well as specific antibodies. Although significant strides have been made in our understanding of TRPV3 channel activation, gating, and physiological function, many areas and questions remain to be explored and answered.

2. EXPRESSION AND FUNCTION OF TRPV3

2.1 TRPV3 in skin

TRPV3 was initially cloned in 2002 by three groups independently (Peier et al., 2002; Smith et al., 2002; Xu et al., 2002). In situ hybridization and immunohistochemical studies show that TRPV3 is abundantly expressed in epidermal keratinocytes, especially in the outer root sheath of the epithelial compartments of hair follicles and epidermal basal layer in mammals (Asakawa et al., 2006; Borbiro et al., 2011; Peier et al., 2002; Xu, Delling, Jun, & Clapham, 2006; Xu et al., 2002). Functional studies using patch-clamp recordings and Ca^{2+} imaging demonstrate that TRPV3 is functionally expressed in cultured primary keratinocytes and that the responses evoked by TRPV3 activators are removed by genetic ablation of TRPV3 function (Cheng et al., 2010; Luo, Stewart, Berdeaux, & Hu, 2012). Genetic studies also show that TRPV3 is closely associated with a variety of physiological and pathological functions of the skin, including hair growth, hair morphogenesis, dermatitis, pain, and pruritus.

2.1.1 TRPV3 and hair growth

Genetic studies have confirmed that TRPV3 is required for normal hair growth, as both global and conditional (epidermal) TRPV3 knockout mice display wavy hair coat and curly whiskers (Cheng et al., 2010). The TRPV3 deficiency-induced hair morphogenesis phenotype is unique and independent of its temperature sensitivity because genetic deletion of TRPV4, another warm-sensitive TRP channel functionally expressed in TRPV3-positive epidermal keratinocytes, does not produce a wavy hair coat or curly whiskers (Chung, Lee, Mizuno, Suzuki, & Caterina, 2004b; Lee, Iida, Mizuno, Suzuki, & Caterina, 2005; Suzuki, Mizuno, Kodaira, & Imai, 2003). Further studies suggest that TRPV3 forms a signaling complex with transforming growth factor alpha (TGF-α)/epidermal growth factor receptor (EGFR). Activation of EGFR signaling in keratinocytes leads to increased TRPV3 channel activity, which in turn stimulates TGF-α release leading to terminal differentiation of suprabasal keratinocytes (Cheng et al., 2010). Although TRPV3 null mice also display misaligned hair follicles and a thin stratum corneum, it is not known if loss of TRPV3 in the hair follicle stem cell pools contribute to the defects of hair morphogenesis in the TRPV3 null mice since the basic mechanisms and signaling pathways in these cells are critical to epithelial morphogenesis in the skin (Blanpain & Fuchs, 2009).

Interestingly, gain-of-function TRPV3 mutations also affect hair growth in rodents, further supporting the importance of TRPV3 in hair morphogenesis. These TRPV3 mutations occur in the DS-Nh (no-hair) mouse and WBN/Kob-Ht (hairless type) rat, both of which are spontaneously hairless (Akimoto, Suzuki, Nakama, & Suzuki, 2000; Watanabe, et al., 2003). The Nh locus has been mapped on mouse chromosome 11 (Watanabe, et al., 2003) and the Ht locus on rat chromosome 10 (Akimoto et al., 2000), and both loci are contiguous with the linkage site. The non-hair Nh and Ht phenotypes are inherited in an autosomal dominant fashion, and the Nh mutation possesses a high efficacy for penetration (Asakawa et al., 2006). Asakawa et al. have identified the region of the Nh gene and performed an extensive sequencing analysis of this genomic region using bacterial artificial chromosome libraries prepared from DS and DS-Nh mice. Their findings revealed a point mutation in the S4–S5 linker of the TRPV3 channel (G573S in DS-Nh mice, and also identified the G573C mutation as a causative factor in the WBN/Kob-Ht rats) (Asakawa et al., 2006). Further studies show that hair growth cycle, a process of cyclic growth (anagen) and regression (telogen) of hair follicles in mice, is altered by the naturally occurring gain-of-function mutations of TRPV3. The DS-Nh mice have decreased levels of keratin-associated protein 16-1, 16-3, and 16-9 genes in the skin, all of which are related to the anagen phase of the hair growth cycle. Consistently, the anagen phase, but not the telogen phase, persists in DS-Nh mice (Imura et al., 2007). These findings strongly support the hypothesis that gain-of-function of TRPV3 disrupts normal hair growth in the anagen phase (Imura et al., 2007).

Interestingly, the no-hair phenotype in the DS-Nh mice and WBN/Kob-Ht rats is also associated with inflammation which is characterized by aggregations of mast cells in the skin lesions where mutated TRPV3 is expressed, but not in lung and intestines (Asakawa et al., 2006). The skin histamine concentration in the DS-Nh mice is also significantly higher than that in the control mice (Asakawa et al., 2006). Furthermore, many other inflammatory mediators are also upregulated in the DS-Nh mice and WBN/Kob-Ht rats, for instance, CCL11, CCL17, IL-13, IL-17, IFN-γ, and MCP-1 (Yoshioka et al., 2009). Since mast cells play a critical role in allergy and anaphylaxis, it is not clear how gain of function of epidermal TRPV3 channels leads to increased number of resident mast cells in the skin. It also remains unclear if these inflammatory mediators also contribute to the defect of hair growth in the DS-Nh mice and WBN/Kob-Ht rats.

Although the cellular basis of the hairless phenotype caused by both G573S and G573C mutations of murine TRPV3 is elusive, in vitro studies show that both mutations are constitutively active in heterologous cells including HEK293 cells and *Xenopus* oocytes (Xiao, Tian, Tang, & Zhu, 2008). Overexpression of G573S and G573C in HEK293 cells also causes cell death which might result from Ca^{2+} overload and cellular toxicity produced by the constitutive activities of the gain-of-function mutant channels. Interestingly, the mutant channels overexpressed in *Xenopus* oocytes are not sensitive to thermal or chemical stimuli, which is likely caused by a large basal current (constitutive activity) that nearly reaches the maximal capacity of the active channels. On the other hand, when the mutant channels are coexpressed with the wild-type TRPV3, the sensitivity to both chemical ligands and temperatures is enhanced (Xiao, et al., 2008). Therefore, it is likely that cell death and/or altered transcriptional activities induced by the constitutively activating TRPV3 mutations in the hair follicles contribute to the hairless phenotype in the DS-Nh mice and WBN/Kob-Ht rats.

2.1.2 TRPV3 and dermatitis

Besides the hairless phenotype, the naturally occurring gain-of-function point mutations in TRPV3 have been linked to spontaneous dermatitis in the DS-Nh mice. Interestingly, both the hairlessness and dermatitis phenotypes are controlled by the Nh mutations and are not separable by backcross breeding. The symptoms of the atopic dermatitis (AD) developed in the DS-Nh mice include erythema, edema, dry skin, skin erosions, and excoriations (Yoshioka et al., 2006, 2009). In addition, the spontaneous dermatitis in DS-Nh mice shares many characteristics of AD in humans. For instance, (1) in both cases hyperkeratosis can be detected in skin lesions; (2) both skin lesions contain *Staphylococcus aureus*; (3) inflammatory cells including high numbers of mast cells and $CD4^{+}$ T lymphocytes infiltrate the lesion sites; and (4) serological analyses show that serum levels of IgE and certain interleukins, including IL-4, IL-13, and IL-17, are markedly increased, suggesting that both allergic and inflammatory responses are involved in pathogenesis of the spontaneous dermatitis (Yoshioka et al., 2009). Furthermore, the spontaneous dermatitis has been recapitulated in the $TRPV3^{Gly573Ser}$ transgenic mice in which the Nh mutation is overexpressed under the control of the TRPV3 promoter (Yoshioka et al., 2009). These transgenic mice can develop characteristic signs of dermatitis after being raised under *S. aureus*-free conditions for 20–25 weeks. Yoshioka et al. (2009) have measured the penetrance of the $TRPV3^{Gly573Ser}$ gene for dermatitis in a newly constructed C57BL-*Nh* mouse model. The C57BL mice are resistant to allergen stimulation due to a T helper 1 response. Interestingly, these mice fail to develop spontaneous dermatitis although they have a thickening of the epidermis, increased numbers of skin mast cells and nerve fibers in the epidermis, as well as scratching behavior (Yoshioka et al., 2009). Therefore, allergic dermatitis induced by the gain of function of TRPV3 is dependent on the genetic backgrounds of mouse strains.

It appears that tissue levels of nerve growth factor (NGF) are markedly increased in skin samples of $TRPV3^{Gly573Ser}$ transgenic mice compared to wild-type animals. NGF has been shown to be a critical factor contributing to human AD and pruritus by increasing the density of itch-related nerve terminals in the skin (Toyoda et al., 2002). Interestingly, the increased production of NGF relies primarily on the temperature sensitivity of TRPV3, i.e., more NGF is produced in the epidermis of the $TRPV3^{Gly573Ser}$ transgenic mice than that in the wild-type mice because of the constitutive activity of the $TRPV3^{Gly573Ser}$ mutation and lowered threshold of temperature activation of the mutant channel. However, neither the hairless nor spontaneous dermatitis phenotypes have been found in the transgenic mice overexpressing TRPV3 in epidermal keratinocytes under the control of the keratin 14 promoter (Huang et al., 2008). The difference between the phenotypes caused by overexpressing the wild-type TRPV3 and a gain-of-function TRPV3 mutation is that the latter has constitutive activity when formed by a homotetramer or heterotetramer with endogenous wild-type TRPV3 (Xiao, et al., 2008).

Similar to rodent TRPV3, human TRPV3 is also expressed by keratinocytes in the outer root sheath, activation of which has been shown to inhibit hair growth (Borbiro et al., 2011). Dermal immunolabeling and gene expression of TRPV3, along with other TRPV channels, are significantly increased in erythematotelangiectatic rosacea- and phymatous rosacea-affected skins in human patients. Therefore, it is hypothesized that dysregulation of TRPV channels in cutaneous non-neuronal cells may be critically involved in the initiation and/or development of rosacea, which is a frequent chronic inflammatory skin disease of unknown etiology (Sulk et al., 2012).

Strikingly, the same Gly573Ser or Gly573Cys mutations are also found in human patients with the Olmsted syndrome (OS), a rare genodermatosis characterized by extensive palmoplantar and periorificial keratoderma, alopecia in most cases, and severe itching (Danso-Abeam et al., 2013; Lai-Cheong et al., 2012; Lin et al., 2012). By using whole-exome sequencing of case-parents trios, Lin et al. have identified the TRPV3 heterozygous mutation Gly573Ser in three individuals, as well as a de novo missense mutation of TRPV3 ($TRPV3^{Gly573Cys}$) in one patient, and a heterozygous mutation ($TRPV3^{Trp692Gly}$) in another. Like the TRPV3 mutations in the DS-Nh mouse, these mutant human TRPV3 channels display constitutive activity when expressed in HEK293 cells and cause cell death of the transfected HEK293 cells and apoptosis in the keratinocytes in palm skin biopsy sections of the affected individuals (Lin et al., 2012). Therefore, upregulation of TRPV3 function impairs hair growth and increases the incidence of dermatitis and pruritus in both humans and rodents, suggesting that either too much or too little TRPV3 activity is detrimental to skin health (Asakawa et al., 2006; Lin et al., 2012; Yoshioka et al., 2009). Thus, maintaining an appropriate balance of TRPV3 activity is essential to normal function of epidermal keratinocytes. Identification of TRPV3 as a causative factor in human skin diseases not only highlights the importance of TRPV3 in human skin physiology and pathophysiology, but also sheds light on therapeutic management of chronic dermatitis, hair growth defects, and chronic pruritus by selective targeting of TRPV3.

2.1.3 TRPV3 and skin barrier function

TRPV3 null mice have defects in skin barrier formation because of the impaired keratinocyte cornification resulting from dysregulation of the activities of transglutaminases, a family of Ca^{2+}-dependent crosslinking enzymes essential for keratinocyte cornification (Cheng et al., 2010). Paradoxically, the TRPV3 activators 2-aminoethyl diphenyl borate (2-APB) and camphor do not affect barrier recovery rate after tape stripping in either hairless mouse skin or human skin, which, interestingly can be accelerated by TRPV4 activation and delayed by TRPV1 activation (Denda, Sokabe, Fukumi-Tominaga, & Tominaga, 2007). Akazawa, Yuki, Yoshida, Sugiyama, and Inoue (2013) have also found that activation of TRPV4, but not TRPV3, strengthens the tight junction-associated barrier of human epidermal cells. Environmental warm temperatures affect epidermal cell–cell junction and skin permeability barrier homeostasis (Denda et al., 2007; Kida et al., 2012). TRPV4 (not TRPV3) seems to be the primary sensor, although both TRPV3 and TRPV4 are activated by warm temperatures (Akazawa et al., 2013; Denda et al., 2007).

2.1.4 TRPV3 and pain

Since skin is the largest sensory organ in mammals, it has long been speculated that skin might serve as a sensor of diverse environmental cues in addition to touch (Lumpkin & Caterina, 2007). Since TRPV3 is a close relative of TRPV1 which plays important roles in many pain conditions, extensive efforts have been made to investigate the role of TRPV3 in pain sensation during the past decade. However, few studies have provided conclusive evidence of TRPV3-mediated pain sensation. As a thermosensitive ion channel, TRPV3 has been suspected to contribute to heat pain sensation because the response to acute noxious heat is disrupted in TRPV3 knockout mice in water immersion (>48 °C) and hot plate (>52 °C) assays (Moqrich et al., 2005). It has been further demonstrated that mice with selective overexpression of TRPV3 in keratinocytes exhibit shorter escape latencies compared with their wild-type littermates under the condition of selective pharmacological inhibition of TRPV1. These studies suggest that the thermosensitive TRPV3 plays very important roles in heat pain sensation although the effect of TRPV3 might be overshadowed by TRPV1 which is a predominant mediator of thermal hyperalgesia (Caterina et al., 2000; Davis et al., 2000).

In addition to thermal pain sensation, TRPV3 also mediates chemical-induced acute nocifensive responses. Farnesyl pyrophosphate (FPP), an intermediate metabolite in the mevalonate pathway, has been shown to be a selective TRPV3 agonist eliciting rapid pain-related behaviors when injected into a mouse hindpaw presensitized with carrageenan, but not in the un-sensitized mice. Local TPRV3 knockdown in mouse hind paws using a TRPV3-shRNA significantly attenuates pain-related behaviors evoked by FPP, but not the TRPV1 activator capsaicin (Bang, Yoo, Yang, Cho, & Hwang, 2010). The FPP-evoked nocifensive response is also significantly reduced by intradermal pretreatment with isopentenyl pyrophosphate (IPP), a potent TRPV3 inhibitor (Bang, Yoo, Yang, Cho, & Hwang, 2011). Furthermore, a recent finding from the same group showed that 17R-resolvin D1, a potent lipid mediator derived from both eicosapentaenoic acid and docosahexaenoic (Hong, Gronert, Devchand, Moussignac, & Serhan, 2003), specifically inhibits TRPV3-mediated cellular responses, and attenuates TRPV3-mediated nocifensive behaviors induced by FPP (Bang, Yoo, Yang, Cho, & Hwang, 2012). In addition to animal studies using chemical agonists of TRPV3, Carstens' group using a split-tongue stimulus paradigm in human subjects also demonstrated that eugenol and carvacrol produced irritation and increased innocuous warmth and heat pain elicited by noxious heat (49 °C) (Klein, Carstens, & Carstens, 2013). On the other hand, the role of TRPV3 in heat pain sensation still remains controversial. Other studies have demonstrated that TRPV3-deficient mice on both C57BL6 and 129S6 background exhibit no change in acute heat pain or inflammatory heat hyperalgesia, suggesting that TRPV3 might not be a major contributor to heat pain sensation (Huang, Li, Yu, Wang, & Caterina, 2011).

2.1.5 TRPV3 and pruritus

One of the prominent behavioral characteristics of the DS-Nh mice is severe scratching, indicating that epidermal TRPV3 may play a causative role in the pathogenesis of chronic pruritus (Steinhoff & Biro, 2009; Yoshioka et al., 2009). Further studies show that although the C57BL-Nh mice, which are resistant to allergen stimulation due to a Th1 response, do not develop dermatitis by 30 weeks of age, their scratching behavior is significantly increased at 15 weeks of age, compared with age-matched C57BL control mice, and this tendency continues for 15 weeks. Therefore, the scratching responses in the DS-Nh mice and the $TRPV3^{Gly573Ser}$ transgenic mice seem to be separable from the hairless and dermatitis phenotypes (Yoshioka et al., 2009).

However, the molecular and cellular mechanisms underlying the scratching responses evoked by gain-of-function TRPV3 mutations remain elusive. In humans TRPV3 is expressed by primary sensory neurons that might contribute to the scratching behavior that occurs in OS

(Danso-Abeam et al., 2013; Lai-Cheong et al., 2012; Lin et al., 2012). On the other hand, there is little evidence that TRPV3 is functionally expressed in mouse dorsal root ganglia (DRG) neurons (Peier et al., 2002). Therefore, it is likely that the scratching response in DS-Nh mice is not directly mediated by activation of primary sensory neurons, but is rather initiated by resident skin cells, for example, keratinocytes, mast cells, and resident immune cells. Consistent with this, the scratching response in the DS-Nh mice is sensitive to anti-histaminergic treatments, suggesting that allergic responses could contribute to pruritus evoked by gain-of-function TRPV3 mutations (Yoshioka et al., 2006). More importantly, the spontaneous scratching response in a mouse dry skin model induced by acetone–ether–water treatment is abolished in the TRPV3 null mice (Yamamoto-Kasai et al., 2012). The decreased itch response is accompanied by a reduced number of nerve fibers in the dermis in the TRPV3 null mice compared with the wild-type mice. The scratching behavior and skin lesions in hairless mice are similar to the chronic itching observed in AD patients (Yoshioka et al., 2009). Furthermore, elevated levels of TRPV3 mRNA transcripts can be detected in AD patients with pruritus compared with those without pruritus. This finding strongly supports the hypothesis that TRPV3 plays a critical role in the development of chronic itch associated with dry skin in both humans and mice (Yamamoto-Kasai et al., 2012). Recently, it has been demonstrated that DS-Nh mice have an increased response to haptens (such as 2,4-dinitrofluorobenzene) compared with the DS mice. This response is mediated by epidermal keratinocytes and dermal dendritic cells (Yamamoto-Kasai, Yasui, Shichijo, Sakata, & Yoshioka, 2013). Also, keratinocytes from DS-Nh mice have significantly higher levels of thymic stromal lymphopoietin (TSLP) compared with the DS mice (Yamamoto-Kasai et al., 2013). TSLP has been shown to be released from keratinocytes through activation of the Ca^{2+} release-activated Ca^{2+} channel protein 1 (Orai1)/nuclear factor of activated T-cells signaling pathway, and it directly activates a subpopulation of TRPA1-positive sensory neurons to produce itch (Wilson et al., 2013).

In regard to the role of TRPV3 in pain and itch sensations, TRPV3 has been designated as the therapeutic target for the treatment of pain and inflammation. Several small molecular TRPV3-specific antagonists have been developed by Hydra Biosciences and Glenmark Pharmaceuticals, which show analgesic effect in some pain conditions in animal studies (Brederson, Kym, & Szallasi, 2013).

2.1.6 Molecules responsible for epidermal TRPV3-mediated responses

Activation of resident skin cells is sufficient to cause a sensory phenotype. For example, mast cell degranulation following IgE-mediated mast cell activation produces histamine-dependent pruritus by activating downstream TRPV1 channels in pruritoceptive sensory neurons (Imamachi et al., 2009; Kim, Lee, Shim, & Oh, 2004). Furthermore, skin-derived neurotrophic factors such as NGF and artemin can also produce sensory hypersensitivity by regulating the structure and function of cutaneous sensory nerve endings (Albers & Davis, 2007; Jankowski & Koerber, 2010) (Figure 12.2). Indeed, Gopinath et al. (2005) have detected increased expression of TRPV3 in keratinocytes of breast skin biopsies from patients associated with breast pain and tenderness (Gopinath et al., 2005). As a consequence, increased expression of TRPV3 might increase the production and release of NGF from epidermal keratinocytes and contribute to the sensory hypersensitivity (Gopinath et al., 2005). Transgenic overexpression of the gain-of-function TRPV3 mutation found in the Nh mice results in a decrease of temperature threshold for NGF release; i.e., it enhances NGF release from epidermal sheets at 33 °C (NGF generally is released at 37 °C in wild-type animals) (Yoshioka et al., 2009).

Although there is evidence supporting a role of epidermal TRPV3 in sensory transduction, how TRPV3-positive skin cells communicate with sensory nerve endings is not clear. TRPV3 has been detected in keratinocytes in mice, but not in DRG neurons, and since keratinocytes do not form synapes with nerve endings, secretion of diffusible molecules must be responsible for thermosensation and pain evoked by activation of the TRPV3-positive keratinocytes (Moqrich et al., 2005; Peier et al., 2002). By using transgenic mice overexpressing TRPV3 selectively in keratinocytes, Huang et al. have shown that activation of overexpressed TRPV3 channels in these mice leads to thermal nociception and hyperalgesia. These phenotypes are mediated at least in part through Ca^{2+}- and cyclooxygenase-1 (COX1)-dependent release of prostaglandin E2 (PGE2), which accounts for the increased escape responses during heat stimulation (Huang et al., 2008). PGE2 causes thermal hyperalgesia in mice by sensitizing TRPV1-positive DRG neurons through activation of G-protein-coupled EP1 and/or EP4 receptors (Lin et al., 2006; Moriyama et al., 2005) (Figure 12.2).

Another plausible candidate molecule is ATP which is released from keratinocytes upon heating in a co-culture system (Mandadi et al., 2009). When TRPV1-deficient DRG neurons were co-cultured with keratinocytes,

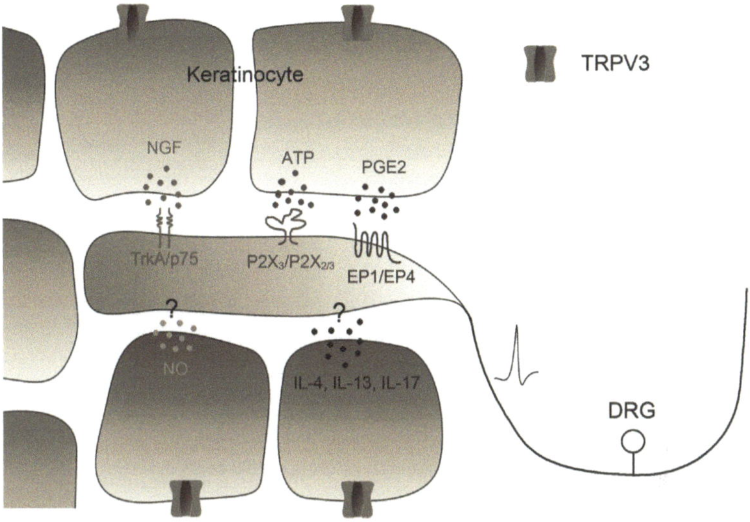

Figure 12.2 TRPV3-mediated crosstalk between keratinocytes and cutaneous nerve endings. TRPV3 activation in the keratinocytes promotes the release of inflammatory mediators including PGE2, ATP, NGF, NO, IL-4, IL-13, IL-17 which act on their corresponding receptors to active sensory nociceptors and initiate pain or itch sensation. (See the color plate.)

increased temperatures evoked an increase in cytosolic Ca^{2+} concentration only in DRG neurons, which was blocked by PPADS and suramin, two P2 purinoreceptor antagonists. When keratinocytes were co-cultured with P2X2-expressing HEK293 cells, heat activated the P2X2-expressing HEK293 cells, however, only if they were co-cultured with the wild-type keratinocytes, but not keratinocytes from TRPV3-deficient mice. Therefore, ATP is a candidate diffusible molecule that potentially mediates TRPV3-mediated thermotransduction in skin (Figure 12.2).

Nitric oxide (NO) is another potential candidate molecule that could contribute to epidermal TRPV3-mediated sensory responses. Indeed, NO plays important roles in wound healing and other biological processes in the skin (Zhu, Wei, Bian, & Murad, 2008). Interestingly, activation of TRPV3 on keratinocytes produces NO through nitrite reduction instead of through de novo synthesis from L-arginine by NO synthase enzymes (NOS) (Moriyama et al., 2005). TRPV3-dependent NO production is involved in migration of keratinocytes, wound healing, and thermosensory behaviors. These findings strongly support the proposal that NO serves as a diffusible molecule mediating communication between keratinocytes and sensory neurons (Moriyama et al., 2005) (Figure 12.2). In addition, keratinocyte-derived cytokines, including IL-4, IL-13, and IL-17, are found in skin lesions from DS-Nh and TRPV3^{Gly573Ser} transgenic mice and might also participate in the severe spontaneous scratching response (Xu et al., 2006; Yoshioka et al., 2009) (Figure 12.2).

2.2 TRPV3 in sensory neurons

Whether TRPV3 is expressed in rodent primary sensory neurons is still controversial although TRPV3 is clearly expressed in human sensory neurons (Smith et al., 2002; Xu et al., 2002). There is no direct evidence for the presence of functional TRPV3 in primary sensory neurons of DRG and trigeminal ganglia in mice (Peier et al., 2002). ^{35}S-radioactive in situ hybridizations and Northern blot analysis have failed to detect TRPV3 in rat DRG (Peier et al., 2002). On the other hand, TRPV3 transcripts were detected in mouse DRG by PCR (Peier et al., 2002) and rat DRG by patch-clamp studies (Stotz, Vriens, Martyn, Clardy, & Clapham, 2008). TRPV3 expression was also detected in DRG, spinal motor neurons, and peripheral nerves of humans and monkeys (Facer et al., 2007; Xu et al., 2002). Interestingly, unlike TRPV1, TRPV3 expression in human DRG is not responsive to a combination of NGF, glial cell (line)-derived growth factor, and neurotrophin3 (Anand et al., 2006). These data suggest that expression of TRPV3 in DRG is species-specific.

Interestingly, TRPV3 may also be able to form heteromeric channels with TRPV1 in human DRG since they could be co-precipitated from HEK293 cells co-transfected with TRPV1 and TRPV3 (Cheng et al., 2012; Smith et al., 2002). Ca^{2+} imaging also showed that co-transfection of TRPV3 increased TRPV1 response to both capsaicin and proton (Smith et al., 2002). Further studies showed that TRPV1/V3 heteromers displayed different single-channel conductance, temperature sensitivity, activation threshold, and heat-induced sensitization (Cheng et al., 2012).

2.3 TRPV3 in the gastrointestinal tract

In addition to keratinocytes, TRPV3 is highly expressed in epithelial cells in the alimental canal, including the oral cavity, tongue, larynx, and distal colon (Hamamoto, Takumida, Hirakawa, Takeno, & Tatsukawa, 2008; Hamamoto, Takumida, Hirakawa, Tatsukawa, & Ishibashi, 2009; Ueda, Yamada, Ugawa, Ishida, & Shimada, 2009; Xu et al., 2006). High expression levels of TRPV3 have been detected in the tongue and nasal epithelium of mice, rats, and humans by in situ hybridization and immunofluorescence (Xu et al., 2006). Furthermore, carvacrol and eugenol, components of spice plants such as oregano, savory, clove, and thyme, evoked intracellular Ca^{2+} responses in the epithelial cells of the tongue, which could be part of the molecular and cellular basis of chemesthesis in oral and nasal epithelium (Xu et al., 2006). TRPV3 mRNA transcripts and proteins were also detected in the epithelial cells of the distal colon, jejunum, and ileum, but not in those of the stomach, duodenum, or proximal colon (De Petrocellis et al., 2012). Carvacrol increased cytosolic Ca^{2+} concentration in a subset of primary epithelial cells cultured from the distal colon and evoked ATP release in primary colonic epithelial cells (Ueda et al., 2009).

Plant cannabinoids such as **Δ** (9)-tetrahydrocannabinol and cannabidiol (CBD) are potent activators of TRPV1 and TRPV2 (Akopian, Ruparel, Jeske, Patwardhan, & Hargreaves, 2009; Neeper et al., 2007; Qin et al., 2008). Interestingly, they also activate TRPV3 and evoke intracellular Ca^{2+} response in TRPV3-expressing heterologus cells. Furthermore, these plant cannabinoids reduce TRPV3 mRNA in the jejunum and ileum of croton oil-treated mice, suggesting that TRPV3 might play a role in gastrointestinal inflammation (De Petrocellis et al., 2012). By examining the association between genetic variability in 43 fatty acid metabolism-related genes and colorectal cancer risk in 1225 colorectal cancer cases and 2032 controls, Hoeft et al. (2010) identified TRPV3 as a high risk factor for colorectal cancer. However, it remains unknown how TRPV3 contributes to the pathogenesis of gastrointestinal inflammation and colorectal cancer. TRPV3 has also been detected in mouse adipocytes (Bishnoi, Kondepudi, Gupta, Karmase, & Boparai, 2013), and acinar and ductal cells in salivary glands (Sobhan et al., 2013) using RT-PCR, although the function and physiological significance of TRPV3 in these tissues are not clear.

2.4 TRPV3 in blood vessels

Messenger RNA transcripts of TRPV1–V4 of the TRPV family have been detected in both pulmonary arteries and aorta by RT-PCR (Yang, Lin, McIntosh, & Sham, 2006). Both TRPA1 and TRPV3 were also detected in the endothelium of intact cerebral arteries. Consistent with TRPV3 being a Ca^{2+}-permeable channel, TRPV3 activators, such as carvacrol and eugenol, increased the intracellular Ca^{2+} concentration in native cerebral artery endothelial cells. Carvacrol also activated TRPV3-like cation currents in cerebral artery endothelial cells, and more importantly, carvacrol elicited an endothelium-dependent vasodilation of intact cerebral arteries. Interestingly, carvacrol-induced vasodilation involves Ca^{2+}-activated K^+ channels, but does not require activation of nitric oxide synthase (NOS) or COX

(Earley, Gonzales, & Garcia, 2010). Since both carvacrol and eugenol are major components of oregano, which is most frequently used in Mediterranean cooking, it is hypothesized that TRPV3-mediated endothelium-dependent vasodilation may account for some of the cardioprotective effects of the Mediterranean diet (Earley et al., 2010).

2.5 TRPV3 in CNS

Although TRPV3 mRNA transcripts have been detected in the brain, expression of TRPV3 protein in the brain has not been reported, at least in part because of the lack of specific TRPV3 antibodies (Smith et al., 2002; Xu et al., 2002). Moussaieff et al. (2008) have shown that incensole acetate (IA), a major active constituent of Boswellia (also known as Indian Frankincense) resin, is a selective TRPV3 activator and induced c-fos expression in several brain areas, including areas that are associated with anxiety and depression, such as the central nucleus of the amygdala and the lateral septum. Interestingly, IA is also used in religious and cultural ceremonies, and is believed to contribute to spiritual exaltation. In fact, application of IA causes anxiolytic and antidepressive activities in wild-type mice, and these activities of IA are not observed in the TRPV3 knockout mice. Therefore, it is likely that TRPV3 might mediate the excitatory effect of IA in the CNS (Moussaieff et al., 2008).

Hypothermia has been shown to be an effective neuroprotective treatment in large randomized trials in patients with acute brain injury and in several animal models of focal cerebral ischemia (van der Worp, Macleod, Kollmar, & EuropeanStroke Research Network for Hypothermis (EuroHYP), 2010). Cooling to below 35 °C produced a significantly improved outcome with the largest benefit obtained at 34 °C in acute ischemic stroke, suggesting that this is the optimal target temperature for therapeutic hypothermia (Yenari & Hemmen, 2010). Hyperthermia, on the other hand, exacerbates ischemic damage (Minamisawa, Smith, & Siesjo, 1990). Lipski et al. (2006) have detected TRPV3 mRNA transcripts in the hippocampus using RT-PCR and proposed that inhibition of TRPV3 might contribute to the beneficial effect of hypothermia on oxygen-glucose deprivation-induced sudden negative shifts of extracellular potential, intracellular Ca^{2+} increase, membrane current, and depolarization in rat CA1 pyramidal neurons. This study revealed an important causative role of TRPV3 in the development of ischemia, which provides a sound basis for treatment of ischemic stroke by inhibition of TRPV3 with hypothermia (Lipski et al., 2006). On the other hand, a recent report showed that activation of TRPV3 by IA protected against ischemic injury by inhibiting the inflammatory responses resulting from the ischemic damage. The neuroprotective effects of IA are partially mediated by TRPV3 channels, since they were significantly attenuated by genetic or pharmacological ablation of the TRPV3 function (Moussaieff et al., 2012).

In addition to pyramidal neurons, TRPV3 also plays a functional role in hippocampal interneurons. For instance, long-term depression in interneurons was absent in slices from TRPV3, but not TRPV4 null mice. Furthermore, high-frequency stimulation-induced long-term potentiation at the CA3-CA1 pyramidal cell synapse was also inhibited because of a reduced inhibition of inhibitory interneurons which innervate the CA1 pyramidal neurons (Brown, Chirila, Schrank, & Kauer, 2013). It was also demonstrated that TRPV3 expression in the hypoglossal nucleus and medial nucleus tractus solitarius was increased during the development of hyperphagia, and infrared heat treatment decreased the number of TRPV3-positive neurons and reduced food intake. Therefore, TRPV3 in the dorsal medulla of the brainstem might play important roles in the control of food intake (Hu, Choo, & Ma, 2011). By using a genetic association strategy, a recent study showed a significant association between single-nucleotide polymorphisms in the TRPV3 gene and migraine in the Spanish population, suggesting that TRPV3 may also contribute to the genetic susceptibility to migraine (Carreno et al., 2012).

3. TRPV3 ACTIVATORS

3.1 Temperature

Our sense of temperature is critical to thermoregulation and protects us from harmful cold or hot temperatures (Bandell, Macpherson, & Patapoutian, 2007). Cloning and characterization of temperature-sensitive TRP channels have significantly extended our understanding of the molecular and cellular mechanisms that allow us to properly sense temperatures in our surroundings (Dhaka, Viswanath, & Patapoutian, 2006). Several members of the family of TRP channels are activated by cold or hot temperatures and serve as molecular sensors in our skin and/or cutaneous sensory neurons. For instance, TRPV1 and TRPV2 are activated by noxious heat; TRPV3, TRPV4, TRPM2, TRPM3, TRPM4, and TRPM5 are sensitive to warm temperatures, while TRPM8 and TRPA1 detect cool and noxious cold temperatures under physiological or pathological conditions (Dhaka et al., 2006; Latorre, Brauchi, Orta, Zaelzer, & Vargas, 2007; Patapoutian, Peier,

Story, & Viswanath, 2003; Talavera et al., 2005; Talavera, Nilius, & Voets, 2008; Togashi et al., 2006; Vriens et al., 2011). TRPV3 is one of the warm-sensitive thermoTRPs that can be activated by warm temperatures with threshold ranging from 31 to 39 °C (Peier et al., 2002; Smith et al., 2002; Xu et al., 2002). Unlike TRPV1 which is desensitized by repeated heat stimulation, heat activation of TRPV3 displays a sensitization property, i.e. the heat-induced current increases upon repeated heat stimulation. This sensitization phenomenon has been recorded in both TRPV3-expressing heterologous cells and primary keratinocytes (Chung, Lee, Mizuno, Suzuki, & Caterina, 2004a; Chung, Guler, & Caterina, 2005; Peier et al., 2002; Smith et al., 2002; Xu et al., 2002).

In a thermal gradient assay, TRPV3 null mice chose their preferred floor temperature more slowly, but settled in the same temperature range as wild-type littermates. In a two-temperature choice paradigm, TRPV3 null mice had a decreased preference for 35 °C over room temperature and longer withdrawal latencies in response to high temperatures. Furthermore, TRPV3 null mice had no deficits in other sensory modalities. Therefore, genetic ablation of TRPV3 function produces selective deficits in responses to innocuous and noxious heat (Moqrich et al., 2005). Overexpression of TRPV3 in keratinocytes under the control of the K14 promoter produced increased escape responses in TRPV3 transgenic mice, but only during pharmacological inhibition of TRPV1 function. This behavioral response was reduced by coadministration of a COX inhibitor, suggesting that skin-derived PGs mediate the response induced by overexpressed epidermal TRPV3. However, further studies have demonstrated that the deficits in thermal sensation in TRPV3 null mice depend on genetic background; i.e., genetic ablation of TRPV3 resulted in a more restrictive range of occupancy centered around cooler floor temperatures on the 129S6 background, whereas TRPV3 null mice on the C57BL6 background displayed no obvious alterations in thermal preference (Huang et al., 2011). Furthermore, TRPV3 and TRPV4 double knockouts showed similar thermal preference behavior as wild-type controls, and little or no change in acute heat nociception or inflammatory thermal hyperalgesia even under the condition of TRPV1 inhibition (Huang et al., 2011). Therefore, TRPV3 likely makes limited and strain-dependent contributions to thermosensation. One possible explanation for this outcome is that there are many other warm-sensing TRP channels expressed in skin and sensory neurons, activation of which could compensate for the loss of TRPV3 in vivo (see earlier). Making things even more complicated is the presence of other temperature-sensitive non-TRP ion channels that could also contribute to warm sensing in rodents, including K^+ channels with two pore domains (K2p) and Ca^{2+}-activated Cl^- channels (TMEM16A or Ano1) (Cho et al., 2012; Noel et al., 2009).

Because of the lack of structural information on TRP channels, the molecular mechanism by which thermoTRPs convert thermal energy (heat or cold) into a protein conformational change that drives the channels into an open state is not understood. Voets et al. (2004) have hypothesized that thermoTRPs are weakly voltage dependent, and that in the resting state their voltage activation curve can only be recorded in the non-physiological positive voltage range. Temperatures and chemical agonists function as gating modifiers to shift voltage-dependent activation curves of thermoTRPs toward more negative, physiological membrane potentials. For instance, heat and capsaicin as well as cold and menthol, modulate voltage dependence of TRPV1, and regulate the voltage dependence of TRPM8, respectively (Voets et al., 2004). On the other hand, it is also proposed that thermoTRP channels are modular proteins with allosteric gating by temperature, voltage, and chemical ligands (Latorre et al., 2007). In this case, temperatures directly gate TRP channels, with the gating separable from activation by other mechanisms including voltages and chemical ligands, although they all allosterically modulate each other's activities (Latorre et al., 2007). For example, both voltage and agonists alter the slope of the temperature-dependent gating curve as well as shifting the midpoint (Yao, Liu, & Qin, 2010). The temperature-dependent activity of thermoTRPs can also be affected by standard-state enthalpy, entropy upon channel gating, gating equilibrium constant, etc. (Jara-Oseguera & Islas, 2013).

Extensive efforts have been made to identify the temperature sensors in TRP channels. By using chimeras between the entire C-terminal domain of TRPV1 and TRPM8, Brauchi, Orta, Salazar, Rosenmann, and Latorre (2006) identified the C-terminal domains of TRPV1 and TRPM8 as critical modules for conferring temperature sensitivity. Indeed, a chimeric channel in which the C-terminal domain of TRPM8 is swapped into TRPV1 conferred cold sensitivity to the TRPV1 pore module. Conversely swapping the C-terminal from TRPV1 into TRPM8 conferred heat sensitivity to the TRPM8 pore module (Brauchi et al., 2006). Interestingly, the TRPV3 channel cloned from the western clawed frog is activated by cold temperatures instead of heat stimuli, which might be critical for the frogs to detect noxious cold temperatures in their surrounding environments (Saito et al.,

2011). However, the frog TRPV3 becomes sensitive to heat stimulation when it is fused to the N- and C-terminal regions of mouse TRPV3 (Grandl et al., 2008). Yao, Liu, and Qin (2011) have also identified an N-terminal region connecting the ankyrin repeats to the first TM segment as a crucial temperature-sensing element by using systemic chimeric analysis between the rat TRPV1 and TRPV2 channels. These findings suggest that the cytosolic N- and C-terminal regions could be critical determinants of temperature sensitivity of TRPV3.

By using a combination of random mutagenesis and high-throughput screening, Grandl et al. (2008) identified five single-point mutations (out of ~14,000 randomly mutated TRPV3 clones) that specifically abolished heat activation without disrupting chemical activation or voltage modulation. Interestingly, all five mutations are located in the extracellular loop within the outer pore region of mouse TRPV3 (Figure 12.1). Heat responses were abolished without affecting responses to the chemical ligands, 2-APB, or camphor, when this domain from the heat-insensitive human TRPV2 was exchanged into the heat-sensitive mouse TRPV3. Therefore, this domain is specifically required for temperature activation in mouse TRPV3, and temperature activation of TRPV3 is separable from all other known activation mechanisms (Grandl et al., 2008). Consistent with the outer pore region being a critical determinant of temperature activation of thermoTRPs, Grandl et al. (2010) also identified point mutations in the outer pore region of rat TRPV1 that impaired temperature activation by using random and site-directed mutagenesis approaches. Further studies with single-channel recordings showed that temperature activation of TRPV1 was associated with sequential occupancy of short and long open states of the single channels. Mutations in the outer pore region of TRPV1 selectively abolished the long channel openings and thus disrupted TRPV1 heat sensitivity (Grandl et al., 2010).

The involvement of TRPV1 pore turret in temperature sensing is further supported by site-directed fluorescence recordings in which temperature change but not ligand- or voltage-dependent channel activation induced a significant rearrangement of TRPV1 pore turret that is coupled to channel opening (Yang, Cui, Wang, & Zheng, 2010). In addition, deletion of the first 10 amino acids of the pore turret of TRPV1 significantly attenuated heat response and increased the heat activation threshold without affecting capsaicin responsiveness. Furthermore, replacement of the TRPV1 turret sequence with an artificial sequence rendered the mutant channel insensitive to temperature without changing ligand-activated responses (Cui et al.,

2012). Kim et al. have used the cysteine accessibility method to probe temperature-dependent conformations of single amino acids in TRP channels focusing on over 50 amino acids in the predicted outer pore domains of TRPV1 and TRPV3 (Grandl et al., 2008, 2010; Kim, Patapoutian, & Grandl, 2013). Residues located within the second predicted extracellular pore loop display temperature-dependent accessibilities to extracellular solvent. These amino acid residues are identical or proximal to residues that were shown to be specifically required for temperature activation, but not chemical activation (Grandl et al., 2008, 2010). Therefore, multiple protein domains including the N- and C-terminal regions as well as the outer pore regions of TRPV3 and TRPV1 channels have been proposed to contribute to temperature activation of thermoTRPs. However, it is not known if these protein domains form a single complex or act independently in response to thermal stimulation.

3.2 2-APB and related analogs

2-APB, a synthetic small molecule is one of the first known chemical activators of TRPV3 (Chung et al., 2004a; Hu et al., 2004). 2-APB was initially found to inhibit store-operated Ca^{2+} channels and inositol 1, 4, 5-trisphosphate (IP_3) receptors, thereby suppressing intracellular Ca^{2+} release (Ma et al., 2000; Maruyama, Kanaji, Nakade, Kanno, & Mikoshiba, 1997). It activates TRPV3 in both heterologous expression and primary mouse keratinocytes (Cheng et al., 2010; Chung et al., 2004a; Hu et al., 2004; Luo et al., 2012). In addition, 2-APB also has a synergistic action with other TRPV3 activators and modulators including temperatures, camphor, arachidonic acid (AA), and many skin sensitizers (Chung et al., 2004a; Hu et al., 2004; Hu et al., 2006; Xu et al., 2006). Two structurally related compounds, diphenylboronic anhydride (DPBA) and diphenyltetrahydrofuran (DPTHF), can also modulate the activity of TRPV3 (Figure 12.3). Interestingly, although structurally similar, DPBA activates whereas DPTHF inhibits TRPV3 (Chung et al., 2005). In a high-throughput mutagenesis screen study, Hu et al. (2009) have identified two amino acid residues in the mouse TRPV3 that are specifically required for activation of TRPV3 by 2-APB but not by camphor or voltage. The cytoplasmic N-terminal mutation H426N severely attenuates 2-APB response in human, dog, and frog TRPV3 without affecting camphor response. The cytoplasmic C-terminal mutation R696K in the TRP box abolishes 2-APB response in a Ca^{2+}-dependent manner. Interestingly, mutating the

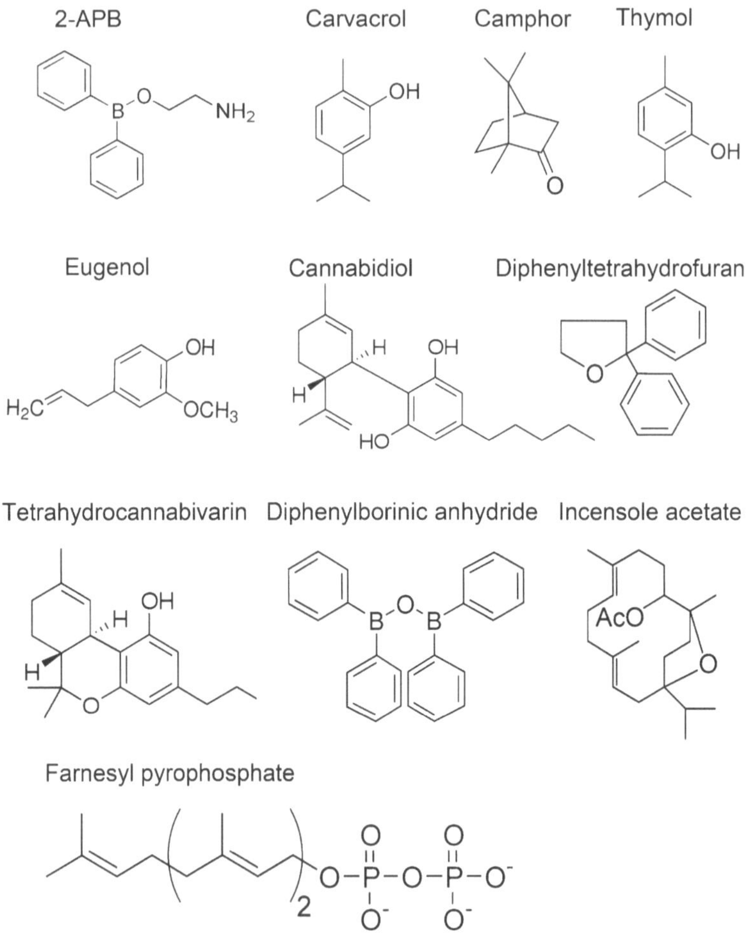

Figure 12.3 Chemical structures of some commonly used TRPV3 activators.

equivalent residue at 426 of the chicken TRPV3 (which is an asparagine) to histidine conferred high 2-APB sensitivity to the chicken TRPV3 which normally is only weakly sensitive to 2-APB. Moreover, mutating two residues in TRPV4 at the equivalent positions to the TRPV3 sequence (N426H and W737R) was sufficient to render the 2-APB-insensitive TRPV4 into a TRPV3-like 2-APB sensitivity. Therefore, 2-APB activation of TRPV3 is critically dependent on these two cytoplasmic residues (Figure 12.1). It has recently been reported that TRPV3 is potently activated by intracellular protons, and the N-terminal histidine residue, H426, known to be critical for 2-APB-mediated TRPV3 activation, is involved in sensing intracellular proton levels (Cao, Yang, Zheng, & Wang, 2012) (Figure 12.1).

Although 2-APB activates TRPV3, it also has many actions on other ion channels; for example, 2-APB increases activities of TRPV1, TRPV2, TRPV6 (Hu et al., 2004; Voets et al., 2001), two-pore K^+ channels (Beltran, Beltran, Aguado, Gisselmann, & Hatt, 2013), store-operated Orai3 channels (Bergsmann et al., 2011; Demuro et al., 2011), but inhibits most of the TRPCs, TRPM2, TRPM8, IP_3 receptors (Hu et al., 2004; Maruyama et al., 1997; Togashi, Inada, & Tominaga, 2008), gap-junctional channels (Harks et al., 2003), and Orai1 channels (Lis et al., 2007). 2-APB also regulates TRPM6 and TRPM7 function in a concentration-dependent manner (Li, Jiang, & Yue, 2006). Therefore, extra care is required to interpret the effect of 2-APB on ion channels, especially in native tissues where many endogenous ion channels may be affected by 2-APB.

3.3 Natural compounds

Like other thermoTRPs, TRPV3 is also activated by natural compounds from plants such as camphor, carvacrol, IA, and eugenol (Moqrich et al., 2005; Moussaieff et al., 2008; Xu et al., 2006). Camphor is a terpenoid with the chemical formula $C_{10}H_{16}O$ found in wood of the camphor laurel (*Cinnamomum camphora*) and is a major active ingredient of balms and analgesic liniments (Figure 12.3). Camphor was identified as the first natural compound to activate TRPV3 (Moqrich et al., 2005). Camphor activated and sensitized TRPV3-mediated current in both TRPV3-expressing Chinese hamster ovary cells and primary keratinocytes, and camphor-induced responses were absent from keratinocytes of TRPV3 null mice (Moqrich et al., 2005). Camphor uses a distinct activation mechanism that is separable from those used by temperatures or 2-APB because mutants carrying mutations disrupting temperature or 2-APB responses had normal camphor responses (Grandl et al., 2008; Hu et al., 2009). Recently, Sherkheli et al. (2013) have shown that a cysteine residue (C619) in the pore region of mouse TRPV3 is likely involved in camphor-mediated activation of TRPV3 (Sherkheli, Vogt-Eisele, Weber, & Hatt, 2013) since the mouse TRPV3-C619S mutant had a severely reduced response to camphor but had a similar response to 2-APB compared with wild-type mouse TRPV3 (Sherkheli et al., 2013) (Figure 12.1). In addition to camphor, TRPV3 can also be activated by many other natural compounds such as carvacrol, thymol, and eugenol, which generally evoke warmth sensation and sensitize skin (Xu et al., 2006). Using Ca^{2+} imaging and patch-clamp recordings, Xu et al. (2006) have demonstrated that these natural compounds strongly activate TRPV3 channels heterologously expressed in HEK293 cells and endogenously expressed in native tongue epithelial cells and keratinocytes. However, like 2-APB, camphor and skin sensitizers are not specific TRPV3 activators because they also activate or modulate activities of many other TRP channels (Macpherson et al., 2006). For instance, camphor activates TRPV1 and produces a rapid desensitization and tachyphylaxis (Macpherson et al., 2006; Xu, Blair, & Clapham, 2005). Camphor also activates TRPM8 expressed in heterologous cells and DRG neurons and sensitizes TRPM8 response to cooling and icilin but it inhibits TRPM8 response to menthol in Ca^{2+} imaging studies (Selescu, Ciobanu, Dobre, Reid, & Babes, 2013). On the other hand, other studies with electrophysiological recordings show that camphor has little or no effect on TRPM8 even at saturating concentrations (Macpherson et al., 2006; McKemy, Neuhausser, & Julius, 2002; Xu et al., 2005). Camphor also inhibits TRPA1 channel function, which might be related to its analgesic property (Macpherson et al., 2006; Xu et al., 2005).

By examining a library of monoterpenoids including many natural plant chemicals, Vogt-Eisele et al. (2007) identified about 30 compounds with agonistic activities on TRPV3, which is consistent with previous findings that many plant monoterpenoids are TRPV3 activators (Macpherson et al., 2006; Moqrich et al., 2005; Vogt-Eisele et al., 2007; Xu et al., 2005). Among these monoterpenoids, six compounds are more potent than camphor for activating TRPV3 (with EC_{50} values up to 16 times lower than that of camphor). It has been hypothesized that a cyclic structure with a secondary hydroxyl group in these monoterpenoids is required for TRPV3 activation (Vogt-Eisele et al., 2007) (Figure 12.3).

Cannabinoids are a group of terpenophenolic compounds in cannabis (*Cannabis sativa*) and are also generated in the nervous and immune systems of mammals. Several members of this family exhibit regulatory effects on several TRP channels. For instance, CBD has been reported to activate TRPV1, TRPV2, and TRPA1 channels (De Petrocellis et al., 2012). TRPV3 can also be activated by cannabinoids, such as CBD and tetrahydrocannabivarin, which stimulate TRPV3-mediated Ca^{2+} influx with higher potency and efficacy (with EC_{50} values in micromolar range) (Figure 12.3) (De Petrocellis et al., 2012).

3.4 Potent endogenous ligands

FPP is an intermediate in the HMG-CoA reductase pathway used to synthesize terpenoids, terpenes, and sterols. FPP is generated via one isomerization and subsequent condensation processes with IPP by IPP isomerase and FPP synthase (Figure 12.3) (Dhar, Koul, & Kaul, 2013). Although FPP is mainly synthesized in the cytosol, it can be detected in human plasma and its steady-state plasma level is around 6.6 ng/ml (17 nM) (Saisho, Morimoto, & Umeda, 1997). FPP is a potent and specific activator of TRPV3 with a nanomolar EC_{50} value, much more potent than monoterpenoids (Bang et al., 2010). FPP evoked intracellular Ca^{2+} responses and activated membrane currents in human TRPV3-transfected HEK293 cells as well as primary keratinocytes. Activation of TRPV3 expressed in epidermal keratinocytes by FPP evoked the release of diffusible signaling molecules that directly excited pain-initiating DRG neurons (Bang et al., 2010). Indeed, paw injection of FPP elicited acute nociceptive behaviors and thermal hypersensitivity, which were reduced in mice when TRPV3 was knocked down using a mouse TRPV3-shRNA (Bang et al., 2010).

4. TRPV3 REGULATORS

4.1 Divalent cations

Channel desensitization is a common phenomenon for TRP channels, which is a protective measure for cells to avoid potential excitotoxicity resulting from sustained activation of the channels. Although the mechanisms of TRP channel desensitization are multifactorial and not completely understood, desensitization of both TRPV1 and TRPA1 following activation requires extracellular Ca^{2+} (Akopian et al., 2009; Touska, Marsakova, Teisinger, & Vlachova, 2011). In marked contrast to TRPA1 and TRPV1, TRPV3 displays a unique property of sensitization characterized by increased channel activity (usually starting with low activity) following repeated stimulation by heat or chemical ligands in both heterologous cells and primary keratinocytes (Chung et al., 2004a; Liu et al., 2011; Peier et al., 2002; Smith et al., 2002; Xiao, et al., 2008; Xu et al., 2002).

Since TRPV3 is highly Ca^{2+}-permeable, it has been hypothesized that intracellular Ca^{2+} plays a critical role in TRPV3 sensitization by regulating interactions between calmodulin and TRPV3 protein (Xiao, et al., 2008). Calmodulin is a common regulator of several thermoTRPs (Gordon-Shaag, Zagotta, & Gordon, 2008). Indeed, there is a conserved binding site on the ankyrin repeat domain in the N-terminus of TRPV1, TRPV3, and TRPV4 involving calmodulin and ATP binding, disruption of which suppresses TRPV3 sensitivity as well as sensitization by repeated stimulation of agonists (Phelps, Wang, Choo, & Gaudet, 2010; Xiao, et al., 2008). It has also been reported that Ca^{2+} binding to calmodulin inhibits TRPV3 function and the inhibitory effect is gradually reduced by repeated stimulation of the channel, which results in the sensitization phenomenon (Xiao, et al., 2008). This is further supported by the finding that sensitization of TRPV3 is mimicked by chelating intracellular Ca^{2+} with a fast Ca^{2+} chelator BAPTA but not EGTA which displays slower buffering kinetics (Xiao, et al., 2008). However, further studies have demonstrated that TRPV3 sensitization is an intrinsic gating property resulting mainly from strong hysteresis following channel activation (Liu et al., 2011). In fact, the sensitization of TRPV3 can also be recorded in excised TRPV3-expressing membrane patches. Furthermore, BAPTA sensitizes TRPV3 by directly interacting with the channel and the sensitizing effect is independent of its Ca^{2+} chelating ability, which is further supported by the finding that synthetic dipeptides that are structurally similar to BAPTA without Ca^{2+} binding capability are also able to potentiate channel gating and accelerate sensitization of TRPV3 (Liu et al., 2011).

Interestingly, the sensitization of TRPV3 channel is irreversible, suggesting that a structural adaptation might occur following sustained channel activation (Liu et al., 2011). Consistent with the irreversible adaptation of TRPV3 gating, Chung et al. (2005) have reported that TRPV3 displays biphasic currents in response to either chemical or thermal activation. The first phase is the sensitizing phase with characteristic outward rectification. The second phase current is much larger than the first sensitizing current and becomes linear without characteristic rectification displayed in the first phase. The second phase current also has alterations in temperature sensitivity, voltage dependence, and ion permeability (Chung et al., 2005). The second phase generally develops abruptly out of the sensitizing phase and the speed of development of the second phase depends highly on the TRPV3 current density and extracellular divalent cations. For instance, removing extracellular Ca^{2+} or mutation of a putative Ca^{2+} binding amino acid residue (D641N) in the channel pore loop facilitates the transition from the first to the second phase following TRPV3 activation (Figure 12.1).

TRPV3 function is also regulated by another divalent cation Mg^{2+}, a critical mineral involved in many aspects of human metabolism. Mg^{2+} homeostasis is critical to skin health, for instance, bathing in a magnesium-rich Dead Sea salt solution can improve skin barrier function and reduce inflammation in atopic dry skin (Proksch, Nissen, Bremgartner, & Urquhart, 2005), whereas magnesium deficiency produces AD associated with robust scratching behavior in rodents (Neckermann, Bavandi, & Meingassner, 2000). Since TRPV3 activity is strongly associated with skin irritation, inflammation, and pruritus, it has been speculated that TRPV3 could be a molecular target for Mg^{2+}. Indeed, TRPV3 function is inhibited by Mg^{2+} through interaction with acidic residues on both sides of the channel pore loop region. Extracellular Mg^{2+} also attenuated the TRPV3-mediated increase of cAMP response element binding (CREB) activity in keratinocytes. The acidic aspartate residue (D641) in the outer pore loop and two acidic residues (E679, E682) in the inner pore region contribute to the inhibition of TRPV3 by Mg^{2+} (Luo et al., 2012). Since the modulatory effects of both Ca^{2+} and Mg^{2+} are mediated by D641, this acidic residue might be part of an extracellular cation sensor for TRPV3 and possibly other structurally related TRPV channels (Figure 12.1).

4.2 Intracellular ATP, calmodulin, and phosphoinositide lipids

Extracellular ATP is an important signaling molecule mediating cell-to-cell communication primarily through activation of membrane-bound purinoceptors (Burnstock, 2012). Interestingly, intracellular ATP is also critical to couple TM signaling cascades with metabolic pathways (Nichols, 2006). Lishko, Procko, Jin, Phelps, and Gaudet (2007) have solved the structure of the N-terminus cytosolic ankyrin repeat domain of TRPV1 and identified a multiligand-binding site for ATP and calmodulin. Disruption of this site regulated TRPV1 tachyphylaxis. Further studies showed that this multiligand-binding site for ATP and calmodulin was conserved in TRPV3 and TRPV4, but not TRPV2 (Phelps et al., 2010). Indeed, ATP directly interacted with the multiligand-binding site at the N-terminal ankyrin repeat domain of TRPV3 channel, blocked the sensitization of TRPV3 to repeated 2-APB applications, and shifted the concentration–response curves for TRPV3 agonists 2-APB and thymol rightward by lowering the sensitivity of TRPV3 (Phelps et al., 2010). ATPγS, the non-hydrolyzable form of ATP, displayed the same effect, supporting a direct interaction between ATP and TRPV3. This has been further confirmed by the finding that mutating a critical residue (K169), the binding site for ATP, abolished the leftward shift of the concentration–response curve in the presence of ATP (Phelps et al., 2010) (Figure 12.1).

As mentioned earlier, intracellular Ca^{2+} plays important roles in the regulation of TRPV3 function. The Ca^{2+}-binding protein, calmodulin has been shown to contribute to the modulation of TRPV1 and TRPV4 by Ca^{2+} (Lishko et al., 2007; Strotmann, Schultz, & Plant, 2003). Calmodulin also directly and competitively binds to the same multiligand-binding site at the N-terminal ankyrin repeat domain that is conserved among TRPV1, TRPV3, and TRPV4 (Lishko et al., 2007; Phelps et al., 2010). Using a calmodulin-agarose pulldown assay, Phelps et al. (2010) have demonstrated that Ca^{2+}-calmodulin directly binds to the ankyrin repeat domain of TRPV3. Using site-directed mutagenesis they also identified a single lysine residue (K169) in the ankyrin repeat domain to be responsible for the interaction between TRPV3 and Ca^{2+}-calmodulin (Figure 12.1) (Phelps et al., 2010). Both BAPTA and anti-calmodulin antibody sensitized TRPV3 function with the current amplitude dramatically increased upon initial application of TRPV3 agonists. In addition, the calmodulin antagonist ophiobolin A mimicked the sensitization effect of BAPTA and anti-calmodulin (Xiao et al., 2008).

Phosphatidylinositol-4, 5-bisphosphate (PIP_2) is an important phospholipid enriched in the plasma membrane, where it plays critical roles in cell signaling (Sun, Thapa, Hedman, & Anderson, 2013). PIP_2 is hydrolyzed into IP_3 and diacyl glycerol (DAG) by phospholipase C (PLC), a membrane-bound enzyme activated upon the activation of G-protein coupled receptors (GPCRs). PIP_2 has been found to regulate functions of many ion channels including TRPV1 (Cao, Cordero-Morales, Liu, Qin, & Julius, 2013; Rohacs, 2007), TRPM8 (Liu & Qin, 2005; Rohacs, Lopes, Michailidis, & Logothetis, 2005), and the M-type K^+ channel (KCNQ2) (Liu, Liang, Liu, & Zhang, 2008). The regulation of the TRPV1 by PIP_2 is controversial. On the one hand, TRPV1 function was shown to be negatively regulated by PIP_2 as evidenced by potentiation of TRPV1-mediated responses by activation of the co-expressed GPCRs (such as bradykinin B_2 receptor) and tyrosine kinase receptors (such as NGF receptor TrkA), activation of which was shown to deplete membrane PIP_2 and release PIP_2 inhibition of TRPV1 (Chuang et al., 2001). Furthermore, reconstitution of purified TRPV1 protein into an artificial lipid bilayer formed a fully functional TRPV1 channel which could be fully activated by several TRPV1 activators in the absence of PIP_2 (Cao et al., 2013). Finally, phosphoinositide lipids including PIP_2 suppressed TRPV1 channel activity in liposomes, which further supports the hypothesis that phosphoinositide lipids negatively regulate TRPV1 function (Cao et al., 2013). On the other hand, other studies have demonstrated that phosphoinositide 3-kinase regulates TRPV1 trafficking to mediate NGF-induced potentiation of TRPV1 and phosphoinositide lipids are indeed endogenous cofactors that modulate TRPV1 activation (Lukacs, Rives, Sun, Zakharian, & Rohacs, 2013; Stein, Ufret-Vincenty, Hua, Santana, & Gordon, 2006; Yao & Qin, 2009).

TRPV3 channel function is potentiated by depletion of PIP_2 because $G_{q/11}$-coupled purinergic and muscarinic receptor-mediated hydrolysis or pharmacological depletion of PIP_2 enhanced TRPV3 channel function (Figure 12.4) (Doerner, Hatt, & Ramsey, 2011). A decrease in PIP_2 caused a negative shift of the half-maximal activation voltage and increased the voltage-independent faction of TRPV3. In addition to the alteration of voltage-dependent activation, PIP_2 also decreased the threshold of 2-APB to activate TRPV3. A decrease of membrane PIP_2 level increased the open probability of TRPV3 channels without altering the single-channel amplitude. TRPV3 mutants carrying mutations that neutralize two positively charged amino acid residues in the TRP domain (R696A or K705A) were insensitive to PIP_2 application, indicating that these two

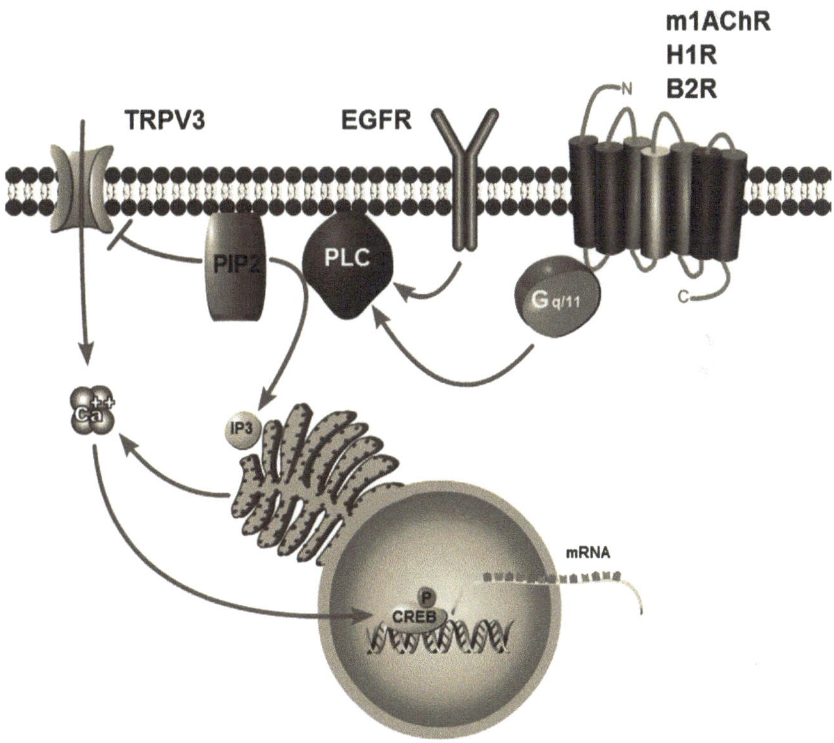

Figure 12.4 TRPV3-mediated signaling is regulated by multiple receptor signaling pathways. G protein-coupled receptors (m1AChR, H1R, and B2R) activate $G_{q/11}$ upon ligands binding. Activated PLC in turn hydrolyzes PIP2 which constitutively suppresses TRPV3, thereby potentiating TRPV3 response. EGFR activation also induces PLC activation and enhances TRPV3 function in a similar manner. TRPV3-mediated Ca^{2+} influx can promote the CREB-initiated gene expression. (See the color plate.)

residues are critical to the interaction between TRPV3 protein and PIP_2 (Doerner et al., 2011) (Figure 12.1).

4.3 Arachidonic acid

AA, a polyunsaturated fatty acid present in the phospholipids of cell membrane, is an important inflammatory mediator involved in many molecular and cellular functions under physiological and pathological conditions (Meves, 2008). Although AA signaling regulates activities of many TRP channels including TRPV1 and TRPV4 (Nilius, Vriens, Prenen, Droogmans, & Voets, 2004), different AA metabolic pathways are involved. For instance, activation of bradykinin B2 receptors produces 12-lipoxygenase metabolites such as 12-HPETE to activate TRPV1 and increase the excitability of sensory nerve endings (Hwang et al., 2000; Shin et al., 2002). On the other hand, the endocannabinoid anandamide and AA use the cytochrome P450 epoxygenase-dependent formation of epoxyeicosatrienoic acids such as 5′, 6′-epoxyeicosatrienoic acid (EET) and 11′, 12′-EET to activate TRPV4 (Sonkusare et al., 2012; Watanabe, et al., 2003). Interestingly, previous studies have suggested that TRPV3 may contribute to AA-mediated inflammatory responses since AA dramatically increased membrane current and intracellular Ca^{2+} response mediated by heterologously or endogenously expressed TRPV3, but not TRPV1 and TRPV2 (Hu et al., 2006). In addition to AA, many other unsaturated fatty acids exhibited similar potentiating effects on TRPV3 channel activity. AA was shown to increase the affinity of TRPV3 to 2-APB as evidenced by a leftward shift of the concentration–response curve of 2-APB in the presence of AA. The potentiating effect of TRPV3 by AA was mediated by an increase of the single-channel open probability but not the channel conductance. Neither AA metabolic blockers nor protein kinase inhibitors against protein kinase C (PKC) or calmodulin-dependent protein kinase II affected the potentiating effect of AA, suggesting that AA directly acts on TRPV3 channels.

4.4 PLC and PKC

PLC plays an important role in mammalian signal transduction pathways by cleaving PIP_2 into DAG and IP_3, which results in activation of PKC and increased intracellular Ca^{2+} levels (Rohacs, 2007). Many GPCRs activated by inflammatory mediators such as bradykinin, ATP, and histamine activate PLC through G proteins ($G_{q/11}$). When TRPV3 was co-expressed with human muscarinic receptor type 1 (HM1), histamine H1 receptor, or bradykinin B2 receptor in HEK293 cells, membrane currents evoked by the

TRPV3 activators carvacrol and 2-APB were dramatically increased by co-application of agonists of the HM1 (carbachol), H1 (histamine), or B2 (bradykinin) receptors, respectively (Figure 12.4). The facilitating response of GPCR activation on TRPV3 function was abolished by a PLC inhibitor U73122 or chelation of intracellular Ca^{2+} (Xu et al., 2006). It is likely that PKC, the critical downstream signaling molecule of PLC signaling, is involved in this process since TRPV3 was also potently sensitized by PKC (Hu et al., 2006). The PKC activator phorbol-12-myristate-13-acetate has been shown to increase TRPV3-mediated whole-cell current in TRPV3-expressing HEK293 cells. The potentiation was abolished when cells were pretreated with selective PKC inhibitors (Ro328110 and Go6976) (Hu et al., 2006). In addition to GPCR signaling, PLC can also be activated by growth factors such as EGF, which activate receptor tyrosine kinases, such as EGFR (Figure 12.4). In fact, loss of functional interaction between TRPV3 and EGFR signaling has been proposed to be a major determinant of defects in hair morphogenesis in the TRPV3 null mice (Cheng et al., 2010).

5. CONCLUSIONS

Great strides have been made to understand the role of TRPV3 under both physiological and pathological conditions. TRPV3 is a versatile receptor integrating diversified sensory signals although it is unlikely to play a major role in sensing warm temperatures in rodents in vivo. TRPV3 is also involved in many cellular functions in addition to its role as a polymodal sensory transducer. Recent findings that both gain-of-function and loss-of-function mutations of TRPV3 produce severe skin disorders and disruption of hair morphogenesis provide compelling evidence that TRPV3 activity is critical to maintain epidermal homeostasis. Future research on investigations into the cellular and systemic functions of TRPV3 may greatly benefit from advent of selective agonists and antagonists as well as specific antibodies. The prominent role of TRPV3 in skin disorders and chronic itch makes it a promising drug target.

ACKNOWLEDGMENTS

We are grateful to Richard Clark and Edgar Walters for critical reading of this manuscript and our colleagues for helpful comments and discussions. This work is supported partly by grants from the National Institutes of Health (RO1RGM101218A), Mission Connect/The Institute for Rehabilitation and Research (TIRR) Foundation (013-108), and The University of Texas Health Science Center to HH.

REFERENCES

Akazawa, Y., Yuki, T., Yoshida, H., Sugiyama, Y., & Inoue, S. (2013). Activation of TRPV4 strengthens the tight-junction barrier in human epidermal keratinocytes. *Skin Pharmacology and Physiology, 26*, 15–21.

Akimoto, T., Suzuki, H., Nakama, K., & Suzuki, K. (2000). Locus of dominant hairless gene (Ht) causing abnormal hair and keratinization maps to rat chromosome 10. *Experimental Animals, 49*, 137–140.

Akopian, A. N., Ruparel, N. B., Jeske, N. A., Patwardhan, A., & Hargreaves, K. M. (2009). Role of ionotropic cannabinoid receptors in peripheral antinociception and antihyperalgesia. *Trends in Pharmacological Sciences, 30*, 79–84.

Albers, K. M., & Davis, B. M. (2007). The skin as a neurotrophic organ. *The Neuroscientist, 13*, 371–382.

Anand, U., Otto, W. R., Casula, M. A., Day, N. C., Davis, J. B., Bountra, C., et al. (2006). The effect of neurotrophic factors on morphology, TRPV1 expression and capsaicin responses of cultured human DRG sensory neurons. *Neuroscience Letters, 399*, 51–56.

Asakawa, M., Yoshioka, T., Matsutani, T., Hikita, I., Suzuki, M., Oshima, I., et al. (2006). Association of a mutation in TRPV3 with defective hair growth in rodents. *Journal of Investigative Dermatology, 126*, 2664–2672.

Bandell, M., Macpherson, L. J., & Patapoutian, A. (2007). From chills to chilis: mechanisms for thermosensation and chemesthesis via thermoTRPs. *Current Opinion in Neurobiology, 17*, 490–497.

Bang, S., Yoo, S., Yang, T. J., Cho, H., & Hwang, S. W. (2010). Farnesyl pyrophosphate is a novel pain-producing molecule via specific activation of TRPV3. *Journal of Biological Chemistry, 285*, 19362–19371.

Bang, S., Yoo, S., Yang, T. J., Cho, H., & Hwang, S. W. (2011). Isopentenyl pyrophosphate is a novel antinociceptive substance that inhibits TRPV3 and TRPA1 ion channels. *Pain, 152*, 1156–1164.

Bang, S., Yoo, S., Yang, T. J., Cho, H., & Hwang, S. W. (2012). 17(R)-resolvin D1 specifically inhibits transient receptor potential ion channel vanilloid 3 leading to peripheral antinociception. *British Journal of Pharmacology, 165*, 683–692.

Beltran, L., Beltran, M., Aguado, A., Gisselmann, G., & Hatt, H. (2013). 2-Aminoethoxydiphenyl borate activates the mechanically gated human KCNK channels KCNK 2 (TREK-1), KCNK 4 (TRAAK), and KCNK 10 (TREK-2). *Frontiers in Pharmacology, 4*, 63.

Bergsmann, J., Derler, I., Muik, M., Frischauf, I., Fahrner, M., Pollheimer, P., et al. (2011). Molecular determinants within N terminus of Orai3 protein that control channel activation and gating. *Journal of Biological Chemistry, 286*, 31565–31575.

Bishnoi, M., Kondepudi, K. K., Gupta, A., Karmase, A., & Boparai, R. K. (2013). Expression of multiple transient receptor potential channel genes in murine 3T3-L1 cell lines and adipose tissue. *Pharmacological Reports: PR, 65*, 751–755.

Blanpain, C., & Fuchs, E. (2009). Epidermal homeostasis: a balancing act of stem cells in the skin. *Nature Reviews: Molecular Cell Biology, 10*, 207–217.

Borbiro, I., Lisztes, E., Toth, B. I., Czifra, G., Olah, A., Szollosi, A. G., et al. (2011). Activation of transient receptor potential vanilloid-3 inhibits human hair growth. *Journal of Investigative Dermatology, 131*, 1605–1614.

Brauchi, S., Orta, G., Salazar, M., Rosenmann, E., & Latorre, R. (2006). A hot-sensing cold receptor: C-terminal domain determines thermosensation in transient receptor potential channels. *Journal of Neuroscience, 26*, 4835–4840.

Brederson, J. D., Kym, P. R., & Szallasi, A. (2013). Targeting TRP channels for pain relief. *European Journal of Pharmacology, 716*, 61–76.

Brown, T. E., Chirila, A. M., Schrank, B. R., & Kauer, J. A. (2013). Loss of interneuron LTD and attenuated pyramidal cell LTP in Trpv1 and Trpv3 KO mice. *Hippocampus, 23*, 662–671.

Burnstock, G. (2012). Purinergic signalling: its unpopular beginning, its acceptance and its exciting future. *Bioessays, 34*, 218–225.

Cao, E., Cordero-Morales, J. F., Liu, B., Qin, F., & Julius, D. (2013). TRPV1 channels are intrinsically heat sensitive and negatively regulated by phosphoinositide lipids. *Neuron, 77*, 667–679.

Cao, X., Yang, F., Zheng, J., & Wang, K. (2012). Intracellular proton-mediated activation of TRPV3 channels accounts for the exfoliation effect of alpha-hydroxyl acids on keratinocytes. *Journal of Biological Chemistry, 287*, 25905–25916.

Carreno, O., Corominas, R., Fernandez-Morales, J., Camina, M., Sobrido, M. J., Fernandez-Fernandez, J. M., et al. (2012). SNP variants within the vanilloid TRPV1 and TRPV3 receptor genes are associated with migraine in the Spanish population. *American Journal of Medical Genetics. Part B, Neuropsychiatric Genetics: The Official Publication of the International Society of Psychiatric Genetics, 159B*, 94–103.

Caterina, M. J., Leffler, A., Malmberg, A. B., Martin, W. J., Trafton, J., Petersen-Zeitz, K. R., et al. (2000). Impaired nociception and pain sensation in mice lacking the capsaicin receptor. *Science, 288*, 306–313.

Caterina, M. J., Schumacher, M. A., Tominaga, M., Rosen, T. A., Levine, J. D., & Julius, D. (1997). The capsaicin receptor: a heat-activated ion channel in the pain pathway. *Nature, 389*, 816–824.

Cheng, X., Jin, J., Hu, L., Shen, D., Dong, X. P., Samie, M. A., et al. (2010). TRP channel regulates EGFR signaling in hair morphogenesis and skin barrier formation. *Cell, 141*, 331–343.

Cheng, W., Yang, F., Liu, S., Colton, C. K., Wang, C., Cui, Y., et al. (2012). Heteromeric heat-sensitive transient receptor potential channels exhibit distinct temperature and chemical response. *Journal of Biological Chemistry, 287*, 7279–7288.

Cho, H., Yang, Y. D., Lee, J., Lee, B., Kim, T., Jang, Y., et al. (2012). The calcium-activated chloride channel anoctamin 1 acts as a heat sensor in nociceptive neurons. *Nature Neuroscience, 15*, 1015–1021.

Chuang, H. H., Prescott, E. D., Kong, H., Shields, S., Jordt, S. E., Basbaum, A. I., et al. (2001). Bradykinin and nerve growth factor release the capsaicin receptor from PtdIns(4,5)P2-mediated inhibition. *Nature, 411*, 957–962.

Chung, M. K., Guler, A. D., & Caterina, M. J. (2005). Biphasic currents evoked by chemical or thermal activation of the heat-gated ion channel, TRPV3. *Journal of Biological Chemistry, 280*, 15928–15941.

Chung, M. K., Lee, H., Mizuno, A., Suzuki, M., & Caterina, M. J. (2004a). 2-Aminoethoxydiphenyl borate activates and sensitizes the heat-gated ion channel TRPV3. *Journal of Neuroscience, 24*, 5177–5182.

Chung, M. K., Lee, H., Mizuno, A., Suzuki, M., & Caterina, M. J. (2004b). TRPV3 and TRPV4 mediate warmth-evoked currents in primary mouse keratinocytes. *Journal of Biological Chemistry, 279*, 21569–21575.

Cui, Y., Yang, F., Cao, X., Yarov-Yarovoy, V., Wang, K., & Zheng, J. (2012). Selective disruption of high sensitivity heat activation but not capsaicin activation of TRPV1 channels by pore turret mutations. *Journal of General Physiology, 139*, 273–283.

Danso-Abeam, D., Zhang, J., Dooley, J., Staats, K. A., Van Eyck, L., Van Brussel, T., et al. (2013). Olmsted syndrome: exploration of the immunological phenotype. *Orphanet Journal of Rare Diseases, 8*, 79.

Davis, J. B., Gray, J., Gunthorpe, M. J., Hatcher, J. P., Davey, P. T., Overend, P., et al. (2000). Vanilloid receptor-1 is essential for inflammatory thermal hyperalgesia. *Nature, 405*, 183–187.

De Petrocellis, L., Orlando, P., Moriello, A. S., Aviello, G., Stott, C., Izzo, A. A., et al. (2012). Cannabinoid actions at TRPV channels: effects on TRPV3 and TRPV4 and their potential relevance to gastrointestinal inflammation. *Acta Physiologica, 204*, 255–266.

Demuro, A., Penna, A., Safrina, O., Yeromin, A. V., Amcheslavsky, A., Cahalan, M. D., et al. (2011). Subunit stoichiometry of human Orai1 and Orai3 channels in closed and open states. *Proceedings of the National Academy of Sciences of the United States of America, 108*, 17832–17837.

Denda, M., Sokabe, T., Fukumi-Tominaga, T., & Tominaga, M. (2007). Effects of skin surface temperature on epidermal permeability barrier homeostasis. *Journal of Investigative Dermatology, 127*, 654–659.

Dhaka, A., Viswanath, V., & Patapoutian, A. (2006). Trp ion channels and temperature sensation. *Annual Review of Neuroscience, 29*, 135–161.

Dhar, M. K., Koul, A., & Kaul, S. (2013). Farnesyl pyrophosphate synthase: a key enzyme in isoprenoid biosynthetic pathway and potential molecular target for drug development. *New Biotechnology, 30*, 114–123.

Doerner, J. F., Hatt, H., & Ramsey, I. S. (2011). Voltage- and temperature-dependent activation of TRPV3 channels is potentiated by receptor-mediated PI(4,5)P2 hydrolysis. *Journal of General Physiology, 137*, 271–288.

Earley, S., Gonzales, A. L., & Garcia, Z. I. (2010). A dietary agonist of transient receptor potential cation channel V3 elicits endothelium-dependent vasodilation. *Molecular Pharmacology, 77*, 612–620.

Facer, P., Casula, M. A., Smith, G. D., Benham, C. D., Chessell, I. P., Bountra, C., et al. (2007). Differential expression of the capsaicin receptor TRPV1 and related novel receptors TRPV3, TRPV4 and TRPM8 in normal human tissues and changes in traumatic and diabetic neuropathy. *BMC Neurology, 7*, 11.

Gopinath, P., Wan, E., Holdcroft, A., Facer, P., Davis, J. B., Smith, G. D., et al. (2005). Increased capsaicin receptor TRPV1 in skin nerve fibres and related vanilloid receptors TRPV3 and TRPV4 in keratinocytes in human breast pain. *BMC Women's Health, 5*, 2.

Gordon-Shaag, A., Zagotta, W. N., & Gordon, S. E. (2008). Mechanism of Ca(2+)-dependent desensitization in TRP channels. *Channels (Austin), 2*, 125–129.

Grandl, J., Hu, H., Bandell, M., Bursulaya, B., Schmidt, M., Petrus, M., et al. (2008). Pore region of TRPV3 ion channel is specifically required for heat activation. *Nature Neuroscience, 11*, 1007–1013.

Grandl, J., Kim, S. E., Uzzell, V., Bursulaya, B., Petrus, M., Bandell, M., et al. (2010). Temperature-induced opening of TRPV1 ion channel is stabilized by the pore domain. *Nature Neuroscience, 13*, 708–714.

Hamamoto, T., Takumida, M., Hirakawa, K., Takeno, S., & Tatsukawa, T. (2008). Localization of transient receptor potential channel vanilloid subfamilies in the mouse larynx. *Acta Oto-laryngologica, 128*, 685–693.

Hamamoto, T., Takumida, M., Hirakawa, K., Tatsukawa, T., & Ishibashi, T. (2009). Localization of transient receptor potential vanilloid (TRPV) in the human larynx. *Acta Oto-laryngologica, 129*, 560–568.

Harks, E. G., Camina, J. P., Peters, P. H., Ypey, D. L., Scheenen, W. J., van Zoelen, E. J., et al. (2003). Besides affecting intracellular calcium signaling, 2-APB reversibly blocks gap junctional coupling in confluent monolayers, thereby allowing measurement of single-cell membrane currents in undissociated cells. *FASEB Journal, 17*, 941–943.

Hoeft, B., Linseisen, J., Beckmann, L., Muller-Decker, K., Canzian, F., Husing, A., et al. (2010). Polymorphisms in fatty-acid-metabolism-related genes are associated with colorectal cancer risk. *Carcinogenesis, 31*, 466–472.

Hong, S., Gronert, K., Devchand, P. R., Moussignac, R. L., & Serhan, C. N. (2003). Novel docosatrienes and 17S-resolvins generated from docosahexaenoic acid in murine brain, human blood, and glial cells. Autacoids in anti-inflammation. *Journal of Biological Chemistry, 278*, 14677–14687.

Hu, J., Choo, H. J., & Ma, S. X. (2011). Infrared heat treatment reduces food intake and modifies expressions of TRPV3-POMC in the dorsal medulla of obesity prone rats.

International Journal of Hyperthermia: The Official Journal of European Society for Hyperthermic Oncology, North American Hyperthermia Group, 27, 708–716.
Hu, H., Grandl, J., Bandell, M., Petrus, M., & Patapoutian, A. (2009). Two amino acid residues determine 2-APB sensitivity of the ion channels TRPV3 and TRPV4. *Proceedings of the National Academy of Sciences of the United States of America, 106*, 1626–1631.
Hu, H. Z., Gu, Q., Wang, C., Colton, C. K., Tang, J., Kinoshita-Kawada, M., et al. (2004). 2-Aminoethoxydiphenyl borate is a common activator of TRPV1, TRPV2, and TRPV3. *Journal of Biological Chemistry, 279*, 35741–35748.
Hu, H. Z., Xiao, R., Wang, C., Gao, N., Colton, C. K., Wood, J. D., et al. (2006). Potentiation of TRPV3 channel function by unsaturated fatty acids. *Journal of Cellular Physiology, 208*, 201–212.
Huang, S. M., Lee, H., Chung, M. K., Park, U., Yu, Y. Y., Bradshaw, H. B., et al. (2008). Overexpressed transient receptor potential vanilloid 3 ion channels in skin keratinocytes modulate pain sensitivity via prostaglandin E2. *Journal of Neuroscience, 28*, 13727–13737.
Huang, S. M., Li, X., Yu, Y., Wang, J., & Caterina, M. J. (2011). TRPV3 and TRPV4 ion channels are not major contributors to mouse heat sensation. *Molecular Pain, 7*, 37.
Hwang, S. W., Cho, H., Kwak, J., Lee, S. Y., Kang, C. J., Jung, J., et al. (2000). Direct activation of capsaicin receptors by products of lipoxygenases: endogenous capsaicin-like substances. *Proceedings of the National Academy of Sciences of the United States of America, 97*, 6155–6160.
Imamachi, N., Park, G. H., Lee, H., Anderson, D. J., Simon, M. I., Basbaum, A. I., et al. (2009). TRPV1-expressing primary afferents generate behavioral responses to pruritogens via multiple mechanisms. *Proceedings of the National Academy of Sciences of the United States of America, 106*, 11330–11335.
Imura, K., Yoshioka, T., Hikita, I., Tsukahara, K., Hirasawa, T., Higashino, K., et al. (2007). Influence of TRPV3 mutation on hair growth cycle in mice. *Biochemical and Biophysical Research Communications, 363*, 479–483.
Imura, K., Yoshioka, T., Hirasawa, T., & Sakata, T. (2009). Role of TRPV3 in immune response to development of dermatitis. *Journal of Inflammation (London), 6*, 17.
Jankowski, M. P., & Koerber, H. R. (2010). Neurotrophic factors and nociceptor sensitization. In L. Kruger, & A. R. Light (Eds.), *Translational pain research: From mouse to man*. Boca Raton, FL: CRC Press.
Jara-Oseguera, A., & Islas, L. D. (2013). The role of allosteric coupling on thermal activation of thermo-TRP channels. *Biophysical Journal, 104*, 2160–2169.
Kida, N., Sokabe, T., Kashio, M., Haruna, K., Mizuno, Y., Suga, Y., et al. (2012). Importance of transient receptor potential vanilloid 4 (TRPV4) in epidermal barrier function in human skin keratinocytes. *Pflügers Archiv – European Journal of Physiology, 463*, 715–725.
Kim, B. M., Lee, S. H., Shim, W. S., & Oh, U. (2004). Histamine-induced Ca(2+) influx via the PLA(2)/lipoxygenase/TRPV1 pathway in rat sensory neurons. *Neuroscience Letters, 361*, 159–162.
Kim, S. E., Patapoutian, A., & Grandl, J. (2013). Single residues in the outer pore of TRPV1 and TRPV3 have temperature-dependent conformations. *PLoS One, 8*, e59593.
Klein, A. H., Carstens, M. I., & Carstens, E. (2013). Eugenol and carvacrol induce temporally desensitizing patterns of oral irritation and enhance innocuous warmth and noxious heat sensation on the tongue. *Pain, 154*, 2078–2087.
Lai-Cheong, J. E., Sethuraman, G., Ramam, M., Stone, K., Simpson, M. A., & McGrath, J. A. (2012). Recurrent heterozygous missense mutation, p.Gly573Ser, in the TRPV3 gene in an Indian boy with sporadic Olmsted syndrome. *British Journal of Dermatology, 167*, 440–442.
Latorre, R., Brauchi, S., Orta, G., Zaelzer, C., & Vargas, G. (2007). ThermoTRP channels as modular proteins with allosteric gating. *Cell Calcium, 42*, 427–438.

Lee, H., Iida, T., Mizuno, A., Suzuki, M., & Caterina, M. J. (2005). Altered thermal selection behavior in mice lacking transient receptor potential vanilloid 4. *Journal of Neuroscience, 25*, 1304–1310.

Li, M., Jiang, J., & Yue, L. (2006). Functional characterization of homo- and heteromeric channel kinases TRPM6 and TRPM7. *Journal of General Physiology, 127*, 525–537.

Lin, C. R., Amaya, F., Barrett, L., Wang, H., Takada, J., Samad, T. A., et al. (2006). Prostaglandin E2 receptor EP4 contributes to inflammatory pain hypersensitivity. *Journal of Pharmacology and Experimental Therapeutics, 319*, 1096–1103.

Lin, Z., Chen, Q., Lee, M., Cao, X., Zhang, J., Ma, D., et al. (2012). Exome sequencing reveals mutations in TRPV3 as a cause of Olmsted syndrome. *American Journal of Human Genetics, 90*, 558–564.

Lipski, J., Park, T. I., Li, D., Lee, S. C., Trevarton, A. J., Chung, K. K., et al. (2006). Involvement of TRP-like channels in the acute ischemic response of hippocampal CA1 neurons in brain slices. *Brain Research, 1077*, 187–199.

Lis, A., Peinelt, C., Beck, A., Parvez, S., Monteilh-Zoller, M., Fleig, A., et al. (2007). CRACM1, CRACM2, and CRACM3 are store-operated Ca^{2+} channels with distinct functional properties. *Current Biology, 17*, 794–800.

Lishko, P. V., Procko, E., Jin, X., Phelps, C. B., & Gaudet, R. (2007). The ankyrin repeats of TRPV1 bind multiple ligands and modulate channel sensitivity. *Neuron, 54*, 905–918.

Liu, B., Liang, H., Liu, L., & Zhang, H. (2008). Phosphatidylinositol 4,5-bisphosphate hydrolysis mediates histamine-induced KCNQ/M current inhibition. *American Journal of Physiology: Cell Physiology, 295*, C81–C91.

Liu, B., & Qin, F. (2005). Functional control of cold- and menthol-sensitive TRPM8 ion channels by phosphatidylinositol 4,5-bisphosphate. *Journal of Neuroscience, 25*, 1674–1681.

Liu, B., Yao, J., Zhu, M. X., & Qin, F. (2011). Hysteresis of gating underlines sensitization of TRPV3 channels. *Journal of General Physiology, 138*, 509–520.

Lukacs, V., Rives, J. M., Sun, X., Zakharian, E., & Rohacs, T. (2013). Promiscuous activation of transient receptor potential vanilloid 1 channels by negatively charged intracellular lipids, the key role of endogenous phosphoinositides in maintaining channel activity. *Journal of Biological Chemistry*, 288, 35003-35013.

Lumpkin, E. A., & Caterina, M. J. (2007). Mechanisms of sensory transduction in the skin. *Nature, 445*, 858–865.

Luo, J., Stewart, R., Berdeaux, R., & Hu, H. (2012). Tonic inhibition of TRPV3 by Mg^{2+} in mouse epidermal keratinocytes. *Journal of Investigative Dermatology, 132*, 2158–2165.

Ma, H. T., Patterson, R. L., van Rossum, D. B., Birnbaumer, L., Mikoshiba, K., & Gill, D. L. (2000). Requirement of the inositol trisphosphate receptor for activation of store-operated Ca^{2+} channels. *Science, 287*, 1647–1651.

Macpherson, L. J., Hwang, S. W., Miyamoto, T., Dubin, A. E., Patapoutian, A., & Story, G. M. (2006). More than cool: promiscuous relationships of menthol and other sensory compounds. *Molecular and Cellular Neuroscience, 32*, 335–343.

Mandadi, S., Sokabe, T., Shibasaki, K., Katanosaka, K., Mizuno, A., Moqrich, A., et al. (2009). TRPV3 in keratinocytes transmits temperature information to sensory neurons via ATP. *Pflügers Archiv – European Journal of Physiology, 458*, 1093–1102.

Maruyama, T., Kanaji, T., Nakade, S., Kanno, T., & Mikoshiba, K. (1997). 2APB, 2-aminoethoxydiphenyl borate, a membrane-penetrable modulator of Ins(1,4,5) P3-induced Ca^{2+} release. *The Journal of Biochemistry, 122*, 498–505.

McKemy, D. D., Neuhausser, W. M., & Julius, D. (2002). Identification of a cold receptor reveals a general role for TRP channels in thermosensation. *Nature, 416*, 52–58.

Meves, H. (2008). Arachidonic acid and ion channels: an update. *British Journal of Pharmacology, 155*, 4–16.

Minamisawa, H., Smith, M. L., & Siesjo, B. K. (1990). The effect of mild hyperthermia and hypothermia on brain damage following 5, 10, and 15 minutes of forebrain ischemia. *Annals of Neurology, 28*, 26–33.

Moqrich, A., Hwang, S. W., Earley, T. J., Petrus, M. J., Murray, A. N., Spencer, K. S., et al. (2005). Impaired thermosensation in mice lacking TRPV3, a heat and camphor sensor in the skin. *Science, 307*, 1468–1472.

Moriyama, T., Higashi, T., Togashi, K., Iida, T., Segi, E., Sugimoto, Y., et al. (2005). Sensitization of TRPV1 by EP1 and IP reveals peripheral nociceptive mechanism of prostaglandins. *Molecular Pain, 1*, 3.

Moussaieff, A., Rimmerman, N., Bregman, T., Straiker, A., Felder, C. C., Shoham, S., et al. (2008). Incensole acetate, an incense component, elicits psychoactivity by activating TRPV3 channels in the brain. *FASEB Journal: Official Publication of the Federation of American Societies for Experimental Biology, 22*, 3024–3034.

Moussaieff, A., Yu, J., Zhu, H., Gattoni-Celli, S., Shohami, E., & Kindy, M. S. (2012). Protective effects of incensole acetate on cerebral ischemic injury. *Brain Research, 1443*, 89–97.

Neckermann, G., Bavandi, A., & Meingassner, J. G. (2000). Atopic dermatitis-like symptoms in hypomagnesaemic hairless rats are prevented and inhibited by systemic or topical SDZ ASM 981. *British Journal of Dermatology, 142*, 669–679.

Neeper, M. P., Liu, Y., Hutchinson, T. L., Wang, Y., Flores, C. M., & Qin, N. (2007). Activation properties of heterologously expressed mammalian TRPV2: evidence for species dependence. *Journal of Biological Chemistry, 282*, 15894–15902.

Nichols, C. G. (2006). KATP channels as molecular sensors of cellular metabolism. *Nature, 440*, 470–476.

Nilius, B., Vriens, J., Prenen, J., Droogmans, G., & Voets, T. (2004). TRPV4 calcium entry channel: a paradigm for gating diversity. *American Journal of Physiology: Cell Physiology, 286*, C195–C205.

Noel, J., Zimmermann, K., Busserolles, J., Deval, E., Alloui, A., Diochot, S., et al. (2009). The mechano-activated K^+ channels TRAAK and TREK-1 control both warm and cold perception. *EMBO Journal, 28*, 1308–1318.

Patapoutian, A., Peier, A. M., Story, G. M., & Viswanath, V. (2003). ThermoTRP channels and beyond: mechanisms of temperature sensation. *Nature Reviews Neuroscience, 4*, 529–539.

Peier, A. M., Reeve, A. J., Andersson, D. A., Moqrich, A., Earley, T. J., Hergarden, A. C., et al. (2002). A heat-sensitive TRP channel expressed in keratinocytes. *Science, 296*, 2046–2049.

Phelps, C. B., Wang, R. R., Choo, S. S., & Gaudet, R. (2010). Differential regulation of TRPV1, TRPV3, and TRPV4 sensitivity through a conserved binding site on the ankyrin repeat domain. *Journal of Biological Chemistry, 285*, 731–740.

Proksch, E., Nissen, H. P., Bremgartner, M., & Urquhart, C. (2005). Bathing in a magnesium-rich dead sea salt solution improves skin barrier function, enhances skin hydration, and reduces inflammation in atopic dry skin. *International Journal of Dermatology, 44*, 151–157.

Qin, N., Neeper, M. P., Liu, Y., Hutchinson, T. L., Lubin, M. L., & Flores, C. M. (2008). TRPV2 is activated by cannabidiol and mediates CGRP release in cultured rat dorsal root ganglion neurons. *Journal of Neuroscience, 28*, 6231–6238.

Rohacs, T. (2007). Regulation of TRP channels by PIP(2). *Pflügers Archiv – European Journal of Physiology, 453*, 753–762.

Rohacs, T., Lopes, C. M., Michailidis, I., & Logothetis, D. E. (2005). PI(4,5)P2 regulates the activation and desensitization of TRPM8 channels through the TRP domain. *Nature Neuroscience, 8*, 626–634.

Saisho, Y., Morimoto, A., & Umeda, T. (1997). Determination of farnesyl pyrophosphate in dog and human plasma by high-performance liquid chromatography with fluorescence detection. *Analytical Biochemistry, 252*, 89–95.

Saito, S., & Shingai, R. (2006). Evolution of thermoTRP ion channel homologs in vertebrates. *Physiological Genomics, 27*, 219–230.

Saito, S., Fukuta, N., Shingai, R., & Tominaga, M. (2011). Evolution of vertebrate transient receptor potential vanilloid 3 channels: opposite temperature sensitivity between mammals and western clawed frogs. *PLoS Genetics, 7*, e1002041.

Selescu, T., Ciobanu, A. C., Dobre, C., Reid, G., & Babes, A. (2013). Camphor activates and sensitizes transient receptor potential melastatin 8 (TRPM8) to cooling and icilin. *Chemical Senses, 38*, 563–575.

Sherkheli, M. A., Vogt-Eisele, A. K., Weber, K., & Hatt, H. (2013). Camphor modulates TRPV3 cation channels activity by interacting with critical pore-region cysteine residues. *Pakistan Journal of Pharmaceutical Sciences, 26*, 431–438.

Shin, J., Cho, H., Hwang, S. W., Jung, J., Shin, C. Y., Lee, S. Y., et al. (2002). Bradykinin-12-lipoxygenase-VR1 signaling pathway for inflammatory hyperalgesia. *Proceedings of the National Academy of Sciences of the United States of America, 99*, 10150–10155.

Smith, G. D., Gunthorpe, M. J., Kelsell, R. E., Hayes, P. D., Reilly, P., Facer, P., et al. (2002). TRPV3 is a temperature-sensitive vanilloid receptor-like protein. *Nature, 418*, 186–190.

Sobhan, U., Sato, M., Shinomiya, T., Okubo, M., Tsumura, M., Muramatsu, T., et al. (2013). Immunolocalization and distribution of functional temperature-sensitive TRP channels in salivary glands. *Cell and Tissue Research, 354*, 507–519.

Sonkusare, S. K., Bonev, A. D., Ledoux, J., Liedtke, W., Kotlikoff, M. I., Heppner, T. J., et al. (2012). Elementary Ca^{2+} signals through endothelial TRPV4 channels regulate vascular function. *Science, 336*, 597–601.

Stein, A. T., Ufret-Vincenty, C. A., Hua, L., Santana, L. F., & Gordon, S. E. (2006). Phosphoinositide 3-kinase binds to TRPV1 and mediates NGF-stimulated TRPV1 trafficking to the plasma membrane. *Journal of General Physiology, 128*, 509–522.

Steinhoff, M., & Biro, T. (2009). A TR(I)P to pruritus research: role of TRPV3 in inflammation and itch. *Journal of Investigative Dermatology, 129*, 531–535.

Stotz, S. C., Vriens, J., Martyn, D., Clardy, J., & Clapham, D. E. (2008). Citral sensing by transient receptor potential channels in dorsal root ganglion neurons. *PLoS One, 3*, e2082.

Strotmann, R., Schultz, G., & Plant, T. D. (2003). Ca^{2+}-dependent potentiation of the nonselective cation channel TRPV4 is mediated by a C-terminal calmodulin binding site. *Journal of Biological Chemistry, 278*, 26541–26549.

Sulk, M., Seeliger, S., Aubert, J., Schwab, V. D., Cevikbas, F., Rivier, M., et al. (2012). Distribution and expression of non-neuronal transient receptor potential (TRPV) ion channels in rosacea. *Journal of Investigative Dermatology, 132*, 1253–1262.

Sun, Y., Thapa, N., Hedman, A. C., & Anderson, R. A. (2013). Phosphatidylinositol 4,5-bisphosphate: targeted production and signaling. *Bioessays, 35*, 513–522.

Suzuki, M., Mizuno, A., Kodaira, K., & Imai, M. (2003). Impaired pressure sensation in mice lacking TRPV4. *Journal of Biological Chemistry, 278*, 22664–22668.

Talavera, K., Nilius, B., & Voets, T. (2008). Neuronal TRP channels: thermometers, pathfinders and life-savers. *Trends in Neurosciences, 31*, 287–295.

Talavera, K., Yasumatsu, K., Voets, T., Droogmans, G., Shigemura, N., Ninomiya, Y., et al. (2005). Heat activation of TRPM5 underlies thermal sensitivity of sweet taste. *Nature, 438*, 1022–1025.

Togashi, K., Hara, Y., Tominaga, T., Higashi, T., Konishi, Y., Mori, Y., et al. (2006). TRPM2 activation by cyclic ADP-ribose at body temperature is involved in insulin secretion. *EMBO Journal, 25*, 1804–1815.

Togashi, K., Inada, H., & Tominaga, M. (2008). Inhibition of the transient receptor potential cation channel TRPM2 by 2-aminoethoxydiphenyl borate (2-APB). *British Journal of Pharmacology, 153*, 1324–1330.

Touska, F., Marsakova, L., Teisinger, J., & Vlachova, V. (2011). A "cute" desensitization of TRPV1. *Current Pharmaceutical Biotechnology, 12*, 122–129.

Toyoda, M., Nakamura, M., Makino, T., Hino, T., Kagoura, M., & Morohashi, M. (2002). Nerve growth factor and substance P are useful plasma markers of disease activity in atopic dermatitis. *British Journal of Dermatology, 147*, 71–79.

Ueda, T., Yamada, T., Ugawa, S., Ishida, Y., & Shimada, S. (2009). TRPV3, a thermosensitive channel is expressed in mouse distal colon epithelium. *Biochemical and Biophysical Research Communications, 383*, 130–134.

Voets, T., Droogmans, G., Wissenbach, U., Janssens, A., Flockerzi, V., & Nilius, B. (2004). The principle of temperature-dependent gating in cold- and heat-sensitive TRP channels. *Nature, 430*, 748–754.

Voets, T., Prenen, J., Fleig, A., Vennekens, R., Watanabe, H., Hoenderop, J. G., et al. (2001). CaT1 and the calcium release-activated calcium channel manifest distinct pore properties. *Journal of Biological Chemistry, 276*, 47767–47770.

Vogt-Eisele, A. K., Weber, K., Sherkheli, M. A., Vielhaber, G., Panten, J., Gisselmann, G., et al. (2007). Monoterpenoid agonists of TRPV3. *British Journal of Pharmacology, 151*, 530–540.

Vriens, J., Owsianik, G., Hofmann, T., Philipp, S. E., Stab, J., Chen, X., et al. (2011). TRPM3 is a nociceptor channel involved in the detection of noxious heat. *Neuron, 70*, 482–494.

Watanabe, A., Takeuchi, M., Nagata, M., Nakamura, K., Nakao, H., Yamashita, H., et al. (2003). Role of the Nh (non-hair) mutation in the development of dermatitis and hyperproduction of IgE in DS-Nh mice. *Experimental Animals, 52*, 419–423.

Watanabe, H., Vriens, J., Prenen, J., Droogmans, G., Voets, T., & Nilius, B. (2003). Anandamide and arachidonic acid use epoxyeicosatrienoic acids to activate TRPV4 channels. *Nature, 424*, 434–438.

Wilson, S. R., The, L., Batia, L. M., Beattie, K., Katibah, G. E., McClain, S. P., et al. (2013). The epithelial cell-derived atopic dermatitis cytokine TSLP activates neurons to induce itch. *Cell, 155*, 285–295.

van der Worp, H. B., Macleod, M. R., Kollmar, R., & European Stroke Research Network for Hypothermis (EuroHYP). (2010). Therapeutic hypothermia for acute ischemic stroke: ready to start large randomized trials? *Journal of Cerebral Blood Flow & Metabolism, 30*, 1079–1093.

Xiao, R., Tang, J., Wang, C., Colton, C. K., Tian, J., & Zhu, M. X. (2008). Calcium plays a central role in the sensitization of TRPV3 channel to repetitive stimulations. *Journal of Biological Chemistry, 283*, 6162–6174.

Xiao, R., Tian, J., Tang, J., & Zhu, M. X. (2008). The TRPV3 mutation associated with the hairless phenotype in rodents is constitutively active. *Cell Calcium, 43*, 334–343.

Xu, H., Blair, N. T., & Clapham, D. E. (2005). Camphor activates and strongly desensitizes the transient receptor potential vanilloid subtype 1 channel in a vanilloid-independent mechanism. *Journal of Neuroscience, 25*, 8924–8937.

Xu, H., Delling, M., Jun, J. C., & Clapham, D. E. (2006). Oregano, thyme and clove-derived flavors and skin sensitizers activate specific TRP channels. *Nature Neuroscience, 9*, 628–635.

Xu, H., Ramsey, I. S., Kotecha, S. A., Moran, M. M., Chong, J. A., Lawson, D., et al. (2002). TRPV3 is a calcium-permeable temperature-sensitive cation channel. *Nature, 418*, 181–186.

Yamamoto-Kasai, E., Imura, K., Yasui, K., Shichijou, M., Oshima, I., Hirasawa, T., et al. (2012). TRPV3 as a therapeutic target for itch. *Journal of Investigative Dermatology, 132*, 2109–2112.

Yamamoto-Kasai, E., Yasui, K., Shichijo, M., Sakata, T., & Yoshioka, T. (2013). Impact of TRPV3 on the development of allergic dermatitis as a dendritic cell modulator. *Experimental Dermatology, 22*, 820–824.

Yang, F., Cui, Y., Wang, K., & Zheng, J. (2010). Thermosensitive TRP channel pore turret is part of the temperature activation pathway. *Proceedings of the National Academy of Sciences of the United States of America, 107*, 7083–7088.

Yang, X. R., Lin, M. J., McIntosh, L. S., & Sham, J. S. (2006). Functional expression of transient receptor potential melastatin- and vanilloid-related channels in pulmonary arterial and aortic smooth muscle. *American Journal of Physiology – Lung Cellular and Molecular Physiology, 290*, L1267–L1276.

Yao, J., Liu, B., & Qin, F. (2010). Kinetic and energetic analysis of thermally activated TRPV1 channels. *Biophysical Journal, 99*, 1743–1753.

Yao, J., Liu, B., & Qin, F. (2011). Modular thermal sensors in temperature-gated transient receptor potential (TRP) channels. *Proceedings of the National Academy of Sciences of the United States of America, 108*, 11109–11114.

Yao, J., & Qin, F. (2009). Interaction with phosphoinositides confers adaptation onto the TRPV1 pain receptor. *PLoS Biology, 7*, e46.

Yenari, M. A., & Hemmen, T. M. (2010). Therapeutic hypothermia for brain ischemia: where have we come and where do we go? *Stroke, 41*, S72–S74.

Yoshioka, T., Hikita, I., Asakawa, M., Hirasawa, T., Deguchi, M., Matsutani, T., et al. (2006). Spontaneous scratching behaviour in DS-Nh mice as a possible model for pruritus in atopic dermatitis. *Immunology, 118*, 293–301.

Yoshioka, T., Imura, K., Asakawa, M., Suzuki, M., Oshima, I., Hirasawa, T., et al. (2009). Impact of the Gly573Ser substitution in TRPV3 on the development of allergic and pruritic dermatitis in mice. *Journal of Investigative Dermatology, 129*, 714–722.

Zhu, H., Wei, X., Bian, K., & Murad, F. (2008). Effects of nitric oxide on skin burn wound healing. *Journal of Burn Care & Research, 29*, 804–814.

Index

Note: Page numbers followed by "f" and "t" indicate figures and tables, respectively.

E

F

G

M

N

O

P

R

S

V

W

X

Z

COLOR PLATES

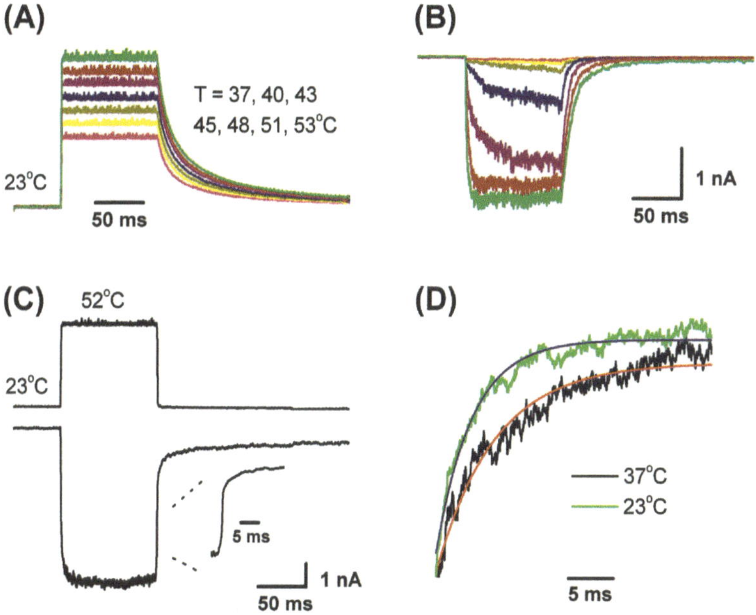

Figure 2.2 The time-resolved activation and deactivation of TRPV1. (A) A family of temperature jumps. (B) Current responses evoked by temperature jumps in (A). (C) Deactivation by rapid cooling. (D) Deactivation slowed by increasing temperature.

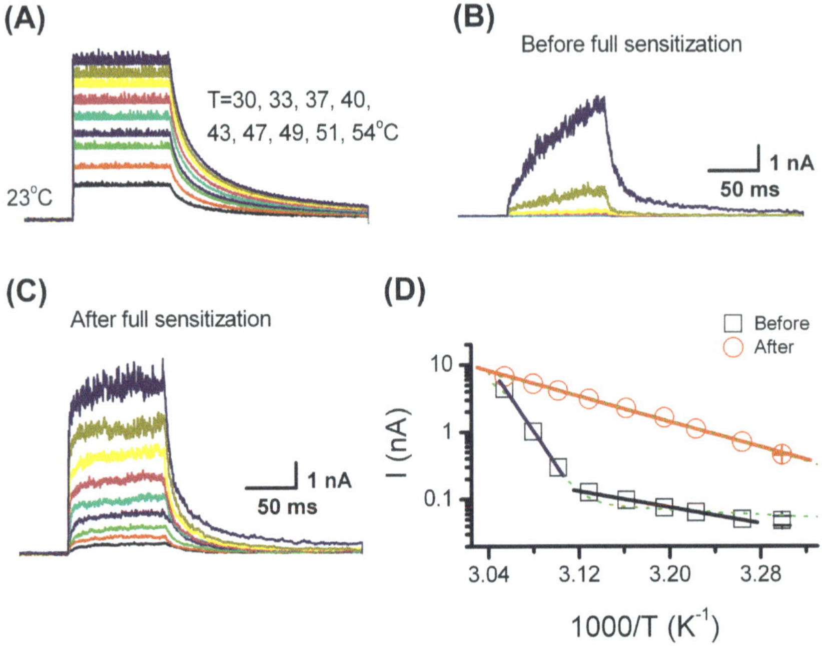

Figure 2.5 Temperature responses of TRPV3 evoked by a family of temperature jumps (A) before (B) and after sensitization (C). Slope sensitivity was decreased after sensitization (D).

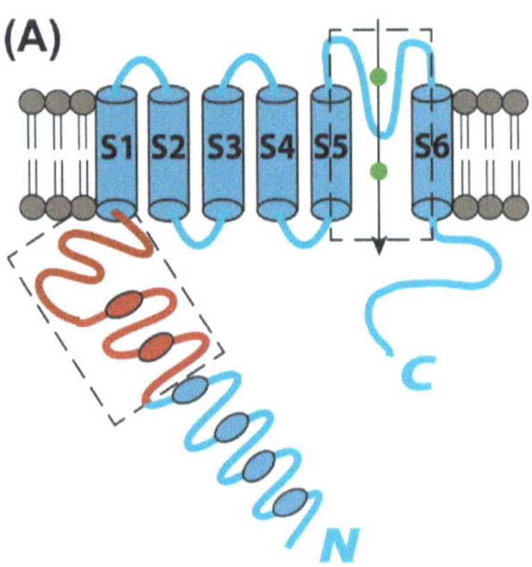

Figure 2.9(A) Proximal N terminus mediates heat sensitivity of vanilloid receptors. (A) Membrane topology of vanilloid receptors. The proximal N terminus meading heat sensitivity is shown in red.

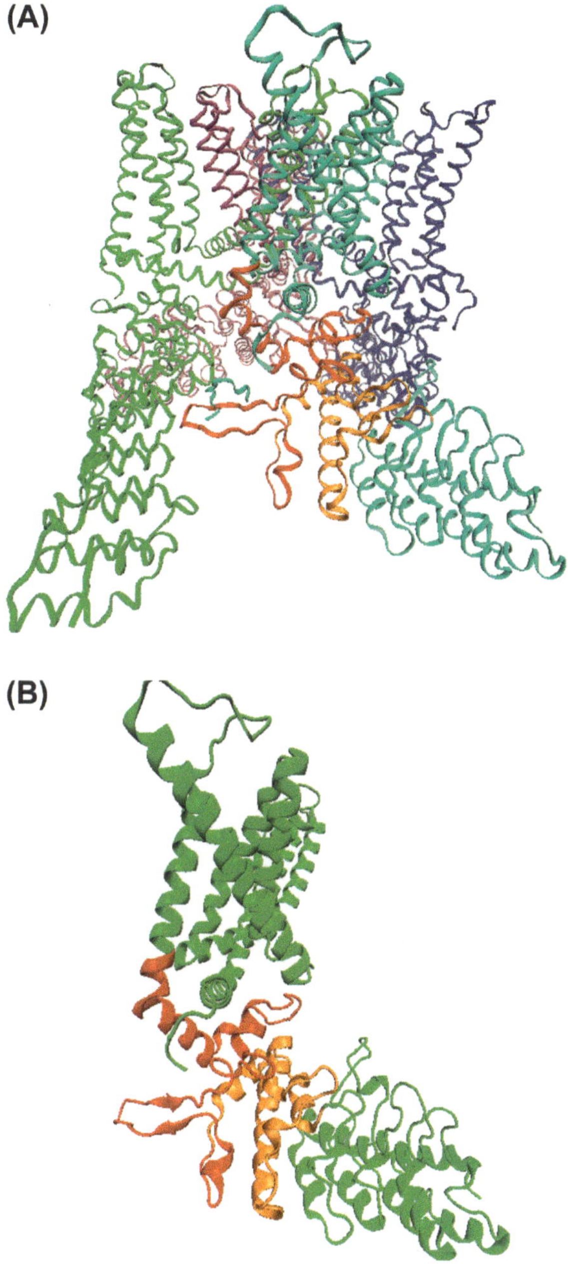

Figure 2.11 Cyo EM structure of TRPV1. (A) Tetramer. (B) Monomer. Regions colored in red corresponds to the linker domain connecting ankyrin repeats to S1. Regions in yellow are the last two ankyrin repeats (Liao et al., 2013).

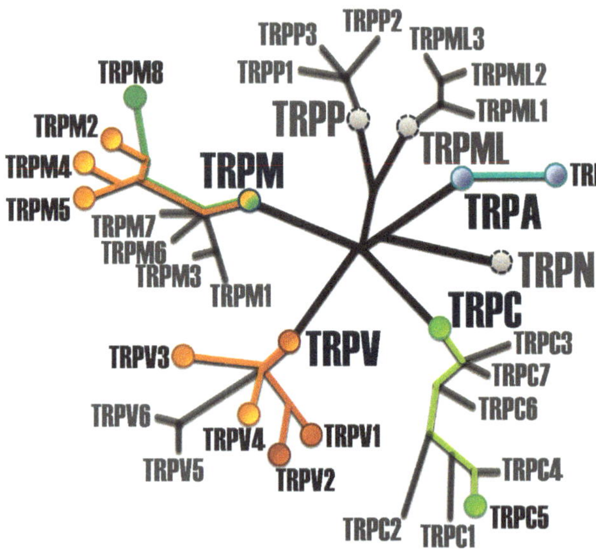

Figure 3.1 TRP superfamily dendogram. Seven subfamilies has been identified and the identity between them is low (~20%). Canonical or classic (TRPC), vanilloid (TRPV), ankyrin (TRPA), nonmechanoreceptor potential (TRPN), mucolipins (TRPML), polycystins (TRPP), and melastatin (TRPM). ThermoTRPs in the TRPM, TRPV, TRPA, and TRPC subfamilies are identified in colors.

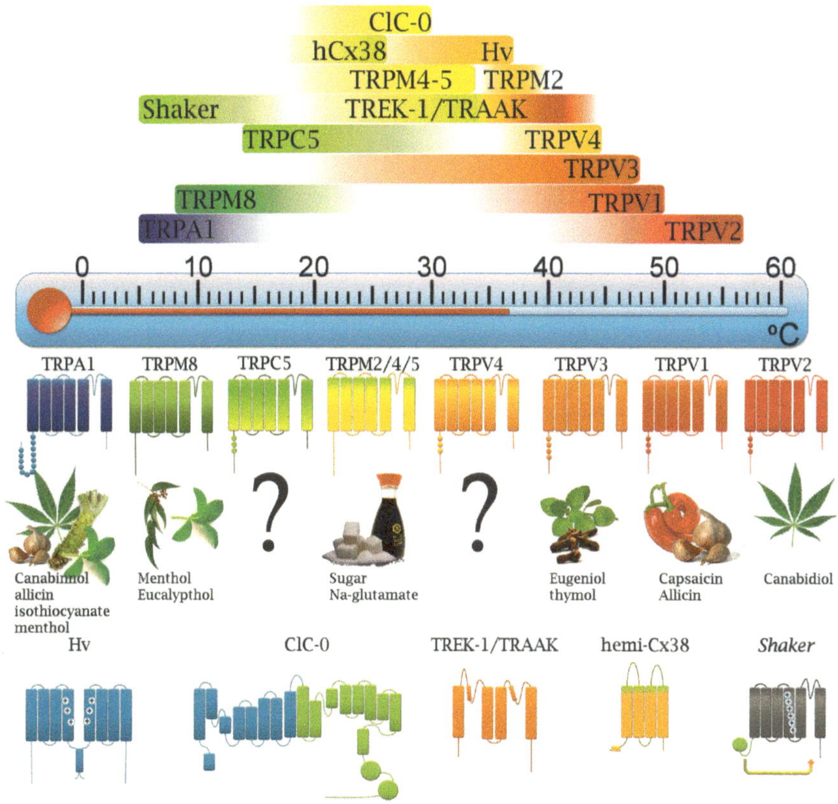

Figure 3.2 The temperature sensitivity of thermoTRPs and other temperature-dependent ion channels. The open probability (P_o) of the different temperature-activated channels increases in different temperature ranges. In this figure the higher the color intensity the larger the P_o. ThermoTRPs share a common organization. They are tetramers where the permeation pathway is located in the center of symmetry of the channel. ThermoTRP are activated by different agonist obtained from different plants. Other highly sensitive ion channels that do not share a common structure are shown in the bottom of the figure. Proton channels (H_v), chloride channel (ClC-0), and connexin 38 (hemi-Cx38). The green ball in the case of the Shaker channel is the inactivation gate that once the channel opens binds to the channel internal vestibule.

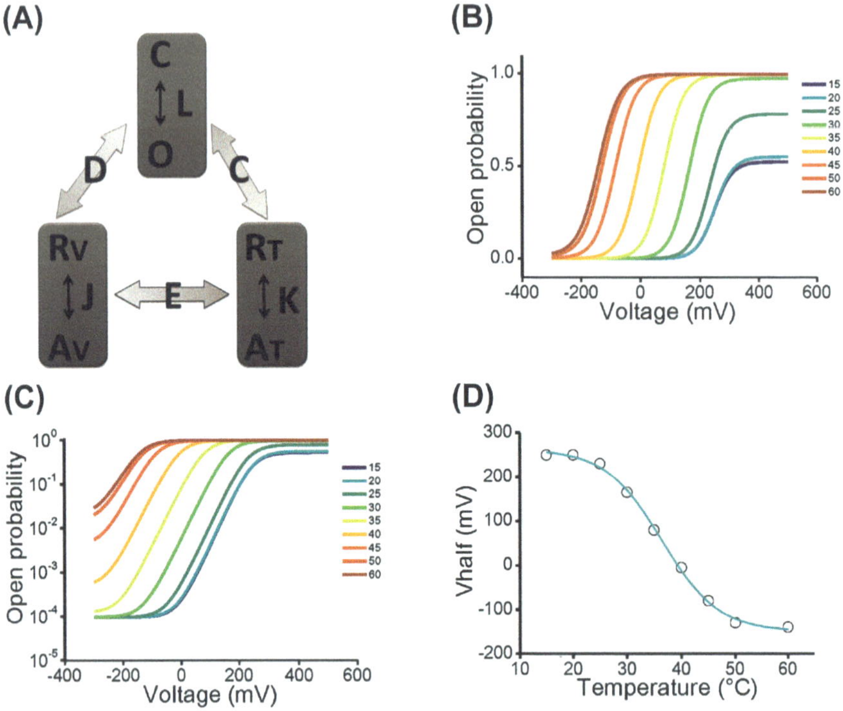

Figure 3.6 Allosteric kinetic model proposed for heat receptor. (A) Schematic representation for the model where the equilibrium constant L governs the open-closed transitions, J the voltage sensor activation and K the temperature sensor activation. Pore opening, voltage and temperature sensor activation are coupled between them through the allosteric constants D, C, and E. D and C connect the pore module with the voltage and temperature sensor, respectively and E represents the strength of the coupling between both sensors. When both sensors are in the model is reduced to the C–O transition only, that is, the channel can open in the absence of activated sensors. When all sensors are activated the equilibrium constant that defines the C–O transition is LDC, where LDC $>>$ L. (B) The allosteric model predicts that an increase in temperature shifts the P_o–V curves to the left along the voltage axis and if the C–O transition is voltage-independent, voltage becomes a partial channel activator (P_o does not reach one). (C) The model predicts that the limiting slope gives the number of apparent gating charges of the C–O reaction, which in this case has been made voltage-independent. Notice that the model also predictas that temperature alone can increase P_o by more than two orders of magnitude. (D) V_{half} saturates at low and high temperatures.

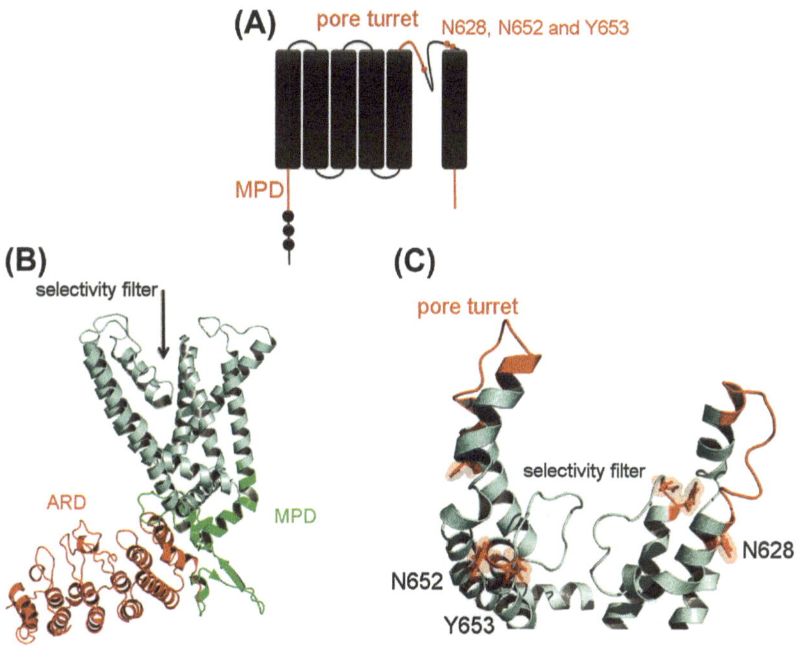

Figure 3.7 TRPV1 molecular determinants associated to temperature gating. (A) Schematic representation of TRPV1 monomer, highlighted in red are the different structures which has an impact on temperature gating. (B) Image of the recently TRPV1 structure determined by crioEM technique (PDB:3J5P) showing the molecular determinants located in the N-terminus claimed to be the temperature sensor structure. MPD is the connection between the ARD and the S1 in TRPV1. This portion of the protein is able to transfer the temperature sensitivity from TRPV1 to TRPV2 (see text). (C) Homology model (see Brauchi et al., 2007) showing the pore turret and the amino acid residues that when mutated reduce TRPV1 temperature sensitivity.

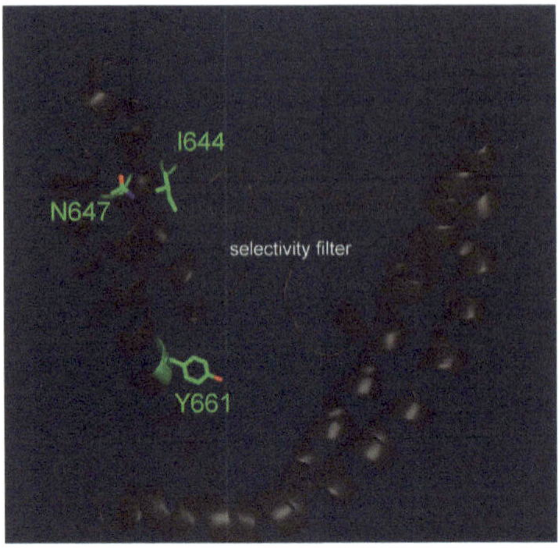

Figure 3.8 TRPV3 amino acid residues involved in temperature detection. The residues I644, N647, and Y661 located in the neighborhood of the pore domain have been identified as the molecular determinants involved in temperature detection with a high throughput mutagenesis technique. The triple mutant in TRPV3 channel abolishes the temperature sensitivity of this channel.

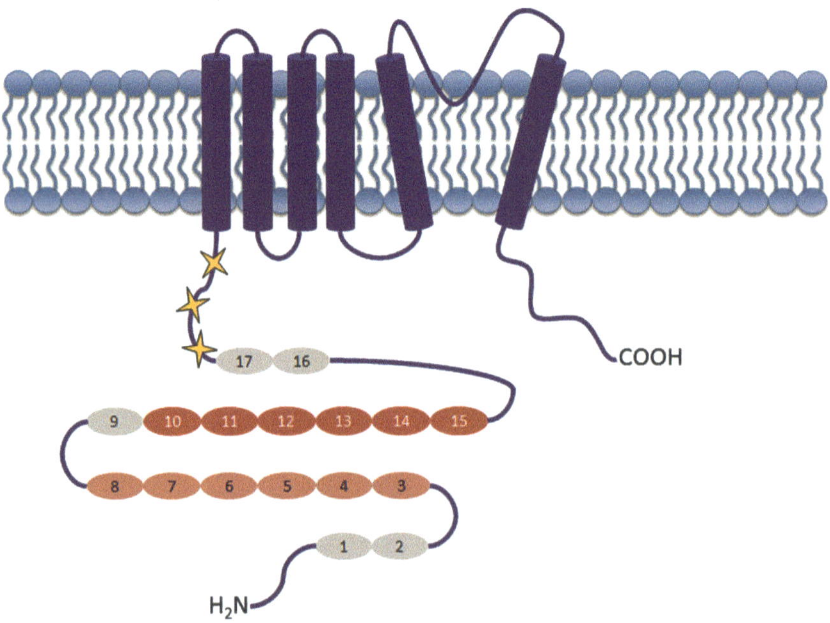

Figure 4.1 Topological organization of a snake TRPA1 channel. Ankyrin repeats are shown as numbered ovals. The primary and enhancer modules important for determining heat sensitivity correspond to repeats number 10–15 and 3–8, respectively. Stars in the linker region between the transmembrane domains and ankyrin repeat domain depict conserved cysteine residues involved in channel activation by electrophiles.

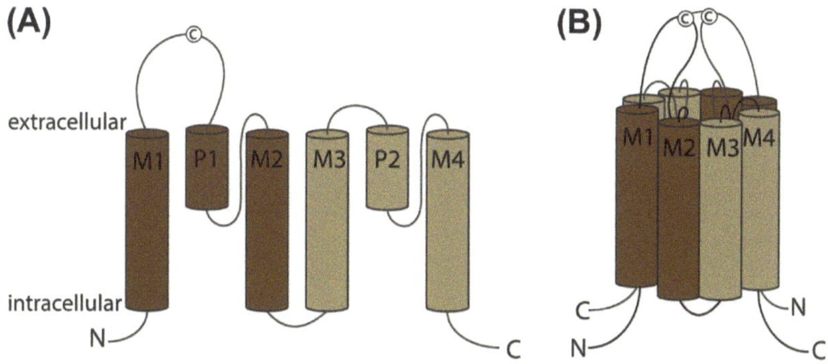

Figure 5.1 Membrane topology of K2P channels. (A) A topology diagram of a single K_{2P} subunit with two pore-forming domains. (B) A mature channel is formed by two subunits covalently linked via the cysteines (C) in the first extracellular loop. M1–M4, transmembrane segment 1–4; P1–2, pore helix 1–2.

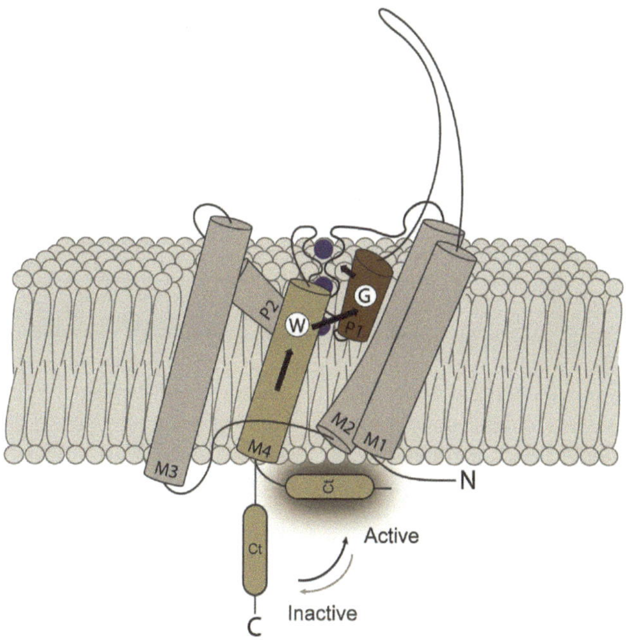

Figure 5.3 A hypothetical cartoon model of how Ct affects TREK-1 activity. A cartoon model of a single TREK-1 subunit showing a hypothetical mechanism of channel activation by temperature. It was proposed that increasing temperature facilitates the transition of Ct from inactive to active conformation, leading to stabilization of an open conformation of the selectivity filter via interaction between Trp275 (W) and Gly137 (G) from a single TREK-1 subunit (Bagriantsev et al., 2012). M1–M4, transmembrane segment 1–4; P1–2, pore helix 1–2; Ct, C-terminal domain. Blue spheres depict potassium ions.

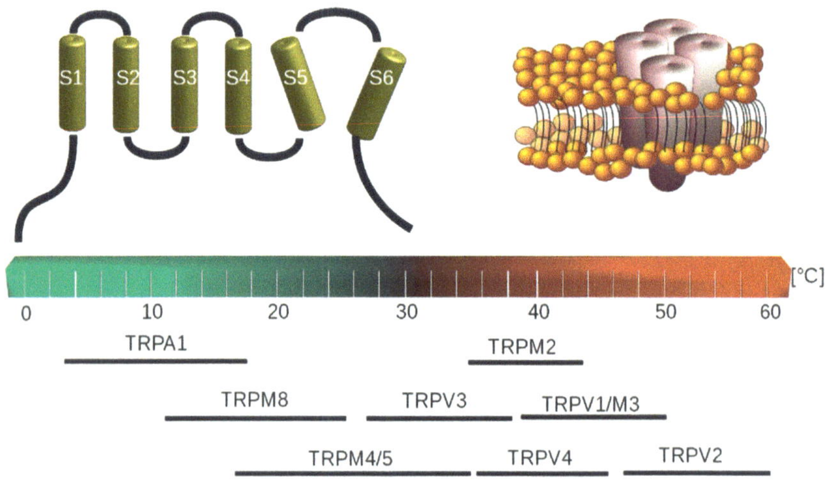

Figure 6.1 General features of thermo-TRP channels. TRP channels are composed of six transmembrane segments (left upper panel), with the S5 and S6 segments and the loop that connects them forming the channel pore. TRP channels are inserted on the membrane (right upper panel) as tetramers. The temperature range to which each thermo-TRP channel responds is depicted as a black line below the thermometer bar.

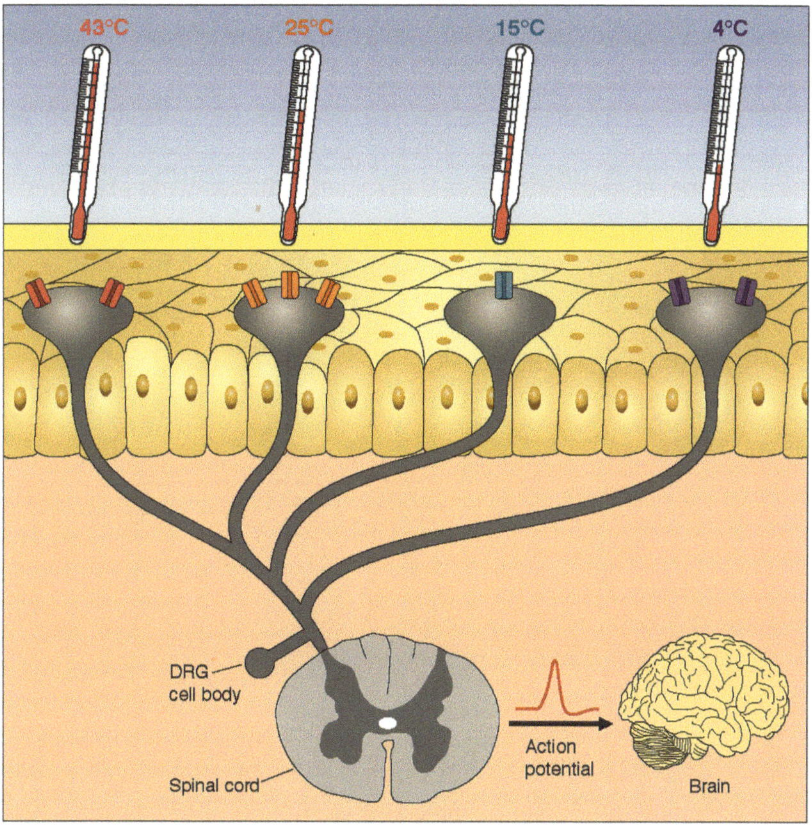

Figure 7.1 Thermosensitive ion channels in sensory neurons. Sensory neurons innervate the skin and contain thermosensitive nonselective cation channels in their terminals that sense a wide range of temperatures. Activation of these channels depolarizes the sensory neuron, leading to propagation of action potentials that are relayed to the spinal cord and eventually reach the brain. *Illustration by Kelly Paralis, Penumbra Design.*

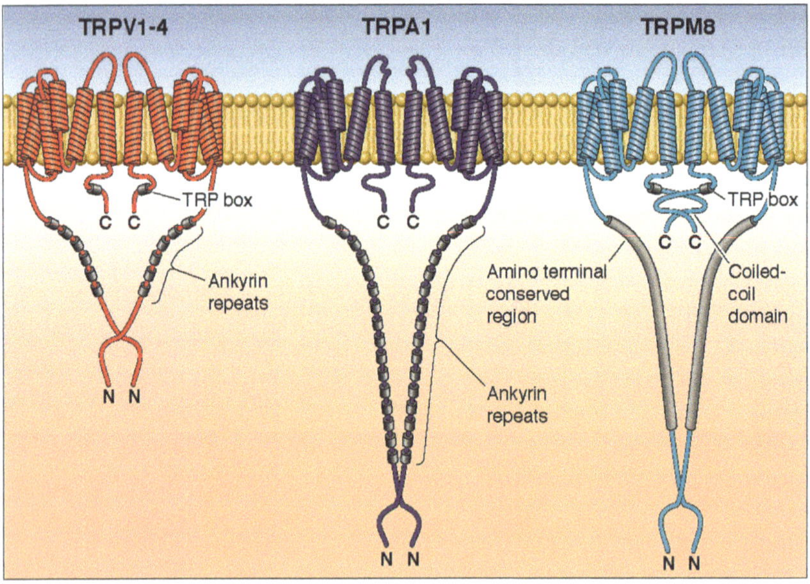

Figure 7.2 Schematic of the domain organization for thermo transient receptor potential (TRP) channels. Representation of the domain organization for TRPV1–4, TRPA1, and TRPM8. For clarity, dimers are shown. A monomer for each channel contains 6 transmembrane segments and large N- and C-termini in the cytoplasm. TRPV1–4 have 6 N-terminal ankryin repeats and a conserved C-terminal TRP box sequence. TRPA1 contains 14–19 N-terminal ankyrin repeats. TRPM8 has an N-terminal conserved region and C-terminal TRP box and coiled-coil domains. *Illustration by Kelly Paralis, Penumbra Design.*

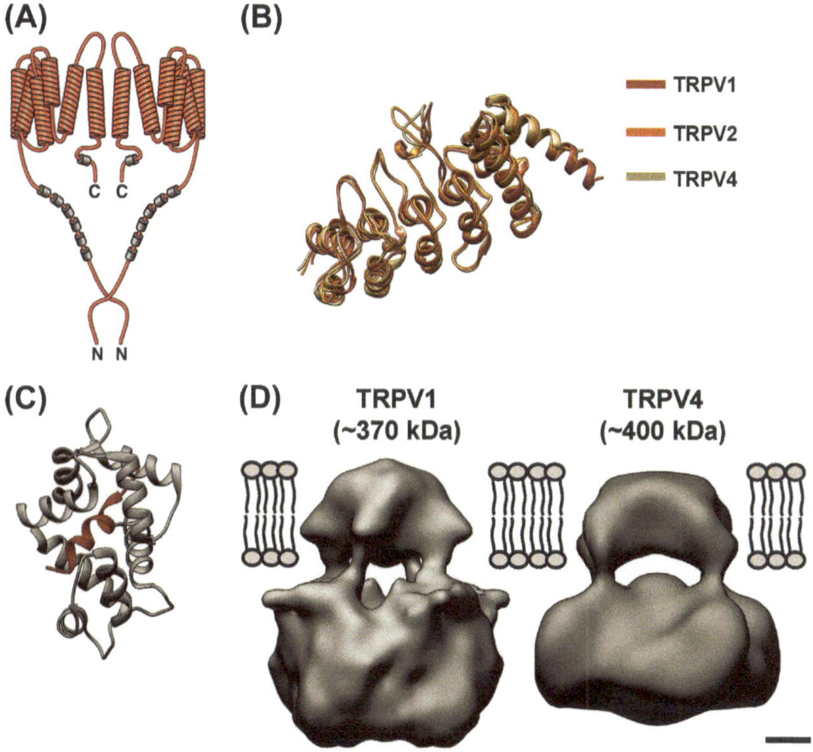

Figure 7.3 Structural analyses of thermoTRPV channels. (A) Domain organization of thermoTRPV channels. (B) Comparison of the crystal structures for the ankyrin repeat domains of TRPV1 (PDB code: 2PNN), TRPV2 (PDB code: 2ETB), and TRPV4 (PDB code: 4DX2) (Phelps, Wang, Choo, & Gaudet, 2010). (C) Crystal structure of a TRPV1 C-terminal peptide (red) in complex with calmodulin (gray) (PDB code: 3SUI) (Lau, Procko, & Gaudet, 2012). (D) Comparison of the cryo-electron microscopy structures for full-length TRPV1 at 19 Å resolution (Moiseenkova-Bell, Stanciu, Serysheva, Tobe, & Wensel, 2008) and TRPV4 at 35 Å resolution (Shigematsu, Sokabe, Danev, Tominaga, & Nagayama, 2010). Scale bar represents 25 Å.

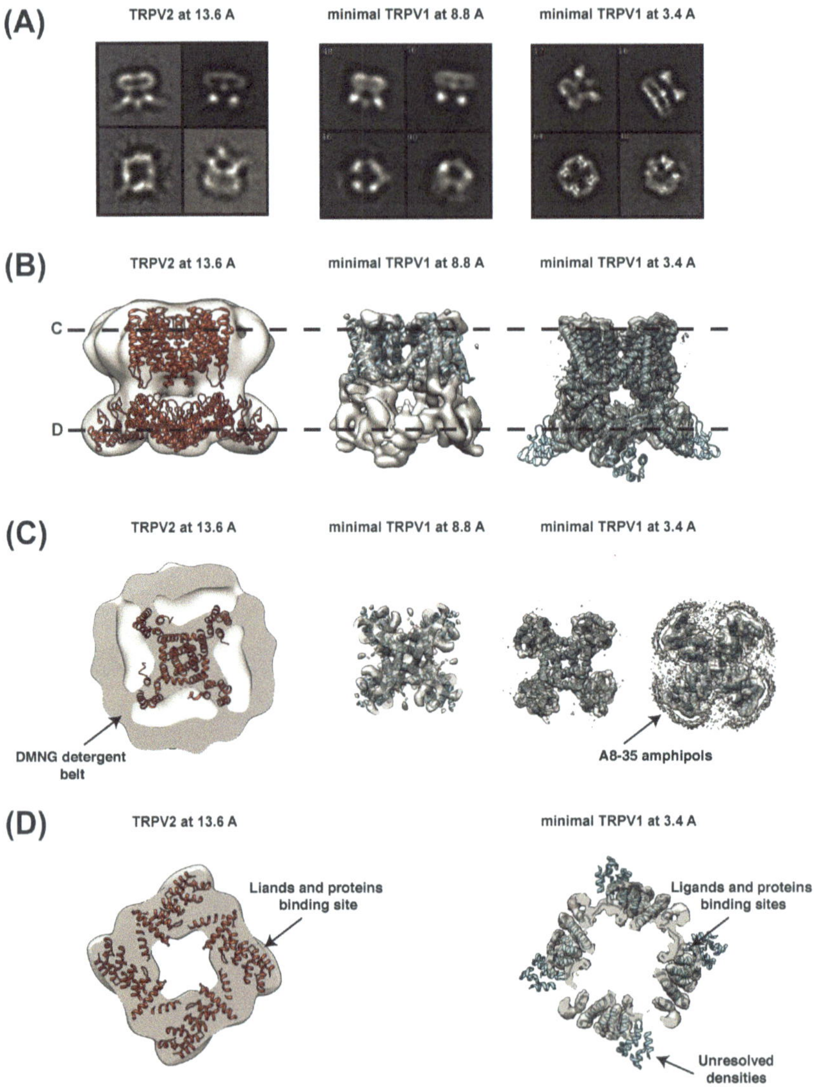

Figure 7.4 Comparison of cryo-electron microscopy (EM) structures of full-length TRPV2 and minimal TRPV1. (A) 2D class averages of TRPV2 at 13.6 Å (Huynh et al., 2014) and minimal TRPV1 at 8.8 Å and 3.4 Å resolution *(Adopted and reprinted from Liao et al. (2013))*. (B) 3D reconstruction of TRPV2 at 13.6 Å resolution (Huynh et al., 2014) and minimal TRPV1 at 8.8 Å and 3.4 Å resolution *(Adopted and reprinted from Liao et al. (2013))*. The TRPV2 EM map is fitted with crystal structures of the MlotiK1 transmembrane (TM) domain (PDB code: 3BEH) and the TRPV2 ankyrin repeat domain (PDB code: 2ETB). Dashed lines correspond to cross-sections represented in (C) and (D). (C) Cross-sectional view through the TM domains of the TRPV2 (Huynh et al., 2014) and minimal TRPV1 EM maps *(Adopted and reprinted from Liao et al. (2013))*. The unfilled densities correspond to the maltose-neopentyl glycol (MNG) detergent belt for TRPV2 and the A8-35 amphipol belt for minimal TRPV1. (D) Cross-sectional view through the cytoplasmic density of the TRPV2 (Huynh et al., 2014) and minimal TRPV1 EM maps. *Liao et al. (2013).*

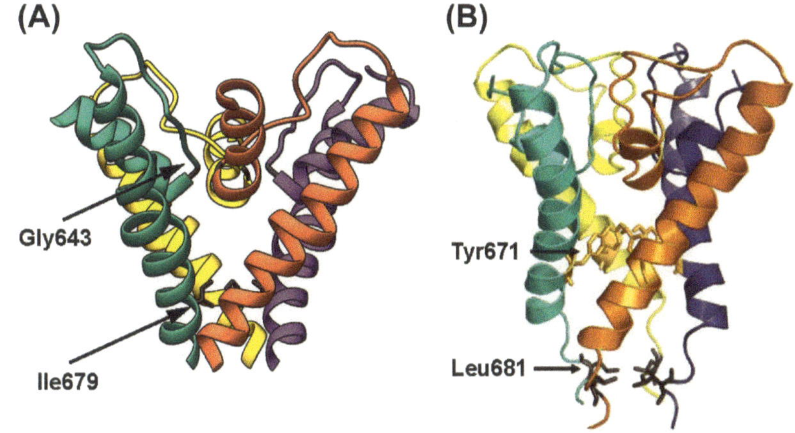

Figure 7.5 Structural analysis of the TRPV1 pore region. (A) Architecture of the TRPV1 pore as determined by cryo-electron microscopy at 3.4 Å (Liao et al., 2013). Residues important for forming the dual gate are indicated by arrows. (B) Predicted pore architecture based on homology modeling and electrophysiological studies using a cysteine accessibility strategy *(Adopted and reprinted from Salazar et al. (2009).)*. A gate for passage of small cations is predicted to begin from the intracellular side at Tyr671. A second gate for larger molecules is predicted to begin from the intracellular side at Leu681 (both residues indicated by arrows).

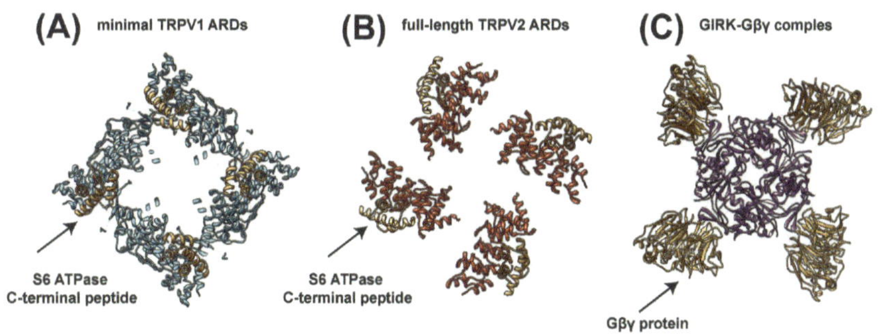

Figure 7.6 Comparison of the ankyrin repeat domain (ARD) orientation in the TRPV2 and minimal TRPV1 electron microscopy (EM) maps. (A) Model of the TRPV1 ARDs as they are arranged in the minimal TRPV1 cryo-EM map and aligned with the crystal structure of the Gankyrin-S6 ATPase C-terminus complex (PDB code: 2DVW). The TRPV1 ARDs are shown in complex with the S6-ATPase protein. (B) Crystal structures of the TRPV2 ARDs (PDB code: 2ETB) as fitted into the TRPV2 cryo-EM map and aligned with the Gankyrin-S6 ATPase C-terminus complex (PDB code: 2DVW). The TRPV2 ARDs are shown in complex with the S6-ATPase protein. (C) Crystal structure of GIRK2 in complex with $G\beta\gamma$ (PDB code: 4KFM).

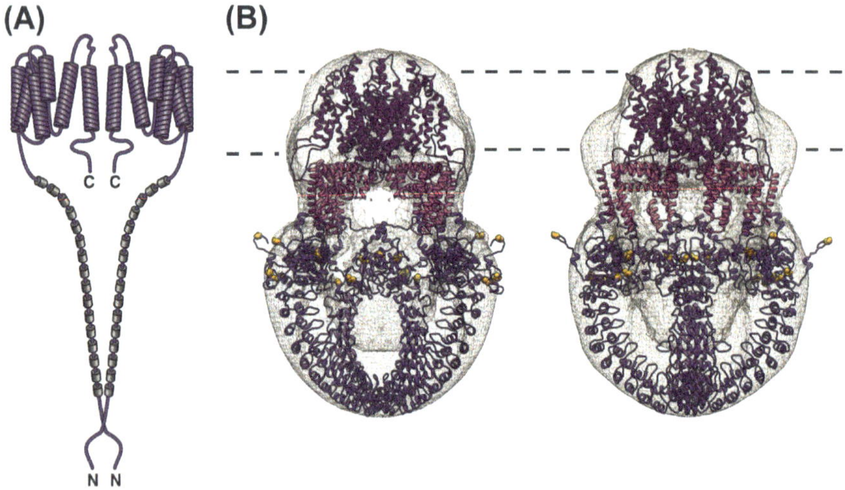

Figure 7.7 Topology model of TRPA1 and TRPA1 electron microscopy (EM) structure at 16 Å resolution. (A) Domain organization of TRPA1. (B) Two side views of the TRPA1 EM structure (Cvetkov et al., 2011). The transmembrane (TM) domain was fitted with a molecular model of the TRPV1 TM domain (purple) (Fernandez-Ballester & Ferrer-Montiel, 2008). A model of the TRPA1 N-terminus was fitted within the cytoplasmic domain of the density map (blue). The cysteine residues critical for electrophilic ligand activation are represented as yellow spheres. A model of the TRPA1 C-terminus was fitted into the 3D density map immediately below the predicted TM domain within the region linking the TM domain to the cytoplasmic domain (light purple).

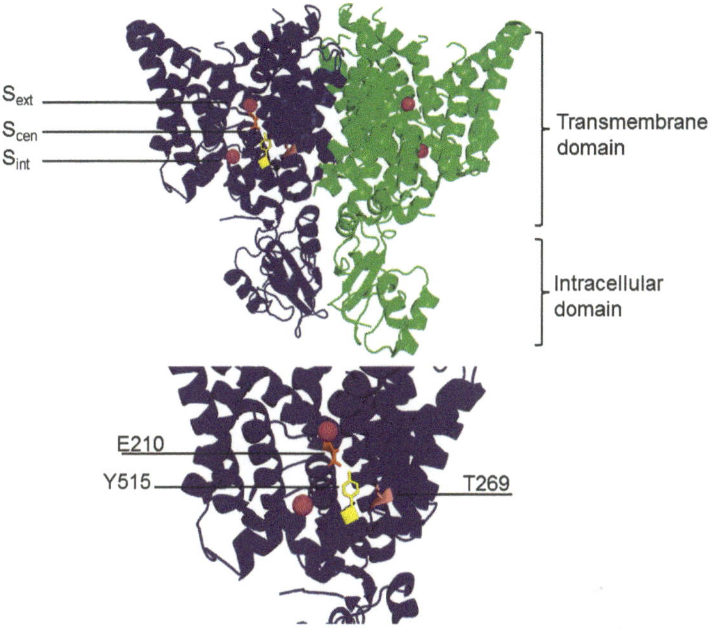

Figure 8.2 Upper panel, dimeric structure of the eukaryotic antiporter CLC-Cm (PDB entry: 3ORG) viewed from the membrane plane (extracellular side above and cytoplasmic side below). The two subunits are shown in green and blue. The residue E210 (also called "gating glutamate") is colored red, T269 is shown in pink, and Y510 in yellow. Cl^- anions bound to S_{ext} and S_{int} are shown in magenta. The position of the three binding sites, S_{ext}, S_{cen}, and S_{int}, is also indicated by horizontal lines. The side chain of E210 occupies S_{cen}. The cytoplasmic C-terminus contains CBS domains. Lower panel, expanded representation of the anion permeation pathway for one of the subunits. In both panels some transmembrane helices were removed for clarity.

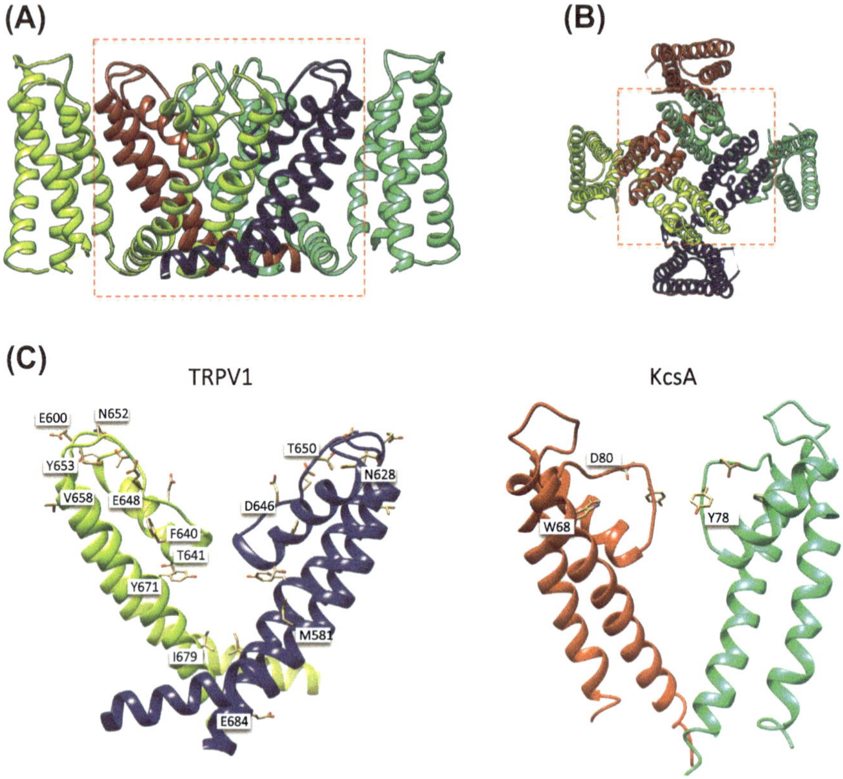

Figure 9.1 Side (A) and top (B) view of the TRPV1 channel transmembrane region in the closed state. Each subunit is shown in a different color. The S1–S4 domains of the red and blue subunits are omitted in (A) to reveal the pore region (highlighted with the box). (C) Comparison of the pore features of TRPV1 (left) and KcsA (right). Some important amino acids discussed in the text are highlighted. Note that 23 amino acids in the turret region are missing in the TRPV1 structure.

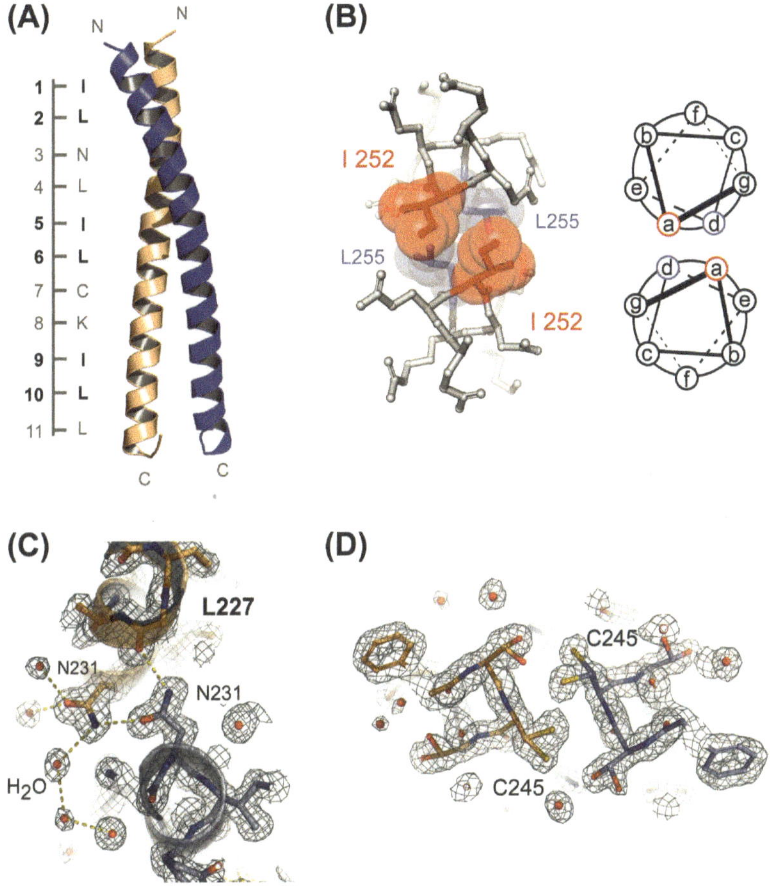

Figure 10.6 Crystal structure of the Hv coiled-coil domain. (A) The coiled-coil domain of mouse Hv1/VSOP (violet, PDB code: 3VMX) (Fujiwara et al., 2012). Layer numbers and the amino acid residues are indicated and Ile and Leu are highlighted. (B) Hydrophobic layers of the coiled-coil core. (B) Representative geometry of the symmetrically packed hydrophobic I/L-type coiled-coil core (left). Van der Waals spheres depicting the side chains of the "a" (red) and "d" (blue) layers on a ribbon backbone (gray) are shown. Highlighted in the hydrophobic core are two Ile residues in layer-9 (red) and two Leu residues in layer-10 (blue). Helical wheel representation of the core-packing arrangements found in the homodimeric coiled coil (right). Heptad repeat positions are labeled a-g. (C) Structure (stick models) and the 2Fo–Fc composite-omit maps in layer-3. Cartoon helix models are also shown to indicate the strand orientation. Red balls depict water oxygen atoms. Maps are contoured at 1.0 σ. Yellow dashed lines depict polar contacts. (D) Structure and 2Fo-Fc maps in layer-7. Maps are contoured at 1.0 σ. Numbers depict occupancies of sulfur atoms in the dual conformations of Cys residues. Neighboring Ser residues also assume dual conformations. *Figures adapted from Fujiwara et al. (2012).*

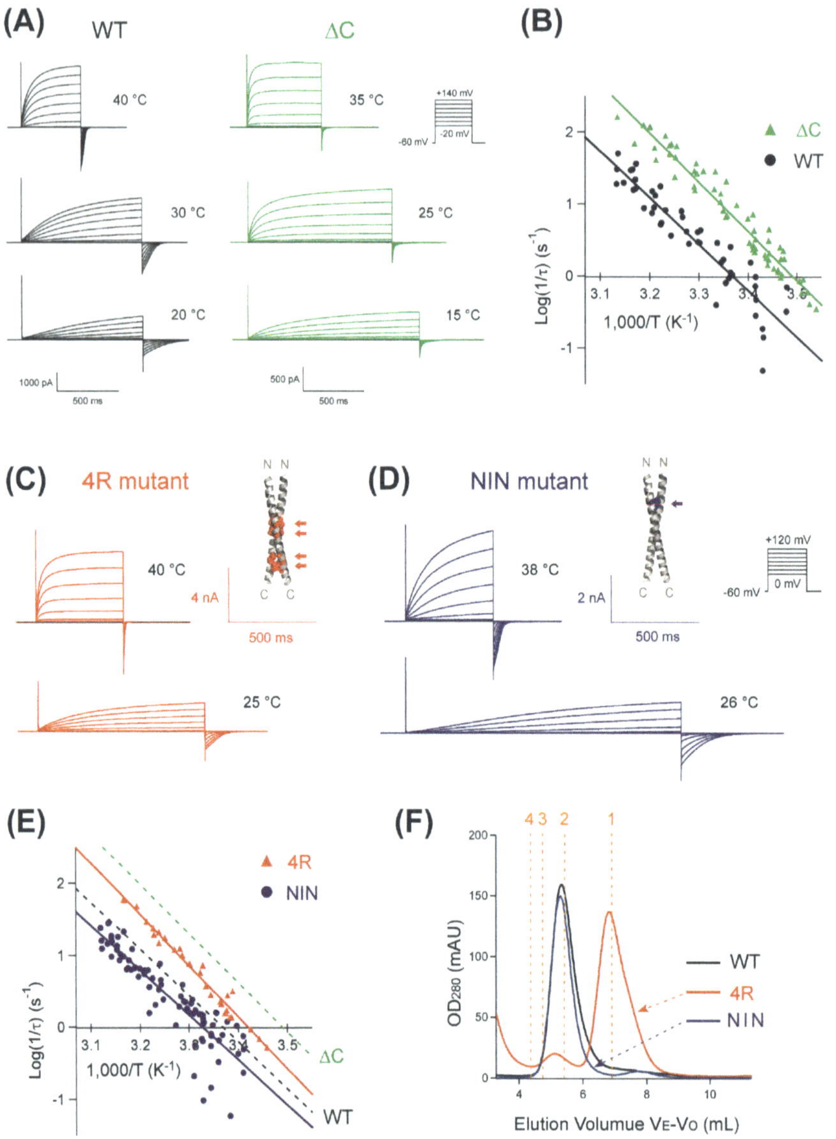

Figure 10.8 Effect of the coiled-coil domain on Hv1/VSOP activation kinetics. (A) Representative currents through WT (black) and ΔC (green) Hv1/VSOP recorded from HEK293T cells at the indicated temperatures (Fujiwara et al., 2012). The holding potential was −60 mV; step pulses from −20 mV to +140 mV were applied in 20-mV increments. The pulse protocol is indicated on the right. (B) Accumulated data from WT (black) and ΔC (green) channels, and linear regression lines showing the correlation of the activation kinetics with temperature. (C,D) Representative currents through 4R (red, in C) and NIN (blue, in D) mHv1/VSOP mutants recorded from HEK293T cells at

the indicated temperatures. The pulse protocol is indicated on the right in D. Structure models are shown to indicate the positions of the mutations. (E) Accumulated data from 4R (red) and NIN (blue) mutant channels and their linear regression lines. Symbols used are shown in the figure. Dotted lines depict the linear regression lines. (F) Coiled-coil stoichiometry determined using size exclusion chromatography monitored at 280 nm. For clear size separation, we used HMT-tagged proteins in these experiments. The horizontal axis shows elution volume (ml), and dotted lines indicate the predicted elution volumes for tetrameric, trimeric, dimeric and monomeric proteins. The elution volume VE was corrected for the void elution volume by subtracting the volume of blue dextran (VO). The oligomeric state of the main component was 2.29 for WT and 1.15 for 4R. *Data are taken from Fujiwara et al. (2012).*

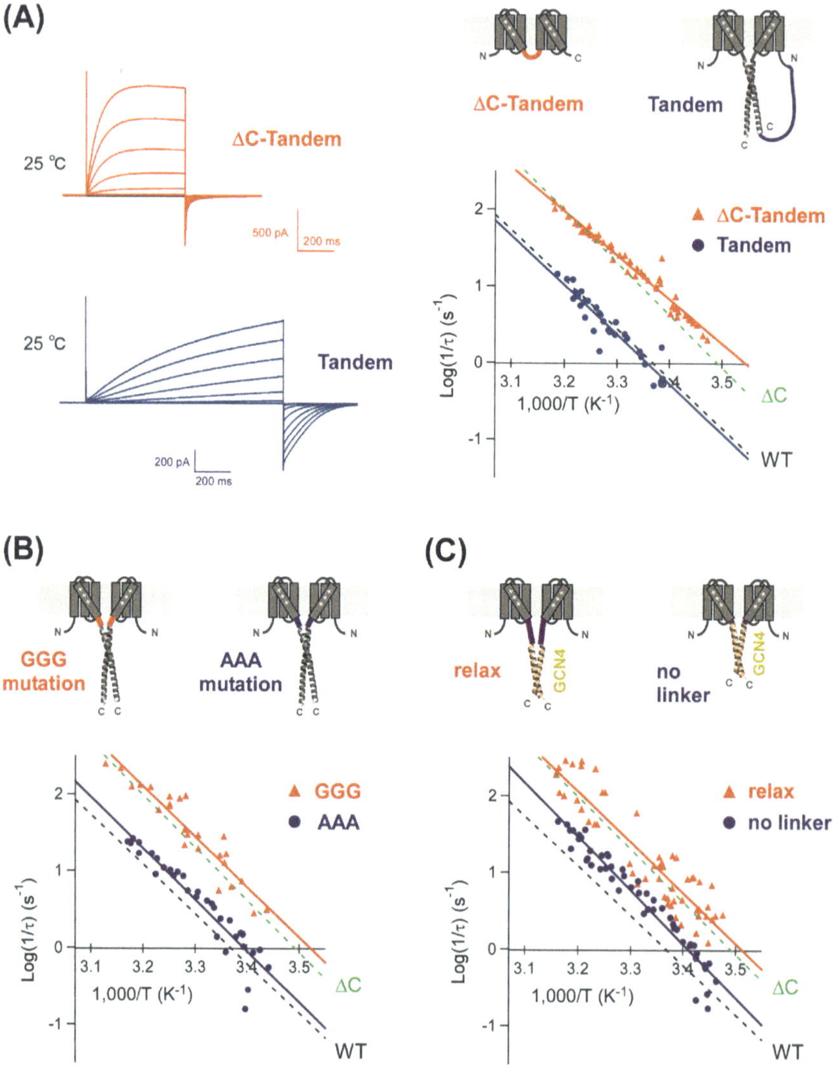

Figure 10.9 The temperature-dependent activation of murine Hv1/VSOP linker mutant channels. (A) Representative currents through murine Hv1/VSOP-Tandem (blue) and ΔC-Tandem (red) channels recorded from HEK293T cells at 25 °C (lower panel). Accumulated data from the Tandem (blue) and ΔC-Tandem (red) channels and the linear regression line (right panel) are shown; the dotted lines depict the linear regression lines for the data from WT (black) and ΔC (green) channels. (B) Accumulated data from the GGG (red) and AAA (blue) mutant channels and their linear regression lines; the dotted lines are the same as in panel (A). (C) Accumulated data from the GCN4 chimeric channel with a long linker (relax) (red) or with no linker (blue) and their linear regression lines; the dotted lines are the same as in panel (A). *Data are taken from Fujiwara et al. (2012).*

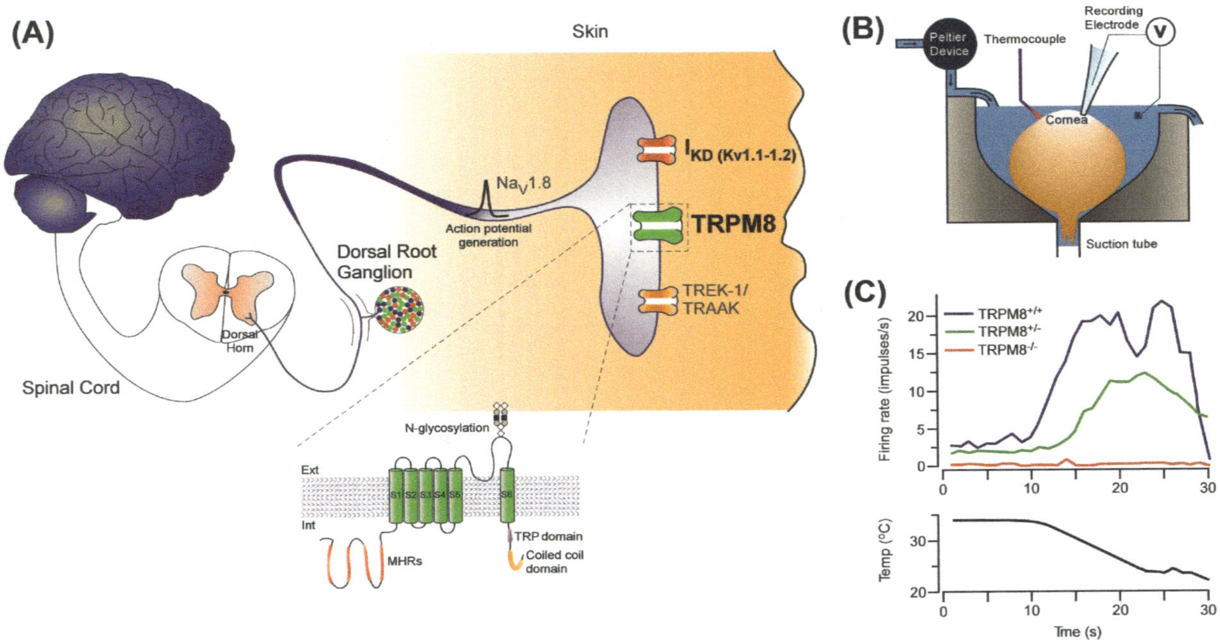

Figure 11.1 TRPM8 is the main molecular determinant of cold sensitivity of primary cold thermoreceptor neurons. (A) The schematic representation of a cold-sensitive primary afferent neuron. The site of cold transduction (enlarged nerve ending at the right) is depicted as separate from the site of action potential generation (axon at left). In this picture, TRPM8 is represented as the most prominent channel. Kv1.1-1.2 (the molecular counterpart of the break current I_{KD}) and TREK-1/TRAAK channels are also important molecular determinants that contribute to the net response to temperature reductions in cold thermoreceptor neurons. (B) Experimental arrangement used to record nerve terminal impulses (NTI) of cold thermoreceptor neurons from isolated mouse eye in vitro. The eye is placed in a small recording chamber and secured by the application of slight suction to a tube connected to the conical bottom. A glass pipette (50 μm in diameter) is brought into contact with the cornea to record NTIs. The temperature of the perfusing solution is adjusted with a Peltier device at the entrance of the chamber, controlled by computer, and is measured at the corneal surface using a thermocouple. (C) The mean firing rate (*top*) of a population of corneal cold-sensitive nerve endings from $TRPM8^{+/+}$, $TRPM8^{\pm}$, and $TRPM8^{-/-}$ mice during cold stimulation (cooling ramp at bottom). Basal firing rate (ongoing activity) and cold responses in these free terminals are proportional to the expression level of TRPM8 channels. *((C) adapted from Parra et al., 2010.)*

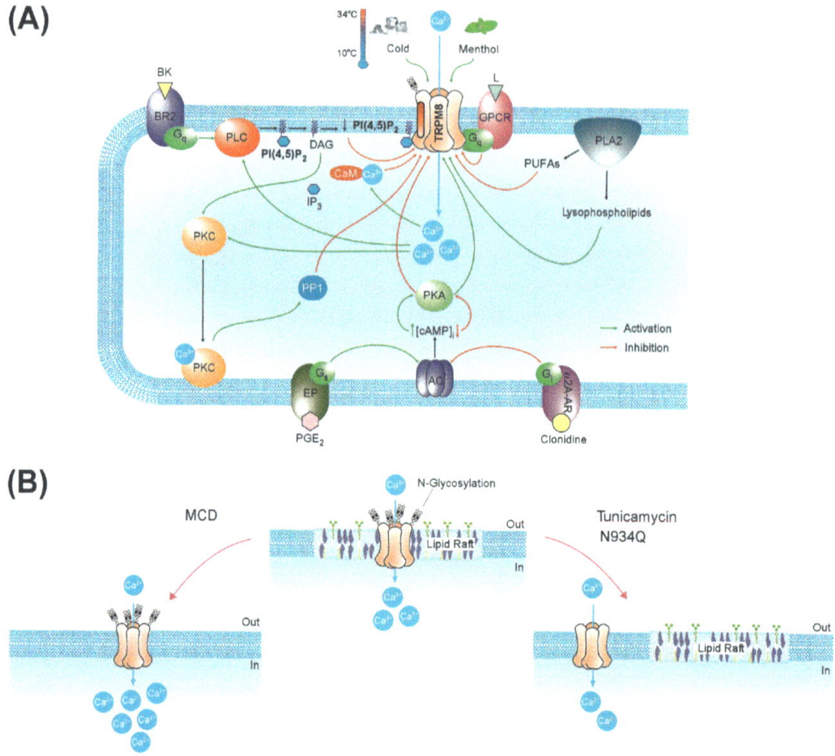

Figure 11.2 The modulation of TRPM8 channel by second messengers, protein–protein interactions and post-translational modifications. TRPM8 is represented as a homotetrameric channel in the scheme. (A) Activation of PLC through the activation of bradykinin (BK) B2 receptor (BR2) induces a reduction in PI(4,5)P_2 levels in the plasma membrane, by the cleavage of PI(4,5)P_2 in diacylglycerol (DAG) and inositol trisphosphate (PI(1,4,5)P_3), which leads to the desensitization of the channel. The activation of prostaglandin receptor (EP) induces an increase in cyclic adenosine monophosphate (cAMP) by the activation of adenylate cyclases (AC). cAMP activates PKA, which can induce a downregulation of the channel by an unknown mechanism. Conversely, the inhibition of α2A adrenoreceptor (α2A-AR) by clonidine induces an inhibition of AC and a reduction of PKA activity that induces a diminution of the phosphorylation level of TRPM8, causing a decrease of channel activity. The activation of TRPM8 channels by cold or menthol can induce an increase in the intracellular Ca^{2+} concentration ($[Ca^{2+}]_i$). This intracellular Ca^{2+} increase can activate PKC, which can modulate protein phosphatase 1 (PP1), leading to the inhibition of the channel by dephosphorylation. The acute Ca^{2+}-dependent desensitization of TRPM8-mediated responses during single sustained application of agonist would be related to Calmodulin (CaM) activation by Ca^{2+}. On the other hand, increases in phospholipase A2 (PLA_2) activity can generate PUFAs and lysophospholipids from glycerophospholipids; PUFAs exert inhibitory effects on TRPM8, and lysophospholipids upregulate the channel function. Gαq subunits bind directly to TRPM8, and through a conformational change of Gαq that follows the activation

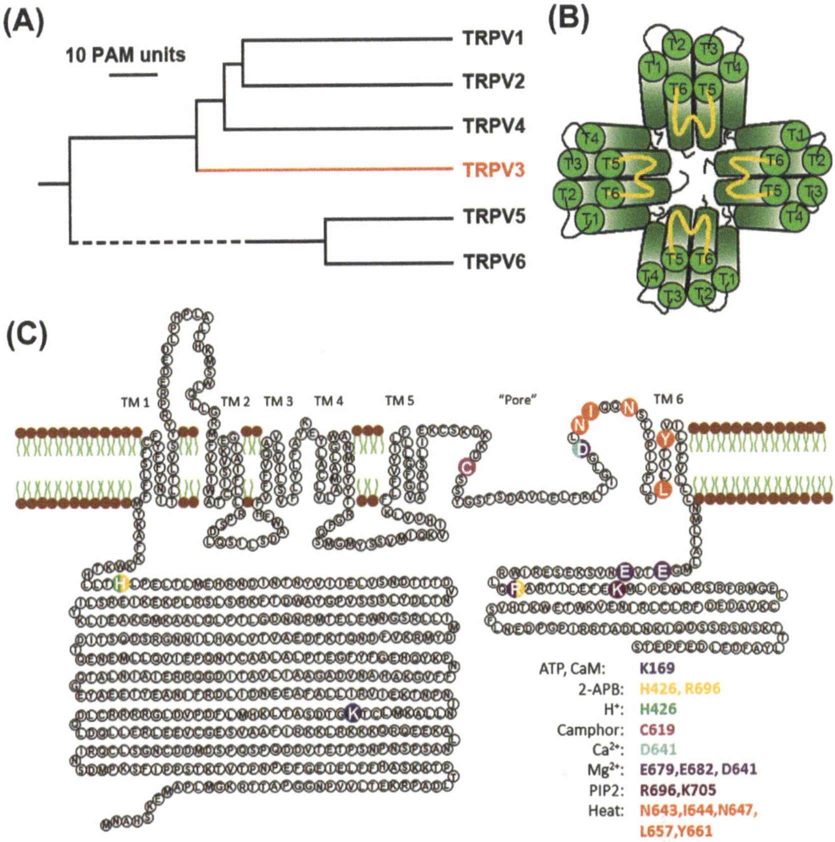

Figure 12.1 (A) The phylogenetic tree of TRPV subfamily. (B) Schematic diagram illustrating the putative tetrameric structure of TRPV3 composed of four identical subunits. The putative pore loops lie between TM5 and TM6 (yellow). (C) Predicted structural topology of TRPV3 with all the amino acid residues indicated. Residues involved in activation/modulation by 2-APB, camphor, heat, H^+, Ca^{2+}, Mg^{2+}, adenosine triphosphate (ATP), CaM, and PIP2 are color coded. (See the color plate.)

of the G protein-coupled receptors (GPCR) by ligand (L) binding, induces a direct inhibition of the channel. PHB modification and N-glycosylation are represented by a red bar and a schematic sugar moiety in one of the subunits of the channel, respectively. (B) The disruption of lipid rafts by methyl-beta-cyclodextrin (MCD) increases the channel activity. Unglycosylation in response to treatment with Tunicamycin and single point mutation of Asp934 to Gln induce a reduction of TRPM8 sensitivity to cold and menthol, despite the main localization of the channel outside of lipid rafts.

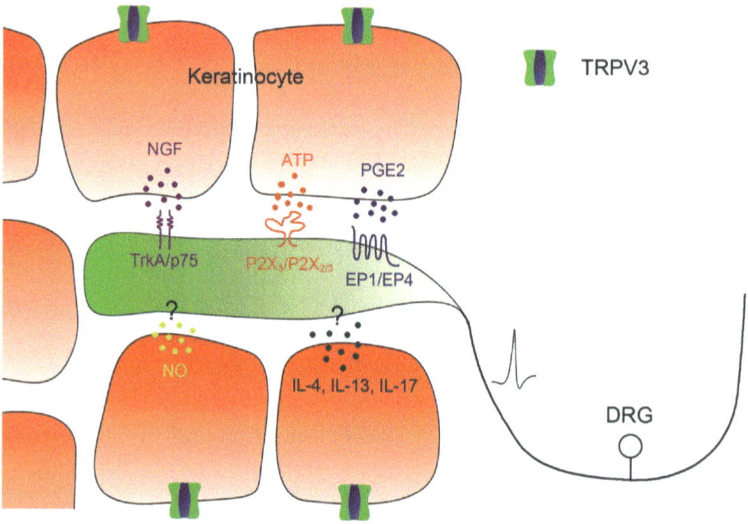

Figure 12.2 TRPV3-mediated crosstalk between keratinocytes and cutaneous nerve endings. TRPV3 activation in the keratinocytes promotes the release of inflammatory mediators including PGE2, ATP, NGF, NO, IL-4, IL-13, IL-17 which act on their corresponding receptors to active sensory nociceptors and initiate pain or itch sensation.

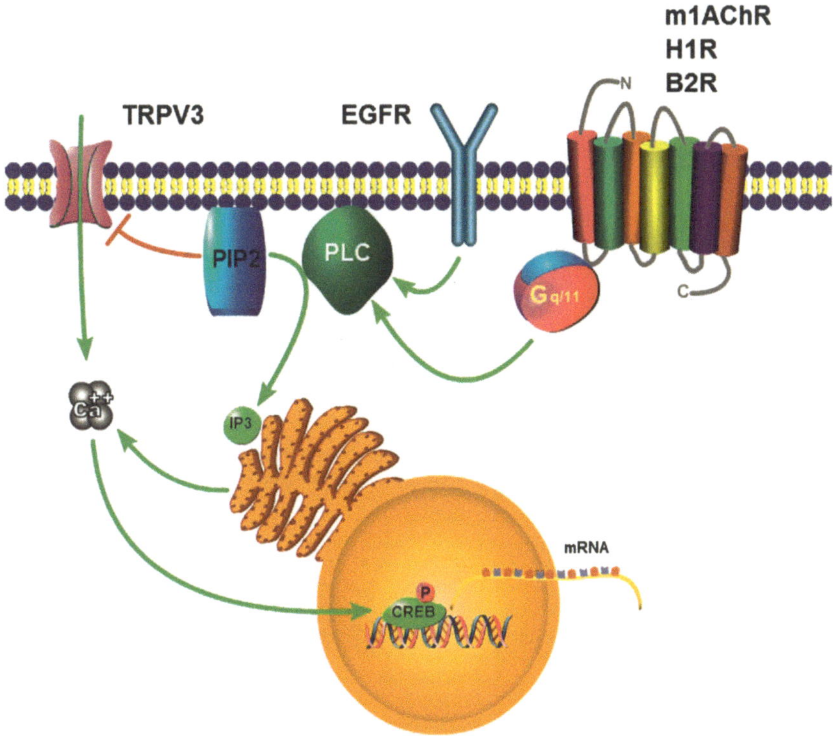

Figure 12.4 TRPV3-mediated signaling is regulated by multiple receptor signaling pathways. G protein-coupled receptors (m1AChR, H1R, and B2R) activate $G_{q/11}$ upon ligands binding. Activated PLC in turn hydrolyzes PIP2 which constitutively suppresses TRPV3, thereby potentiating TRPV3 response. EGFR activation also induces PLC activation and enhances TRPV3 function in a similar manner. TRPV3-mediated Ca^{2+} influx can promote the CREB-initiated gene expression.

FSC
www.fsc.org
MIX
Papier | Fördert gute Waldnutzung
FSC® C083411

Kolibri 360 GmbH
Ferdinand-Jühlke-Straße 7
99095 Erfurt, Deutschland
produktsicherheit@kolibri360.de